经典中国国际出版工程
China Classics International

化工过程强化关键技术丛书
CHEMICAL PROCESS INTENSIFICATION SERIES

中国化工学会组织编写

Heat Transfer Enhancement in Chemical Processes

化工过程强化传热

孙丽丽（Lili Sun）　　吴德飞（Defei Wu）
高莉萍（Liping Gao）　　魏志强（Zhiqiang Wei）　等著

ELSEVIER

化学工业出版社
Chemical Industry Press
·北京·

内容简介

Heat Transfer Enhancement in Chemical Processes systematically analyzes the basic principles of heat transfer, summarizes the theories and practices of heat transfer enhancement, and develops new methods for heat transfer enhancement in chemical processes on the basis of integrating process systems with heat exchange equipment technologies. The book focuses on the representative processes in the petrochemical industry, i.e., oil refining, aromatics, and ethylene production, and systematically details the system optimization and integrated application of heat transfer enhancement in the petrochemical industry from the equipment components, and the process units to the whole plant.

This book provides scholars, engineers, students, and technicians in oil refining, petrochemical, fine chemical, and coal chemical industries, chemical and biochemical pharmacy, food processing, and waste gas and waste water treatment, with a comprehensive reference.

图书在版编目（CIP）数据

化工过程强化传热 = Heat Transfer Enhancement in Chemical Processes : 英文 / 中国化工学会组织编写 ; 孙丽丽等著. -- 北京 : 化学工业出版社，2025. 6. -- ISBN 978-7-122-47735-4

Ⅰ. TQ02

中国国家版本馆CIP数据核字第20259C8P97号

责任编辑：任睿婷　杜进祥　马泽林　吴　刚　装帧设计：张　辉
责任校对：边　涛

出版发行：化学工业出版社
　　　　　（北京市东城区青年湖南街13号　邮政编码100011）
印　　装：北京建宏印刷有限公司
710mm×1000mm　1/16　印张28¼　字数603千字
2025年8月北京第1版第1次印刷

购书咨询：010-64518888　　　　　售后服务：010-64518899
网　　址：http://www.cip.com.cn
凡购买本书，如有缺损质量问题，本社销售中心负责调换。

定　　价：598.00元　　　　　　　　版权所有　违者必究

About the author

Lili Sun is an academician of the Chinese Academy of Engineering, a national engineering survey and design master, and an expert in refining and petrochemical engineering technology and management. She is working at the forefront of engineering technology according to the needs of national development, and has completed the engineering design of more than 100 refining and petrochemical units and more than 10 modern refining and petrochemical bases.

Ms. Sun led her team in making major breakthroughs in the highly efficient and eco-friendly aromatics technology, and led the construction of China's first aromatics complex with proprietary technology, reaching the world's leading level, leading the high-quality development of the aromatics industry, and playing an important role in solving the arable land allocation contradiction between grains and cotton. She has made significant breakthroughs in the integration and innovation of efficient and clean conversion technology for crude oils, and led the construction of China's first refinery with a single-train capacity of 10 million tons per year, simultaneously solving a series of major problems such as efficient utilization of low-quality crude oils and pollution source control. This technology has been applied to more than 10 refining and petrochemical bases in Huizhou, laying a solid foundation for securing China's energy supply. She has developed the key technology for ultra-large-scale high-acid natural gas purification and safe and high-efficiency sulfur recovery, and led the construction of the second largest high acid natural gas purification plant in the world, filling gaps in key technologies for high acid natural gas purification. The cumulative gas supply exceeds 100 billion cubic meters, providing clean energy to more than 70 cities and thousands of companies in the Yangtze River economic zone. Harmful H_2S can be converted into valuable products, and the annual sulfur production accounted for 30% of the national total production of that year. She has led the high-quality construction of Yanbu Aramco Sinopec Refining Company Ltd., China's largest overseas joint venture refining and petrochemical project, creating multiple new records in the history of local engineering construction, achieving world-class production, and winning high praise from both governments.

She is the winner of two Special National Prizes and two Second National Prizes for Progress in Science and Technology; one Gold Award of National Engineering Survey and Design; two Awards of Merit for Outstanding Project of the Year by FIDIC; the Science and Technology Innovation Award by Ho Leung Ho Lee Foundation; and the Hou Debang Chemical Science and Technology Achievement Award. She has 48 patents and has published 7 books. She was awarded the National Excellent Scientific and Technical Worker, and her team was honored as an advanced state-owned enterprise by the government and a role model for state-owned enterprises.

Preface

Promoting a revolution in energy production and consumption and building an energy sector that is clean, low carbon, safe, and efficient are the essential conditions for building a beautiful China. The petrochemical industry is not only a large energy producer but also a large energy consumer, and making innovations in its energy-saving technologies is particularly important. Heat transfer enhancement, as a new energy-saving technology that can significantly improve the process heat transfer performance, has been widely used in chemical industry and other process industries in recent years because it can reduce equipment size, improve thermal efficiency, and reduce energy consumption and waste discharge. It has been widely valued by numerous university researchers, engineers, technicians, and production management personnel. Relevant research results keep mushrooming, including both complete theoretical research and partial practical application. However, no monograph systematically elaborates on the application and implementation of heat transfer enhancement in chemical industry based on case analysis. In order to comprehensively summarize the practical experience and application achievements of heat transfer enhancement in chemical processes, especially in petrochemical processes, and jointly push the application of heat transfer enhancement technologies to a new level, Professor Lili Sun, a national engineering survey and design master, took the lead in preparing this monograph.

This monograph analyzes the basic principles of heat transfer enhancement, sorts out the theories and methods of this technology, and briefly discusses the heat transfer enhancement methods for typical energy-consuming equipment. On this basis, the comprehensive practical application of heat transfer enhancement in petrochemical industry is systematically summarized from three application levels of equipment components, processes and plants as a whole, focusing on the three representative sectors in petrochemical industry: oil refining, aromatics, and ethylene. The first level is the heat transfer enhancement of equipment components. Heat exchange equipment includes general heat transfer equipment in petrochemical and other fields. Over half a century, it has evolved from the first-generation plain tube shell and tube heat exchanger with only a heat transfer function to the fourth generation with compound heat transfer enhancement technology and microchannel heat transfer enhancement functions. By adopting enhanced heat transfer elements and improving the structure of heat exchange equipment, the fourth-generation heat exchange equipment has achieved the purpose of economical and efficient use of energy. The second level is the heat transfer enhancement of processes. "Pinch point" technology is applied to process flow, and the characteristics of the heat energy during production are analyzed from the perspective of plants.

The energy bottleneck of the system is identified, and specific measures are taken to achieve process enhancement and energy integration. The third level is the heat transfer enhancement of the plant as a whole. Based on the heat transfer enhancement of equipment components and processes mentioned above, the system energy consumption of the plant is macroscopically constructed and designed from the aspects of plant-wide process flow optimization, integration of process units, matching of utilities and auxiliary facilities, and comprehensive utilization of low-temperature waste heat, so as to build high-quality energy basis for the plant.

This book is the crystallization of more than 40 national and provincial Science and Technology Progress Awards, National Excellent Engineering Design Awards, National EPC Awards, and other achievements of the Sinopec Engineering Incorporation (SEI), including the Development and Application of Efficient and Eco-Friendly PX Suite Technology, which won the grand prize of the National Science and Technology Progress Award, and the Development and Industrial Application of VRFCC Suite Technology, which won the first prize of the National Science and Technology Progress Award. This book is the first monograph in China that comprehensively, systematically, specifically, and graphically discusses the theory, application methods, and effects of heat transfer enhancement in chemical processes. It is hoped that this book will provide strong guidance and reference value for engineers, researchers, and production management personnel in related fields as well as for teachers and students in colleges and universities.

This book is rewritten from *Heat Transfer Enhancement in Chemical Process* in Chinese, compiled by Lili Sun. Lili Sun was responsible for designing the structure of the book, drawing up an outline, setting the writing requirements, and coordinating and finalizing the manuscript. Lili Sun, Defei Wu, Liping Gao, Jiangning Ying, Zhiqiang Wei, and Yabiao Wang revised, translated and proofread the book, and Xiaohua Jiang, Zhao Yang, Bo Cao, Yiru Li, Jiayi Zhu, Wenxin Zhang, Liwei Cao, and Jianguang Cai participated in the translation.

Except for Professor Bofeng Bai from the National Key Laboratory of Multiphase Flow in Power Engineering at Xi'an Jiaotong University, the authors of the book in Chinese are all from SEI. With a high theoretical level and rich practical experience, they have made great efforts for the manuscript, showing a rigorous and meticulous attitude and professionalism. We owe them many thanks.

This book strives to achieve the integration of theory with practice, the integration of process technologies with process enhancement technologies, the integration of process units with utilities, and the integration of local with overall plant, so as to ensure that it is academic, systematic, original, novel, and practical. However, due to the limitation of our knowledge, there may be inadequacies in this book. Corrections from readers are more than welcome and greatly appreciated.

Lili Sun

Acknowledgments

Na An, Yan Bai, Gang Chen, Kaibei Chen, Maochun Chen, Wensong Dai, Wenyou Ding, Xiaoyuan Guo, Jian Han, Xi'ou He, Rongxing Jiang, Jianghai Jian, Changli Li, Chuhe Li, Guanghua Li, Jinfeng Li, Kaixiang Liu, Chang'ai Liu, Jiangfeng Liu, Jie Liu, Yongfang Liu, Liguo Ma, Yiqiang Nie, Jing Ning, Yuhong Ruan, Li Sha, Jian Sima, Zhibo Song, Yi Sun, Enmin Wang, Jianbo Wang, Qingchuan Wang, Shaohua Wang, Songtao Wang, Tianyu Wang, Xinquan Wang, Yanli Wang, Zhenwei Wang, Lei Wu, Yao Xu, Zhao Yang, Jianbin Yi, Chengzhi Yuan, Yifu Yuan, Zhongxun Yuan, Fangfang Zhang, Jianhua Zhang, Lixin Zhang, Weiqian Zhang, Yinrong Zhang, Bairen Zhao.

Contents

CHAPTER 1

Introduction

Contents

1.1 Significance of heat transfer enhancement

1.1.1 Heat transfer enhancement: An important means of energy conservation and emission reduction

Global ecological and environmental problems have become increasingly prominent in recent years; especially major issues such as global warming, environmental degradation, and frequent disasters have not only affected the sustainable development of the global economy and society but also threatened the basis of human existence at an increasingly rapid rate, garnering great attention from the international community. The United States, the European Union, and Japan are all actively committed to the

1

reduction of fossil energy use. China, in its *13th Five-Year Plan for Energy Development,* also proposes that the total energy consumption should be controlled within 5 billion tons of standard coal by 2020.

There are two ways to reduce the use of fossil energy. One is to adjust the energy structure by vigorously developing renewable energy to increase its proportion in energy consumption. The other is to improve the energy utilization efficiency of large energy consumers such as the process industry so as to save fossil energy consumption. Improving the comprehensive utilization efficiency of process energy and slowing down the degradation of energy in its utilization are the key links and source measures for energy conservation and emission reduction. Moreover, heat transfer enhancement is an important key to achieving the reduction in fossil energy use and has important significance for energy conservation and emission reduction.

1.1.2 Positive effects of heat transfer enhancement on cost saving

For the chemical industry, the investment in heat exchangers accounts for about 5%–15% of the whole plant investment. For a crude oil distillation unit, the investment in heat exchangers accounts for about 11%–16% of the total unit investment and 30%–40% of the static equipment investment. For an ethylene plant, the investment in heat exchangers accounts for about 6% of the unit investment and about 30%–40% of the static equipment investment. For a residual oil hydrogenation unit, the investment in heat exchangers accounts for about 5%–8% of the unit investment and about 34%–40% of the static equipment investment. In a thermal power plant, if boilers are regarded as heat exchange equipment, the investment in heat exchangers accounts for about 70% of the total investment in the overall plant [1]. In a refrigerator, the weight of the evaporator accounts for 30%–40% of the total weight of the refrigerator, and its power consumption accounts for about 20%–30% of the total consumption. Therefore, the reasonable design, operation, and improvement of heat exchangers are very important for saving investment, material use, energy consumption, and layout space.

1.1.3 Important contribution of heat transfer enhancement to energy conservation and emission reduction in chemical industry

The chemical industry is not only an important supplier of energy but also a large consumer of energy. The energy consumed by the chemical industry is mainly fuel gas and electricity, in which all the fuel gas is converted into heat energy to drive the processes. There is a large amount of heat transfer between process flows in the chemical processes in which the amount of heat energy circulation in the process of energy consumption is about 1.5–3 times the input heat energy.

Since the beginning of this century, petrochemical enterprises have made remarkable progress in energy conservation. On the one hand, traditional energy conservation technologies have gradually matured and been widely used. Especially in recent years, newly built oil refining enterprises have taken into consideration and applied a large number of mature energy conservation technologies during engineering design, bringing the energy consumption of main refining units such as crude oil distillation, catalytic cracking, continuous reforming, hydrotreating, and hydrocracking to an advanced level. On the other hand, the quality of oil products has been constantly upgraded. On January 1, 2023, National ⅥB Standard oil products were produced and supplied nationwide. Some indexes stipulated in the *Emission Standard of Pollutants for Petroleum Refining Industry* (GB 31570—2015) and the *Emission Standard of Pollutants for Petroleum Chemistry Industry* (GB 31571—2015) jointly issued by the Ministry of Environmental Protection and relevant ministries and commissions in 2015 met the world's most stringent standards. With the quality upgrading of oil products and the increasingly stringent environmental protection requirements, additional production units, processing technologies, and energy consumption are needed, which slow down the decreasing trend of oil refining energy consumption to a certain extent.

In the future, the energy conservation of chemical enterprises will be in a critical stage for a long time and require energy conservation planning, diagnosis, optimization, and evaluation from a systematic and intelligent perspective. Heat transfer enhancement is one of the important means to realize energy conservation and consumption reduction in chemical enterprises and plays an important role in energy conservation in the chemical industry.

1.2 Progress in heat transfer enhancement technology

Traditional heat transfer enhancement technologies in chemical processes focus on the heat transfer enhancement of equipment (such as heat exchanger) and components, which plays a positive role in the enhanced utilization of heat energy in chemical processes. However, in the engineering practice of today's chemical processes, the enhanced utilization of heat energy is a complex system engineering, which is decisively affected by factors such as process and the whole plant. Heat transfer enhancement in chemical processes should systematically consider various factors of enhanced utilization of heat energy. Up to now, heat transfer enhancement technology in chemical processes has been vigorously developed and extended. The first level is the strengthening of heat transfer equipment and components, the second level is the strengthening of processes, and the third level is the strengthening of the overall plant. The first level is the traditional heat transfer enhancement technology, which has been developed to the fourth generation. The second and third levels are the extension of heat transfer enhancement technology in chemical processes based on system engineering theory. This section introduces the main progress of heat transfer enhancement technology from the aspects of heat exchanger heat transfer enhancement and heat exchanger network optimization.

1.2.1 Heat transfer enhancement technology of heat exchanger

Heat exchanger is the equipment that transfers part of heat from hot fluid to cold fluid. Heat exchanger family products occupy an important position in process industries such as petrochemical, electric power, and metallurgys, and are widely used [2]. With the application of heat exchanger in various fields, heat transfer enhancement technology has been vigorously developed from the first generation to the fourth generation. For heat exchange tube, the first-generation heat transfer technology is represented by plain tube, which mainly achieves heat transfer function. The second-generation heat transfer technology is represented by parallel-fin and two-dimensional fin, which mainly change the structure of heat exchanger, improve process efficiency, and reduce energy consumption. The third-generation heat transfer technology is represented by three-dimensional rough element and three-dimensional fin, which mainly improve heat transfer efficiency. The fourth-generation heat transfer technology is also called compound heat transfer enhancement technology, which mainly uses compound enhancement method to achieve better heat transfer effect.

Heat transfer enhancement technology mainly focuses on four objectives [1,3–5]: the first one is to reduce the heat transfer area of initial design, reduce the volume of equipment, and reduce the metal consumption of equipment; the second is to improve the heat transfer capacity of the existing heat exchanger; the third is to make the heat exchanger work at a lower temperature difference; and the fourth is to reduce the heat exchanger pressure drop and reduce power consumption. Around the above objectives, researchers have carried out a lot of research [1,6,7], promoting the continuous progress of heat transfer enhancement technology of heat exchanger.

Generally, the heat transfer capacity of heat exchanger can be strengthened from three aspects. The first one is to increase the heat exchange area, such as low-finned tube, corrugated tube, and porous surface tube, to not only increase the effective heat exchange area but also improve the flow pattern of fluids, improving the heat transfer performance of heat exchanger. The second is to increase the effective average temperature difference and increase the temperature difference between the inlet and outlet of cold and hot fluids as much as possible if the process conditions permit. If the change in fluid temperature is not allowed by the process conditions, the effective average temperature difference can be increased by countercurrent heat transfer. The third is to improve the total heat transfer coefficient of heat exchanger. By improving the membrane heat transfer coefficient and reducing the fouling thermal resistance, the total heat transfer coefficient can be increased and the performance of heat exchanger can be improved.

Table 1.1 shows the heat transfer enhancement technology of typical heat exchangers in chemical industry [8].

Table 1.1 Typical heat transfer enhancement technology in chemical industry.

Shell and tube heat exchanger	Shell side enhancement		Structure features and mechanism of heat transfer enhancement	Suitable conditions
		Corrugated tube	The outer thread structure of the heat exchange tube enlarges the secondary heat transfer area	No phase transition, boiling, condensation
		Porous surface tube	The porous structure increases the bubble core	Boiling
		T-shaped finned tube	The T-shaped channel increases the bubble core	Boiling
		Serrated fin tube	The discontinuous fins make the condensate film thinner	Condensation
		Longitudinal groove tube	The V-shaped grooves make the condensate film thinner	Condensation
		Rod baffle	Rod baffle changes the flow direction of the medium, and the longitudinal flow reduces the pressure drop and eliminates vibration	Working condition of phase transition
		Helical baffle	Spiral flow reduces flow dead zone and vibration tendency	No phase transition, boiling, condensation
	Tube side enhancement	Porous surface tube	The porous structure increases the bubble core	Boiling
		Inserts	Inserts change flow state, destroy boundary layer, strengthen heat transfer, and increase turbulence	No phase transition

Continued

Table 1.1 Typical heat transfer enhancement technology in chemical industry—cont'd

Double-sided enhancement	Transversally corrugated tube	One time processing with double-sided forming and double-sided strengthening	No phase transition, boiling, condensation
	Twisted tube	No baffle structure, spiral flow of fluid at tube and shell side	No phase transition, boiling, condensation
	Corrugated pipe	Increase fluid turbulence	No phase transition, boiling, condensation
	High-efficiency tube with inner porous surface and external longitudinal groove	The porous surface inside the tube strengthens boiling heat transfer, and the vertical groove outside the tube strengthens condensation	Vertically mounted reboiler and working conditions of boiling in tube and condensation outside tube
Others	Coil-wound heat exchanger	Compact equipment, multiflow heat exchange in a shell	No phase transition, boiling, condensation
Plate heat exchanger	Spiral plate	Flexible channel width adjustment, pure countercurrent heat exchange	Heat transfer of high viscosity fluid containing particles
	Corrugated plate	Compact heat exchanger, high turbulence	No phase transition, boiling, condensation
	Plate-fin type	Compact heat exchange equipment, multiflow heat exchange in a shell	No phase transition, boiling, condensation

1.2.2 Optimization technology of heat exchanger networks

The optimization design of heat exchanger networks to improve the comprehensive utilization efficiency of energy has been widely used in high energy consumption industries such as the oil refining and petrochemical industry, garnering continuous attention from researchers. Heuristic and mathematical programming methods are typical comprehensive optimization methods of heat exchanger networks.

1.2.2.1 Heuristics

Based on the principles of thermodynamics and engineering design experience, the heuristic method forms the initial network structure according to some design objectives and heuristic rules and then adjusts the initial structure until it reaches the optimal structure. The heuristic method is simple and easy to use and is widely used in engineering practice. Pinch point technology is the most widely used and most influential heuristic method nowadays.

In 1978, Linnhoff et al. first proposed the pinch point technology. After application and development, it was promoted from heat exchanger network synthesis to energy analysis and optimization of the whole process system [8–10]. Pinch point design method firstly draws the composite curve of cold and hot flows, determines the pinch point position through the temperature-enthalpy diagram (T-H diagram) as shown in Fig. 1.1, and calculates the minimum amount of cold and hot utilities. Three basic rules should be followed when applying pinch point technology to heat exchanger network design. Firstly, no heat shall pass through the pinch point. Secondly, no cold utilities should be used above the pinch point. Thirdly, no thermal utilities should be used below the pinch point. Pinch point technology provides direct guidance and suggestions for engineering designers and has important practical value.

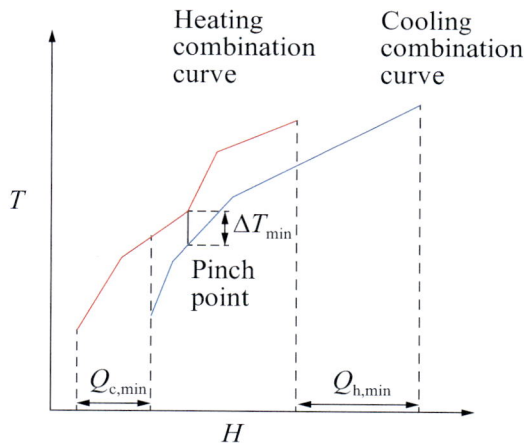

Fig. 1.1 Composite curve of heat on T-H diagram.

1.2.2.2 Mathematical programming method

The mathematical programming method is to establish the mathematical model of the heat exchanger network synthesis problem and obtain the optimal heat exchanger network structure by solving the objective function that satisfies certain constraints. According to the characteristics of objective functions and constraints, commonly used mathematical models can be divided into Linear Program (LP), Nonlinear Program

(NLP), Mixed Integer Linear Program (MILP), and Mixed Integer Nonlinear Program (MINLP). The mathematical programming method generally includes three steps to solve the problems in heat exchanger network synthesis [8,11–14]:

Firstly, a superstructure model that contains all possible alternatives is constructed.

Secondly, the superstructure model is written into a mathematical expression and its basic form is:

$$\begin{cases} \min \; Z = f(x, y) \\ \text{s.t.} g_i(x, y) \leq 0 \qquad i \in I \\ h_j(x, y) = 0 \qquad I \in J \end{cases} \qquad (1.1)$$

Thirdly, an appropriate algorithm solution, Eq. (1.1), is selected to obtain the optimal heat exchanger network structure.

The heat exchanger network involves many factors and is very complicated. In the establishment of mathematical models of various heat exchanger networks, simplification and assumptions are inevitable, so the mathematical model established deviates from the actual process described, and the optimal solution obtained will deviate from the actual optimal solution.

In engineering applications, the optimization of heat exchanger networks based on pinch point technology still needs to solve many specific problems, such as threshold problems, the layout of the rectification tower, the layout of the heat pump, and multi-unit or plant-wide thermal integration. The main functions of heat exchanger network optimization technology are to lay the frame of optimal utilization of heat energy in the whole plant and the process units and to clarify the direction of heat transfer enhancement at the macroscopic level. Heat transfer enhancement technology of the heat exchanger is a specific enhancement measure to realize optimal utilization of heat energy. Both of them complement each other and are important ways to enhance heat transfer in chemical processes.

1.3 Engineering applications of heat transfer enhancement

1.3.1 Engineering strategies of innovative heat transfer enhancement

Fig. 1.2 shows the engineering strategy of innovative heat transfer enhancement in refining and petrochemical enterprises, including the objectives of innovative heat transfer enhancement, related enhancement principles, application background and some enhancement technologies, etc.

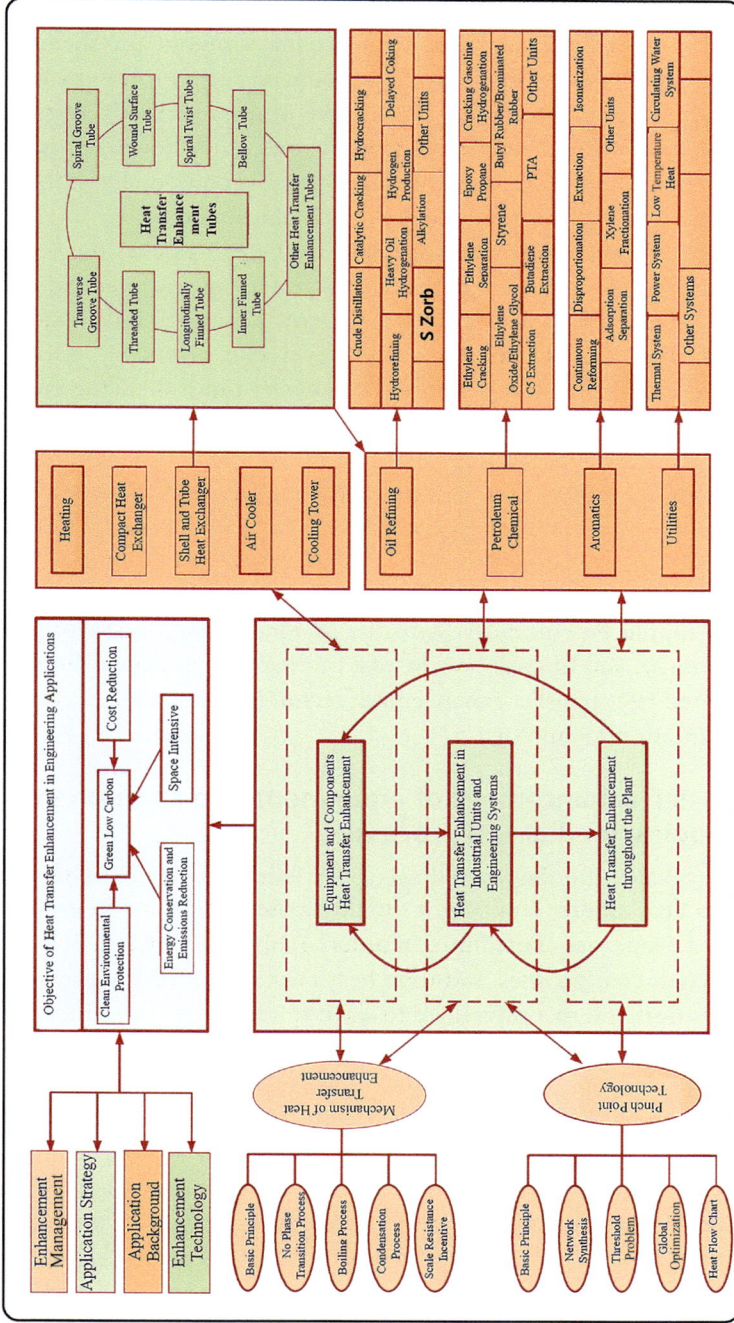

Fig. 1.2 Engineering strategy diagram of innovative heat transfer enhancement for refining and petrochemical enterprises.

Energy saving and emission reduction, clean production and environmental protection, spatial intensification, investment saving, and green and low carbon are not only the overall goals of modern refining and petrochemical enterprises but also the new goals of improving the competitiveness of refining and petrochemical enterprises by heat transfer enhancement. The implementation of the innovative heat transfer enhancement objective is a complex system engineering, which involves three levels of heat transfer enhancement. Heat transfer enhancement of equipment and components is the first level, which is the common basis for refining and petrochemical enterprises to achieve heat transfer enhancement. Heat transfer enhancement in the processes is the second level, which is an important link connecting the heat transfer enhancement of equipment and components with the plant-wide heat transfer enhancement. The plant-wide heat transfer enhancement is the third level. The second level relies on and limits the first level. The third level relies on the first level and the second level and at the same time provides feedback and regulation on both levels.

The innovative heat transfer enhancement in refining and petrochemical enterprises requires a global tradeoff optimization within the entire plant, rather than simply adding up the heat transfer enhancements of equipment and components or specific processes. Among them, the heat transfer enhancement of equipment and component should be combined with the integrated innovation of unit technology, the heat transfer enhancement of processes should be combined with the integrated application of process and engineering technology, and the plant-wide heat transfer enhancement should be combined with integrated heat transfer enhancement technology and collaborative application of plant optimization technology.

1.3.2 Heat transfer enhancement of equipment and components for special technologies, media, and processes

There are many special technologies and media in refining and petrochemical enterprises, so enhanced heat transfer equipment should be used according to process characteristics and physical properties of medium. In order to solve the problem of low shell side film heat transfer coefficient in shell and tube heat exchanger, for the process medium without phase transition, it is recommended to use heat transfer enhancement tubes such as low-finned tube, transversally corrugated tube and corrugated low-finned tube; for the process medium condensation, it is recommended to use heat transfer enhancement tubes such as longitudinal groove tube and serrated fin tube; and for process medium boiling, it is recommended to use heat transfer enhancement tubes such as T-shaped finned tube and tube with porous surface. In order to solve the problems of high pressure drop and vibration caused by traditional structures, it is recommended to use double segmental baffle, triple segmental baffle, helical baffle, rod baffle, etc. In order to solve the contradiction between improving the film heat transfer coefficient on the shell side and the increased pressure drop and vibration, the rod baffle can be integrated with low-finned tube or the helical baffle can be integrated with low-finned tube. In order to solve the problem of poor film heat transfer coefficient caused by laminar flow in tube, it is recommended to use inserts to strengthen fluid disturbance and destroy the boundary layer on the tube wall surface. In order to further enhance the film heat transfer coefficient under turbulent flow in the tube, it is recommended to adopt corrugated tube, finned tube, rough ribbed tube, and porous surface tube. For heat exchangers with similar film heat transfer coefficients both on the tube side and shell side, double-sided heat transfer enhancement elements are recommended, such as corrugated tube, corrugated low-finned tube, spiral-twisted tube, and porous surface inside tubes with longitudinal groove tubes outside.

In addition, in order to enhance boiling heat transfer at a small temperature difference, porous surface tube heat exchanger was developed and applied in the bottom reboiler of the raffinate column in aromatics complex. In order to solve the problems of low film heat transfer coefficient and high pressure drop of high viscosity fluid, a spiral plate heat exchanger was developed and applied in the hot high-pressure separation steam generator of the slurry bed hydrogenation unit. In order to solve the problems of the small temperature difference and narrow space, plate and frame heat exchanger was applied in the jacket water cooler of high-density polyethylene unit. In order to solve the problems of large flow rate, large load, deep temperature overlap of hot and cold fluid, and small allowable pressure drop, fully welded plate and shell heat exchanger was applied in the reactor feed-effluent heat exchanger of the continuous reforming unit. In order to solve the problems of small space and small allowable pressure drop, the plate-type air cooler was developed and applied. In order to solve the problem of the low outlet temperature of the process medium, an evaporative air cooler was developed and applied. In order to solve the problem of multi-flow heat transfer in one equipment, the cold box integrated with a plate-fin heat exchanger was applied in the ethylene plant. All these heat transfer enhancement technologies are for the heat transfer enhancement of equipment and components for special technologies, media, and processes.

For example, 8 streams need to exchange heat with each other in an ethylene unit, which requires 21 sets of shell and tube heat exchangers, and the equipment weighs 266 tons. If the cold box shown in Fig. 1.3 is used, multiple streams can be collected in a single equipment for heat exchange. The weight of the equipment can be reduced to 20 tons, effectively reducing the weight and floor space of the equipment and achieving the goal of intensive design.

For example, the evaporation air cooler can directly cool the process medium to below the dry bulb temperature, and the cooling water cooler can be canceled, not only saving the metal consumption but also saving the corresponding investment of the cooling water tower, cooling water pump, cooling water pipe network, and the steel framework platform.

The rod baffle is mainly used to eliminate vibration and decrease shell side pressure drop, and the film heat transfer coefficient of the shell side decreases accordingly.

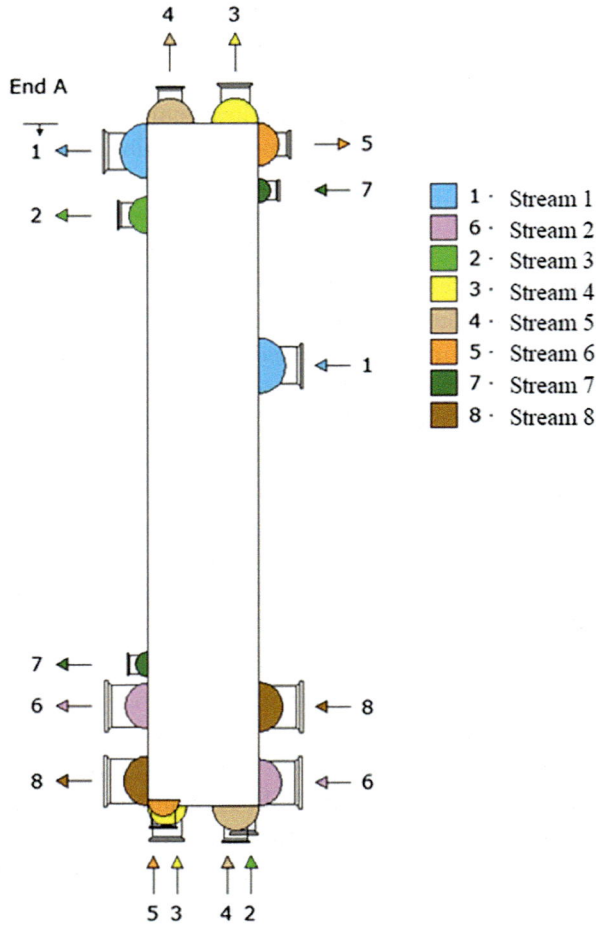

Fig. 1.3 Schematic drawing of cold box.

Vibration can be easily induced when the gas phase medium flows across the tube bundle. When the vibration cannot be eliminated by adjusting the baffle, the resonant frequency can be avoided by enlarging the shell diameter to decrease the velocity on the shell side. Taking a compressor interstage cooler as an example, the vibration warning can be completely eliminated only when the shell diameter is enlarged to 1200 mm for a shell and tube exchanger with a segmental baffle. If the rod baffle is used instead of the segmental baffle, for the same length of heat exchange tube, a diameter of 700 mm would be sufficient to meet the requirements of heat exchange and eliminate the hidden vibration.

The application of heat transfer enhancement technology can not only reduce the weight of equipment but also affect the process, achieving the purpose of energy conservation and consumption reduction. If the porous surface tube is used for a light hydrocarbon reboiler, the temperature level requirement of hot fluid can be reduced so that lower-grade utilities or low-temperature medium can be used in the unit. Plate heat exchanger or coil-wound heat exchanger can reduce the temperature difference in the hot end; when used for raw material heaters, it can reduce the furnace load and fuel consumption. Meanwhile, it can reduce the subsequent power consumption and cooling water consumption of the air cooler.

Fluid inside the flow channel of the spiral plate heat exchanger shows a spiral flow, with severe turbulence, and fouling is hard to settle down. When used for heat transfer of high viscosity and easy scaling medium, it can obtain a higher heat transfer coefficient, and its self-cleaning function can prolong the operation period.

1.3.3 Heat transfer enhancement of processes

Heat transfer enhancement of processes is of great significance to the energy conversation, emissions reduction and investment saving. Some typical processes have special characteristics of heat utilization and heat transfer enhancement.

1.3.3.1 Crude oil distillation unit—The best application demonstration of pinch point technology

Crude oil distillation is the first production process in the refining industry. After heat exchange, crude oil is fractionated into products with different boiling ranges. After heat exchange of products, the entire production process is completed. Only physical changes are involved. The energy consumption is high because a great amount of medium is heated and cooled in the unit. Fuel consumption accounts for 70%–85% of the energy consumption of the unit. Making full use of the heat energy released by fuel combustion through the complex heat exchanger network is the key to rational utilization of energy in the unit.

The crude oil distillation unit has a complex heat exchanger network. Let's take a 3.5 Mt/a crude oil distillation unit in China as an example, as shown in Fig. 1.4. There are 33 heat exchange stations with about 69 sets of heat exchangers merely in the crude oil preheating section, as shown in Fig. 1.4. During the engineering design of the heat exchanger network, the pinch point technology is fully applied, and the crucial factors affecting the heat exchanger network are evaluated and selected comprehensively according to engineering experience.

The first is the optimization of crude oil routes. Multiroute design can realize the reasonable matching of cold and heat flows to a certain extent and improve the flexibility of the heat exchanger network. In this case, the crude oil is divided into three heat transfer routes, while the desalted crude oil and topped oil are divided into two heat transfer routes. Equal flow is designed, and the heat transfer quantity and pressure drop of each route are basically equal.

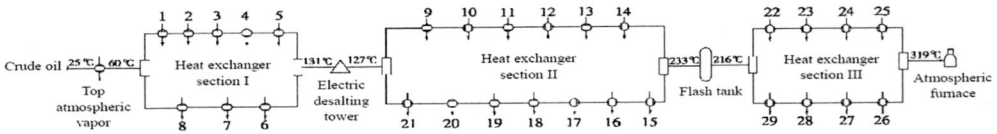

Fig. 1.4 Schematic diagram of heat exchanger network flow.

The second is the reasonable heat transfer sequence of hot flow in the heat exchanger network. First of all, heat sources with small heat capacity and low temperatures, such as primary fractionator top pump around and atmospheric top pump around, are used to pre-heat crude oil in the heat exchange section I. Secondly, heat sources with small heat capacity and high temperature level also participate in the preheating of crude oil in the heat exchange section I. Because the temperature drops rapidly during heat exchange, the temperature difference at the outlet is small. Thirdly, heat sources with large heat capacity and high temperature level exchange heat for the first time in the rear section and then return to the front section for heat exchange with cold flow. For example, high-temperature vacuum residue exchanges heat with topped oil in heat exchange section III, and then exchanges heat with desalted oil in section II and crude oil in section I successively.

Table 1.2 lists some parameters of the 10-million-ton crude oil distillation units designed and constructed by Sinopec Engineering Incorporation (SEI). The energy consumption of these units was found to achieve an international advanced level with heat exchanger network optimization and enhanced heat transfer equipment. After the analysis and evaluation of the crude oil distillation unit in Huizhou Phase II, Aspen Tech believed that the final heat exchange temperature reached a level that is extremely difficult to reach.

Table 1.2 Key data of 10-million-ton crude oil distillation units in China.

Enterprise	Processing capacity (Mt/a)	Final temperature of heat exchange (°C)	Energy consumption (kgoe/t)	Remark
Hainan Refining	8	290	8.5	Atmospheric residue sent out of the unit
Qingdao Refining	10	320	9.0	Vacuum deep cut
Huizhou Phase I	12	310	8.2	
Huizhou Phase II	10	300	8.0	Supply heat to light hydrocarbon units
Quanzhou Petrochemical	12	320	8.5	Vacuum deep cut, supply heat to light hydrocarbon units, heat input from FCC unit

1.3.3.2 FCC unit—Heat exchanger network for hot end threshold problem

The FCC unit is the core unit for converting heavy oil into light oil in refineries. The unit produces a lot of coke as a by-product, which will release a large amount of heat from combustion in the regenerator, exceeding its own heat exchange demand. At the same time, the unit needs cold utilities. It is a typical heat exchanger network for hot end threshold problems, as shown in Fig. 1.5. The key to strengthening the utilization of heat energy is to make good use of the heat released by coke combustion. The thermal integration between the FCC unit and other refining units is possible. The rational design of the heat exchanger network can realize the effective thermal integration of the FCC unit with other units, which is beneficial to reducing the energy consumption of the whole refinery.

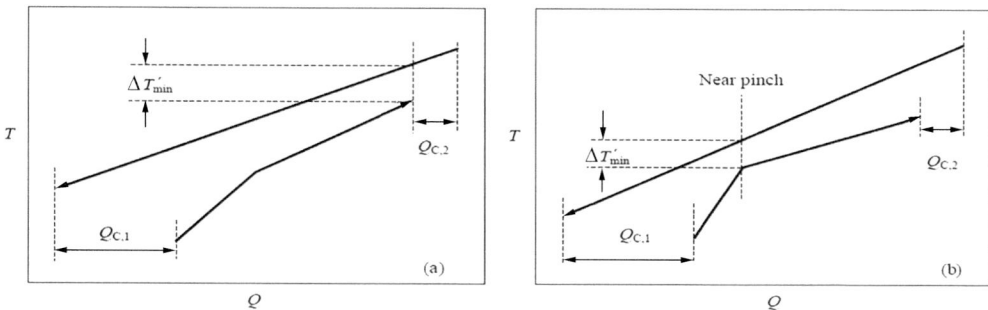

Fig. 1.5 Diagram of heat exchanger network for hot end threshold problem. (A) Basic heat exchanger network. (B) Optimized heat exchanger network.

As for the specific heat transfer process, for high temperature level heat energy, the heat released by coke combustion in the regenerator is partly removed by the heat recovery unit inside and outside the regenerator for the generation of medium-pressure or subhigh-pressure steam and the other part is brought into the fractionation system by the reaction oil and gas, to generate medium-pressure steam by the circulating oil slurry or thermal integrate with heat from other units. The medium temperature level heat energy of the unit is brought into the fractionation system by the reaction oil and gas, i.e., the absorption oil circulation reflux, the first midpump around and the second midpump around, and is mostly used to preheat the catalytic cracking raw oil and integrate with the heat from the absorption stabilization system, providing a heat source for the reboilers of desorption tower and stabilization tower. Low temperature level heat energy of the unit, such as the top pumparound and overhead oil gas from the tower, is mostly recovered by hot medium water or directly integrated with the heat from gas fractionation unit.

1.3.3.3 Hydrogenation unit—Coordinated heat transfer enhancement of processes and equipment and components

Hydrogenation units in refineries include the hydrotreating unit and hydrocracking unit, which play an important role in the oil refining process. The heat energy utilization of hydrogenation process units has two characteristics: the first is to make full use of the heat generated by the reaction by enhancing the raw oil-effluent heat exchanger and the relevant processes, and the second is to divide the heat exchanger network into the high-pressure part and low-pressure part, which are optimized and integrated respectively. The optimization and integration of the heat exchanger network are combined with the product separation process.

As for the heat transfer enhancement process of hydrogenation feed preheating, there are two kinds of heating processes for the feed and hydrogen: one is mixing hydrogen with feed at the inlet of the heater process, and the other is mixing feed with hydrogen at the outlet of the hydrogen heater. For the "mixing hydrogen with feed at the inlet of heater" process, the feed and hydrogen are mixed into a two-phase flow and then heat exchange with the reaction products, and then enter the feed heater. For "mixing feed with hydrogen at the outlet of hydrogen heater," the feed and hydrogen respectively exchange heat with reaction products in single phase, and the feed is mixed with hydrogen at the outlet of the feed heater. The two heat exchange processes are shown in Fig. 1.6. Compared with "mixing feed with hydrogen at the outlet of hydrogen heater," "mixing hydrogen with feed at the inlet of heater" strengthens the heat transfer coefficient of cold fluid and can save more than 40% of the heat transfer area.

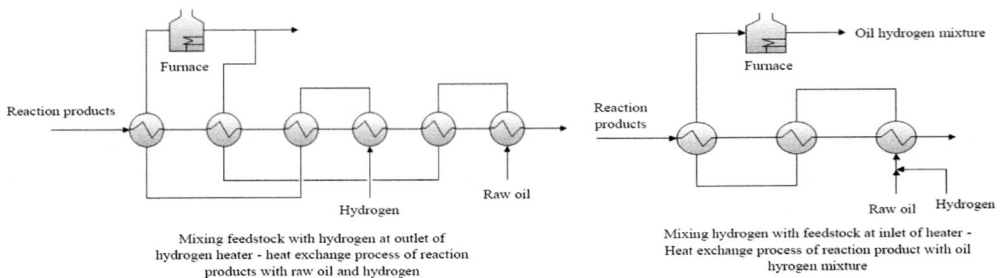

Fig. 1.6 Schematic diagram of different heat transfer processes.

There are two processes for cooling and separation of hydrogenation reaction products: one is the "hot high-pressure separation" process and the other is the "cold high-pressure separation" process. In the "hot high-pressure separation" process, the reaction product is cooled to 240°C or higher temperature and then enters the hot high-pressure separator for gas-liquid separation. The separated liquid phase is depressurized and fractionated, while the gas phase continues to be cooled to 50°C and enters the cold high–pressure separator. In the "cold high-pressure separation" process, the reaction product is cooled to 50°C and then enters the cold high-pressure separator for gas–liquid separation, and the cold oil is depressurized and then exchanges heat with the reaction products to above 230°C before entering the fractionation part. The heat exchange flow of the two processes is as shown in Fig. 1.7. The "hot high-pressure separation" process strengthens the heat transfer of the hot flow (reaction products), requires 40% less heat transfer area than that of the "cold high-pressure separation" process, and the fuel consumption saved is about 5%–10%.

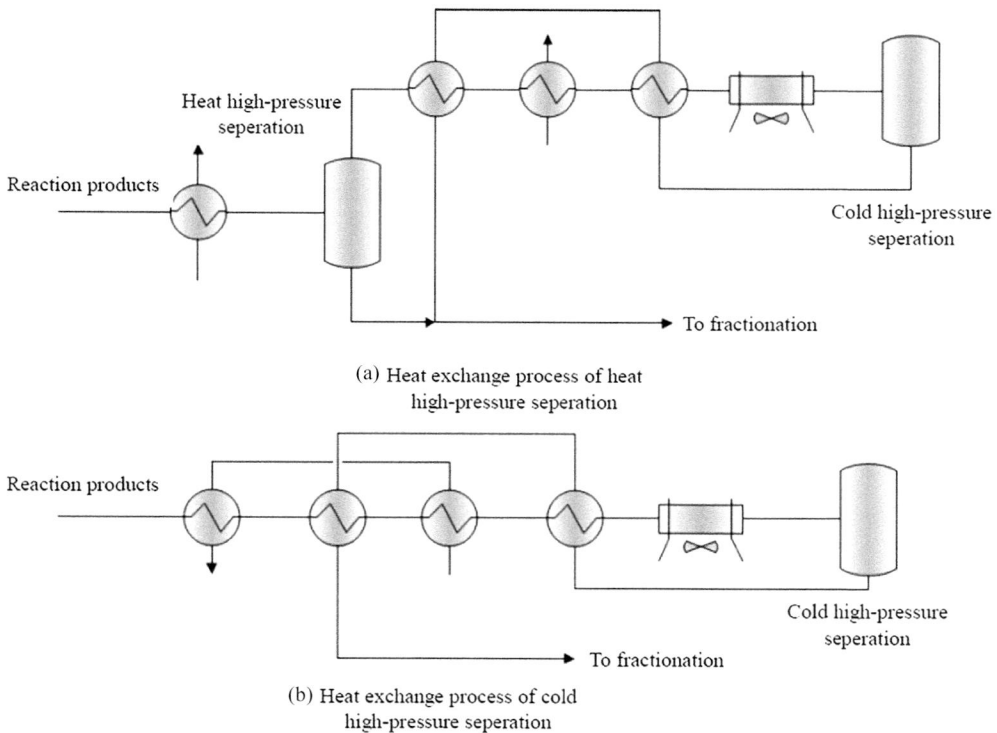

(a) Heat exchange process of heat high-pressure seperation

(b) Heat exchange process of cold high-pressure seperation

Fig. 1.7 Schematic diagram of "hot high-pressure" and "cold high-pressure" separation process. (A) Hot high-pressure separation process. (B) Cold high-pressure separation process.

1.3.3.4 Ethylene unit—Optimal utilization of cold energy

The ethylene unit consists of ethylene cracking and ethylene separation, and ethylene separation includes cracking gas drying, component precutting, hydrogen separation, methane separation, ethylene/ethane separation, C_3 separation, etc. The ethylene unit requires a large amount of cold energy as a cooling medium because more distillation and separation of low-carbon hydrocarbons are involved. The energy consumption of cold utilities in ethylene plant accounts for about 25% of the total energy consumption of the plant. Rational and optimal utilization of cold energy is key to heat energy utilization and heat transfer enhancement in the ethylene plant.

As for the cold energy required for ethylene separation, room temperature grade is provided by cooling water, 40°C to −37°C grade is provided by propylene (or methane/ethylene/propylene ternary) refrigeration system, −37°C to −98°C grade is provided by ethylene (or methane/ethylene binary, or methane/ethylene/propylene ternary) refrigeration system, and below −98°C grade is provided by methane, binary, ternary, or methane tail gas expander refrigeration system.

Fig. 1.8 is the schematic diagram of the ethylene separation system.

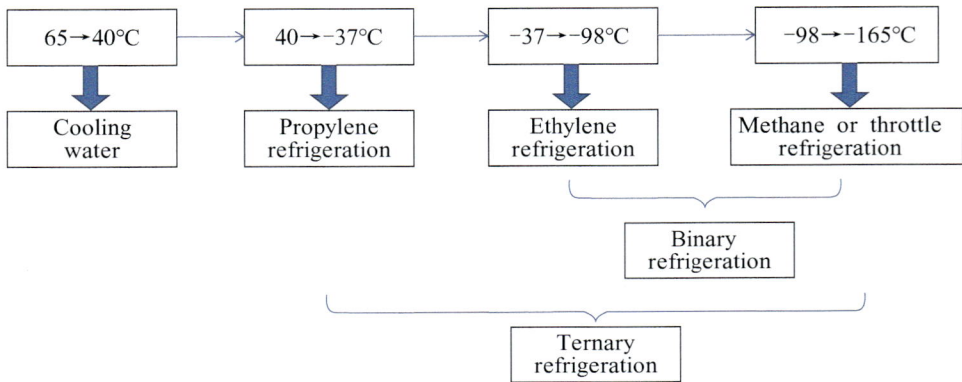

Fig. 1.8 Cooling principle diagram of ethylene separation system.

There are two ways to strengthen cold energy utilization in the ethylene plant. The first is the wide application of pinch point technology, such as giving priority to recycle cold energy by process flows, rationally selecting cold and heat utilities, decreasing heat transfer temperature difference, using multitemperature-level refrigeration according to process requirements, and reducing the power of the refrigerator. The second is to widely use a cold box and plate-fin heat exchanger for the demethane tower heat exchanger, deethane tower heat exchanger, ethylene rectifying tower heat exchanger, and methanation heat exchanger in the cryogenic area of the ethylene cold separation system.

Sinopec is one of the world's top five ethylene technology licensors. After more than 30 years of independent research and development of ethylene technology, its overall technical level is advanced. With energy conservation and consumption reduction measures such as heat transfer enhancement, the energy consumption per unit product of its ethylene plant has reached the international leading level.

1.3.3.5 Aromatics complex unit—Heat source integration and optimal utilization of low-temperature heat

The heat energy enhancement integration scheme of typical heat source integration is adopted in the aromatics complex unit. The xylene column is the material and heat center of the aromatics complex unit. The xylene column bottom reboiler heater indirectly provides a heat source for the reboiler of six columns, including the raffinate column, extract column, deheptanizer column, heavy aromatics column, desorbent rerun column, and o-xylene column. The thermal integration process of the aromatics complex unit is as shown in Fig. 1.9. When the heat integration scheme is implemented, in order to recover the condensation heat of overhead materials from the xylene column, the heat integration of the xylene column and the distillation column such as the raffinate column and the extract column is usually carried out. The xylene column is operated under moderate pressure, and the raffinate column and extract column are operated under atmospheric pressure. The condensation latent heat of overhead gas phase flow from the xylene column acts as the heat source of the bottom reboiler of the raffinate column and extract column. Taking a unit as an example, about 80 MW more heat is recovered after the implementation of the scheme. In order to further recover the condensation from the overhead of the raffinate column and extract column, the operating pressure of the raffinate column and extract column is increased from normal pressure to 0.25–0.31 MPa

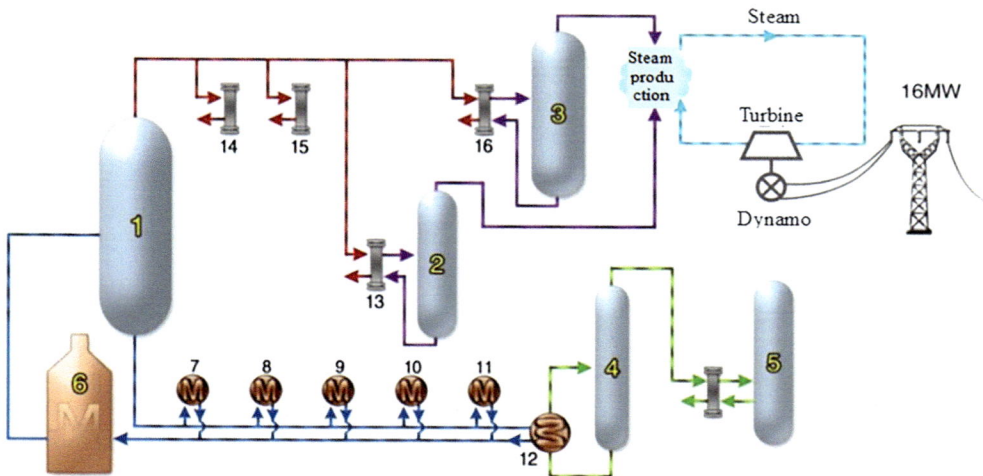

Fig. 1.9 Schematic diagram of heat energy enhancement of heat source integration of aromatics complex unit. (1) Xylene column; (2) extract column; (3) raffinate column; (4) heavy aromatics column; (5) reformate splitter; (6) reboiler heater; (7) tatoray stripper reboiler; (8) extract column no. 2 reboiler; (9) desorbent rerun column reboiler; (10) deheptanizer column reboiler; (11) xylene clay treater feed heater; (12) heavy aromatics column reboiler; (13) extract column no. 1 reboiler; (14) PX column reboiler; (15) OX column reboiler; (16) raffinate column reboiler.

through process optimization. The condensation heat temperature at the top of the column was around 140–150°C, and 0.5 MPa saturated steam was generated at the top of the column. The pressure of the corresponding xylene column is raised to 1.2 MPa, and the overhead material provided heat for the reboiler of the raffinate column and extract column. Taking a 0.6 Mt/a aromatics complex unit as an example, about 90 MW heat is further recovered after the implementation of the scheme. The condensation heat from the raffinate column and extract column can generate 0.5 MPa saturated steam at a rate of 180 t/h, and the net power generation is 18,000 kW. The energy consumption per ton of p-xylene product can be reduced by about 50 kgoe.

The heat source integration of "one furnace and seven columns" and the heat integration of multiple columns through steam grade upgrade are typical cases of heat transfer enhancement and rational energy use through process optimization. In the aromatics project in Hainan Refinery, the heat transfer enhancement integrated scheme reduced the energy consumption of the plant by 15% compared to the same period last year, which was comparable with the international advanced process and effectively opened up international market for domestic aromatic technologies.

1.4 Heat transfer enhancement of the whole plant

The heat transfer enhancement of the whole plant depends on the heat transfer enhancement of equipment and components and the heat transfer enhancement of processes, and also gives feedback and coordination strategies, so as to achieve global optimization at the enterprise level. The ways to achieve heat transfer enhancement in the refining and chemical industry in the future include optimizing and improving the direct heat supply and storage temperature level of materials in the whole plant, carrying out thermal integration and intensive design of process units, carrying out centralized refinery gas processing, generating higher-pressure grades of steam for cascade utilization and to reduce fuel consumption, recycling and optimizing the utilization of low-temperature heat resources in the whole plant, and implementing cascade utilization and cold transfer enhancement of cooling water system.

Since the 21st century, Sinopec Engineering Incorporation has undertaken the design and construction of a number of 10-million-ton refineries. After the start-up of Hainan Refinery, large-scale single-line, 10-million-ton, localization, technology integration, and technology innovation have gradually become the keywords and themes in China's oil refining field. After that, a series of large refineries were built and put into operation, including 10 Mt/a of Qingdao Refinery, 12 Mt/a of Huizhou, and 12 Mt/a of Quanzhou Petrochemical, etc. Since the engineering strategy of innovative heat transfer enhancement has been applied and improved during the design and construction of the above refinery, this section takes the above refinery as an example to make corresponding explanations.

1.4.1 Direct heat supply and storage of materials in the whole plant

The direct heat supply of materials in the whole plant is very important for the efficient and reasonable utilization of heat energy in refineries and petrochemical enterprises. On the one hand, the materials are effectively prevented from being cooled in upstream process units and then heated in downstream process units. Repeatedly cooling down and heating up the materials among process units wastes not only the heat energy needed for heating up but also the cold utilities needed for cooling down. On the other hand, after the realization of the direct heat supply of materials, the thermal conditions of process units change, and the secondary optimization of the heat exchanger network of the process unit is needed. The secondary optimization of the heat exchanger network is an important opportunity to improve the energy consumption level of refining and chemical enterprises.

In recent years, with the progress of technology and equipment, direct heat supply of materials should be promoted comprehensively. The recommended target temperature of direct heat supply materials is as follows:

(1) For the feed directly supplied to the hydrocracking unit, such as atmospheric residue, vacuum gas oil, and vacuum residue, the recommended temperature of direct heat supply materials is 140–180°C.

(2) For the feed directly supplied to the FCC unit, such as hydrotreated residue and hydrotreated gas oil, the recommended temperature of direct heat supply materials is 180–210°C.

(3) For the feed directly supplied to the delayed coking unit, such as vacuum residual, deasphalted oil, and FCC slurry oil, the recommended temperature of direct heat supply materials is 180–220°C.

(4) For the feed directly supplied to the hydrogenation unit, the recommended temperature of direct heat supply of materials is 130–160°C.

(5) For the feed directly supplied to the diesel hydrogenation unit, the recommended temperature of direct heat supply materials is 130–160°C.

(6) For the feed directly supplied to the kerosene hydrogenation unit, the recommended temperature of direct heat supply materials is 110–130°C.

(7) For the feed directly supplied to the S Zorb unit, the recommended temperature of direct heat supply materials is 100–150°C.

During the direct heat supply of materials in the whole plant, it is necessary to pay attention to the steam generation of the feed to downstream process units in upstream process units. Steam generation is also a typical cooling process of flows. For example, hydrotreated residue and hydrotreated gas oil in refineries are mostly used for the generation of 1.0 and 0.4 MPa steam and then sent to the FCC unit as the feed for secondary heating of circulating oil slurry. From a global perspective, after the generation of 1.0 and 0.4 MPa steam in the upstream residue hydrogenation unit and gas oil hydrogenation unit, the heat of circulating oil slurry in the FCC unit is required, while the circulating oil slurry can generate 3.5 MPa steam, so the heat at high-quality temperature level is not rationally utilized. Such a process should be optimized and combined with the direct heat supply of materials in the whole plant. At the same time, for the intermediate storage of materials as per process requirements, the storage temperature of materials should be reasonably increased to avoid heat loss caused by storage.

1.4.2 Intensive design of process units

In order to meet the needs of rational utilization of resources and energy conservation and consumption reduction, the concept of a complex plant should be changed from the traditional combination of individual units to the whole refinery as a "large complex plant". Hainan Refinery has carried out overall optimization of its process units, utilities, and systems. The traditional direct heat supply of materials between units has been changed to cross mutual supply, so as to optimize the comprehensive utilization of heat and greatly reduce the mechanical energy and footprint. The main process units are divided into four functional areas, i.e., heavy oil processing, distillate oil processing, gas processing, and environmental protection, which are relatively concentrated to facilitate the transportation of large material, reduce transportation distance, and avoid mechanical energy loss and heat loss. The average storage days of intermediate raw materials are 2 days. After reasonable division of the functional area, in order to realize the mutual supply of heat between units, the units are integrated into seven complexes with cross-mutual supply of materials between units. For example, heat integration of the FCC unit and gas fractionation unit is achieved to use the waste heat in the FCC unit to heat hot water, which is provided to the gas fractionation unit as a heat source; residue hydrotreating is arranged together with the hydrocracking unit to realize the shared use of some compressors, high-pressure pumps, and utilities; continuous reforming unit and isomerization unit share the stabilization column system to realize the close integration of the processing process; and integration of materials is achieved via the recovery and centralized treatment of light hydrocarbons in refinery and the centralized optimization of sour water and solvent regeneration and sulfur recovery system.

The recovery and centralized treatment of light hydrocarbon in the whole plant adopts integrated technology to recover effective components such as slight hydrocarbon, hydrogen, ethylene, and ethane on the basis of less energy consumption by means of adsorption, absorption, refrigeration, and separation, so as to realize effective utilization of resources. At the same time, the refinery gas recovery process expands new heat trap for the large-scale system to utilize low-temperature heat. On the one hand, 70–90°C hot-medium water is a good heat source for the reboiler during gas distillation and separation. On the other hand, the gas freezing separation process needs a lot of low-temperature cooling medium, and lithium bromide low-temperature cooling water can meet the production needs. It is a good way for the high-value-added utilization of hot medium water in the refinery. For a 10 Mt/a hydrogenation refinery, the setup of a hydrogen recovery system can recover about 7000 tons of hydrogen every year. At the same time, hydrogen-rich pressure swing adsorption (PSA) tail gas can be used as raw material for hydrogen plants after pressure boost. This innovation provides an outlet for PSA tail gas with low heat and also saves raw material for hydrogen plants. In addition, it can consume the excess hot medium water obtained by recovering process low-temperature heat, reduce process cooling load, and save power consumption and cooling water consumption.

The integrated setup of environmental protection units facilitates the improvement of low-pressure steam energy utilization. Environmental protection units in refineries include sour water stripping unit and amine regeneration unit, which are major consumers of low-pressure steam in refineries. Taking a 10 million ton refinery as an example, the sour water stripping unit and amine regeneration unit consume about 120–150 t/h of 0.4 and 1.0 MPa steam, which accounts for 5.0%–10.0% of the energy consumption of the whole refinery. The integrated setup of the environmental protection units is mainly because of the different types and quantities of impurities such as ammonia, phenols, and cyanide in sour water produced by different units. Separate treatment can reduce water pollution, reduce the additional energy consumption caused by mixed processing, and save the consumption of low-pressure steam.

1.4.3 Generation of higher pressure grade steam for cascade optimization and reduction of fuel consumption

In steam utilization, the waste heat of oil and flue gas in the unit should be fully utilized to generate steam or heat the feed water. In the waste heat recovery design of each unit, steam generation equipment generates high parameter steam as far as possible according to heat source temperature level, while steam consuming unit consumes low parameter steam as far as possible when process conditions permit. Between steam pipe networks of various pressure grades, back-pressure steam turbine should be selected as far as possible to drive high-power mechanical equipment. The steam discharged from the turbine can be used by the next level of the pipe network to realize the cascade utilization of steam energy. The steam condensate of the whole plant reduces the supplementary water, so the energy consumption of the unit and the whole refinery is fully recovered and utilized to reduce the make-up water so that the energy consumption of the unit and the whole refinery is reduced accordingly. The realization of thermal combination in some units closely related with each other during operation not only reduces the cooling water consumption of the upstream units but also reduces the steam consumption of the down-stream units, thus reducing the steam supply of boilers and saving fuel consumption. For example, the power boiler of traditional refinery mainly generates 3.5 MPa steam. The first reason is to meet the production needs of the process, and the second reason is the pressure rating restrictions of engineering equipment. With the development of engineering equipment technology, the main way to optimize the utilization of heat energy in the steam power system of refinery is to generate higher pressure grade steam and cascade utilization. For instance, the heat required to generate 10.0 MPa steam is similar to that required to generate 3.5 MPa steam. While 10.0 MPa steam is reduced to 3.5 MPa steam, the power generated expands the cascade utilization of steam (Fig. 1.10).

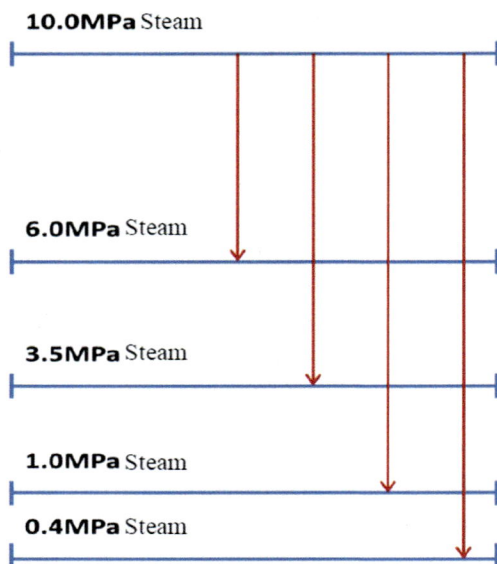

Fig. 1.10 Schematic diagram of generating steam of higher pressure grade and cascade optimization of utilization.

In terms of fuel utilization, the proportion of fuel used in furnace is larger. It is estimated that if H_2S content in fuel gas is not more than 30 ppm. The furnace exhaust temperature can be reduced to below 120°C. In the furnace design, by taking enhancement measures such as a highly efficient burner and air preheater, the design thermal efficiency can exceed 92%, thus greatly saving fuel consumption.

1.4.4 Recycle and optimize the utilization of low-temperature heat resources in the whole plant

Refining and chemical enterprises have a large number of low-temperature heat resources. A 0.5 Mt/a refinery can generate more than 80 MW of low-temperature heat resources per hour. These low-temperature heat resources need to be balanced comprehensively in the whole plant to achieve recovery and optimal utilization in the whole plant. Low-temperature heat resources of refining and petrochemical enterprises refer to oil products at 50–200°C or gas or flue gas at 100–400°C. According to different temperatures, low-temperature heat resources can be divided into three categories: oils at 150–200°C, and gas or flue gas at 300–400°C are high-grade low-temperature heat resources. These heat resources have been basically utilized via direct heat transfer. Oils at 80–150°C and gas or flue gas at 200–300°C are medium-grade, low-temperature heat resources. These heat resources are the key points in the research and development of comprehensive utilization technologies for low-temperature heat in refining and chemical enterprises nowadays. Oils at 50–80°C and gas or flue gas of 100–200°C are low-grade, low-temperature heat resources. Nowadays, such heat resources are mostly cooled and discharged directly. Under the current technological development situation, economic irrationality is the key to restricting such low-temperature heat resources. There are two ways to utilize low-temperature heat resources [15]. The first is matching and utilizing by grades, that is, heat exchanging with lower temperature grade flows, mainly including providing a heat source for low-temperature reboiler, air preheating, temperature keeping and heat tracing of storage tanks and pipelines, crude oil preheating, winter heating, and domestic hot water. The second is upgrade and utilization, specifically in the form of thermal upgrade, refrigeration, and work, including absorption heat pump, lithium bromide refrigeration, low-temperature power generation, and seawater desalination.

For the recovery and optimized utilization of low-temperature heat resources in the whole plant of refining and petrochemical enterprises, it is necessary to optimize the low-temperature thermal system of circulating hot-medium water by fully considering the low-temperature heat source, the quantity of heat trap, and temperature level and plane layout in the whole refinery and then realize the matching optimization of hot and cold flows and maximum heat recovery of heat through the circulation of hot medium water flow in series and/or in parallel [15]. At the same time, the actual situation of energy and heat consumption in the enterprises are considered, and the matching by grades and the upgrade and utilization of low-temperature heat resources are reasonably optimized to realize the optimal utilization of low-temperature heat resources in refining and chemical enterprises.

1.4.5 Cascade utilization of cooling water systems

Cascade utilization of cooling water refers to the realization of multiple cascade utilization according to the temperature rise requirements of cooling water. According to the cooling load of process flows, when the primary cooling water has a small temperature rise, it can be used to cool other flows to realize the cascade utilization of cooling water and reduce the consumption of cooling water. The diagram of the traditional cooling water system and cascade cooling water system is shown in Fig. 1.11. After cascade utilization of cooling water is realized, the total consumption of cooling water in refining and petrochemical enterprises can be reduced by about 30%–50%.

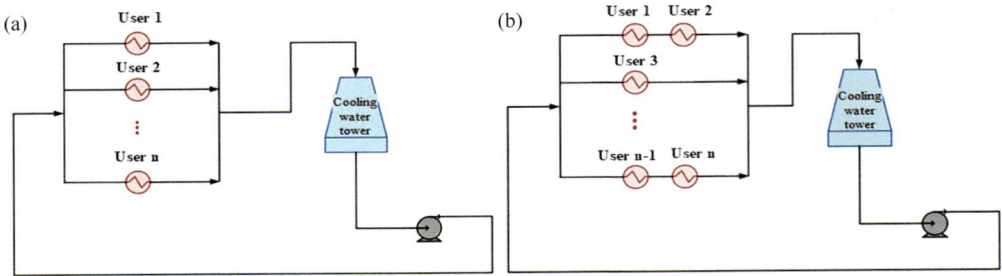

Fig. 1.11 Schematic diagram of cooling water system. (A) Traditional cooling water system. (B) Cascade cooling water system.

The engineering strategy of innovative heat transfer enhancement has been success-fully applied in refineries, such as Hainan Refinery and Qingdao Refinery, supporting them to achieve green and low-carbon, clean and environmental protection, energy conservation and emission reduction, space intensification, and investment saving. The average values of comprehensive energy consumption, single factor energy consumption, energy consumption of fuel and power, profit per ton of oil, and energy consumption per unit of output value in the last 3 years are selected as the evaluation indicators to draw the radar chart of the core competitiveness indicators of the enterprises.

The radar chart of Qingdao Refinery's core competitiveness is shown in Fig. 1.12, where the indicator for the domestic average level is defined as 10. As can be seen from Fig. 1.12, the comprehensive energy consumption and single factor energy consumption

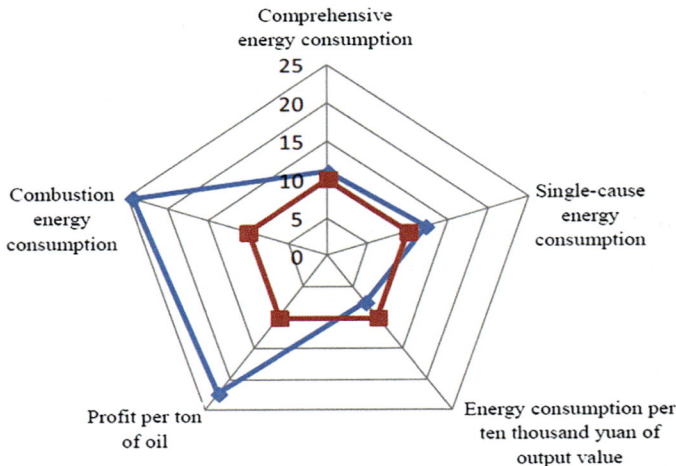

Fig. 1.12 Radar chart of Qingdao Refinery's core competitiveness.

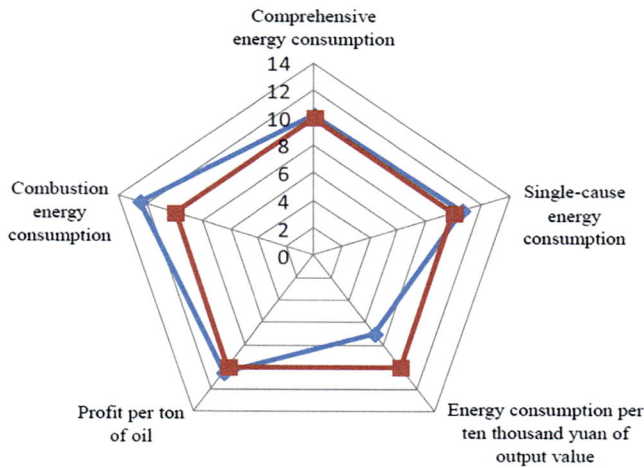

Fig. 1.13 Radar chart of Hainan Refinery's core competitiveness.

are at the domestic advanced level in China, the fuel energy consumption of fuel and power and profit per ton of oil are at the leading level, the energy consumption per 10,000 yuan of output value is at the advanced level, and the comprehensive competitiveness is at the advanced level in China.

The radar chart of Hainan Refinery's core competitiveness is shown in Fig. 1.13. It can be seen that the comprehensive energy consumption, single factor energy consumption, energy consumption of fuel and power, and profit per ton of oil are all at the advanced level; the energy consumption per 10,000 yuan of output value is at the general level, and the comprehensive competitiveness is at the advanced level.

References

[1] Lin ZH, Wang J, Li RY, et al. Heat transfer enhancement technology. Beijing: Chemical Industry Press; 2007.
[2] Lanzhou Petroleum Machinery Research Institute. Heat exchanger. 2nd ed. Beijing: China Petrochemical Press; 2013.
[3] Cui HT, Peng PY. Heat transfer enhancement technology and its application. Beijing: Chemical Industry Press; 2006 [in Chinese].
[4] Zhu DSH, Qian SW. Heat transfer enhancement technology and its design and application. Chem Equip Technol 2000;21(6):1–9.
[5] Lin ZH. Heat transfer enhancement technology of single phase fluid in tubular heat exchanger. J Nat 2013;35(5):313–9.
[6] Lin WZH, Cao JH, Fang XM, et al. Research progress in heat transfer enhancement of shell and tube heat exchangers. Chem Ind Eng Prog 2018;37(4):1276–86.
[7] Xiao W, Shi ZHX, Jiang XB, et al. Comprehensive research progress of heat exchanger network considering heat transfer enhancement of shell and tube heat exchangers. Chem Ind Eng Prog 2018;37 (4):1267–75.

[8] Kemp IC. Pinch analysis and process integration: a user guide on process integration for the efficient use of energy. 2nd ed. UK: Butterworth-Heinemann; 2007.

[9] Feng X. Principle and technology of energy saving in chemical industry. 4th ed. Beijing: Chemical Industry Press; 2015.

[10] Yao PJ. Process systems engineering. Shanghai: East China University of Science and Technology Press; 2009.

[11] Yee TF, Grossmann IE. A screening and optimization approach for the retrofit of heat-exchanger networks. Ind Eng Chem Res 1991;30(1):146–62.

[12] Sreepathi BK, Rangaiah GP. Review of heat exchanger network retrofitting methodologies and their applications. Ind Eng Chem Res 2014;53(28):11205–20.

[13] Smith R, Jobson M, Chen L. Recent development in the retrofit of heat exchanger networks. Appl Therm Eng 2010;30(16):2281–9.

[14] Zhang BJ, Luo XL, Chen QL, et al. Heat integration by multiple hot discharges/feeds between plants. Ind Eng Chem Res 2011;50(18):10744–54.

[15] Hua B. Research and application of oil refinery energy system optimization technology. Beijing: China Petrochemical Press; 2009.

CHAPTER 2

Basic principles and methods of heat transfer enhancement

Contents

2.1 Basic principles of heat transfer enhancement

For heat exchangers, heat transfer enhancement technology achieves the purpose of optimal production by using heat transfer enhancement components, improving the structure of heat exchanger, improving the heat transfer per unit time and per unit volume of heat exchanger, improving the heat transfer efficiency, and reducing the cost [1]. Its main task is to improve the heat transfer efficiency by transferring the specified heat with the most economical equipment, protecting the safety of high-temperature components with the most economical cooling way, and achieving the rational utilization of energy with the highest heat transfer efficiency [2]. So, its research is very significant. Heat transfer enhancement technology has developed vigorously in various industrial fields from the first generation to the current fourth generation. The first-generation heat transfer technology, represented by plain tube, mainly achieves heat transfer between hot and cold fluids. The second-generation technology is represented by flat fin and two-dimensional fin, and compact heat transfer equipment is applied. The third-generation technology, represented by three-dimensional roughness element and three-dimensional fin, mainly improves heat transfer efficiency. The fourth-generation technology, also known as compound heat transfer enhancement technology, mainly uses the compound enhancement method to achieve better heat transfer effect [3].

2.1.1 Basic theories of heat transfer enhancement of heat exchangers

There are many theories describing the heat transfer enhancement of heat exchange elements, mainly including boundary layer theory, core flow heat transfer enhancement theory, entransy theory, thermal-mass theory, and field synergy principle.

Boundary layer theory [4] is the basis of traditional in–tube heat transfer enhancement technology theory. The boundary layer theory established by Prandtl divides the convective heat transfer temperature field into the thermal boundary layer region and main flow region. The thermal boundary layer is a resistance concentration area of heat transfer, and the traditional convective heat transfer enhancement technology is based on the study of thermal boundary layer, such as extended surface, rough surface, and swirl generator. By changing the structure and heat transfer surface of the heat exchanger, the flow boundary layer is destroyed, the thermal boundary layer is compressed, the secondary flow is produced, and the fluid turbulence is improved, so as to achieve enhanced heat transfer. Its disadvantage is the increased flow resistance.

Convective heat transfer temperature field is divided into the thermal boundary layer region and main flow region. The flow in the tube is divided into boundary flow and core flow. The region outside the boundary layer is called the core flow region. When the fluid flows in a confined space, the boundary layer merges at the central axis with the full development of the flow, and the temperature gradient exists in the entire flow field of the fully developed convective heat transfer inside the tube. Professor Liu proposed the theory of core flow heat transfer enhancement [5] and verified its correctness by experiments. To enhance the core flow heat transfer, the most direct method is to homogenize maximally the fluid temperature in the core flow region, in order to form an equivalent thermal boundary layer with a large temperature gradient near the tube wall, thus achieving a significant enhancement effect (e.g., the application of metal porous media). Four basic principles of heat transfer enhancement theory of core flow are summarized, that is, enhancing the temperature uniformity of core flow as much as possible, maximizing the flow disturbance in the core flow zone, reducing the enhancement elemental area of core flow zone as much as possible, and minimizing the flow disturbance near the boundary of core flow zone. Core flow heat transfer enhancement is a fluid-based enhanced heat transfer method. Its biggest difference from boundary layer theory is that there is no convective heat transfer between the enhanced heat transfer surface and the fluid, which emphasizes not only enhanced heat transfer but also the reduction of flow resistance increment and energy dissipation.

Academician Guo and others put forward the theories of entranspy and thermal-mass [6]. Considering that irreversibility is the essence of heat transfer under finite temperature difference, entranspy, a new core physical quantity of heat transfer formerly known as heat transfer potential capacity, is defined from the perspective of heat transfer using inductive and deductive methods. Entranspy represents the ability of objects to transfer heat outward. In the process of heat transfer, entranspy is not conserved and there is dissipation. Entranspy dissipation can measure the irreversibility of heat transfer process and can be used to define the efficiency of heat transfer process. The ratio of entranspy dissipation to heat flow square is defined as the entranspy dissipation thermal resistance in heat transfer process, and the principle of minimum entranspy dissipation thermal resistance for heat transfer process optimization is proposed; i.e., under given constraints, when the entranspy dissipation thermal resistance is the smallest, the heat transfer efficiency is the highest.

Based on the nature of heat, the theory of thermal-mass is established and the energy-mass duality theory of heat is put forward, which points out that the heat transfer patterns can be studied by analytical methods such as Newtonian mechanics and analytical mechanics. For example, Newton's law of motion can be used to describe the process of heat transfer, so as to establish a universal law of heat conduction. Thermal-mass theory reveals the physical meaning of entranspy and makes it clear that entranspy is a simplified expression of thermal-mass transfer potential energy.

The field synergy principle proposed by Academician Guo et al. [7] reviews the physical mechanism of heat transport from the perspective of two-dimensional laminar boundary layer energy, the convective heat transfer is likened to the heat conduction process with internal heat source, and it is pointed out that the heat source intensity depends not only on the velocity and physical properties of the fluid but also on the synergy between velocity and heat flux vector. The smaller the synergy angle of the two fields, the higher the synergy degree and the greater the heat transfer intensity. According to the field synergy principle, the best convective heat transfer enhancement effect can be achieved by improving the synergy degree of velocity and heat flow field. The synergy degree can be improved by decreasing the included angle of velocity and heat flow vector and increasing the uniformity of velocity and heat flow field. The field synergy principle provides a new idea for developing heat transfer enhancement technology. Many scholars have studied the field synergy principle and its application and have made some achievements. For example, Academician Tao applied the field synergy principle to the research on refrigerator and optimized the related parameters of pulse tube refrigerator. The field physical quantity synergy principle has made new contributions to the development of heat transfer enhancement technology by recognizing and understanding the mechanism of enhanced heat transfer from a new perspective.

2.1.2 Mechanism analysis of convective heat transfer enhancement

The mechanism of heat transfer enhancement has been described in many literatures, which mainly analyze various factors affecting heat transfer according to the basic formula of heat transfer. The basic calculation formula of heat transfer is

$$Q = KS\Delta T_{\mathrm{m}} \tag{2.1}$$

where,

Q = heat load, W;

K = overall heat transfer coefficient, $W/(m^2 \ K)$;

S = heat transfer area, m^2; and

ΔT_{m} = effective mean temperature difference between hot and cold flow, K.

According to the basic formula of heat transfer, the heat transfer rate per unit heat transfer area is proportional to the driving force of heat transfer and inversely proportional to the thermal resistance. Improving the heat transfer capacity of heat exchanger can be achieved by improving the driving force of heat transfer or reducing the thermal resistance of heat transfer. Therefore, increasing the heat transfer of heat exchanger can be achieved by increasing the heat transfer area, increasing the mean heat transfer temperature difference, or improving the heat transfer coefficient [2,8].

Increasing heat transfer area per unit volume to enhance heat transfer is the most studied and effective way to enhance heat transfer, such as the use of ribbed tube, transversally corrugated tube, finned tube, plate rib type heat transfer surface and porous medium structure, which not only increase the heat transfer area per unit volume, but also improve the flow state of the fluid. The core of heat transfer enhancement is to increase heat transfer area or improve heat transfer coefficient by changing the structure of heat exchanger.

There are two ways to increase the mean temperature difference. One way is to increase the inlet temperature of the hot flow or lower the inlet temperature of the cold flow to increase the difference between the inlet and outlet temperature of the cold and hot flows while the process conditions permit. The other way is to change the arrangement of heat transfer surface to change the temperature difference, so as to achieve heat transfer enhancement. When there is no phase transition in the cold and hot flows in the heat exchanger, countercurrent or a flow pattern close to countercurrent can be used in the structure to increase the mean heat transfer temperature difference, and the number of shell passes of heat exchanger can also be increased to increase the mean temperature difference. When increasing the heat transfer temperature difference, the technical feasibility and economic rationality should be taken into consideration, and the rational utilization of energy of the whole system should be taken into account.

Process conditions limit the increase of heat transfer temperature difference. In practical engineering, enhancing heat transfer by increasing heat transfer coefficient is the most effective way. The equation of heat transfer coefficient K is:

$$K = \frac{1}{\frac{1}{h_i} + R_i + \frac{\delta}{\lambda} + R_o + \frac{1}{h_o}} \tag{2.2}$$

where,

h_i, h_o = film heat transfer coefficient of the flow inside and outside the tube, $W/(m^2\,K)$;

δ = wall thickness of heat exchange tube, m;

λ = wall thermal conductivity of heat exchange tube, $W/(m\,K)$; and

R_i, R_o = fouling thermal resistance of the flow inside and outside the tube, $(m^2\,K)/W$. Thermal resistance has three types, i. e., flow thermal tesistrance, wall thermal resistance, and fouling thermal resistance on both sides of the wall. Under normal circumstances, convective thermal resistance is the main factor hindering heat transfer, the wall thermal resistance of metal heat exchanger is a secondary factor, and fouling thermal resistance is a variable factor. As time passes by, fouling thermal resistance will change from a nonmajor factor to a major factor hindering heat transfer, and the maximum thermal resistance should be analyzed according to the actual application. According to Eq. (2.2), to improve the heat transfer coefficient, the most influential thermal resistance must be reduced and measures should be taken to enhance heat transfer. Increasing the flow velocity, changing the structure of heat exchanger, or changing the heat exchanger wall material can reduce the thermal resistance.

2.1.3 Evaluation methods of heat transfer enhancement performance

The key to energy saving of heat exchange equipment is to improve its comprehensive performance. With the extensive application of heat transfer enhancement technology in engineering, it is necessary to evaluate the technology and judge its level of advancement, so as to facilitate the application and research guidance of high-performance heat transfer technology. In the early stage, single parameter was used for evaluation, such as the heat transfer coefficient and pressure drop, friction coefficient, efficiency, and heat transfer enhancement ratio Nu/Nu_0 at a specific flow rate. The resistance coefficient increases significantly with the increase of heat transfer coefficient because the flow resistance is increased while the irreversible loss of temperature difference heat transfer is reduced. Considering heat transfer and fluid flow comprehensively, the criterion number of $(Nu/Nu_0)/(f/f_0)$ is proposed. The higher the criterion number, the higher the energy efficiency. For different heat exchangers, there are certain differences in criterion numbers.

Evaluation methods and indicators [9] based on the first law of thermodynamics include specific pressure drop *J*, energy coefficient, and area mass factor, with clear physical meaning. However, evaluation results vary greatly with different heat exchange areas and different working conditions. Webb et al.[10] defined a general performance evaluation criterion (*PEC*) as shown in Eq. (2.3), which is used to evaluate the overall heat transfer performance and the increase of pressure drop of heat transfer unit. Under the same *Re* number, when the physical properties and heat transfer area of the fluid remain unchanged, the heat transferred is compared based on the same pump power consumption, as shown in Eq. (2.4).

$$PEC = \frac{hA/(h_0A)}{(f/f_0)^{1/3}(A/A_0)^{2/3}} = \frac{Nu/Nu_0}{(f/f_0)^{1/3}(Re/Re_0)(Pr/Pr_0)^{1/3}} \tag{2.3}$$

$$PEC = \frac{Nu/Nu_0}{(f/f_0)^{1/3}} \tag{2.4}$$

where,

f = the fluid resistance coefficient of the enhanced heat transfer surface, which is a dimensionless characteristic number; and

f_0 = the fluid resistance coefficient of the plain tube, which is the benchmark.

Based on the above theories, Ji et al. conducted in-depth research and development [11]. According to the contrast between enhanced surface and reference surface, namely the relationship between Nu/Nu_0 and f/f_0, the comprehensive performance evaluation chart of heat transfer resistance for the purpose of energy saving is summarized, which can be divided into enhanced heat transfer nonenergy saving zone, enhanced heat transfer zone with equal pump power, enhanced heat transfer zone with the same pressure drop, and the optimal enhanced heat transfer zone with the same friction coefficient.

Methods based on the second law of thermodynamics mainly include entropy and exergy evaluation methods. These two evaluation methods have been successfully applied in the analysis of flow and heat transfer process of heat exchanger and the optimization of the overall structure [12]. Methods based on entropy and exergy can reflect the energy efficiency characteristics of heat exchanger under specific thermal parameters, but they cannot determine the energy efficiency level among all heat exchangers of the same type.

Through theoretical analysis of the comprehensive characteristics of water-water no-phase-transition flow and heat transfer of heat exchangers, Professor Bai led his research team to deduce and analyze the energy efficiency index (*EEI*) of plate heat exchangers [13], as shown in Eq. (2.5). *EEI* reflects the inherent energy efficiency attributes of heat exchanger as a whole and represents the heat transfer coefficient of heat exchanger consumption unit converted into flow pressure drop. When the temperature difference and fluid flow rate are the same, for the heat exchangers with the same heat exchange area and the same fluid flow length, the higher the energy efficiency index, the higher the flow rate that will be obtained per unit of pump work. By selecting a reasonable coefficient n, EEI can maintain good stability.

$$EEI = \frac{K}{\nabla P^n} \qquad (2.5)$$

where,

K = overall heat transfer coefficient, W/(m^2 K); and

∇P = pressure gradient.

The comprehensive evaluation index of heat transfer enhancement performance can also reflect the multifield synergetic relationship of convective heat transfer by synergetic angle. The better the synergetic angle, the better the effect of heat transfer enhancement [12]. At present, there are many evaluation methods for heat transfer enhancement performance, so appropriate evaluation methods should be selected according to specific requirements in practical application.

2.2 Heat transfer enhancement technology

According to the characteristics of heat transfer process, heat transfer enhancement technology can be classified from different angles [3].

According to the heat transfer process, it can be divided into heat conduction process enhancement, radiation heat transfer process enhancement, and convective heat transfer enhancement. Heat conduction process enhancement tries to reduce contact thermal resistance at high heat flow. Radiation heat transfer process enhancement can be realized by changing the material and surface roughness that affect the radiation. Convective heat transfer enhancement can be realized by changing the flow state and heat exchanger structure. The most widely used convective heat transfer enhancement is the focus of research in various application fields.

According to the phase transition of fluid heat transfer process, it can be divided into heat transfer enhancement process with or without phase transition. Different enhancement technologies are used for different heat transfer processes. Heat transfer enhancement process without phase transition means that the fluid does not have phase transitions such as evaporation or condensation in the heat transfer process, and the heat transfer can be enhanced by improving the fluid turbulence. Heat transfer technology with phase transition includes condensation and boiling heat transfer enhancement processes. In the process of condensation, the thermal resistance is mainly in the condensation liquid film. Thinning the liquid film is the key to enhancing the condensation heat transfer. For forced convective boiling and pool boiling, there are differences in heat transfer enhancement methods. Heat transfer surface roughness and extended surface are helpful to enhance convective boiling. Adding foaming core is the main means to enhance pool boiling heat transfer.

From the source of enhancement technology, it can be divided into active, passive, and composite heat transfer enhancement technologies. Active enhancement technology requires external power input, also known as active power technology, and includes heat transfer enhancement technologies such as electromagnetic field action, electrostatic field method, vibration method, mechanical method, jet impingement, jet, or suction. Passive enhancement technology refers to the enhancement of heat transfer by changing the shape and structure of heat exchanger without external power input, also known as reactive power technology. It mainly includes extended surface, irregular surface, rough surface method, in-tube insert, bionic optimization, and additives. Composite enhancement technology refers to the technology that applies two or more active or passive heat transfer enhancement methods to heat exchangers at the same time in order to obtain greater heat transfer enhancement effect. The research scope of composite heat transfer enhancement technology is very wide, such as the insert of bond or cyclone device in corrugated low-fin tube, the insert of helix in corrugated tube, the insert of bond in rough tube, etc.

2.2.1 Heat transfer enhancement process without phase transition

2.2.1.1 Artificial rough surface and irregular surface method

There are many factors affecting the convective heat transfer of no–phase-transition fluid, such as flow state, fluid physical properties, and the size and shape of flow channel. To enhance the convective heat transfer of no–phase-transition fluid is to reduce the thickness of laminar sublayer, increase the turbulent intensity of the fluid, and reduce the heat transfer resistance of laminar sublayer, so as to improve the uniformity of fluid velocity field and temperature field and improve the coordination between velocity and temperature gradient. The principle of field coordination is universal for enhancing single–phase convective heat transfer and points out the direction for the development of efficient heat transfer surface [14].

Local backflow zone and local separation zone will appear when the fluid flows through rough surface, increasing disturbance and thinning boundary layer, thus enhancing heat transfer. In practical engineering applications, proper roughness should be selected to achieve the ideal enhancement effect. If the roughness is too small, the low-speed fluid flows smoothly close to the rough tube wall without vortex, and the heat transfer improvement effect is not good. If the roughness is too large, it is easy to form a vortex with close stagnation, and the thermal resistance is too high, which has an adverse effect on the laminar flow heat transfer. The irregular surface not only increases the heat transfer surface but also constantly changes the flow state of the fluid in the flow channel, increases the disturbance, and reduces the thickness of the thermal boundary layer, so as to enhance the heat transfer. The irregular tubes mainly used include transversally corrugated tubes, swirl tubes, corrugated tubes, spiral-twisted tubes, inner rib tubes, etc. [15].

Garcia et al. [16] compared the experimental results of corrugated tubes, swirl tubes, and tubes inserted with metal coils in laminar flow zone, transition zone, and turbulent zone, and it was found that when $Re < 200$, the enhancement effect of rough tubes is not obvious. When $200 < Re < 2000$, the shape of rough structure has a great influence on the enhancement effect. The surface roughness method with metal coil inserted into the tube has the best enhancement effect and the fluid state can be predicted. When $Re > 2000$, the effect of corrugated tubes and swirl tubes is better.

2.2.1.2 Extended surface [1,2]

Extended surface mainly increases fluid disturbance, destroys laminar boundary layer, increases turbulence degree, and promotes convective heat transfer through its special structure, which has significant enhancement effect on laminar flow and turbulence. Meanwhile, it increases the heat transfer area and increases the heat transfer, enhancing the effect of heat transfer from two aspects. At the same time, fluid disturbance will prevent the accumulation of fouling and achieve the effect of scale inhibition. It is suitable for heat transfer enhancement of gas phase medium. However, the extended surface method will increase the pressure drop while increasing the heat transfer area, which may also bring noise and vibration.

Fin is the most effective way to expand the surface. There are many kinds of fins, mainly including straight fin, corrugated fin, zigzag fin, oblique needle fin, porous fin, shutter fin, and nail head fin. Compared with the plain tube, finned tube has larger surface area with the same metal consumption, not only improving the equipment compactness and heat transfer efficiency but also improving the strength and pressure bearing capacity of heat exchanger, so it is widely used in process units in the chemical and petrochemical industries. Finned tube has strict preparation process requirements, and its quality directly affects the performance of heat exchanger. When choosing the structure and material of base tube and fin, comprehensive consideration should be given so as to select the best scheme.

2.2.1.3 Vibration methods

Vibration methods include the vibration of heat transfer surface and the fluid vibration or pulsation. The mechanism is to enhance fluid disturbance through vibration, so as to enhance heat transfer. For natural convection, Park and Kim [17] found that during experiments the heat transfer coefficient increases with the increase of frequency or amplitude. When the vibration intensity reaches the threshold value to achieve resonance, the enhancement effect is the best. For forced convection, studies show that the heat transfer coefficient can be increased by 20%–400%, but the vibration of the heat transfer surface may reduce the pressure in local areas to the saturated pressure of the liquid, resulting in the risk of cavitation. Ultrasonic vibration descaling method has been applied in the process. Fluid induced vibration is very destructive and is prevented as much as possible in heat exchanger design. With the development of technology, fluid–induced vibration can be controlled to decontaminate heat exchange tubes and improve heat transfer efficiency.

2.2.1.4 Fluid rotation method (inserts)

Heat transfer enhancement with inserts produces the radial flow of fluid by using inserts to enhance fluid mixing and improve convective heat transfer coefficient, especially for enhancing the heat transfer of gas, low Reynolds number fluid, or high viscosity fluid. There are many kinds of inserts, such as twisted strip, helix, spiral piece, shutter, and static mixer.

2.2.1.5 Other heat transfer enhancement technologies

(1) Mechanical stirring method. It relies on mechanical equipment to stir the fluid, to rotate the heat transfer surface, or to scrape the surface. Surface scraping is widely used in viscous fluids in chemical production. Bagajewicz et al. [18] devised an innovative solution for automatic cleaning inside shell and tube heat exchangers. The scraping element is fully mounted on the reciprocating connecting rod to clean the tube wall, not only preventing scaling but also improving heat transfer. When the scraper moves at the average speed of the fluid, the heat transfer efficiency increases by 140% and the pressure drop increases by 150%.

(2) Additive method. In order to meet the needs of heat transfer enhancement, some additives such as solid particles, liquid, bubbles, or polymer solid particles are added to the fluid working medium; this method is called additive method. The effect of millimeter and micron particles on liquid heat transfer enhancement is not very obvious. The flow resistance increases a lot, but the heat transfer coefficient only increases by 40%–50%, which greatly limits its application in industry. With the development of nanotechnology, compared with water, the heat transfer efficiency of nanofluid is increased by 60%, and the enhancement effect is obvious. Nanotechnology has garnered attention from people. The irregular movement of nanoparticles in nanofluids destroys the laminar sublayer of fluid and increases disturbance and turbulence, thus enhancing heat transfer. The preparation method, stability, heat transfer characteristics, mechanism of heat transfer enhancement, nanofluid system, and its application have been extensively studied [19].

(3) Jet impingement. Jet impingement heat transfer is one of the most efficient and effective methods of forced convective heat transfer. Gas or liquid is sprayed directly (or at an angle) to the cooled or heated surface through a circular or slit nozzle under differential pressure, resulting in a strong heat transfer effect in the area directly impinged. It is characterized by the direct impingement of fluids on the surface that needs cooling or heating. The process flow is very short, and a very thin boundary layer is formed near the stagnation point, so it has a very high heat transfer efficiency and saves a lot of space. It is suitable for local heat transfer and has been widely used in some industrial technologies and production fields, such as the drying of textiles and paper, the glass tempering, the cooling and heating of steel, the cooling of aviation engines, and the cooling of high load microelectronic components for computers. With the development of high science and technology, jet impingement technology will be more widely used.

2.2.2 Heat transfer enhancement of boiling process

Boiling is a phase transition phenomenon in which a liquid is heated beyond its saturation temperature and vaporizes violently both inside and on its surface.

According to the uniformity of boiling, boiling heat transfer can be divided into uniform boiling and nonuniform boiling. For uniform boiling, there is no fixed heating surface inside the liquid during the boiling process, and bubbles are generated by the movement and aggregation of high-energy molecular groups of the liquid with relatively concentrated energy. Nonuniform boiling refers to the formation and growth of bubbles on the liquid–solid contact surface during the boiling process. Compared with the former, the latter requires lower superheat. Nonuniform boiling is usually used in industrial and commercial applications. Nonuniform boiling can be divided into pool boiling and flow boiling according to whether the liquid flows during the boiling process.

Heat transfer enhancement methods in boiling process mainly include:

2.2.2.1 Enhanced nucleation site activity

In the actual boiling process, bubbles can form on the wall when the superheat reaches a few degrees or tens of degrees. This is because there are cavities of different sizes on the heating wall. When the liquid enters and exits the cavity, it is guaranteed that there is still some gas in the cavity as the vaporization core. Taking measures to enhance the nucleation site activity can make the single-phase convective heat transfer enter the nucleate boiling section in advance, which can reduce the superheat required for boiling and obtain better heat transfer effect.

Various enhancement surfaces have been developed over the past few decades to enhance nucleation site activity. It can be divided into three main categories: inherent surface enhancement, additional surface enhancement, and mixed enhancement. The following are some surface enhancement methods that are widely used in industrial and commercial applications.

(1) Inherent surface enhancement. Sandpaper grinding and other mechanical methods can increase the heating surface roughness, so as to enhance heat transfer. El-Genk

and Suszko [20] found in their pool boiling experiment that the heat transfer coefficient is 150% higher than that of the smooth surface and the critical heat flux is enhanced by 37%.

Material is removed from the work piece using discharge or electric spark to obtain the desired solid structure and geometry. Yu and Lu [21] found in their experimental results that the maximum heat transfer coefficient reaches $10,000 \, W/(m^2 \, K)$ and the maximum critical heat flux reaches $9.8 \times 10^5 \, W/m^2$, which is about five times that of smooth surface. In the past two decades, micro-/nano-electromechanical system (MEMS/NEMS) processes such as deposition, photolithography, and etching that have been commonly used for semiconductor device fabrication can also be applied to the production of boiling enhancement surfaces. These processes can produce highly ordered micro-/nano-surface features. Yu et al. [22] used the MEMS/NEMS method to prepare cavity arrays with diameters ranging from 50 to 200 μm on the heated surface. Pool boiling experiment was carried out with saturated FC-72 working medium at atmospheric pressure. The maximum heat transfer coefficient was found to be $11,000 \, W/(m^2 \, K)$, and the critical heat flux was enhanced by up to 2.5 times.

Selective laser melting (SLM) is an additive manufacturing technology that uses a high-power laser source to melt layer by layer and melt matrix metal powder to form three-dimensional geometric shapes. Wong and Leong [23] recently prepared porous structures using SLM. Pool boiling experiment results show that the maximum heat transfer coefficient is enhanced by 70% and the maximum critical heat flux is enhanced by 100%.

(2) Additional surface enhancement. Porous structure can be formed on smooth surface by means of resin bonding, impregnation drops, spraying, and electrochemical deposition.

In the fabrication of enhancement surfaces, one of the first methods used was to directly bond the particles to the heated surface using an adhesive epoxy resin. Diamond grits are often used as coating materials because of their excellent thermal conductivity. The coating forms a porous cavity, which increases the effective nucleation density of the surface.

Wu et al. [24] left a 1-μm-thick silicon dioxide coating on the substrate by immersion drip method. Because of its great hydrophilicity, it can increase the adhesion of the liquid on the heated surface, thus enhancing the heat transfer coefficient of the heated surface and delaying the occurrence of drying conditions. The experimental results show that the heat transfer coefficient and critical heat flux are enhanced by 91.2% and 38.2%, respectively, by wire harness.

Spraying method is to directly spray the coating particles on the surface at high speed, and the particles undergo plastic deformation and adhere to the surface. Sahu et al. [25] prepared nanocoatings on copper plates in this way, and verified by experiments that the critical heat flux is increased by about 33% and the heat transfer coefficient is also greatly improved.

In free particle bonding method, particles move freely on a heated surface. Kim et al. [26] used free copper particles to enhance the boiling process. Deposition of copper particles provides additional nucleation sites and rapid recovery after bubble release. Experimental results show that the heat transfer coefficient increases by 76.3% and the critical heat flux reaches $160,000 \, W/m^2$.

El-Genk and Ali [27] used electrochemical deposition method to electroplate copper substrate as a cathode in sulfuric acid and copper sulfate electrolytic cell. Experimental results show that when the average thickness of porous surface is $171.1 \, \mu m$, the critical heat flux is $278,000 \, W/m^2$.

2.2.2.2 Nanofluid

Nanofluid refers to a dilute suspension of nanoparticles with a diameter of less than 100 nm. Since 2002, researchers have attempted to use nanofluids to enhance pool boiling heat transfer. However, there are contradictions and conflicts in the experimental results. These inconsistencies are generally attributed to the fact that the results of pool boiling experiments are highly dependent on the specific test conditions because there are so many variables involved, including the type of nanofluid, heating material, surfactant use, heater geometry, saturation pressure, and measuring equipment.

2.2.2.3 Surfactant

Surfactant is a class of functional organic compounds, and its molecular structure generally consists of nonpolar hydrocarbon chain (also known as lipophilic group) and polar groups (also known as hydrophilic group), so it is also called amphiphiles. After surfactant is added into the solution, the change in the physical properties of the solution will affect the formation, growth, and movement of bubbles in the boiling solution, thus affecting the boiling heat transfer characteristics of the solution. It mainly affects the properties of solution including static surface tension, dynamic surface tension, viscosity, wetting performance, solubility, and thermal conductivity.

In addition to the above widely used boiling heat transfer enhancement methods, there are also some methods that are not commonly used or are under exploration at present, such as enhancing boiling heat transfer through electrostatic field, ultrasonic vibration, and eddy current. In the process of natural convection and subcooled boiling, the cavitation phenomenon formed by ultrasonic wave has great influence on the enhancement degree of heat transfer. Cavitation does not occur in the saturated boiling process, so the size of bubble separation and acoustic streaming are the main factors to improve the heat transfer effect. Kwon et al. [28] studied through experiments the effect of ultrasonic vibration on critical heat flux under natural convection. It was found that the formation of large bubbles or steam film was inhibited under the effect of ultrasonic wave, which delayed the occurrence of drying conditions and enhanced heat transfer. The critical heat flux was enhanced by 10%–15%.

2.2.3 Heat transfer enhancement of condensation process

Condensation is the transformation of a substance from metastable gas phase to steady-state liquid phase, usually accompanied by jumps in physical properties such as density and specific heat. It can be divided into two processes of nucleation and condensate (droplet, liquid film) growth. Due to pressure and temperature changes, the system is in the metastable state of steam supersaturation. Gas molecules in metastable state collide with each other and gather into thermodynamic stable microscopic molecular clusters (condensation nuclei), that is, condensation nucleation. After nucleation, gas molecules collide and adhere to the surface of the condensation nucleus under the action of chemical potential, which makes the condensation nucleus grow into larger droplets or liquid films. In the nucleation process, if there is no material such as walls and dust for condensation, the gas molecules will spontaneously form condensation nuclei by the agglomeration of their own molecules; it is called homogeneous nucleation. The theoretical description of homogeneous nucleation process is very complex. After the establishment of classical nucleation theory, many scholars have made continuous improvement and innovation, forming a variety of theories such as self-consistent classical nucleation theory, density functional theory, dynamic nucleation theory, and semiphenomenological theory. Homogeneous condensation usually occurs in devices where the gas state parameters vary dramatically, such as turbines, nozzles, diffusers, and ejectors. If the gas molecules nucleate on the surface of the existing condensation core, it is called heterogeneous nucleation. During heterogeneous nucleation, the state parameters of the system change slowly, and nucleation usually occurs near the equilibrium state, and the nucleation process can be regarded as a near equilibrium process. According to the geometry of the heat transfer process, heterogeneous condensation heat transfer can be divided into the following three modes: direct contact condensation, surface condensation, and condensation in channel. The condensation of heat exchangers in the process industry is mostly surface condensation and condensation in channel.

2.2.3.1 Surface condensation heat transfer and enhancement

Surface condensation can be divided into dropwise condensation and film condensation. Although the film heat transfer coefficient of dropwise condensation can be as high as dozens of times that of film condensation, film condensation is usually adopted for condensation heat transfer in industrial processes due to the limitation of equipment manufacturing technology. It is of great significance to correctly understand the process of film condensation and enhance the heat transfer of film condensation. Scholars at home and abroad have studied dropwise and film coexisting condensation. Since 1990s,

Professor Ma of Dalian University of Technology has conducted a systematic study on the enhancement of condensation heat transfer on dropwise and film coexisting surface [29].

(1) Dropwise condensation and enhancement. Steam condenses and forms dispersed droplets after contacting the supercooled wall surface. Steam condenses on the surface of the droplets, and adjacent droplets coalesce, resulting in droplet growth. Then the droplets fall off from the surface under the action of gravity or shear force, exposing the supercooled surface, and steam continues to condense into droplets on the surface. The process by which droplets form, grow, coalesce, and fall off the supercooled surface is called dropwise condensation. During the whole process, the droplets remain dispersed, and the condensed surface is in direct contact with the steam. Dropwise condensation is an ideal condensation heat transfer process with low thermal resistance.

When steam condenses on the surface of droplets, it is necessary to overcome the thermal conductivity resistance of condensate and the thermal resistance of gas–liquid and solid–liquid interfaces. The thermal conductivity resistance of large droplet is much larger than that of small droplet, and the heat is mainly transferred through the wall and the small droplets. Therefore, the main ways to enhance the heat transfer of dropwise condensation include increasing nucleation rate, decreasing droplet size, accelerating droplet shedding, and decreasing interfacial thermal resistance. The main factors affecting the heat transfer of dropwise condensation include steam pressure, physical and chemical properties of droplet and surface, surface inclination angle (gravity), and noncondensing gas. The contact angle reflects the wettability of liquid to the solid wall. The larger the contact angle, the smaller the wall area occupied by droplets of the same volume, and the more conducive to droplet shedding. Taking water as an example, when the contact angle is less than 90 degrees, the solid can be wetted and the solid surface is called hydrophilic surface; when the contact angle is greater than 90 degrees, the solid cannot be wetted and the solid surface is called hydrophobic surface. When the solid wall surface is rough, there are different calculation methods for the contact angle under different droplet morphology.

In order to reduce the droplet size and speed up droplet shedding, it is necessary to reduce the surface energy of solid wall and increase the surface roughness to reduce the wettability of wall. When the contact angle is greater than 150 degrees, the solid surface is called superhydrophobic surface. Superhydrophobic surface has been widely paid attention to in academic circles for its high condensation heat transfer. Boreyko and Chen found in 2009 that condensation droplets on superhydrophobic surfaces can spontaneously bounce during coalescence [30]. Different from the traditional gravity shedding, the excess surface energy released when the droplets coalesce on the superhydrophobic surface is converted into kinetic energy to drive the droplets to bounce. Spontaneous bounce can promote droplet shedding. Some experiments have compared the condensation heat transfer performance of superhydrophobic copper surface of nanocopper oxide, smooth hydrophilic copper surface, and silane-coated hydrophobic copper surface. It is found that in droplet bouncing mode, the heat transfer performance of superhydrophobic surface is 25% and 150% higher than that of hydrophobic copper surface and hydrophilic copper surface, respectively [31]. However, with the increase of undercooling degree, the surface nucleation density of copper nanoparticles is too high, and the superhydrophobic surface is continuously wetted by droplets, leading to the decrease of condensation heat transfer performance.

The microstructure of superhydrophobic surface is easy to be destroyed, which is the bottleneck for industrial application of superhydrophobic surface in enhancing condensation heat transfer. Torresin et al. found in experiments that the copper oxide nanofibers of superhydrophobic surface gradually peel off as the experiment goes on, and the condensation heat transfer effect is worse than the smooth hydrophilic surface of copper oxide in the film condensation mode [32].

(2) Film condensation and enhancement. Film condensation is the formation of a liquid film by steam contacting the supercooled wall. When the wettability of wall is high or the condensation rate is high, the condensation droplets coalesce and form a film. For vertical or inclined surfaces, the film flows under the action of gravity or steam shear and thickens as the condensation proceeds. The phase change heat released by condensation needs to pass through the liquid film to reach the cooling wall, which increases the heat transfer resistance, so the heat transfer coefficient of film condensation is usually one order of magnitude lower than that of dropwise condensation.

Because the main thermal resistance of film condensation is in the condensation liquid film, heat transfer enhancement of film condensation should focus on breaking and thinning the condensation liquid film. Commonly used condensation heat transfer enhancement structure outside the tube is low fin tube, including two-dimensional circumferential smooth trapezoidal or rectangular fin tubes, three-dimensional serrated fin, discontinuous fin, pin fin, and petal fin as well as corrugated low-fin tube. The mechanism of enhanced condensation heat transfer lies in that the condensate flows from the top of the fin to the root of the fin under the action of surface tension, converges with the root groove of the fin, and is discharged under the action of gravity. The liquid film thickness of tube wall is reduced, the condensation thermal resistance decreases, and the condensation heat transfer coefficient increases by 1.5–4 times. However, due to the surface tension, the fin at the bottom of tube can hold the condensate, so it is necessary to design the fin structure reasonably. The ideal fin structure should be that the fin tip has a small radius of curvature, which is beneficial to reduce the liquid film thickness; the curvature radius from fin tip to fin root increases gradually, and the pressure gradient of liquid film is maintained by surface tension, which is conducive to the movement of condensate from fin tip to fin root; the large space at fin root is conducive to the convergence of condensate along the tube wall to the bottom [33]. Because the lower tube in the bundle is affected by the condensate of the upper tube bundle, the condensation heat transfer enhancement result of single tube may not be applicable to the tube bundle. Therefore, it is necessary to analyze the specific application objects and conditions when designing the condensation heat transfer structure.

2.2.3.2 Characteristics and enhancement of condensation heat transfer in channel

Condensation heat transfer in channel is a key process in the electric power, air conditioning, and chemical industries. Different from the relatively stable gas–liquid distribution of surface dropwise condensation and film condensation, the contents of gas and liquid are constantly changing during the condensation in channel and the flow structure is complex, so the condensation in channel has many unique properties. After entering the channel, steam condenses on the wall of the channel to form a liquid film. As the flow moves on, steam condenses continuously, and the flow pattern changes constantly, including annular flow, stratified flow, wavy flow, and annular fog flow. The phase interface area and the thickness of condensate liquid film also change accordingly, which greatly affects the condensation heat transfer and flow resistance in channel. Therefore, correctly understanding the characteristics and occurrence conditions of the flow pattern in channel and developing the flow heat transfer model of each flow pattern is important for the study of condensation heat transfer in channel. In addition, unlike surface condensation, friction resistance is an important parameter for condensation in channel. Excessive frictional pressure drop will increase pump power loss and also reduce local saturation temperature and reduce condensation heat transfer.

When annular flow occurs during condensation heat transfer in horizontal tube, the gas phase flow rate is high, the shear liquid film promotes the liquid drainage, the liquid film on tube wall is thin, and the condensation heat transfer effect is good. At higher gas phase velocities, condensate may be sucked from the wall film to form a mist flow. At this time, the temperature of condensate droplets dispersed in the tube is the same as that of the steam, and no surface condensation occurs. However, the condensation heat transfer is further enhanced due to the thinning of the liquid film on tube wall. Annular laminar flow is characterized by continuous gas and liquid phase. The difference between annular laminar flow and annular flow is that both the gas and liquid phase flow velocities are low when annular laminar flow occurs. Under the action of gravity, the condensate accumulates downward and the liquid film at the bottom of the tube is thick. If the gas phase flow rate is very low, the gas–liquid interface may be very smooth, forming a stratified flow. With the increase of gas phase flow rate, the interface wave is formed and the condensation heat transfer is enhanced, which is wavy flow now. As condensation occurs everywhere around the tube, the top wall of the tube is always covered with liquid, which is the difference between stratified and wavy flows in the condensation process and stratified and wavy flows in adiabatic and boiling processes. When the liquid volume gradually increases and the wave peak of liquid film touches the top of the tube, the gas phase is dispersed into discontinuous gas bombs or bubbles, and intermittent flow is formed at this time.

The main factors affecting the condensation flow pattern in tube are tube diameter, gas phase mass fraction, mass velocity, gas–liquid density, gas–liquid viscosity, and surface tension. When the saturation temperature rises with the pressure, the stratified flow changes little and the range of wavy flow increases. The reason is that when the pressure rises, the gas–liquid density difference decreases and the gas–liquid velocity difference also decreases, so intermittent flow and annular fog flow are easy to occur.

There are two heat transfer mechanisms in circular tube condensation, i.e., the film condensation heat transfer caused by the movement of condensate from top to bottom under gravity and the convective condensation heat transfer caused by the axial movement of working medium under pressure difference. The film condensation heat transfer occurs at the thin film of stratified flow and wavy flow, as indicated by the 1 in Fig. 2.1, and the flow state is laminar flow. The convective condensation heat transfer occurs at the thick film of annular flow, intermittent flow, fog flow, stratified flow, and wavy flow, as indicated by the 2 in Fig. 2.1, and the flow state is turbulent.

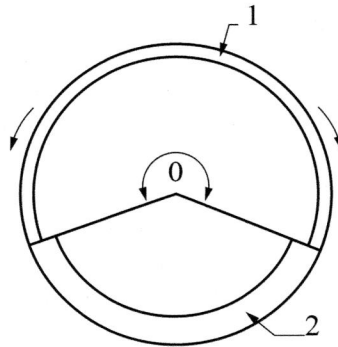

Fig. 2.1 Diagram of condensing mechanism in horizontal circular tubes.

Microfin tube is commonly used to enhance condensation heat transfer inside the tube. Microfin tube has triangular or trapezoidal microfin on the inner wall of plain tube. The microfin tube increases the disturbance of condensate near the wall and destroys the boundary layer. The thickness of liquid film at the bottom is reduced to achieve a more uniform distribution of liquid film, which promotes the movement of condensate toward the top, increases the mixing of condensate, and disturbs the gas–liquid interface. The surface tension promotes the movement of condensate from the top of microfin toward the microfin, reducing the thickness of the condensate at the top of microfin. The area of tube wall is increased. Compared with plain tube, the condensation heat transfer can be enhanced by 80%–180% at the same working conditions.

Shaped tubes and tubes with inserts are also used to enhance condensation heat transfer. Shaped tubes include corrugated tube, corrugated low-finned tube, spiral groove tube, etc. The main mechanism of condensation heat transfer enhancement is that the curved wall promotes boundary layer disturbance, flow separation, and mixing. The area of tube wall is increased. Corrugated tube can also constantly change the axial velocity of the fluid in tube by using the periodical change of static pressure, thus producing strong vortex to disturb the liquid film flow and achieving heat transfer enhancement. Compared with plain tube, the condensation heat transfer of shaped tube can be increased by 20%–95%. Commonly used tubes with inserts include tubes with inserted coils and tubes with porous inserts. Inserted coils can periodically destroy the laminar sublayer of liquid, enhance the turbulent flow, and greatly reduce the thermal resistance. Porous inserts with large voids can also effectively enhance the condensation heat transfer in tube. The mechanism of the enhancement is that the porous skeleton in tube forces the mainstream molecular microclusters to continuously produce complex three-dimensional macroscopic mixed flows in the direction of heat transfer and flow, namely mechanical dispersion effect, which increases the turbulence of the fluid. It is better to enhance laminar film condensation by tubes with porous inserts.

Some scholars have studied the vertical downward flow pattern distribution of air and water mixture in a chevron–type plate channel under adiabatic conditions [34]. Due to the different structures of plates used in the study, the flow pattern distribution and transition boundary summarized by different scholars are quite different, and the subjective judgment of flow pattern is also an important factor.

2.2.4 Fouling and cleaning

Fouling refers to solid substance, usually a mixture, deposited on heat exchanger surface after contact with a fluid containing impurities. It is an accumulation of unexpected materials on the surface of heat exchanger. During the operation of heat exchanger, fouling deposition increases thermal resistance and surface roughness of heat exchanger, which leads to an increase in friction coefficient and average flow rate, and finally an increase in fluid pressure drop. Therefore, fouling deposition seriously affects the operation effect of heat exchanger [35]. Fouling has a significant impact on the mechanical properties of heat exchangers, which increases the overall thermal resistance of heat exchangers, aggravates corrosion, increases pressure drop, and seriously affects the safe operation of heat exchange equipment. According to Malayeria et al. [36], the economic loss caused by heat exchanger fouling is about 0.25% of the gross domestic product of industrialized countries.

Fouling growth is affected by equipment and operation factors. Taking measures to restrain fouling growth can improve the heat transfer performance of heat exchange equipment and prolong the cleaning cycle.

(1) Changing the surface properties of materials. The surface performance of heat exchange surface has an important influence on the formation of fouling, so the surface performance of original materials can be improved by surface treatment technologies such as surface modification technology and coating technology.

(2) Changing the shape of heat exchange element. Zhang et al. [37] studied the flow, heat transfer, and fouling characteristics of corrugated tube and plain tube and found that corrugated tube has the advantages of slow scaling speed, long induction period, and small asymptotic value of fouling thermal resistance at the same flow rate. Moreover, corrugated tube has good heat transfer enhancement performance before and after scaling. Professor Xu et al. [38] studied the heat transfer performance and fouling resistance of transversally corrugated tube, and the results showed that the heat transfer performance and fouling resistance of transversally corrugated tube were better than those of plain tube and low-fin tube, the asymptotic fouling thermal resistance of transversally corrugated tube was about 83% of that of plain tube, and the convective heat transfer coefficient of transversally corrugated tube was 1.4 times of that of plain tube in scaling state.

(3) Adding inserts. Helix inserts, rotor assembly inserts, and vortex generator can not only meet the demand for enhanced heat transfer but also play a role in scale inhibition to a certain extent.

During actual operation, fouling is difficult to avoid after a long time of operation. When the fouling on heat exchange surface accumulates to a certain extent, it needs to be cleaned. Cleaning technology is mainly divided into online cleaning and offline cleaning. The heat transfer performance of heat exchanger is improved by cleaning, and the resistance characteristic is reduced.

Online cleaning is particularly important for towers and shell and tube equipment because equipment disassembly is not required. Online cleaning devices include cleaning elements, power pumps, supporting pipelines, valves, and control cabinets. The cleaning elements in tube can remove the sediment in tube by sliding, rubbing against, and colliding with the inner wall of the tube. By using online cleaning technology, there is no need for the heat exchanger to shut down and disassemble, so it is easy to operate and easy to clean with short cleaning time and obvious cleaning effect. Online cleaning has reliable operation and low failure rate and will not produce corrosion and other adverse effects on heat exchange equipment. While improving the efficiency of heat exchanger, it can prolong the service life and maintenance cycle of heat exchanger, reduce the loss caused by maintenance shutdown, and reduce the maintenance cost.

Online mechanical cleaning depends on the flow of fluid or the mechanical action to overcome the adhesion of fouling to make the fouling fall off. Mechanical cleaning is mainly aimed at hard fouling such as carbide fouling. Mechanical cleaning includes spring, twisted belt, helix and rubber ball, etc. Online physical cleaning removes sediment by acoustic, thermal, and light effects. Common physical cleaning methods include ultrasonic cleaning technology, magnetic field descaling technology, electric field descaling technology, and PIG pigging technology. Online chemical cleaning method is an effective cleaning method to reduce fouling by adding scale remover, acid, or alkali to the fluid. Chemical cleaning method cannot be used in the heat exchangers of nuclear power plants because the corrosion effect of the cleaning agents cannot be predicted accurately.

Offline cleaning of heat exchanger needs to shut down the heat exchanger for disassembly and cleaning, and the cleaning cycle is long, but good cleaning effect can be achieved. It can be divided into mechanical cleaning and chemical cleaning.

Mechanical cleaning refers to the cleaning by mechanical actions such as scraping and vibration. It is effective for fouling with high viscosity and hardness, causing no corrosion and no pollution, but it is easy to damage the tube wall. Commonly used methods include high-pressure water jet, thermal impact, and ice pigging technology. Offline chemical cleaning technology reduces the adhesion between fouling and tube wall through chemical reaction, so as to achieve the purpose of cleaning. Chemical cleaning has low cleaning intensity; the cleaning is more thorough and can avoid damage to the heat exchanger, but attention should be paid to the dangerous problems such as burns and poisoning that may occur during the cleaning process as well as the anticorrosion treatment, such as rust prevention and passivation, after cleaning.

2.3 Commonly used enhanced heat transfer elements and equipment

When adopting a heat transfer enhancement technology, the safety of equipment in transportation, installation, and long period operation should be considered, the requirements of process technology and manufacturing technology should be considered comprehensively, and the economic comparison should be made.

2.3.1 Shell and tube heat exchangers

Shell and tube heat exchanger can enhance the heat transfer effect from both the shell side and tube side. The shell side heat transfer can be enhanced by changing the baffle plate and improving the outer surface performance of heat exchange tube, and the tube side heat transfer can be enhanced by improving the inner surface performance of heat exchange tube and the technologies of inserts. In the 1980s, Tan et al. sorted out the effective and simple heat transfer enhancement technologies and published works of literature [39,40]. Suitable heat transfer enhancement technologies have been identified in the works of literature based on fluid flow state and phase state, and their correctness has been verified by subsequent decades of industrial application.

2.3.1.1 Shell side heat transfer enhancement technology

Literature [40] points out that low-fin tube and corrugated low-fin tube are suitable for the heat transfer enhancement of shell side medium laminar flow, corrugated low-fin tube, low-fin tube, transversally corrugated tube, corrugated tube, and longitudinal groove tube are suitable for turbulent flow; low-fin tube, T-shaped finned tube, and porous surface tube are suitable for boiling; and low-fin tube, longitudinal groove tube, and serrated finned tube are applicable to shell side medium condensation.

Enhanced heat transfer tube

① Low-fin tube. Low-fin tube is simple in processing, stable in performance, and highly replaceable. After successful development, it has been widely applied. Low-fin tube can be used in shell side medium of shell and tube heat exchanger without phase transition and condensation.

Low-fin tube can effectively enlarge the heat transfer area and improve the heat transfer capacity of heat exchanger. When it is used for condensation, under the action of capillary force, the condensate will stay in the fin groove and affect the heat transfer efficiency. The effect depends mainly on the geometry of the low-fin tube and the surface tension of the condensate. For most light hydrocarbons and organic solvent media, the effect of condensate retention is negligible, but the use of low-fin tube is not recommended for water vapor condensation and the condensation of mixtures with very different surface tension values. The effect of the low-fin tube with small fin spacing and high fin height is much more serious than that of the low-fin tube with large fin spacing and low fin height. The greater the surface tension of the condensate, the more serious the effect.

② Porous surface tube. Porous surface tube is high-efficiency heat exchange tube used to enhance boiling heat transfer. At present, there are three main types used in industry. The first one forms a porous layer on the outer wall of heat exchange tube by mechanical processing, which is called machined porous surface tube, and the representative product is Thermoexcel-E tube [41]. The second one forms a porous layer of metal particles on the outer or inner wall of heat exchange tube through sintering technology, which is called sintered porous surface tube, and the representative product is the high flux tube introduced by UOP in the 1970s [42]. The third one forms a porous layer of metal particles on the outer or inner wall of heat exchange tube by flame spraying technology, which is called sprayed porous surface tube, as shown in Fig. 2.2.

Fig. 2.2 Porous surface tube.

Due to the tool size limitation, the hole diameter of machined porous surface is generally 0.1–0.3 mm, and the porosity is lower than that of sintered porous surface. The sprayed porous surface is similar to the sintered porous surface. At present, sintered or sprayed porous surface is often used in industry to enhance boiling heat transfer at a small temperature difference. Experimental studies on micro- and nanoscale have been reported. With the decrease of porous surface size, the enhancement of boiling heat transfer will be greatly improved [43]. Geometric parameters such as average pore diameter (or average particle diameter), porosity, and average thickness are commonly used to characterize the structural characteristics of porous layers. Porosity, generally 30%–50%, indicates the relative size of the space available for steam and liquid two-phase movement and heat transfer in the porous layer and represents the volume proportion between the medium and particles involved in heat transfer. It will have a great influence on the heat transfer performance of the porous layer. The effect of porous layer thickness on boiling heat transfer is related to pore diameter. According to the experiments of Beijing Groundsun Technology Co., Ltd., it is found that the optimal thickness corresponding to aperture 50–150 μm is 0.4 mm.

Boiling heat transfer of media in porous layer includes heat and mass transfer between vapor and liquid. In boiling heat transfer analysis, the flow characteristics of vapor-liquid two-phase flow under the action of internal friction resistance, gravity, surface tension, and inertia force in porous layer should be considered, and the gas-liquid heat transfer characteristics inside and outside porous layer should be considered as well, namely, the effect of strong evaporation, heat conduction and convection, and the mutual influence of flow, heat transfer, and mass transfer. There are a lot of concave and convex holes on the surface of porous tube, which can produce a lot of bubble cores. Under the action of surface tension, the small bubble core in the liquid phase will be trapped in the porous layer. Due to the temperature difference between the top and tail of bubble, the surface tension will change slightly, prompting the heat flow to leave the surface and producing upward jet flow. The higher the emission frequency of bubbles, the more the heat taken away and the higher the heat transfer efficiency. After the bubbles are released from the surface, a large number of air columns form a honeycomb natural convection on the surface of the tube, which reduces the laminar flow layer between the liquid and the tube surface and enhances the heat transfer. The porous surface tube can promote bubble growth in each stage of bubble formation, thus improving the heat transfer efficiency.

Beijing Groundsun Technology Co., Ltd. has cooperated successively with Xi'an Jiaotong University and Heat Transfer Research Inc. (HTRI) to conduct in-depth research on porous surface tubes and successfully applied porous surface tubes in ethylene plant, catalytic cracking unit, gas fractionation unit, aromatics plant, MTO plant, PDH plant, PX unit, and EO/EG plant. According to the experimental data obtained by the company, the comparison results of tube with outside porous surface and plain tube are shown in Fig. 2.3. As can be seen from the figure, under the same heat flux condition, the external boiling heat transfer coefficient of porous surface tube is about 3–6 times that of smooth carbon steel tube. The enhancement rate varies with the material system and can be more than 10 times for light hydrocarbons.

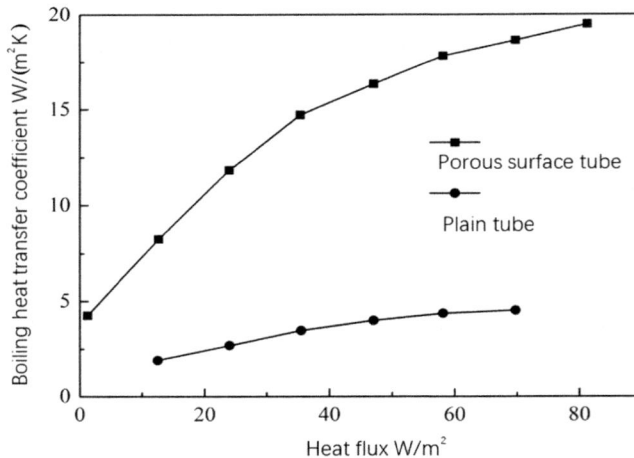

Fig. 2.3 Comparison of shell side boiling of sintered porous surface tube and plain tube.

It is also found that the porous tube has no obvious enhancement effect on medium with wide boiling range but has very good effect on light hydrocarbons with narrow boiling range. With the increase of successful application cases of porous surface tube, the research combined with industrial application is gradually deepened [44,45]. Because the heat transfer characteristics of porous surface tubes are closely related to their pore structure and size, heat transfer models vary from manufacturer to manufacturer, and the published calculation methods are not applicable to all products. Literature [46] reported a method for simulating porous surface tubes using commercial software developed by HTRI, which can be used for simple estimation of equipment heat transfer and reboiler piping scheme.

③ T-shaped finned tube. T-shaped finned tube is a kind of high-efficiency heat exchange tube formed by rolling of plain tube. Its structure is characterized by a series of spiral ring T-shaped tunnels formed on the outer surface of the tube. The appearance of T-shaped finned tube is similar to that of spiral grooved tube, but the root of the fin is circular groove. In the longitudinal section of the tube, the fin resembles the letter "T"; hence it has the name of "T-shaped finned tube."

When the liquid is boiling outside the T-shaped finned tube, the medium outside the tube is heated and forms a series of bubble nuclei in the tunnel. Being heated all around in the tunnel cavity, the bubble nuclei rapidly expand and fill the inner cavity, and the continuous heating rapidly increases the pressure in the bubbles, ejecting the bubbles from the slit on the surface of the tube. The bubbles ejected have large scouring force and produce a certain local negative pressure, which makes the surrounding low-temperature liquid pour into the T-shaped tunnel and form a continuous boiling. The liquid phase continuously penetrates through the narrow slit at the top of the fin to the bottom of the fin, and the vapor phase continuously flows upward and escapes through the narrow slit between the fins, thus achieving the effect of enhanced boiling heat transfer. When the temperature of the hot medium is 12–15°C higher than the boiling point or bubble point of the cold medium, the cold medium will bubble and boil on the plain tube. However, for T-shaped finned tube, a temperature difference of 2–4°C can make the cold medium boil [47].

Since the 1980s, T-shaped finned tubes have been fully studied and successfully used in the reboilers of crude oil distillation, catalytic cracking, and delayed coking units with good economic benefits.

④ Serrated finned tube (thermoexcel-C tube). Because the fin shape is similar to serration, thermoexcel-C tube is vividly called serrated finned tube. The main mechanism of enhanced heat transfer with serrated fins is that the three-dimensional discontinuous fins enhance the effect of surface tension, which can effectively reduce the thickness of condensate film and facilitate the rapid discharge of condensate under shear stress. In addition, the condensate flooded area of C tube is smaller than that of low-fin tube with the same fin spacing, and the liquid film is evenly distributed in the flooded area. The condensation heat transfer effect of C tube is better than that of low-fin tube [48].

Although serrated finned tubes are not as widely used in industry as spiral grooved tubes, there are still some successful cases.

⑤ Longitudinal groove tube. Longitudinal groove tube is used for vertical condenser. Driven by surface tension, the condensate flows rapidly from the top to the bottom of groove and is discharged along the longitudinal groove under the action of gravity so that the liquid film at the groove top and its vicinity is very thin. For straight tubes, it is the same from top to bottom so that the thermal resistance of the whole tube is significantly reduced from top to bottom. The longitudinal groove tube is especially suitable for fluid with high surface tension and cannot be used for material that is easy to scale. Vertical high-flux pipes usually have V-shaped longitudinal groove outside the pipe to enhance the condensation heat transfer in the shell side.

Nonsegmental baffle

Conventional shell and tube heat exchangers use single segmental baffle, which can obtain better heat transfer effect caused by the transverse flushing of fluid, but also have shortcomings such as large flow dead area, high pressure drop, and easy vibration. Double-segmental baffle and triple-segmental baffle can reduce the pressure drop, but sometimes it is difficult to avoid vibration. Rod baffle, helical baffle, and helical blade baffle can solve the vibration problem by changing the direction of fluid flow.

① Rod baffle. Using rod baffle instead of segmental baffle can reduce the pressure drop at shell side of heat exchanger and solve the vibration problem of heat exchanger. However, the film heat transfer coefficient of shell side decreases. The heat transfer effect can be maintained while solving the vibration problem if heat transfer tube with enhanced shell side heat transfer and rod baffle are adopted at the same time.

Rod baffle heat exchanger has been successfully applied in the process industry. It is suitable for the gas phase heat transfer and condensation of shell side medium which is easy to produce vibration, such as compressor interstage cooler, and also suitable for shell side medium with low allowable pressure drop, such as overhead condenser.

② Helical baffle. Helical baffle was developed by ABB Lummus Global and applied to engineering practice with good application effect. Since the 1990s, there have been more and more papers on the research results of helical baffle, and the structure of helical baffle has also changed. Currently, there are two commonly used structures, as shown in Fig. 2.4. One has a cylindrical core in the middle, which can form a continuous helicoid. The other is a discontinuous helical baffle with several baffles lapped to form an approximate helicoid [49]. Because of its simple manufacture, the lapped helical baffle has become a hot topic of research and application. Overlapped helical baffle [50], leak-proof helical baffle [51,52], and helical baffle with discharge groove [53] have been also developed.

Fig. 2.4 Diagram of continuous helical baffle and discontinuous helical baffle.

Different from the zigzag flow state of shell side fluid in segmental baffle heat exchanger, the baffles in helical baffle heat exchanger are arranged as approximate or complete helicoid, in which the shell side fluid flows in an approximate or continuous spiral shape, in order to effectively reduce shell side flow resistance, reduce vibration, and enhance heat transfer.

Since the core occupies the effective heat transfer space, the helical baffle heat exchanger with core is far less widely used than the lapped helical baffle heat exchanger. There are great differences in shell side heat transfer of lapped helical baffle heat exchanger under different factors which affect shell side medium flow and heat transfer, such as lapping degree and leakage prevention measures. The performance of helical baffle heat exchanger designed according to the literature method may be different from that of the equipment provided by the manufacturer. It is recommended that the manufacturers conduct thermodynamic design and ensure the performance.

The helical blade baffle heat exchanger and segmental baffle structure with staggered window patented by Sinopec and the curved baffle structure developed by Beijing University of Chemical Technology have their own characteristics, which can be used in suitable working conditions to solve practical engineering problems. The computational models for these structures are in the hands of the developers, and the methods for engineering design have not yet been published.

2.3.1.2 Enhanced heat transfer technology in tube

Tube side heat transfer can be enhanced by inserts and heat transfer tube. Literature [2] pointed out that the heat transfer enhancement technologies suitable for laminar flow of medium inside tube include corrugated low-fin tube, corrugated tube, and inserts. Corrugated low-fin tube, inner grooved tube, corrugated tube, and inserts are suitable for tube side turbulent flow. Spiral groove tubes are suitable for medium condensation in

tube. Among them, inner grooved tube, corrugated low-fin tube, and corrugated tube
are double-sided enhanced heat transfer tubes formed by one-step processing, which can
effectively enhance heat transfer both inside and outside the tube. It will be detailed in the
double-sided enhanced heat transfer part.

(1) Sintered porous surface tube

A remarkable feature of flow boiling is the bulk motion of liquid, due to the natural
circulation of the liquid caused by the density difference of liquid in boiling loop or the
forced motion caused by an external force. The heat transfer mechanism of flow
boiling is more complicated than that of pool boiling because the heat transfer process
in flow boiling is accompanied by various types of gas–liquid two-phase motion.

When the liquid flows and boils in tube, nucleate boiling coexists with convective boiling, and the heat transfer characteristics are closely related to the flow pattern
of two-phase mixture. The porous surface tubes have obvious enhancement effect
on nucleate boiling and less enhancement effect on convective boiling.

In 2008, Beijing Groundsun Technology Co., Ltd. and Xi'an Jiaotong University conducted a joint test, using 20% and 30% glycol aqueous solutions as test media,
and obtained the film heat transfer coefficient of boiling in porous surface tube and
the test results of plain tube, as shown in Fig. 2.5. It can be seen from the experimental results that the enhanced heat transfer coefficient of boiling in porous tube
is 3–6 times that of the plain tube, which is consistent with the results of the joint test
conducted by the company and American Heat Transfer Research Institute.

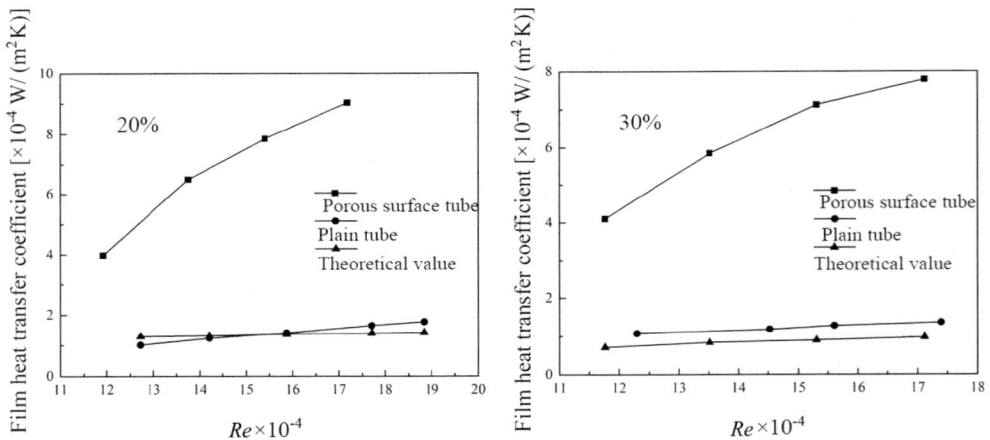

Fig. 2.5 Comparison of tube side boiling of plain tube and tube with sintered porous surface inside.

Researchers [54] studied the properties of inner porous surface by computational
fluid dynamics and proved that it greatly improved the performance of boiling heat
transfer. The inner porous surface structure has been successfully applied in catalytic
reforming and isomerization units.

(2) Inserts

Inserts are a kind of tabulator. The fluid in tube produces spiral flow, and its tangential velocity causes the secondary flow near the wall surface of the heat exchange tube, which enhances the fluid disturbance and destroys the boundary layer on the wall surface of the tube, so as to achieve the purpose of enhancing heat transfer. Also it has the functions of scale prevention and descaling. Although inserts can significantly improve the film heat transfer coefficient in tube, the tube side resistance also increases correspondingly, so inserts are mostly used in the laminar flow region with low Reynolds number of single-phase medium.

As for inserts, there are spiral thread, twisted tape, spiral piece, and static mixer. Commonly used structure is the spiral twisted tape of different sizes. In cooperation with South China University of Technology, Sinopec has developed a cross-serrated bond, which has achieved good results in industrial experiments.

HiTRAN is a patented product developed by Cal Gavin Ltd. In the heat transfer tube enhanced by HiTRAN, due to the disturbance of the HiTRAN in tube, the turbulence of the fluid in the dyeing area is enhanced, the flow boundary layer is destroyed, the radial distribution of the fluid is changed, and the heat and mass transfer process is fully enhanced. At the same time, the disturbance can inhibit the wall nucleation, increase the wall shear rate, and slow down the scaling. The company has successfully applied it in working conditions without phase transition, boiling, and condensation of medium in tube.

2.3.1.3 Compound heat transfer enhancement technology

Compound heat transfer enhancement refers to the technical means by which several heat transfer enhancement technologies play a role simultaneously; for example, by using rod baffle instead of segmental baffle, the vibration problem of shell side gas flow is solved, but the disadvantage is that the film heat transfer coefficient at shell side of heat exchanger decreases sharply. At this time, replacing plain tube with transversally corrugated tube and other heat transfer tubes that enhance heat transfer through secondary extended heat transfer surface can eliminate the hidden vibration while maintaining the heat transfer performance of the heat exchanger. If the overall heat transfer effect of the heat exchanger is not good and both the film heat transfer coefficients at tube side and shell side are low, the heat transfer tube which can simultaneously improve the film heat transfer coefficients at tube side and shell side is selected according to the characteristics of the medium, so as to improve the heat transfer performance of the whole equipment. The compound heat transfer enhancement technology described in this section is mainly the heat exchange tube with double-sided heat transfer enhancement.

Double-sided enhanced heat exchange tube is a single heat transfer element which can simultaneously enhance the heat transfer on tube side and shell side. Spiral grooved tube and transversally corrugated tube mentioned in the literature and twisted tube, widely used in recent years, have the characteristics of double-sided heat transfer enhancement. The porous surface tube with sintered porous surface inside and longitudinal groove tube outside is an efficient heat exchange tube which can enhance boiling and condensation simultaneously.

(1) Transversally corrugated tube. Since the former Soviet Union scientists published research reports on transversally corrugated tube in the 1970s, the advantages of transversally corrugated tube heat transfer enhancement have been recognized, and great progress has been made in the research on transversally corrugated tube.

Transversally corrugated tube is a double-sided enhanced heat transfer tube formed by one-step rolling. Please see Fig. 2.6 for its structure diagram. The mechanism of heat transfer enhancement is that the film heat transfer coefficient can be improved by increasing the turbulence degree when single-phase fluid is in the tube. When the fluid in the heat exchange tube flows through the raised rib of the transverse ring, axial vortex is formed nearby and fluid disturbance increases, destroying the boundary layer. When the vortex is about to disappear, the fluid passes through the next ring rib and the boundary layer is destroyed again. Over and over again, it is difficult to form a stable boundary layer at the metal wall surface in the tube, so as to ensure that the medium in the tube is always in the high heat transfer performance zone. The experimental results show that the main structural parameters affecting the transversally corrugated tube are rib spacing and rib shape. When the medium outside the tube condenses, with reasonable choice of pitch, the surface tension of condensate plays a controlling role. The condensate drops from below the trough, and the pressure gradient generated by the surface tension of the condensate reduces the thickness of condensate film and enhances the condensation heat transfer outside the tube. Researchers [55] have carried out experiment with water inside the tube to cool the ammonia vapor outside the tube, and the overall heat transfer coefficient of transversally corrugated tube condenser is 1.65 times that of plain tube.

Fig. 2.6 Structure diagram of transversally corrugated tube.

Since the vortex generated by the fluid in transversally corrugated tube is near the wall, it only destroys the boundary layer and has little influence on the central fluid. The fluid resistance is smaller than that of the corrugated low-fin tube with the same pitch and groove depth [56].

Test results show that [57] the asymptotic fouling thermal resistance of transversally corrugated tube is about 0.83 times that of plain tube, indicating the good scale inhibition performance of transversally corrugated tube.

(2) Twisted tube. Round tubes are flattened and twisted into twisted tubes. There is no baffle in twisted tube bundle. The heat exchange tubes depend on each other in different directions to form a self-supporting form, as shown in Fig. 2.7.

The shell side medium of twisted tube heat exchanger has no transverse scouring process but rotates and flows forward, similar to spiral flow. Unlike the spiral flow of shell side medium with centrifugal force in helical baffle heat exchanger, the shell side medium in twisted tube heat exchanger flows spirally around each heat exchange tube. This spiral flow effectively thins the boundary layer outside the heat exchange tube, enhances the heat transfer performance, avoids the violent collision between the fluid and the heat exchange tube, and eliminates the pressure loss caused by the transverse scouring of shell side fluid and the constant change of flow direction when baffle structure is adopted.

Fig. 2.7 Diagram of self-supporting structure of twisted tubes.

Since the 1990s, domestic scholars have done a lot of research on the performance of twisted tubes. Professor Si et al. conducted experimental research on the heat transfer and fluid resistance performance of twisted tubes with different structures [58]. In 2000, Professor Qian wrote an article on the research and application of twisted tube and mixed tube bundle heat exchangers, claiming that their enhanced heat transfer can save 26.5%–51% of heat transfer area [59]. It was proved that twisted tube is better for enhancing laminar flow or transition flow [60]. In cooperation with East China University of Science and Technology, Sinopec Engineering Incorporation studied the twisted tubes manufactured by Fushun Chemical Machinery Manufacturing Co., Ltd. The research group has published papers on the experimental and numerical simulation results [61–64].

Brown Fintube company from the United States has continuously improved its twisted tube heat exchanger manufacturing technology and used it in no-phase transition heat exchangers, condensers, and reboilers in the oil refining and chemical industries.

(3) Corrugated tube. Corrugated tube was first used in shell and tube heat exchangers in the 1970s, and its structure is shown in Fig. 2.8. With further research and industrial applications, corrugated tubes with different waveforms have been developed, and good economic benefits have been obtained in engineering applications over the years.

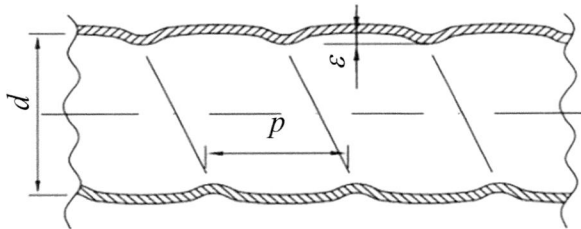

Fig. 2.8 Structure diagram of corrugated tubes.

There is a certain height difference between the wave crest and wave trough of corrugated tube. When fluid flows inside and outside corrugated tubes, the fluid velocity decreases and the static pressure increases at the wave crest of corrugated tubes, while the velocity increases and the static pressure decreases at the wave trough. The flow of fluid occurs under the condition of repeatedly changing the axial pressure gradient, forming small axial eddies near the wave-shaped wall, disturbing the fluid in the boundary layer, separating the boundary layer, and thus thinning the boundary layer. Turbulent flow can easily occur even at lower flow rates. Heat transfer is enhanced inside and outside the tube at the same time so that the overall heat transfer coefficient of the heat exchanger is doubled. At low Reynolds number, the heat transfer and resistance performance ratio of corrugated tubes is obviously better than that of plain tubes. At high Reynolds numbers, due to the rapid increase in pressure drop, the heat transfer and resistance performance ratios of corrugated tubes and plain tubes are very close. Because the tube bundle interface of corrugated tubes is constantly changing, the tube side and shell side fluids are always in a highly turbulent state, which scours the tube wall of the heat exchange tube, thus destroying the thickness of boundary layer and fouling layer, and fouling layer is not easy to form on the tube wall. Turbulence also makes it difficult for fouling particles in the fluid to deposit. Since the curvatures of the inner surface and outer surface of corrugated tube change greatly, and the tube has scalability, even if scale is formed, hard scale is easy to fall off. Its antiscale performance is better than that of plain tube [65]. No fouling was found in a heating station after five heating periods [66].

With regard to corrugated tube technology, a large number of papers have been published on thermodynamic studies and laboratory work as well as on industrial applications and strength design. Liu W also introduced the calculation method of corrugated tubes in his "Process Design of Heat Exchanger" [67]; if corrugated tubes are used when the medium is in the region of high Reynolds number, power loss should be considered comprehensively.

(4) High flux heat exchanger tube. The inside of the heat exchange tube used in vertical reboiler is processed as sintered porous surface to enhance the boiling heat transfer in the tube. The outer surface of the tube is longitudinal groove, which enhances the condensation heat transfer outside the tube. UOP first named it as high flux heat exchange tube.

This kind of heat exchange tube is generally used in vertical reboiler, suitable for vertical reboiler with tower top materials condensed at shell side. The hot tower top materials condense on the outside of heat exchange tube, and the condensate flows down along the longitudinal groove. The reboiling circulating materials at the bottom of tower vaporize on the sintered surface of metal inside the tube, and heat transfer is enhanced on both sides.

2.3.1.4 Other heat transfer enhancement technologies

At the end of the 19th century, Linde Company from Germany successfully developed coil-wound heat exchanger, which is a shell and tube heat exchanger that can realize multistream heat transfer. The heat exchange tubes are winded on the center cylinder, which plays a supporting role in manufacturing, as shown in Fig. 2.9.

Tube side fluid enters the tube box from the tube side entrance, passes through and enters the outlet tube box in a spiral way along the winded tube, and is finally discharged from the outlet. Shell side fluid enters through the shell side inlet, passes through the shell space filled with winded tube bundles in a countercurrent and transverse crossing way,

Fig. 2.9 Core of coil- wound heat exchanger.

and is discharged from the outlet, to avoid the disadvantages of baffle heat exchanger such as the heat transfer dead zone on the back of baffle and the scale accumulation. At the same time, the fluid separation and convergence between adjacent tubes and between layers strengthens the turbulence of the fluid outside the tube and greatly improves the heat transfer coefficient. The heat exchanger is compact in structure, and the heat transfer area per unit volume can reach more than two times that of ordinary tubular heat exchanger, which makes it suitable for large-scale units. The length of heat exchange tube is not limited, and the pressure resistance is good. It has more reliable operation and easier maintenance than plate heat exchanger.

Professor Zhang of Dalian University of Technology instructed his student Q.Y. Yu to complete the research on the calculation method of spiral wound heat exchanger [68], and the film heat transfer coefficient is closely related to the tube winding mode and structure. A simple method can be used to calculate the heat transfer performance of spiral wound heat exchanger [69,70]. Researchers [71] studied the thermodynamic performance of spiral wound heat exchanger by computational fluid dynamics.

Compared with tubular heat exchanger, spiral wound heat exchanger has small temperature difference at hot end. If it is used for heat exchange between reaction feed and discharge, it can effectively recover the heat of reaction products, reduce the load of heating furnace, reduce the air cooling and water cooling load, and reduce the energy consumption of reaction system. Up to now, it has been successfully applied in hydrocracking unit, aromatics unit, and low-temperature methanol scrubbing unit.

2.3.2 Plate heat exchangers

Plate heat exchanger is an efficient and compact heat exchanger which takes plate surface as heat transfer surface. Compared with shell and tube heat exchangers, plate heat exchangers have smaller plate spacing, which easily produces turbulence, and can enhance heat transfer effect. The thin plate reduces the wall's thermal resistance. The flow dead zone between plates is small and the effective heat transfer area is large. The wall surface is smooth and the shear force is large, which makes it not easy to scale, and the fouling heat resistance is small.

The heat transfer of plate heat exchanger is mainly enhanced by changing plate spacing, corrugations, spirals, and adding fins between plates. The high efficiency and compact characteristics of plate heat exchanger can realize heat transfer under small temperature difference, making the process more energy saving.

There are many structure forms of plate heat exchanger. Detachable gasket plate frame heat exchanger is suitable for low-pressure and low-temperature occasions due to the limitation of gasket. All-welded plate heat exchanger has good high-temperature and high-pressure resistance and is widely used because its plates are sealed by welding. Two flow channels are formed after the plates of welded tube-plate heat exchanger are combined. One side is the tube-type channel in the direction of plate width, and the other side is the wavy channel in the direction of plate length. The tube-type channel can achieve 100% mechanical cleaning, similar to the tube side of shell and tube heat exchanger. There is no support between the plates on one side of welded wide channel plate heat exchanger and the flow state is similar to that in rectangular channel. The channels on the other side are supported by dimples or pins in between. The rectangular channel side of such product is specially equipped with cleaning tools, so it is suitable for occasions with large granularity, easy adhesion, and easy coking. All-welded plate-shell heat exchanger adopts square corrugated plate as heat transfer element and installed it in cylindrical shell, truly achieving large-scale equipment. Circular plate shell heat exchanger adopts circular corrugated plate, the plates are welded to form plate bundle, and the plate bundle is placed in the pressure shell. It can be used in high-temperature and high-pressure occasions.

2.3.2.1 Corrugated plate heat exchanger

There are many types of corrugated plate. Chevron corrugated plate is a typical plate with reticular flow with good heat transfer performance, but the flow resistance is large, and it is not suitable for medium-containing particles or fibers. Qiu [72] studied the effect of structural parameters of Chevron plate on heat transfer. The heat transfer effect increases gradually with the increase of corrugated angle and reaches the maximum at an angle of 60 degrees. The heat transfer effect increases with the increase of Chevron height and the decrease of pitch, and the flow pattern of fluid gradually changes from zigzag flow to cross-flow.

Corrugated structure increases the wall shear force between fluid and heat exchange surface so that fouling is not easily attached to the heat exchange surface for growth and it has a certain self-cleaning effect. The fouling thickness is reduced and the heat transfer is further enhanced. In engineering design, the fouling coefficient of plate heat exchanger is generally 1/10 to 1/5 of that of conventional shell and tube heat exchanger.

In corrugated plate heat exchangers, the corrugated structure forces the fluid to constantly change its flow direction, effectively promoting fluid turbulence. When $Re > 500$, it can reach the turbulent state and maintain high heat transfer performance. The flow pattern of fluid varies in plates of different structures. Therefore, the heat transfer models of corrugated plate heat exchanger differ greatly, and the heat transfer correlation formula obtained from experiments should be used as the basis of heat transfer design of plate heat exchanger.

2.3.2.2 Spiral plate heat exchanger

Spiral plate heat exchanger is composed of shell, helix, seal, and inlet and outlet nozzles. The helix is made of two flat steel plates with sealed edges. Both sides of the helix are sealed by cover plates, which can be detachable or nondetachable. Regularly distributed spacing columns welded to plates can ensure plate spacing. It has the advantages of low temperature stress, simple manufacture, and reliable sealing. However, the processing capacity of a single equipment is limited and the pressure-bearing capacity of equipment is limited, so maintenance and mechanical cleaning are difficult.

According to flow channel type, spiral plate heat exchanger is generally divided into type I, type II, type III, and type G [73], as shown in Fig. 2.10.

Normally, no phase change fluid flows along the spiral channel, while phase change fluid flows axially in the spiral channel. The fluids flow along a compact, narrow, and single channel between two plates with large flow turbulence. In spiral plate heat exchangers, cold and hot fluids can realize countercurrent flow with small heat transfer temperature loss.

2.3.2.3 Plate-fin heat exchanger

Plate-fin heat exchanger is a compact plate heat exchanger with fins between parallel plates. Multiflow heat transfer can be achieved in the heat exchanger, and the heat transfer rate is 10–30 times that of ordinary shell and tube heat exchanger. It is suitable for heat transfer enhancement of clean medium under small heat transfer temperature difference, small allowable pressure drop, and other process conditions. It can realize heat transfer under a temperature difference between 0.5°C and 1°C, so it is more energy saving. However, due to structural limitations, it is easy to plug and has poor strength and limited maximum temperature difference, so it is not suitable for high-temperature and high-pressure occasions and also not suitable for corrosive medium.

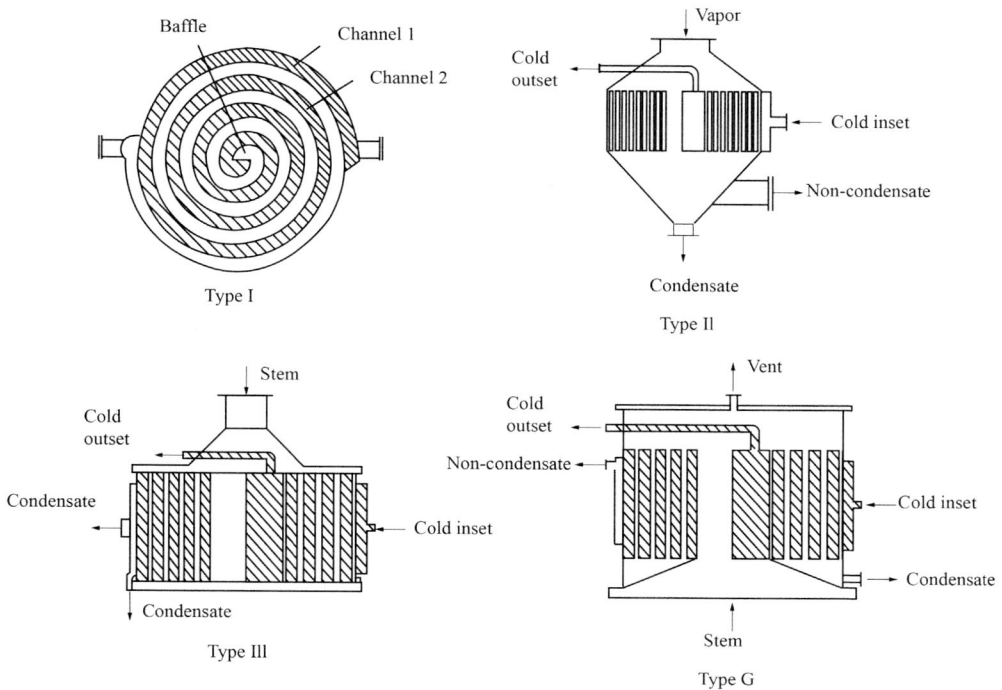

Fig. 2.10 Spiral plate heat exchanger.

As the basic element of plate-fin heat exchanger, fin plays multiple important roles. Fin promotes fluid turbulence and destroys the development of heat transfer boundary layer, thus enhancing heat transfer. Fin also supports the plates and improves the strength of plate-fin heat exchanger.

There are many types of fin used in plate-fin heat exchangers, including straight fin, serrated fin, perforated fin, and corrugated fin. According to the demand for heat transfer enhancement, different fin structures and parameters are selected. High and thin fins should be selected on the side with small heat transfer coefficient, aiming at enhancing heat transfer and increasing heat transfer area. Low and thick fins should be used on the side with large heat transfer coefficient.

Straight fin is made by stamping and rolling thin metal sheet and has high pressure strength. Its heat transfer and hydrodynamics characteristics are similar to those of in-tube flow. Since it has less disturbance effect on the medium and its heat transfer coefficient and pressure drop are relatively small, it is mainly used in situations where flow resistance is strictly required. It can also be used in occasions where fluid contains small particles and need to avoid its precipitation. Compared with other fins, straight fins have higher strength. It can be used to add an empty channel to the outermost layer of the tube bundle of plate-fin heat exchanger to increase the strength of plate-fin heat exchanger.

Serrated fin can be regarded as a straight fin cut into many short segments and staggered at a certain interval to form a discontinuous fin. It is characterized by many tiny grooves along the length of the fin, forming a zigzag channel. Serrated fin is very effective in promoting fluid turbulence and destroying boundary layer and can also realize fluid turbulence in the range of low Reynolds number. Its heat transfer coefficient and pressure drop are greater than those of straight fin. Serrated fin is the most widely used efficient fin in plate-fin heat exchanger. It is suitable for gas channel and liquid channel with high viscosity.

Perforated fin is made by punching holes in thin metal sheet and then stamping or rolling it into shape. There are many holes in the fin, and the opening rate of the perforated fin is generally 5%–20%. The existence of pores can cause the thermal resistance boundary layer to continuously break, so as to improve the heat transfer efficiency and make the medium distributed more evenly in the fin, and also facilitate the washing and elimination of impurity particles. Punching reduces the heat transfer area and weakens the fin strength. This kind of fin is mostly used for heat transfer of medium with phase transition, such as reboiler and condenser.

Corrugated fin is a metal sheet stamped or rolled into a certain corrugated shape, forming a curved channel. This structure can make the medium change its flow direction continuously and promote turbulence, separation, and destruction of thermal resistance boundary layer, thus improving heat transfer efficiency. The denser the corrugation, the greater the amplitude, the better the heat transfer performance and the bigger the resistance. This type of fin has high compressive strength and can be used for gas with high pressure.

By studying the influence of fin height, pitch, and thickness on heat transfer effect and resistance drop, a variety of new fins with different characteristics were developed, including high-strength and high-density fin, high heat transfer factor, and low friction factor fin, which met the high reliability requirements of heat transfer in large-scale petrochemical units.

Plate-fin heat exchanger is widely used in low-temperature processes in the air separation industry and petrochemical industry. Taking ethylene plant as an example, there are about 20 streams in its large cold box, with a total heat load of tens of megawatts, of which the refrigerant cooling capacity accounts for 60%–70% of the total cooling capacity in the cold box. The temperature and pressure of ethylene, propylene, and methane in the cold box are closely related to their composition. The condensation and evaporation processes of refrigerant in the cold box are complicated, and two-phase flow exists in a wide temperature range. When gas-liquid two-phase fluid enters the feeder of plate-fin heat exchanger, gas-liquid separation will occur at the turning point due to centrifugal force, resulting in uneven distribution of two-phase fluid to each channel. Studies show that in low-temperature double-phase transition plate-fin heat exchanger, as long as there is less liquid distribution in an evaporation channel, dryout will immediately appear, the heat transfer temperature difference will sharply decrease, the heat transfer in plate-fin heat exchanger will immediately deteriorate and the fluids in adjacent channels cannot achieve effective heat exchange, resulting in the failure of heat exchange group components. So how we evenly distribute the two-phase flow into each channel of plate heat exchanger becomes the key factor of cold box design. It is particularly important to set up effective distribution structure and uniform mixing structure of gas–liquid two-phase flow at the inlet of gas–liquid two-phase fluid and between each layer of channel in the heat exchanger. Reasonable design of vapor–liquid uniform distribution structure can effectively improve the heat transfer efficiency, and the lowest heat transfer temperature difference can reach $0.57°C$ at present.

2.3.3 Tubular fired heaters

The tubular fired heater commonly used in refinery can be divided into two categories, i.e., general fired heater without chemical reaction in tube and hydrogen reformer with reaction in tube. The type of tubular fired heater has developed from stacker furnace and pure convection furnace to box furnace and cylindrical furnace. Now a tubular fired heater is generally composed of radiation chamber, convection chamber, waste heat recovery system, burner, and ventilation system.

In tubular fired heater, the high-temperature flame and flue gas generated by the combustion of fuel in the radiation chamber serve as the heat source to heat the flowing medium in the tubes via the tube wall, so as to achieve the heat transfer temperature required by the process. There are many complicated heat transfer processes, such as combustion process, flame and high-temperature flue gas radiation heat transfer process, flue gas convective heat transfer process, solid heat transfer process of tube wall, and heat absorption process of medium in tube. Therefore, radiation heat transfer, convective heat transfer, and heat conduction coexist, and enhancing the heat transfer inside and outside of tubes is of great significance to improve heat transfer efficiency, improve the heat efficiency of fired heater, and save energy.

2.3.3.1 Enhanced heat transfer technology of tube outside

Because the high-temperature flame and flue gas of fired heater transfer heat to tubes by radiation and convection, the heat transfer enhancement technology of tube outside is studied based on this.

1. Radiation heat transfer enhancement

    The study on radiation heat transfer enhancement in fired heater focuses on the flow field and temperature field of flue gas and the formation rule of pollutants as well as the optimization design. At present, radiation heat transfer enhancement is mainly achieved through the following aspects.

    ① Installation of enhanced radiation element in radiation chamber. The essential reason for the low heat transfer efficiency of traditional technology is that the heat rays are diffuse, and eventually part of the heat rays are lost by heat dissipation of fired heater wall. Li et al. [74] proposed a method to increase both blackness and area, that is, to enable the blackbody element. Without increasing the heating power of fired heater and reducing the emissivity, the blackbody element can increase the area of fired heater, improve the blackness of fired heater, and increase the radiation ability. According to Lambert's law, among all the radiation energy emitted from all directions in heat source space, the maximum energy is in the normal direction, so the best heat transfer enhancement effect can be achieved by installing the blackbody element on the top of the radiation chamber.

② Installation of reflection board at the outlet of radiation chamber. The high-temperature flue gas directly flows out of radiation chamber, takes away a considerable amount of heat, and reduces the thermal efficiency of the radiation chamber. Dou et al. [75] proposed to install a reflection board at the outlet of the radiation chamber to make the flue gas flow back and increase the residence time of flue gas in the chamber. CFD numerical simulation method is used to compare the effects of different structures with reflection board and without reflection board on the radiation heat transfer. The results show that the installation of reflection board can promote radiation heat transfer, especially the baffle board with slope on both sides, can improve the heating condition in the upper part of the radiation chamber, reduce the outlet temperature of flue gas, and improve the radiation thermal efficiency.

③ Application of high radiation coating on fired heater lining. The efficiency of radiation heat transfer can be increased by coating the wall of radiation chamber with high radiation coating. Coatings used at present can be roughly divided into high-temperature infrared radiation coating and high–radiation ceramic coating. The infrared radiation coating can increase the blackness of substrate surface, enhance the radiation heat transfer of the substrate surface after absorbing heat from heat source, change the radiation spectrum distribution in the heat transfer zone, and change the discontinuous heat source spectrum into a continuous form, thus promoting the heat absorption of the heated object. For example, the far-infrared radiation coating [76] based on high-quality far-infrared radiation composite powder such as Al_2O_3 developed and produced by a domestic research institute has good impact resistance and shock resistance and energy saving of 15%–30%. High-radiation ceramic coating can greatly increase the radiation coefficient of wall and achieve obvious energy-saving effect. There is no need to make any changes to the original structure of fired heater with convenient implementation. It can be widely applied to any tubular fired heater that can spray radiation coating, and the fired heater efficiency is significantly improved.

④ Application of nanocoating on the surface of fired heater tube. In addition to the coating on the lining inside the fired heater, the nontoxic water-based coating with rare earth oxide as the main raw material developed by a company has the characteristics of high emissivity, high temperature resistance, protectiveness, and nonwettability. It is suitable for metal and nonmetal material substrate. It can shrink and expand with the substrate and has strong performance of thermal shock resistance. It's easy to construct. It can be sprayed like finishing paint and can be used for the surface coating of fired heater tube, so as to improve the blackness of the heating surface and improve the heat transfer efficiency.

2. Convective heat transfer enhancement. The heat absorption ratio between radiation chamber and convection chamber of tubular fired heater is related to the heat efficiency and metal consumption of the fired heater. The control thermal resistance in the convection chamber is on the flue gas side, and enhanced heat transfer technology is needed to improve the heat transfer efficiency of the convection chamber.

   Fin tube can not only increase the area of convective heat transfer but also enhance the effect of flue gas turbulence, which can effectively increase the convection heat. Commonly used fin forms are strip fin, ring fin, and stud. The fluid temperature rise curve of the enhanced heat transfer fired heater tube with fins is obviously higher in the initial section of fired heater tube than that of the plain tube; however, when it comes to the outlet section, there is little difference between the two fluid temperature curves [77]. The application of fin tube can increase the production capacity and prolong the operation cycle of the fired heater without significantly affecting the product yield [78]. Using shaped steel tube with concave thread groove as fired heater tube, its heat transfer coefficient is 1.4–1.5 times that of the plain tube.

   The flue gas baffle is set on the wall along the flow path of flue gas in convective section to prevent the flue gas from drifting and to disturb the flue gas at the same time, so as to enhance the heat transfer effect of flue gas. Flue gas spoiler is set in a smooth flue to destroy the boundary layer near the wall and enhance the turbulence of flue gas, so as to enhance the convective heat transfer.

   A soot blower is set in the convection section of the oil–fired or oil–gas fired heater to regularly clean the ash deposited on the convection tube and extended surface, so as to reduce the thermal resistance of scaling on the outer wall of tube, reduce the flue gas pressure drop, and improve the heat transfer effect of the convection chamber.

2.3.3.2 Enhanced heat transfer technology in tube

The heat transfer in tube is mainly convective heat transfer, and the thermal resistance in the tube is mainly concentrated in the coking layer and the boundary layer. The commonly used method is to improve the flow rate in the tube to thin the boundary layer, enhance the heat transfer process in the tube, and weaken the thermal resistance influence of the boundary layer, such as the new type tube and internals with new structures.

Inner ribs and inserts can also be used to enhance heat transfer inside the tube. At present, more researchers focus on plum blossom tube, MERT tube, and twisted-tape-enhanced heat transfer.

There are two types of plum blossom tubes, namely spiral type and straight fin type, which can effectively increase the heat transfer area and improve the flow of materials in the tube. The spiral tube is the earliest enhanced heat transfer used in industry. Compared with plain tube, the heat transfer coefficient of MERT tube is increased by 20%–50%, and the inner surface area is increased by 2%. Under the same reaction depth, the operation period of the unit is twice that of plain tube, but the pressure drop is increased by 2–3.5 times.

Zeng et al. [79] invented the internal twisted-tape heat exchange tube. In addition to improving the heat transfer coefficient and reducing the tube wall temperature, the material in the tube rotates forward, and the resulting transverse flow forms a strong scouring of the tube wall, which not only thins the boundary layer and improves the convective heat transfer coefficient but also makes the surface less prone to coking and prolongs the operation cycle of equipment. The twisted-tape tube can reduce the coking amount by 40%, significantly prolong the operation cycle, increase the tube capacity by 10%, and reduce the temperature of tube wall by about 20°C, thus enhancing the heat transfer [80,81].

2.3.3.3 Enhanced heat transfer technology for waste heat recovery

The temperature of flue gas leaving the convection chamber of fired heater is usually 180–400°C. Flue gas/air preheater is a common and effective way to recover the waste heat from flue gas. Common air preheaters are tube bundle preheater, gravity heat pipe preheater, plate preheater, bidirectional fin cast iron plate preheater, water heat medium preheater, or a combination of the above forms.

Tube bundle preheater is a recuperative air preheater. In order to enhance heat transfer, the spoiler is arranged inside the tube, and the extension form of plain tube, fin tube, or stud tube is set outside the tube according to the medium.

Gravity heat pipe preheater transfers heat by means of liquid–vapor phase transition of heat transfer medium, and its heat transfer performance is very superior. In order to enhance the heat transfer performance of heat pipe, extension forms, such as fin or stud, are usually adopted outside the cold section and hot section of heat pipe. Based on different working conditions, the specific parameters of the extended tube can be adjusted. On the other hand, increasing the vacuum degree in tube is helpful to enhance the vapor–liquid two-phase transition of the working medium in tube, which is conducive to effective heat transfer. For preheater, according to the temperature drop of flue gas, the row spacing of heat pipes and the number of heat pipes per row are adjusted so that the flow rate of flue gas outside the tube is always in the high range, which is beneficial to improving the heat transfer coefficient outside the tube.

Because of the large overall heat transfer coefficient and the small heat exchange area required to recover the same heat, plate air preheater has the advantages of small volume and small pressure drop. This type of air preheater consists of stamped corrugated sheet forming channels of cold and hot fluids spaced from a few millimeters to tens of millimeters. Cold and hot fluids pass through the channels in pure countercurrent or cross-flow. Heat exchange plate is composed of convex and concave in various forms. On the one hand, it increases fluid disturbance to enhance heat transfer. On the other hand, the convex parts between plates have a supporting and strengthening effect.

The principle and structure of cast iron bidirectional fin air preheater and plate preheater are exactly the same. The difference is that the heat exchanger plate is relatively thick, usually 5–6 mm. There are many fins arranged in a certain pattern on both sides of the heat exchanger plate to enhance the heat transfer performance on both sides. The heat exchange element is a cast iron special-shaped plate with a base plate molded by casting and discontinuous fins vertically arranged on both sides of the base plate. There are discontinuous fins on the air side and the flue gas side, which is beneficial to enhance the heat transfer efficiency. The direction of fins is consistent with the direction of fluid flow. While making the fluid turbulent, smooth fins can reduce the pressure loss of fluid flow, reduce the formation of sediment and scale, and play a certain role of self-cleaning.

The water heat medium preheater is mainly composed of flue gas-water heat exchanger, air-water heat exchanger, hot water circulating pump, and the corresponding pipes. It is an air preheating system with softened water circulating in closed piping as the heat transfer medium. In order to enhance heat transfer, the heat exchange tubes on flue gas side and air side are extended with external continuous fins. The internal water is always liquid, and high flow rate is adopted to enhance the heat transfer effect inside the tube. On the flue gas side, the high-temperature flue gas transfers heat to the softened water in the pipe. On the air side, the hot water transfers heat to the combustion air. The hot water circulation pump on the piping is used to maintain the flow of softened water to achieve continuous heat transfer.

2.3.3.4 Case of heat transfer enhancement of heating

Radiation chamber is the main heat transfer part of tubular fired heater. More than 80% of the heat in the radiation chamber is transferred by thermal radiation. Good radiation chamber structure, burner arrangement, and radiation tube arrangement are very important to improve the uniformity of flue gas distribution in fired heater, the average flux density of tube, and the uniform absorption of radiation heat.

Delayed coking fired heater and hydrogenation reaction feed fired heater are usually box type with double-sided radiation horizontal tube, in which the radiation side wall is designed to tilt at a certain angle and a number of low power against wall burners are used. Combustion flame clings to the wall and stays in stable combustion, which can better keep the rigidity and stability of the flame. The high-temperature flue gas jet can flow steadily upward along the fired heater wall, less affected by the flue gas flow outside the jet, forming a flue gas backflow zone between the burner and the tube. The backflow zone makes the flame stick to the wall stably, and the high-temperature flue gas forms a relatively uniform plug flow along the height direction of fired heater, which results in small temperature distribution gradient of flue gas in the upper and lower parts of the side wall and is conducive to the uniform heat absorption of tube. The volume of radiation with inclined wall structure decreases from bottom to top, and the flow rate of flue gas increases gradually, which improves the uniformity of the volumetric flux density of the upper and lower fired heater and enhances the convective heat transfer between the upper radiation tube and flue gas. The against-wall combustion and the inclined wall structure of radiation chamber have the advantages of overcoming combustion flame deflection, optimizing flue gas distribution, increasing average flux density of tube, reducing flue gas distribution gradient in upper and lower parts of fired heater, and preventing local overheating of tube, which is beneficial to the uniform distribution of flux density on the surface of tube along length direction and height direction, the improvement of average flux density on the surface of tube, and the enhancement of heat transfer.

According to the flow rate, flow pattern, and oil film temperature of the medium in tube, the tube diameter of vacuum distillation fired heater is reasonably selected and expanded step by step. If conditions permit, appropriately increasing the flow rate of medium can improve the film heat transfer coefficient in tube, enhance the heat transfer, and reduce the outer wall temperature of tube.

2.4 Heat transfer network synthesis technology

Based on thermodynamics, pinch point technology analyzes the distribution of energy flow along temperature in a process system from a macroscopic point of view, finds out the bottlenecks in energy consumption of the process system, and provides a debottlenecking method. From 1978 to 1993, Linnhoff et al. [82–86] proposed the problem of temperature pinch point in heat exchanger network and made a comprehensive summary of pinch point technology.

The world's famous engineering companies such as Bayer, Monsanto, Dupont, ICI, and Mitsubishi have adopted pinch point technology to design new plants and revamp old plants at earlier times, which has achieved remarkable results in energy consumption reduction, investment reduction, and environmental protection. In the current engineering design, engineering companies such as Technip, Fluor, KBR, Foster Wheeler, and SEI are also using pinch point technology to guide the design of heat exchanger networks and the energy integration design of process systems, which has achieved good economic and social benefits. In recent years, some methods including thermal integration strategy of multiple refining units based on pinch point technology have been proposed to guide engineering design and technical improvement [87], which strongly promoted the application of pinch point technology.

2.4.1 Basic concepts of pinch point technology

2.4.1.1 Temperature enthalpy graph

The temperature enthalpy diagram (*T-H* diagram) can describe the thermal characteristics of process flows and utility flows in a process system. The vertical axis of the diagram is temperature *T* in °C and the horizontal axis is enthalpy *H* in kW. Flows are represented by line segments (straight lines or curves) on the *T-H* diagram. When the mass flow W, initial temperature T_s, and target temperature T_t of a flow are given, it can be plotted on the *T-H* diagram.

2.4.1.2 Composite curve

An industrial process often contains multiple hot flows and cold flows. In practical use, the hot and cold flows need to be combined together, and the matching between hot and cold flows is to be considered. Multiple hot flows are represented by hot composite curves on the *T-H* diagram, and multiple cold flows are represented by cold composite curves. The core of curve construction is to sum up the heat loads of all flows in the same temperature range and then represent it by a virtual flow with cumulative heat load value in this temperature range.

2.4.1.3 Meaning of pinch point

To understand the meaning of pinch point, it is necessary to start with the two characteristics of pinch point.

The first characteristic is the smallest heat transfer temperature difference between hot and cold flows at the pinch point, equaling $\Delta T_{\min}$. This characteristic limits the further energy recovery of the process system and constitutes the bottleneck in the energy consumption of the system. In order to increase the energy recovery of the process system, it is necessary to improve the pinch point to solve the bottleneck.

The second characteristic is the zero heat flow of process system at the pinch point. From the angle of heat flow (or from the angle of potential temperature), the pinch point divides the process system into two independent subsystems. The upper one is the hot end (high potential temperature), with only heating utilities required, known as heat sink, while the lower one is the cold end (low potential temperature), with only cooling utilities required, known as heat source.

To ensure maximum energy recovery in a process system, three basic principles should be followed, i.e., heat flow shall not pass through the pinch point, cooling utilities shall not be introduced above the pinch point, and heating utilities shall not be introduced below the pinch point.

2.4.1.4 Determination of pinch point of process system

The minimum temperature difference ΔT_{min} should be determined before determining the pinch point of the process system. ΔT_{min} can be determined based on experience or can be optimized and selected with the target of minimizing the total cost including equipment cost and operation cost in the process system.

According to the mass flow, composition, pressure, initial temperature, and target temperature of hot and cold flows, the composite curve of hot flows and the composite curve of cold flows are made on T-H diagram, respectively.

The composite curve of hot flows is placed above the composite curve of cold flows, and the two curves are close to each other in the horizontal direction on the T-H diagram. When the vertical distance between the closest points of the two composite curves is exactly equal to ΔT_{min}, this closest point is the pinch point.

Due to the difference of process systems, the two composite curves may have more than one closest point. In this case, all the closest points are the pinch points of the process system, which is called the multiple pinch point problem. When a process system requires only one utility flow, such as heating utilities only or cooling utilities only, such a system does not have a pinch point, which is called the threshold problem.

2.4.1.5 Pinch point design method of heat exchanger network

The flows that need heating and the flows that need cooling during the production of process industries are reasonably matched together, and the hot flows are fully used to heat the cold flows, so as to improve the heat recovery efficiency of the process and reduce the heating and cooling loads of utilities as much as possible. This is the structure of matched heat transfer between flows and the corresponding load distribution problem [88].

Aiming at these problems, heat exchanger network design using pinch point technology is optimized with the goal of minimizing equipment (heat exchangers, heaters and coolers) investment and operating (hot and cold utilities) cost. Each flow is heated or cooled from the initial temperature to the target temperature, while the network operation is stable, is easy to adjust, and has certain operational flexibility [89].

2.4.2 Optimal design of threshold problem heat exchanger network

The threshold problem [90] heat exchanger network refers to the heat exchanger network that only needs one kind of utilities. The heat exchanger network that only needs cold utilities is called the hot end threshold problem heat exchanger network, while the heat exchanger network that only needs hot utilities is called the cold end threshold problem heat exchanger network. The research on the design and application of threshold problem heat exchanger network is relatively scarce, and the research on the optimization and synthesis method of threshold problem heat exchanger network lags behind the research on conventional heat exchanger network.

The hot end threshold problem heat exchanger network is considered to exist only below the pinch point, while the cold end threshold problem heat exchanger network is considered to exist only above the pinch point. The design principle of hot end threshold problem heat exchanger network is to cancel the hot utilities in the heat exchanger network, start the design from the high-temperature side of the heat exchanger network to ensure that the cold flows at a higher temperature can obtain heat from the hot flows, and replace the cooling process with process flow heat exchange as far as possible. The design principle of cold end threshold problem heat exchanger network is to cancel the cold utilities in the heat exchanger network, start the design from the low temperature side to ensure that the heat of the hot flows at a lower temperature can be transferred to the cold flows, and replace the heating process with process flow heat exchange as far as possible.

2.4.2.1 Analysis of hot end threshold problem heat exchanger network

Fig. 2.11 shows the composite curves of cold and hot flows for the hot end threshold problem. Hot end threshold problem heat exchanger network has two forms; i.e., the minimum heat transfer temperature difference between composite curves of cold and hot flows (ΔT_{min}) appears at the end that does not require utilities, as shown in Fig. 2.11A, and there is pinch point between the composite curves of cold and hot flows, as shown in Fig. 2.11B. When the minimum heat transfer temperature difference drops from ΔT_{min} to $\Delta T'_{min}$, the composite curve of cold flows moves to the left, and the composite curves of cold and hot flows further overlap, as shown in Fig. 2.11C and D. In this process, the total consumption of cold utilities remains unchanged, but the required temperature of cold utilities changes. At this time, the sum of low-temperature cold utility load $Q_{C,1}$ and high-temperature cold utility load $Q_{C,2}$ is equal to the original minimum cold utility load $Q_{C,min}$. High-temperature cold utility load $Q_{C,2}$ provides the possibility of heat integration between the threshold problem heat exchanger network and other heat exchanger networks. In the design of heat exchanger network, the minimum heat transfer temperature difference is optimized and determined with the target function of minimizing the sum of equipment investment and operating cost, or the minimum allowable heat transfer temperature difference is determined considering the constraint of heat transfer coefficient. Therefore, when $\Delta T'_{min}$ is equal to the set minimum allowable heat exchange temperature difference, $Q_{C,2}$ is the maximum heat integration load of threshold problem heat exchanger network.

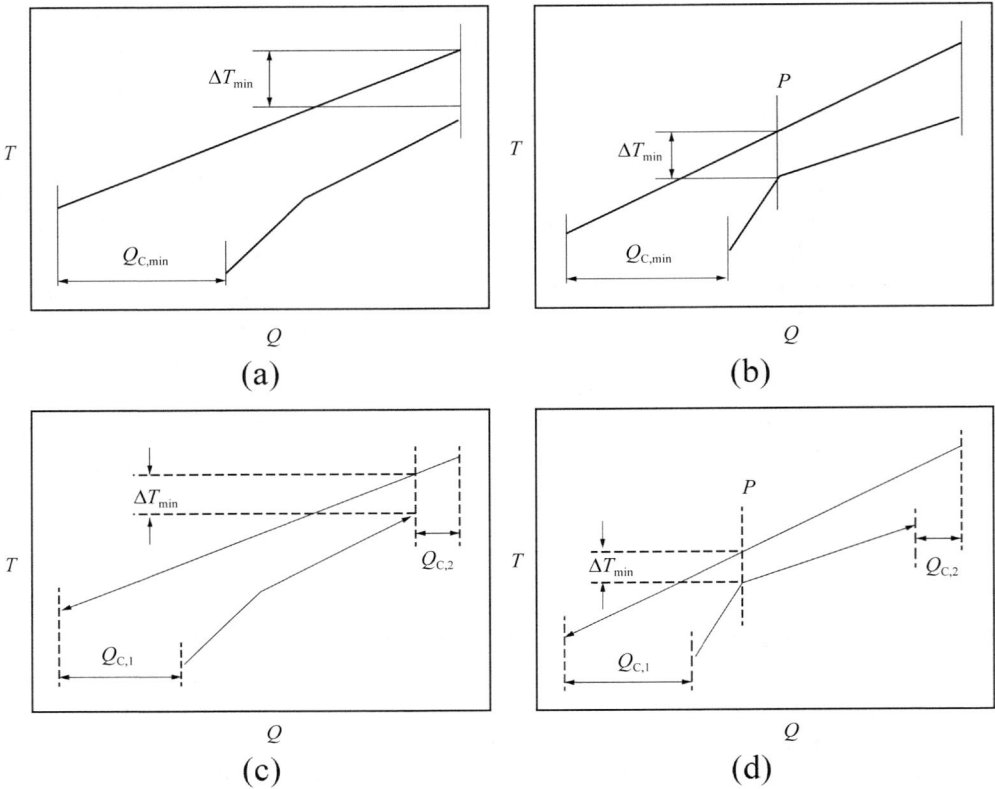

Fig. 2.11 Cold and hot composite curves for hot end threshold problem.

In general, there are two ways of heat integration [91]. One is indirect heat integration, which is realized through the transfer of energy-consumed medium such as steam generation. The other is direct heat integration, which is realized through the direct heat transfer between flows. Heat integration only changes the way of heat utilization and has nothing to do with the maximum heat load of heat integration. Therefore, the threshold problem heat exchanger network can be divided into two parts for optimization and improvement, i.e., the self-matching part used to meet the heat transfer demand of the unit itself and the heat integration part for the steam generation or the direct heat integration with other units. No matter whether the near pinch point appears in the self-matching part or the heat integration part, it should be designed outward from the near pinch point. As shown in Fig. 2.11D, a near pinch point appears in the self-matching part of the unit; in this case, the heat exchanger network of the self-matching part should be designed and matched from the near pinch point.

2.4.2.2 Optimization design block diagram for hot end threshold problem heat exchanger network

Design or production operation data should be collected and checked first when the optimization and improvement of hot end threshold problem heat exchanger network is carried out. Then, process simulation software is used for process simulation calculation to reproduce the design or operation conditions and obtain the data of cold and hot flows of heat exchanger network. Then, the temperature enthalpy diagram is drawn and the pinch point analysis is conducted to determine the correlations between the minimum heat transfer temperature difference and the heat recovery rate and cold utilities. The optimal heat transfer temperature difference is determined by minimizing the sum of equipment investment and operation cost or based on engineering constraints, or the minimum allowable heat transfer temperature difference is determined based on engineering knowledge. Based on the optimal heat transfer temperature difference or the minimum allowable heat transfer temperature difference, the maximum heat load that the threshold problem heat exchanger network can integrate with the cold flows of other heat exchanger networks and the temperature range corresponding to the maximum heat load are determined. The threshold problem heat exchanger network is divided into self-matching part and heat integration part for optimization and improvement, including flow connection mode, heat exchange sequence, increase and decrease in the number of heat exchangers, and heat integration with other heat exchanger networks. Finally, the improved heat exchanger network is analyzed and evaluated, and the economically feasible optimal and improved scheme with the lowest energy consumption is obtained.

Based on the above analysis and combined with the conventional heat exchanger network optimization ideas, the optimization calculation block diagram for hot end threshold problem heat exchanger network is presented, as shown in Fig. 2.12.

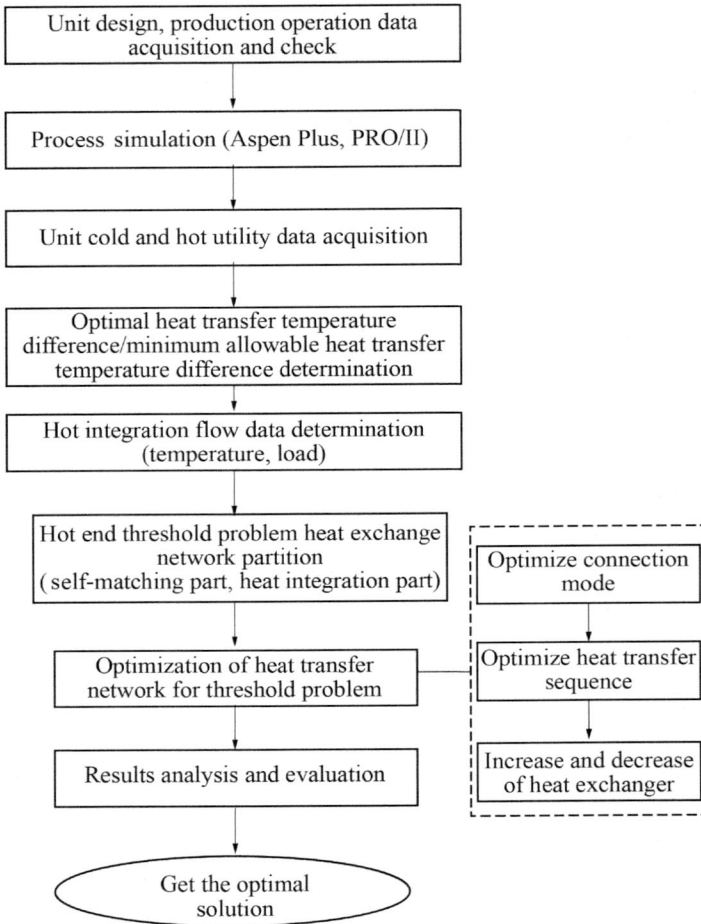

Fig. 2.12 Block diagram of heat exchanger network design for hot end threshold problem.

2.4.3 Example analysis of heat exchanger network in FCC unit

In real industrial processes, the heat exchanger network in FCC unit in refinery is a typical hot end threshold problem heat exchanger network. The FCC unit has a great possibility of heat integration with other refining units. Reasonable design of heat exchanger network in FCC unit to achieve the effective heat integration of FCC unit with other units can lower the energy consumption of the whole refinery.

2.4.3.1 Overview of heat exchanger network in FCC unit

According to the optimization design block diagram for hot end threshold problem heat exchanger network, the basic design data of a refinery was collected and checked, and the process simulation software Aspen Plus was used to complete the process simulation of main fractionator and absorption stabilization system. The design conditions were reproduced, and the data of cold and hot flows of heat exchanger network in FCC complex were extracted, as shown in Table 2.1.

Table 2.1 Cold and hot flow data of FCC unit.

Flow number	Flow name	T_s (°C)	T_t (°C)	Q (MJ)
H1	Overhead gas in main fractionator	120	40	16,155
H2	Light diesel oil	200	50	62,790
H3	Overhead circulating oil	140	80	96,275
H4	First midpump around	270	160	77,440
H5	Second midpump around	320	260	75,345
H6	Circulating slurry	330	275	179,990
H7	Residue slurry	275	70	12,560
H8	Stabilized gasoline	170	70	83,720
H9	Absorption oil	70	40	10,465
H10	Stabilizer overhead gas	50	40	75,345
C1	Raw oil	180	200	23,020
C2	Absorption tower cold feed	40	60	20,930
C3	Intermediate reboiler flow of desorption tower	85	95	23,020
C4	Reboiler flow of desorption tower	130	140	71,160
C5	Deethanized gasoline	136	145	18,835
C6	Reboiler flow of stabilizer	165	175	75,345
C7	Demineralized water	30	110	138,135
C8	3.8 MPa steam feed water	240	260	156,970
C9	1.0 MPa steam feed water	140	145	6280
C10	Rich absorption oil	40	120	17,160
C11	Hot medium water	30	95	83,720

Fig. 2.13 shows the matching diagram of the base case heat exchanger network of FCC unit. The unit generates about 70.0 t/h of 3.8 MPa steam, the heat consumption is about 157,000 MJ/h, and the cold utility load of the unit is 200,500 MJ/h.

According to the temperature enthalpy diagram of the ground state FCC unit as shown in Fig. 2.14, the heat exchanger network of the FCC unit is a typical hot end threshold problem heat exchanger network, and *P* point is the near pinch point. At this point, the actual heat exchange temperature difference between the composite curves of cold flows and hot flows of the unit is about 30°C, greater than the minimum allowable heat exchange temperature difference, and the cold utility of the unit is 200,500 MJ/h.

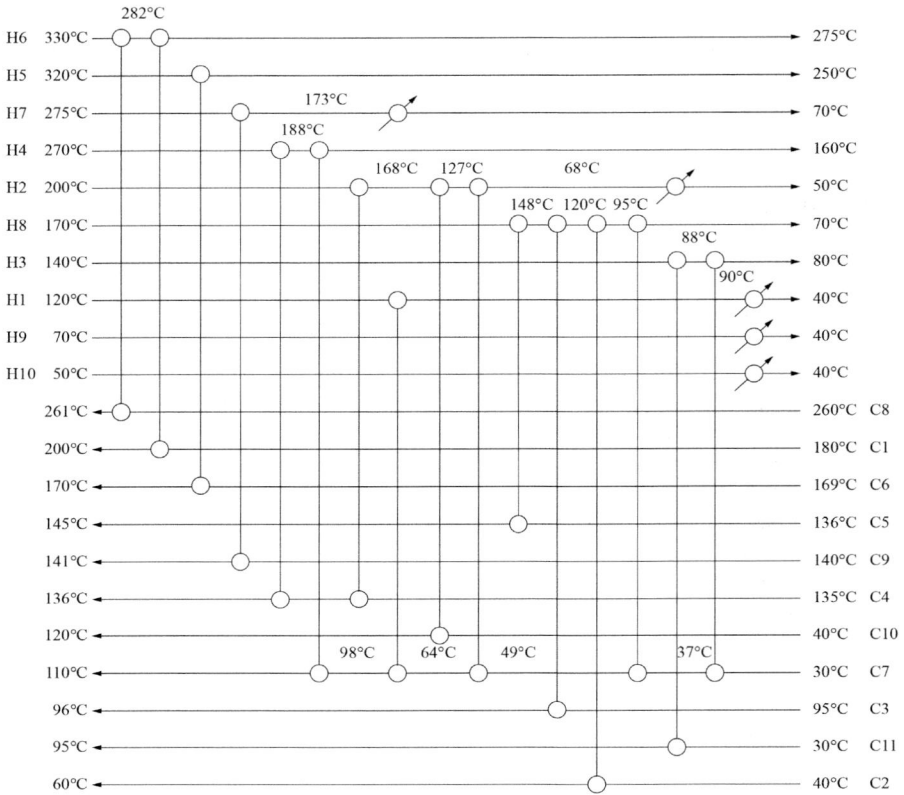

Fig. 2.13 Base case heat exchanger network of FCC unit.

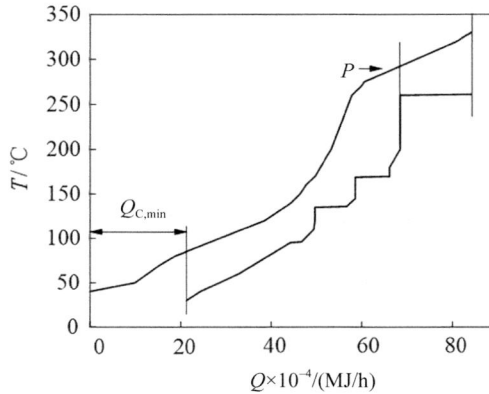

Fig. 2.14 Base case temperature-enthalpy diagram of FCC unit.

2.4.3.2 Improvement of heat exchanger network with the goal of maximum steam generation

(1) Determination of maximum heat from heat integration

The surplus heat of FCC unit in oil refineries is mainly used to generate 3.5–4.0 MPa medium-pressure steam. Therefore, the maximum heat from heat integration of FCC unit, i.e., the maximum heat from indirect heat integration of FCC unit, is determined in order to maximize the generation of 3.8 MPa steam. Combined with engineering knowledge, the minimum allowable heat transfer temperature difference of FCC unit is determined to be 15°C. In Fig. 2.14, the composite curve of cold flows is moved left so that the actual heat exchange temperature difference equals 15°C, as shown in Fig. 2.15. At this time, the cold utility of the unit is 174,500 MJ/h, and the maximum heat from indirect heat integration of the unit is 183,500 MJ/h, which can be used to generate 3.8 MPa medium-pressure steam.

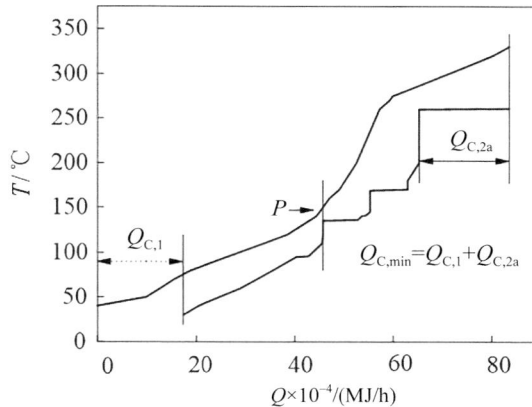

Fig. 2.15 Temperature-enthalpy diagram of FCC unit at maximum steam production.

(2) Improvement of heat exchanger network

The maximum heat from indirect heat integration of the unit is about 183,500 MJ/h, which is similar to the circulating slurry heat of 180,000 MJ/h. In the actual engineering design, all the circulating slurry heat is used to generate 3.8 MPa medium-pressure steam, and the remaining cold and hot flows are used to match the heat exchanger network of the unit. That is, flows H6 and C8 constitute the heat integration part of the hot end threshold problem heat exchanger network. The remaining flows form the self-matching part of the heat exchanger network. Due to the simple heat transfer process in the heat integration part, the focus of heat exchanger network optimization lies in the self-matching part of heat exchanger network.

Fig. 2.16 shows that there are near pinch points in the self-matching part of heat exchanger network. Therefore, the self-matching part should be matched and designed from the near pinch point. Based on this, the self-matching part of heat exchanger network of the unit, including flow connection mode, heat exchange sequence, and increase and decrease in number of heat exchangers, is optimized according to the optimization design block diagram of hot end threshold problem heat exchanger network. The process flow of the improved heat exchanger network is shown in Fig. 2.17. After the heat exchanger network is improved to maximize the steam generation, the cold utility load of the unit decreases by about 11.5% to 177,480 MJ/h. The unit can generate about 80.0 t/h of 3.8 MPa steam, with an increase of 10.0 t/h. Engineering changes include adding oil slurry and desalinization water heat exchanger, increasing the heat exchange area of 3.8 MPa steam generator, and adjusting corresponding pipelines. It is estimated that the additional investment cost is about 3 million RMB, and the investment payback period is about 0.25 years.

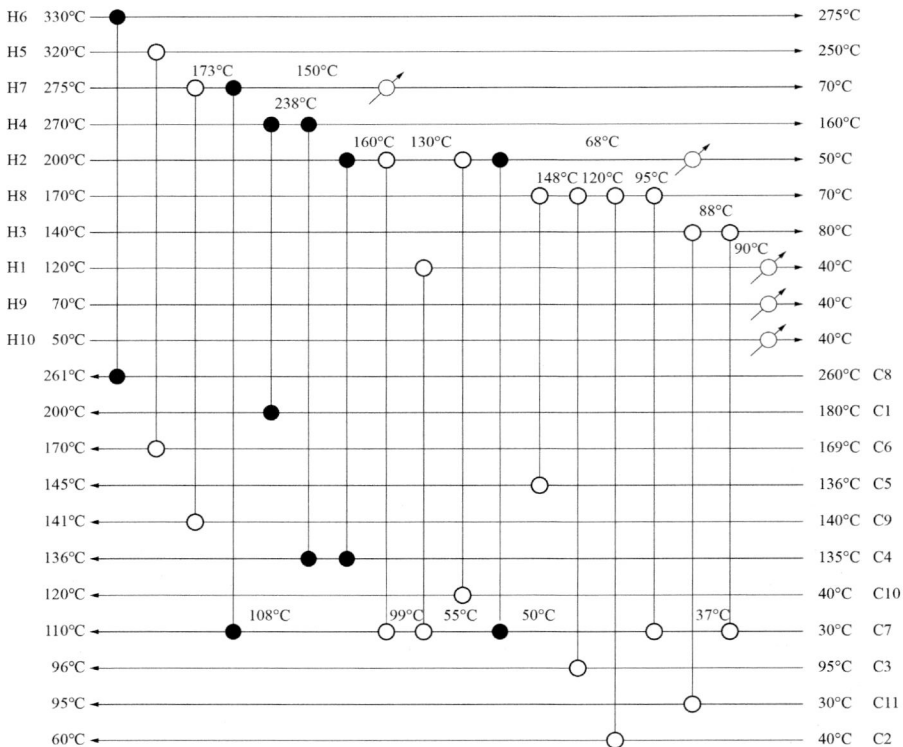

Fig. 2.16 FCC heat exchanger network at maximum steam production.

2.4.3.3 Improvement of heat exchanger network with the goal of maximum heat integration of FCC unit and atmospheric and vacuum unit

(1) Determination of maximum heat from heat integration

　　Direct heat transfer between circulating slurry in FCC unit and topped crude oil in atmospheric and vacuum unit is one of the important ways for rational utilization of excess high-temperature level heat in FCC unit. Taking a refinery as an example, the temperature and flow rate of topped crude oil of atmospheric and vacuum unit after heat exchanger network is 290°C and 1600 t/h, respectively. In Fig. 2.14, the composite curve of cold flows is moved left, and the cold flows of topped crude oil and steam generation feed water are added at the same time, so the actual minimum allowable heat exchange temperature difference is equal to 15°C, as shown in Fig. 2.17, where point P is the near pinch point. At this time, the cold utility of the unit is 174,500 MJ/h, and the maximum heat from heat integration between the units is still 183,500 MJ/h, of which the heat from direct heat integration is 100,500 MJ/h and the heat from indirect heat integration is 83,000 MJ/h.

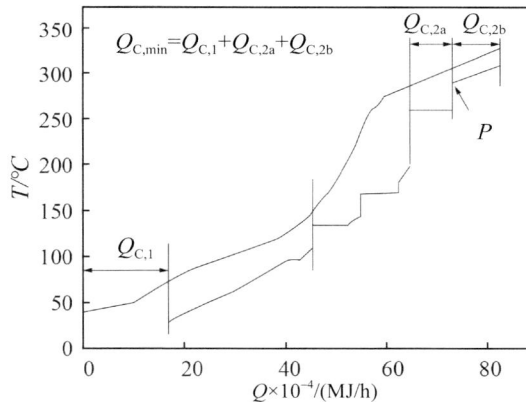

Fig. 2.17 Temperature-enthalpy diagram of FCC unit at maximum direct heat integration.

(2) Improvement of heat exchanger network

　　Whether the goal is to maximize the steam generation or maximize the heat integration of FCC unit and atmospheric and vacuum unit, the cold utility of the unit determined based on pinch point analysis is 174,500 MJ/h; that is, the heat integration method only changes the way of heat utilization and has nothing to do with the maximum heat load of the heat integration. Therefore, same as when the heat exchanger network is improved to maximize the steam generation, considering that the heat of circulating slurry is close to the excess high-temperature level heat of the unit, in the process of improving the heat exchanger network, the circulating slurry heat is used for the heat integration with the atmospheric and vacuum unit and the generation of 3.8 MPa medium-pressure steam, and the remaining cold and hot flows are used to match the heat exchanger network of the unit itself. At this point, flows H6, C8, and atmospheric and vacuum primary oil (C12) form the heat integration part of the heat exchanger network of the threshold problem at the hot end, and the other flows form the self-matching part of the heat exchanger network. There are near pinch points in both the self-matching heat exchanger network and the heat integration part. Therefore, the heat exchanger network should be matched and designed from the near pinch point. The heat exchanger network flow is improved according to the heat exchanger network optimization design block diagram for the hot end threshold problem, as shown in Fig. 2.18. After the improvement, the cold utility load of FCC unit decreased to 177,480 MJ/h, which is about 11.5% lower. About 35.5 t/h of 3.8 MPa steam was generated, and the steam generation decreased by 24.5 t/h. At the same time, the temperature level of primary oil in the atmospheric and vacuum unit is increased from 290°C to 310°C, which saves about 2.4 t/h of

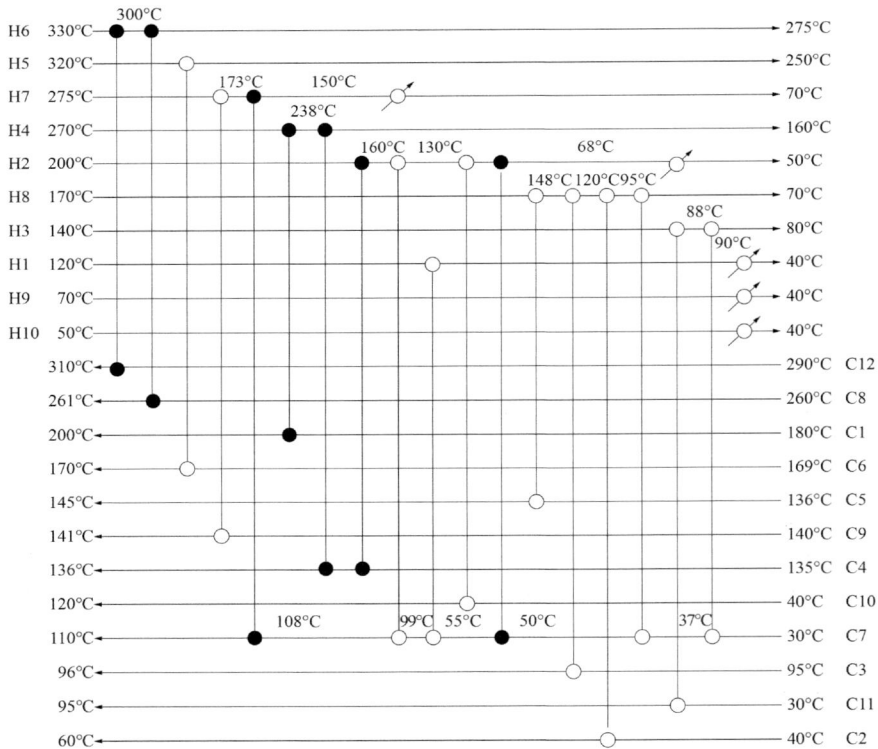

Fig. 2.18 Heat exchanger network of FCC unit at maximum direct heat integration.

fuel gas. The engineering changes include the addition of circulating oil slurry/ atmospheric and vacuum primary oil heat exchanger, the addition of product oil slurry/desalinization water heat exchanger and the corresponding pipeline adjustment, etc. It is estimated that the additional equipment investment cost is about 8 million yuan, and the investment payback period is about 0.83 years.

For hot end threshold problem heat transfer network, under the condition that the minimum heat transfer temperature difference and heat trap temperature are determined, the heat output of the heat integration will not be changed either indirectly or directly. However, from the perspective of the second law of thermodynamics and the global optimization of energy system in oil refining enterprises, priority should be given to fuel conservation. After analysis, although an additional investment of about 3 million yuan is needed to maximize the amount of steam generated, and about 8 million yuan is needed to maximize the thermal integration of FCC and atmospheric and vacuum units, the improvement of heat integration of FCC and atmospheric and vacuum units still has certain advantages from the perspective of fuel saving. At the same time, the heat exchanger network with the threshold problem at the hot end is an important steam-generating device, and its heat integration mode also has a certain influence on the balance of the steam system. Therefore, the choice of heat integration mode of heat exchanger network for hot end threshold problem should be determined from the dual perspective of energy saving or steam balance regulation.

References

[1] Bergles AE. ExHFT for fourth generation heat transfer technology. Exp Therm Fluid Sci 2002;26 (2-4):335–44.
[2] Cui HT, Peng PY. New technology and application of heat transfer enhancement. Beijing: Chemical Industry Press; 2006.
[3] Webb RL, Kim NH. Principles of enhanced heat transfer. London: Taylor & Francis; 1994.
[4] Schlichting H, Gersten K. Boundary-layer theory. New York: McGraw-Hill Book Company; 1979.
[5] Liu W, Yang K. Mechanism and numerical analysis of heat transfer enhancement by core flow in tube. Sci Sin Technol 2009;4:661–6.
[6] Guo ZY, Liang XG, Zhu HY. Entransy—a physical quantity describing the heat transfer ability of objects. Prog Nat Sci 2006;16(10):1288–96.
[7] Guo ZY, Tao WQ, Shah RK. The field synergy (coordination) principle and its applications in enhancing single phase convective heat transfer. Int J Heat Mass Transf 2005;48(9):1797–807.
[8] Bergles AE. Recent developments in enhanced heat transfer. Heat Mass Transf 2011;47(8):1001.
[9] Hesselgreaves JE. Compact heat exchangers: selection, design and operation. Oxford: Pergamon Press; 2001.
[10] Webb RL. Performance evaluation criteria for use of enhanced heat transfer surfaces in heat exchanger design. Int J Heat Mass Transf 1981;24(4):715–26.
[11] Ji WT, Jacobi AM, He YL, et al. Summary and evaluation on single-phase heat transfer enhancement techniques of liquid laminar and turbulent pipe flow. Int J Heat Mass Transf 2015;88:735–54.
[12] Manjunath K, Kaushik SC. Second law thermodynamic study of heat exchangers: a review. Renew Sustain Energy Rev 2014;40:348–74.

[13] Zhang Y, Jiang C, Shou B, et al. A quantitative energy efficiency evaluation and grading of plate heat exchangers. Energy 2017;142:228–33.

[14] He YL, Tao WQ. The basic mechanism of enhanced monophasic convective heat transfer. J Mech Eng 2009;45(3):27–38.

[15] Xu ZHM, Zhang ZB, Zhan HB, et al. Experimental study on mixing fouling characteristics of corrugated tube. J Eng Thermophys 2008;29(2):320–2.

[16] García A, Solano JP, Vicente PG. The influence of artificial roughness shape on heat transfer enhancement: corrugated tubes, dimpled tubes and wire coils. Appl Therm Eng 2012;35(1):196–201.

[17] Park HS, Kim SJ. Thermal performance improvement of a heat sink with piezoelectric vibrating fins. In: International heat transfer conference; 2010. p. 495–501.

[18] Bagajewicz MJ, Rodera H. Multiple plant heat integration in a total site. AIChE J 2002;48(10): 2255–70.

[19] Sadik K, Anchasa P. Review of convective heat transfer enhancement with nanofluids. Int J Heat Mass Transf 2009;52(13-14):3187–96.

[20] El-Genk MS, Suszko A. Saturated nucleate boiling and correlations for PF-5060 dielectric liquid on inclined rough copper surfaces. J Heat Transf 2014;136:081503.

[21] Yu CK, Lu DC. Pool boiling heat transfer on horizontal rectangular fin array in saturated FC72. Int J Heat Mass Transf 2007;50:3624–37.

[22] Yu CK, Lu DC, Cheng TC. Pool boiling heat transfer on artificial micro-cavity surfaces in dielectric fluid FC-72. J Micromech Microeng 2006;16:2092–9.

[23] Wong KK, Leong KC. Pool boiling enhancement of porous structures fabricated by selective laser melting. In: Proceedings of the international symposium of heat transfer and heat powered cycles. Nottingham, United Kingdom; 2016.

[24] Wu W, Bostanci H, Chow LC, et al. Nucleate boiling heat transfer enhancement for water and FC-72 on titanium oxide and silicon oxide surfaces. Int J Heat Mass Transf 2010;53:1773–7.

[25] Sahu RP, Sinha-Ray S, Sinha-Ray S, et al. Pool boiling of Novec 7300 and self-rewetting fluids on electrically-assisted supersonically solution-blown, copper-plated nanofibers. Int J Heat Mass Transf 2016;95:83–93.

[26] Kim TY, Weibel JA, Garimella SV. A free-particles-based technique for boiling heattransfer enhancement in a wetting liquid. Int J Heat Mass Transf 2014;71:808–17.

[27] El-Genk MS, Ali AF. Enhanced nucleate boiling on copper micro-porous surfaces. Int J Multiphase Flow 2010;26:780–92.

[28] Jeong JH, Kwon YC. Effects of ultrasonic vibration on subcooled pool boiling critical heat flux. Heat Mass Transf 2006;42(12):1155–61.

[29] Ma XH, Xu DQ, Lin JF, et al. Condensation heat transfer enhancement on co-existing surface of drop film. J Chem Ind 1999;50(4):535–40.

[30] Boreyko JB, Chen CH. Self-propelled dropwise condensate on superhydrophobic surfaces. Phys Rev Lett 2009;103(18):184501.

[31] Miljkovic N, Enright R, Nam Y, et al. Jumping-droplet-enhanced condensation on scalable superhydrophobic nanostructured surfaces. Nano Lett 2012;13(1):179–87.

[32] Torresin D, Tiwari MK, Del CD, et al. Flow condensation on copper-based nanotextured superhydrophobic surfaces. Langmuir 2013;29(2):840–8.

[33] Lanzhou Petroleum Machinery Research Institute. Heat exchanger. 2nd ed. Beijing: China Petrochemical Press; 2012.

[34] Grabenstein V, Polzin AE, Kabelac S. Experimental investigation of the flow pattern, pressure drop and void fraction of two-phase flow in the corrugated gap of a plate heat exchanger. Int J Multiphase Flow 2017;91:155–69.

[35] Mikko OL, Harold WW, Linda KW. Mechanisms and factors influencing the ultrasonic cleaning of particle-fouled ceramic membranes. Membr Sci 2004;237(1-2):213-23.

[36] Malayeri MR, Müller-Steinhagen H, Watkinson AP. 11th International Conference on Heat Exchanger Fouling and Cleaning 2015, Enfield, Republic of Ireland. Heat Transfer Eng 2016;38 (7-8):667–8.

[37] Zhang ZB, Xu ZM, Shao TC. Experimental investigation on the heat transfer and fouling characteristics of corrugated tube. J North China Electr Power Univ 2007;34(5):68–71.

[38] Xu ZM, Yang SR, Gan YH. Experimental investigation on the fouling performance of the transversally corrugated tube. Proc CSEE 2005;25:159–63.

[39] Tan YK, Luo YL, Yang XX. Heat transfer enhancement technology. Chem Eng 1985;5:1–5.

[40] Tan YK. Heat transfer enhancement technology. Mod Energy Conserv 1986;4:24–32.

[41] Chen Y. Research and application of high efficiency "E" tube. Energy Conserv 2006;1:38–41.

[42] Tan ZHM, Deng SJ. Progress in research of enhanced boiling heat transfer on porous surface. Chem Ind Eng Prog 1994;1(9-14):41.

[43] Zhao HX, Han JT, Xu YT, et al. Research progress of enhanced boiling heat transfer surface. In: The 21st national symposium on hydrodynamics and the 8th national symposium on hydrodynamics and cross-strait ship and ocean engineering hydrodynamics; 2008. p. 1009–16.

[44] Xu H, Dai YL, Xia XM, et al. Development of high flux heat exchanger and its energy-saving application in large petrochemical plant. J Taiyuan Univ Technol 2010;41(5):577–80.

[45] Lin XJ. Application of high efficiency heat exchanger in one-million-tons ethylene plant. Ethylene Ind 2009;21(1):25–7.

[46] Yu J, Zhao P, Yang WL, et al. How to calculate heat transfer of high flux heat exchanger with HTRI. Ind Metrol 2017;27(4):96–9. 121.

[47] An WH, Cui B, Zhu J. Research and application of characteristics of T-fin tube. Brand Standard 2012;4:48–50.

[48] Lin ZH, Wang J, Li RY, et al. Enhanced heat transfer technology. Beijing: Chemical Industry Press; 2007. p. 222–5.

[49] Zhang Y. Review on technological innovation of helical baffle heat exchanger at home and abroad. Pet Chem Equip 2015;18(10):94–6.

[50] Wang QW, Luo LQ, Zeng M, et al. Heat transfer and resistance performance of staggered helical baffle shell and tube heat exchanger. J Chem Ind 2005;56(4):598–601.

[51] Song XP, Pei ZHZH. Short circuit proof helical baffle shell and tube heat exchanger. Petrochem Equip Technol 2007;S 28(3):13–4. 17.

[52] Wang SM, Wen J. Experimental study on heat transfer performance of novel helical baffle heat exchanger without short circuit. J Xi'an Jiaotong Univ 2012;46(9):12–5. 42.

[53] Wang W, Zhang GF, Song TM, et al. Experimental study on heat transfer coefficient and pressure drop of helical baffle heat exchanger with discharge groove. Energy Conserv Technol 2008;26(6):528–9. 574.

[54] Han K, Liu AL, Peng DH, et al. Numerical simulation of flow boiling heat transfer in sintered porous tube. Chem Mach 2011;38(1):104–9.

[55] Lu YSH, Zhuang LX, Chen GH, et al. Experimental study on heat transfer enhancement of horizontal ammonia condenser. Refrigeration 1992;2:8–13.

[56] Jia T, Lu YSH, Zhuang LX, et al. Study on heat transfer and hydrodynamics of transverse tube. J Chem Ind 1990;5:612–7.

[57] Liu K. Experimental study on heat and mass transfer by enhancing the absorption of steam by lithium bromide with transversally corrugated tube. Zhengzhou: Zhengzhou University; 2010.

[58] Si Q, Xia Q, Liang LH, et al. Heat transfer and resistance of spiral flat tube heat exchanger. J Chem Ind 1995;46(5):601–8.

[59] Qian SW, Fang JM, Jiang N. Heat exchanger with twisted tube and mixed tube bundle. Chem Equip Pipes 2000;37(2):20–1.

[60] Huang DB, Deng XH, Wang YJ, et al. Study on heat transfer enhancement of spiral elliptic flat tube. Petrochem Equip 2003;32(3):1–4.

[61] Liu QL, Yang L, Zhu DSH. Simulation analysis of heat transfer enhancement in twisted tube. Petrochem Equip Technol 2010;31(6):19–22. 28.

[62] Yang L. Numerical simulation and experimental study on heat transfer performance of twisted tube double shell side heat exchanger. Guangzhou: South China University of Technology; 2010.

[63] Zhu DSH, Guo XCH, Liu QL. Numerical simulation of heat transfer and flow characteristics in twisted tubes. Fluid Mach 2012;40(2):63–7.

[64] Song D, Jian JH, Zhang YK. Research and performance analysis of twisted tube double shell side heat exchanger. Petrochem Equip Technol 2012;33(5):1–3.

[65] Wang H, Song JM, Wang RX. Discussion on the characteristics of high efficiency and antifouling of corrugated tube heat exchanger. Chem Equip Corros Prev 2000;2:21–3.

[66] Qiu GT, Feng YCH. Corrugated tube heat exchanger (part I)—origin, present situation and development. Pipe Technol Equip 1998;1:43–5.

[67] Liu W. Process Design of Heat Exchanger. Beijing: China Petrochemical Press; 2003.

[68] Yu QY. Study on calculation method of winding pipe heat exchanger. Dalian: Dalian University of Technology; 2011.

[69] Qu P, Wang CHY, Yu YG. Simple calculation of winding pipe heat exchanger. Big Nitrogen Fertil 1998;21(3):178–81.

[70] Yang DQ, Zhou HT. Numerical simulation of flow and heat transfer in shell side of tubular heat exchanger. Pressure Vessel 2015;32(11):40–6.

[71] Xu CHL, Ding GZH. A simple calculation method for heat transfer area of tubular heat exchanger. Low Temp Special Gas 2015;33(1):1–4.

[72] Qiu XL. Study on the influence of structural parameters of herringbone plate on heat transfer and flow resistance characteristics of plate heat exchanger. Guangzhou: South China University of Technology; 2013.

[73] Wang ZZ. Petrochemical design manual (I). Volume 3. Beijing: Chemical Industry Press; 2015. p. 988.

[74] Li ZHM, Wei YW. A new mechanism of blackbody strengthening radiation heat transfer and energy saving—A new way of reenergy saving in heating furnace. In: National symposium on energy saving, emission reduction and low-carbon technology development in metallurgy; 2011. p. 495–501.

[75] Dou CC, Mao Y, Wang J, et al. Effect of new structure reflector plate on flow and heat transfer in radiation chamber of cylindrical furnace. Ind Furnace 2009;31(6):9–12.

[76] Gao XF, Guo Y. Development and application of high quality middle-high temperature far-infrared coatings. Mod Tech Ceram 2000;1:22–5.

[77] Nicolantonio ARD, Spicer DB, Wei VK. Pyrolysis furnace with an internally finned U—shaped radiant coil. US 6419885 B1.1999-01-25.

[78] Wang SH. Ethylene process and technology. Beijing: China Petrochemical Press; 2012. p. 273–82.

[79] Zeng QQ, Zhang SL, Xu HB, et al. Enhanced heat transfer tube. CN 2144807Y.1992-11-21.

[80] Wang GQ, Zeng QQ. Heat transfer enhancement of ethylene cracking furnace tube. Petrochem Ind 2001;30(7):528–30.

[81] Song YK, Zhao LK. Experimental study on heat transfer enhancement technology of passive spoiler twisted sheet tube in cracking furnace. Ethylene Ind 2000;2:8–13.

[82] Linnhoff B, Flower JR. Synthesis of heat exchanger networks: part I: systematic generation of energy optimal networks. AIChE J 1978;24:633–42.

[83] Linnhoff B, Flower JR. Synthesis of heat exchanger networks: part II: evolutionary generation of networks with various criteria of optimality. AIChE J 1978;24:642–54.

[84] Umeda T, Itoh J, Shiroko K. Heat-exchange system synthesis. Chem Eng Prog 1978;74:70–6.

[85] Linnhoff B, Hindmarsh E. The pinch design method for heat exchanger networks. Chem Eng Sci 1983;38:745–63.

[86] Linnhoff B. Pinch analysis—a state-of-the-art overview. Chem Eng Res Des 1993;71:503–22.

[87] Wei ZHQ, Sun LL. Research and application of heat integration strategy for multiple units in refining process based on pinch technology. J Petrol Sci (Petrol Process) 2016;32(2):221–9.

[88] Douglas JM. A hierarchical decision procedure for process synthesis. AIChE J 1985;31:353–62.

[89] Nish N, Stephanopoulos G, West AW. A review of process synthesis. AIChE J 1981;27:321–51.

[90] Feng X. Principle and technology of chemical energy saving. 3rd ed. Beijing: Chemical Industry Press; 2012. p. 1.

[91] Zhang BJ, Luo XL, Chen QL. Hot discharges/feeds between plants to combine utility streams for heat integration. Ind Eng Chem Res 2012;51(44):14461–72.

CHAPTER 3

Heat transfer enhancement in typical oil refining units

Contents

3.1 Crude oil distillation unit

3.1.1 Brief description of process

Crude oil distillation is the first process of fresh crude processing in oil refineries. The main purpose of this unit is to provide qualified feed for downstream secondary processing units. The crude oil distillation technology usually adopts the methods of atmospheric distillation and vacuum distillation to cut crude oil into fractions that meet the feed requirements of downstream units or be used directly as blending components. The process of crude oil distillation mainly involves physical changes, which means the crude oil, after being heated, is fractionated to cut different boiling point products and then cooled by heat exchanging with feed oil. There are various types of crude oil, including sulfuric crude oil, sour crude oil, light crude oil, medium crude oil, and heavy crude oil, as well as special oils such as condensate and ultrathick crude oil. In addition, the light component and fractions of various crude oils are very different. The process and heat exchanger network of crude oil distillation units are also different due to various compositions of production units in oil refineries.

Nowadays, the energy consumption of crude oil distillation units usually accounts for about 15% of the total energy consumption of refinery. For example, the average energy consumption of crude oil distillation units of Sinopec was 9.15 kgoe/(t feed) in 2013 and 8.97 kgoe/(t feed) in 2015. Therefore, crude oil distillation unit is now facing severe challenges in energy saving.

The crude oil distillation process can be divided into fuel type, fuel–lube type, and fuel–chemical type according to different target products of the refinery. A typical crude oil distillation process is shown in Fig. 3.1 [1]. This is a three-stage process consisting of initial distillation, atmospheric distillation, and vacuum distillation. It is characterized by its simple process, strong adaptability to feed oil, large equipment volume, and flexible operation. Such a process is adopted in most of China's domestic crude oil distillation units, such as the Shanghai Gaoqiao 8.0 Mt/a crude oil distillation unit of Sinopec and the Dalian Petrochemical Company 10.0 Mt/a crude oil distillation unit of PetroChina.

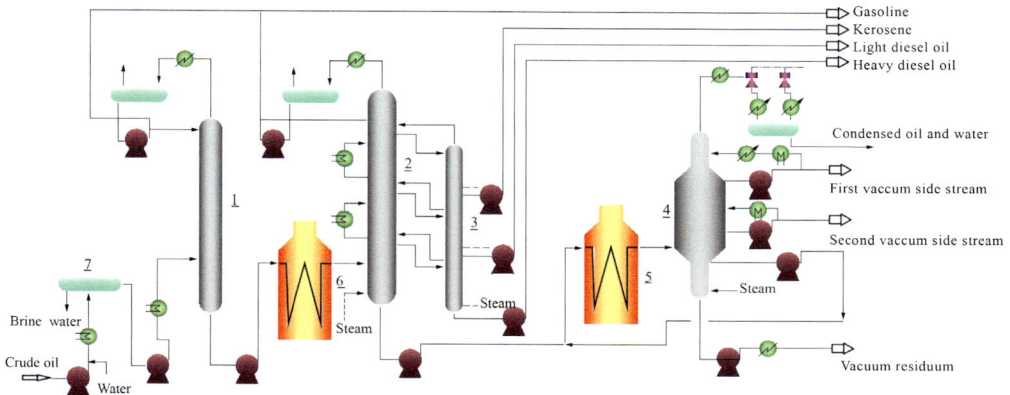

Fig. 3.1 Typical crude oil distillation process. (1) Predistillation column; (2) atmospheric column; (3) stripper; (4) vacuum column; (5) vacuum furnace; (6) atmospheric furnace; (7) demineralizing tank.

3.1.2 Analysis of heat energy characteristics

The crude oil distillation unit involves unit operation processes of distillation, material flow, heat transfer, stripping, and vacuuming. The fuel consumption accounts for 70%–85% of the energy consumption of the whole unit. Therefore, the energy saving of the crude oil distillation unit is mainly on reducing fuel consumption. In the production process of crude oil distillation, there are three characteristics of energy consumption:

① The main forms of energy used in the process are heat, flow work, and steam. Heat, work, and steam are obtained from fuel or electricity conversion equipment such as furnace, pump, and compressor.

② The energy of heat, work, and steam provided by the conversion equipment as well as the recovered recycling energy enters the fractionator to complete the fractionation process from crude oil to the atmospheric column and the vacuum column side-draw products. Apart from a small part of the energy being transferred to the product and energy loss, most of it enters the energy recovery system.

③ After the energy completes its mission in fractionation section, the quality of this "core" energy decreases, but it still has a certain temperature and pressure, which can be recycled by heat exchange equipment. Due to the constraints of technical and economic conditions, the heat is unable to be recovered completely and will eventually be discarded by cooling, heat dissipation, and other ways. The discarded energy and the loss of energy during conversion together constitute the energy consumption of the unit.

According to the three-link energy flow model of the process system (see Fig. 3.2 [2]), the evolution of energy in the process system of crude distillation unit is divided into three different functional links: energy conversion link, energy utilization link, and energy recovery link. The main equipment of energy recovery link includes heat exchangers and coolers. The key to improving the energy recovery link is to optimize the heat exchanger network and apply heat transfer enhancement equipment to improve the final temperature of crude oil. At the same time, the low-temperature heat recovery and utilization system should be designed to effectively recover the low-temperature waste heat and reduce the heat discharge of the unit.

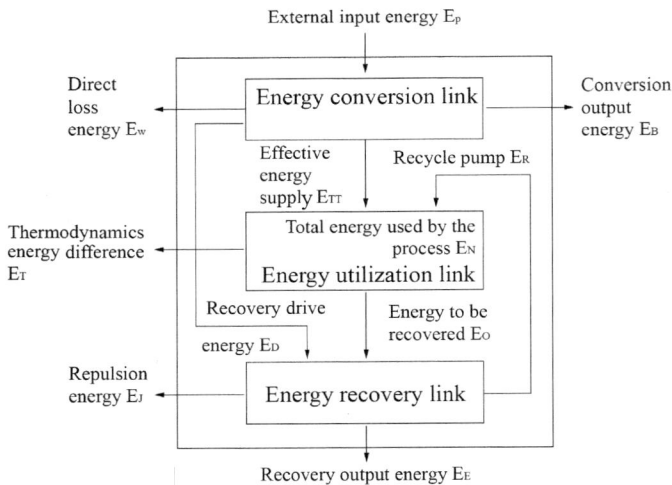

Fig. 3.2 Three-link energy flow structure model.

3.1.3 Optimization of process energy consumption and heat exchanger network

3.1.3.1 Process energy optimization

According to the three-link energy flow model, the energy utilization link of crude oil distillation unit usually includes the predistillation column, atmospheric column, and vacuum column. The energy recovery link includes crude oil preheating heat exchanger network. The energy conversion link includes furnaces, pumps, and compressors. In order to improve the energy utilization efficiency of crude oil distillation unit, it is necessary to optimize the core process firstly by adopting efficient, energy-saving technology and process which can reduce the total energy consumption of distillation process from the source. Secondly, based on the determined process, the heat transfer process of the unit is optimized and enhanced. It includes the optimization of self-produced steam grade and steam production volume after considering the steam consumption requirements of the unit. The corresponding heat exchanger network pinch point analysis and optimization are also considered. During the process of heat exchange enhancing, heat integration, heat exchange process, and self-produced steam interact and cooperate with each other, which requires overall balance to achieve the overall optimization. Finally, the unit achieves the purpose of improving the energy conversion efficiency and reducing the external energy consumption.

Nowadays, the commonly used effective process energy optimization methods are as follows:

① The flash column process is used to flash the light components of crude oil before entering the furnace and directly sent to the appropriate section of the atmospheric column. This will reduce the fuel consumption of the furnace. On the other hand, it will also reduce the temperature of crude oil, which is conducive to the recovery of heat in the unit.

② Multistage vaporization process, such as "predistillation-flash process" and "multistage pressure reduction process," is used to save energy mainly by providing less heat energy to complete the process. But too many vaporization processes will lead to complex processes and increase the difficulty of waste heat recovery in the unit.

③ Optimize the pump around (PA) ratio of atmospheric column and vacuum column and reduce the ratio of low-temperature heat as much as possible, to improve the energy grade of heated materials of the unit, which is conducive to the recovery of waste heat. However, excessive heat removal in the middle and lower part of the column will affect the separation accuracy of the products.

④ Optimize the stripping steam at column bottom to reduce steam consumption. Stripping steam is to reduce the light components of the column bottom heavy oil and reduce the partial pressure of vapor in the feed section of the fractionator. This will reduce the outlet temperature of the furnace while completing the process. Appropriately reducing the consumption of the steam, although the fuel consumption of the furnace may be increased, will save the steam consumption and the energy consumption of the overhead (OH) system.

⑤ The reflux ratio/tray number/fractionation accuracy of fractionator are optimized. The number of trays is increased appropriately, and new types of high-efficiency trays and packings are adopted to achieve energy saving effect.

⑥ The "dry" vacuum distillation can be used to achieve energy saving effect, whenever possible.

3.1.3.2 *Optimization of heat exchanger network*

The optimization of heat exchanger network can improve the final temperature of heat exchange, effectively reduce the duty of atmospheric furnace, and reduce fuel consumption. According to the pinch point theory, when the final heat exchange temperature is increased, the temperature of the product can therefore be reduced. The cooling duty and the cooling water or electric energy can also be reduced. According to the calculation and experience, the fuel consumption can be reduced by 1 kgoe/(t feed) when the final temperature is increased by about 13°C.

There are many streams involved in the heat exchange of crude oil distillation unit (typically around 20 streams). The flowrates and temperatures of the streams are quite different. By using pinch point technology, based on the heat exchanger network optimal goals, $\triangle T_{min}$ can be determined. Then, according to the pinch point design principle and engineering experience, the optimized heat exchanger network of crude oil distillation unit can be designed to achieve the optimal matching between cold and hot fluids. It will also achieve the effective utilization of temperature and achieve the goals of high final heat exchange temperature, with fewer heat exchangers and less heat exchange area. In the design of heat exchanger network, the efficient heat exchanger is used so that the heat exchanger network is more reasonable and the heat exchanger investment is lower. At the same time, according to process optimization and unit heat integration, low-temperature streams are used to generate low-pressure steam and produce hot water. This will fully recover the low-temperature heat and improve the thermal utilization rate of the unit.

1. Optimization of heat exchanger network for new units

   A newly built crude oil distillation unit with a processing capacity of 10 Mt/a processes mixed fresh crude of Arabian light:Arabian heavy (1:1). As the first newly built single-line 10-million-ton-per-year crude oil distillation unit in China, the advantages and disadvantages of its heat exchanger network are of great significance to the energy consumption and the economic benefits of the unit or even the economic benefits of the whole plant. Through the optimization of process conditions and the adoption of pinch point technology to optimize the heat exchanger network design, and according to the engineering experience, the following factors that have a significant impact on the heat exchanger network are comprehensively evaluated and selected.

   ① Optimization of the number of crude oil series. Multiseries design can achieve reasonable matching of cold and hot fluid to a certain extent and increase the flexibility of heat exchanger network. However, with more series of crude oil, more heat exchangers will be needed, which increases the floor area as well as the number of corresponding pipelines and valves in the network. The project investment will be hugely increased. Two-series heat exchanger network refers to the two series of crude oil heat exchange in two series, before and after electric demineralized and flash column. Two-series heat exchanger network has high heat exchange efficiency, simple process, relatively small investment, and low operating costs. After weighing various factors, in this unit, the crude oil is divided into two series with same flow rate for heat exchange. The heat transfer and pressure drop of the two series are basically the same. The outstanding advantage of two-series heat exchanger network is that the number of heat exchangers is reduced to 64, after the selection of large shell diameter heat exchangers, while other crude oil units of the same type use more than 90 heat exchangers. This does not increase the cost of heat exchangers, save space, or reduce the cost of utility and related investments in pipelines, valves, and instruments.

② Reasonable heat flow exchange sequence of heat exchanger network. In the design of heat exchanger network, in order to simplify the structure and reduce the position and number of heat exchangers, the heat exchange sequence of heat flow and cold fluid is optimized:

(A) Heat sources with small heat capacity and low temperature are arranged to preheat crude oil in the heat exchange section I, such as kerosene $(147 \rightarrow 94°C)$ and vacuum top pump around (vacuum top PA) $(125 \rightarrow 93°C)$.

(B) The heat source with small heat capacity and high temperature is also involved in the preheating of crude oil in the heat exchange section I due to the rapid temperature drop and small temperature difference at the outlet in the heat exchange process. For example: light diesel $(206 \rightarrow 143°C)$ and MVGO $(181 \rightarrow 144°C)$.

(C) Heat source with large heat capacity and high temperature is arranged firstly in the later section to exchange heat with hot streams, and then to the front and exchange heat with cold streams. For example: vacuum residue is divided into two series to participate in heat transfer. The part with high temperature $(360 \rightarrow 264°C)$ exchanges heat with the topped crude in heat exchange section III. After the temperature decreased (below 264°C), the stream exchanges heat with demineralized oil in heat exchange section II.

The final design of heat exchanger network of a 10.0 Mt/a crude oil distillation unit is shown in Fig. 3.3 [3]. The calculation results are shown in Table 3.1 [3].

As can be seen from Table 3.1, the final heat exchange temperature of the heat exchanger network of this unit can reach as high as 319°C, ranking among the top of the heat exchanger networks of crude oil distillation units around the world. After the unit was started up, the results of the unit's heat exchanger network were verified, and the final heat exchange temperature of the unit was kept at a high level after the change of fresh crude varieties, indicating that the heat exchanger network of the unit has good adaptability.

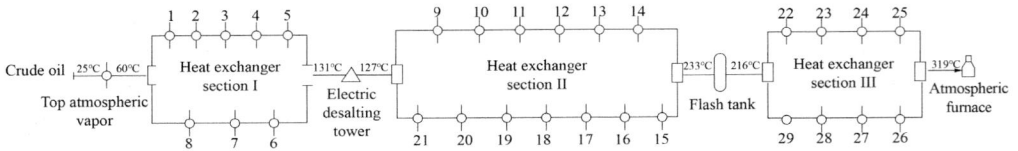

Fig. 3.3 Schematic diagram of heat exchanger network. (1) Atmospheric top pump around (2); (2) vacuum top pump around; (3) kerosene; (4) atmospheric first pump around (2); (5) MVGO; (6) light diesel; (7) atmospheric top pump around (1); (8) heavy diesel (3); (9) atmospheric 1st pump around (1); (10) vacuum residue (5); (11) atmospheric 2nd pump around (2); (12) vacuum residue (4); (13) atmospheric 2nd pump around (1); (14) vacuum residue II (3); (15) vacuum residue I (3); (16) heavy diesel (1); (17) HVGO (1); (18) vacuum 1st pump around (1); (19) vacuum 1st pump around (2); (20) HVGO (2); (21) heavy diesel (2); (22) vacuum 2nd pump around I (2); (23) vacuum residue I (2); (24) vacuum 2nd pump around I (1); (25) vacuum residue I (1); (26) vacuum residue II (1); (27) vacuum 2nd pump around II (1); (28) vacuum residue II (2); (29) vacuum 2nd pump around II (2).

Table 3.1 Summary of calculation results of heat exchanger network.

Item	Data	Number	Data
Total number of units (set)	64	Temperature after heat change (°C)[a]	
Total duty (MW)	220.5	Atmospheric (ATM) top vapor	78
Actual heat transfer area (m^2)	35,132	Kerosene	94
Thermal strength (W/m^2)	6276.3	Light diesel	123
Final heat exchange temperature (°C)	319	Heavy diesel	88
		LVGO	93
Electric demineralizing inlet temperature (°C)	131	MVGO	144
Electric demineralizing outlet temperature (°C)	127	HVGO	146
Temperature of crude oil after heat transfer before entering flash column (liquid phase) (°C)	233	Vacuum residue	164
		0.35 MPa (gauge pressure) steam generated (t/h)	11.23
Flash column bottom temperature (°C)	216	Total pressure drop (MPa)	1.31

[a]Products are sent outside the battery limit (OSBL) as hot streams.

2. Optimization of heat exchanger network of modified unit

The energy-saving optimization and modification of existing heat exchanger networks can be carried out according to the idea of "simulation, analysis, improvement and economic evaluation", as shown in Fig. 3.4. When optimizing the existing heat exchanger network, the production and operating data of crude oil distillation unit and heat exchanger network should be obtained and checked first. The process simulation software is then used to simulate the process. The data of cold and hot streams of heat exchanger network is then obtained. On this basis, the temperature enthalpy map can be drawn and pinch point analysis can be carried out to determine the relationships of the minimum heat exchange temperature difference, the heat exchange area, heat recovery, and the amount of cold and hot utilities. The optimal heat exchange temperature difference is determined by minimizing the total operating cost. The heat exchanger network can be then optimized based on the pinch point analysis. Therefore, the flow connection mode, heat exchange sequence, and new heat exchanger can be successfully improved. Finally, the improved heat exchanger network is analyzed and evaluated to obtain the energy consumption optimization and economic and feasible heat exchange network improvement scheme.

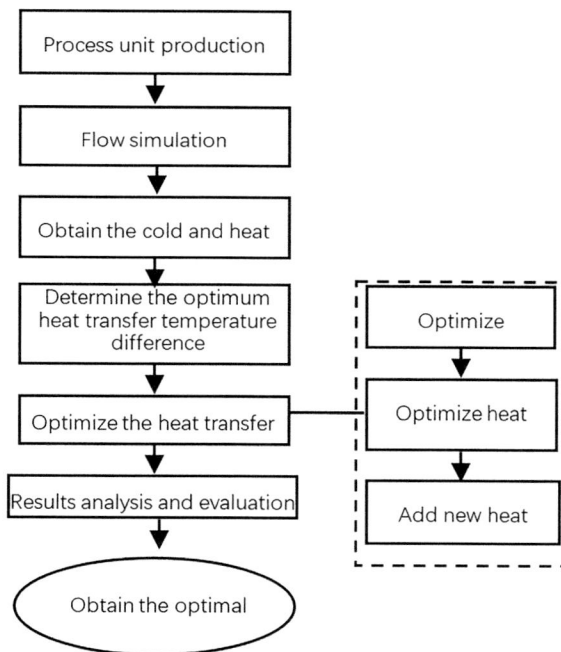

Fig. 3.4 Heat exchanger network optimization method.

For a crude oil distillation unit of 8.0 Mt/a capacity, a four-column process of predistillation column, atmospheric column, first stage, and second stage vacuum column was adopted. The final heat exchange temperature before the modification of heat exchanger network is 284°C, as shown in Fig. 3.5 [4].

After optimization, the minimum pinch point temperature is determined to be 15°C, and the heat transfer network after modification is shown in Fig. 3.6. The final heat exchange temperature of crude oil reaches about 295°C, saving fuel of 400 kg/h and reducing energy consumption of the unit by about 0.36 kgoe/(t feed), and the annual benefit is nearly 10 million Yuan.

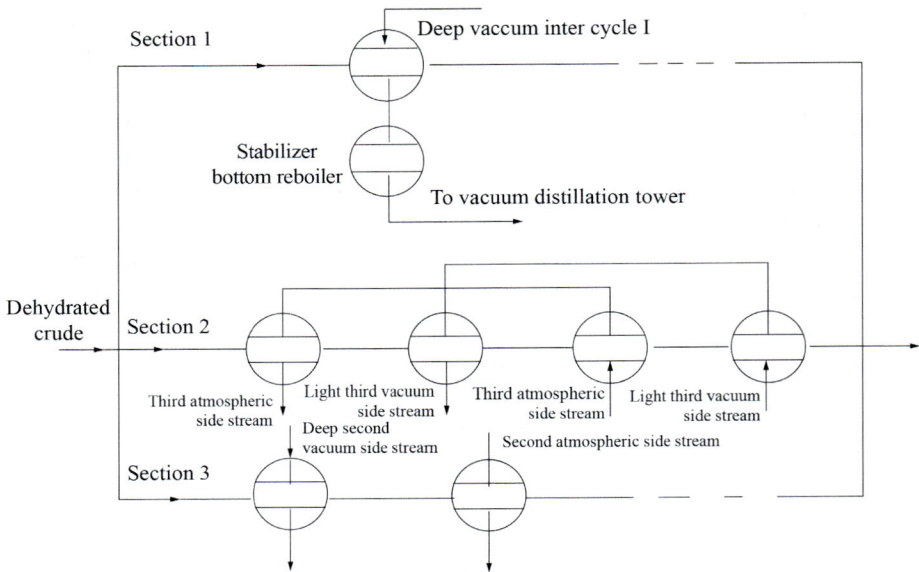

Fig. 3.5 Original heat exchanger network.

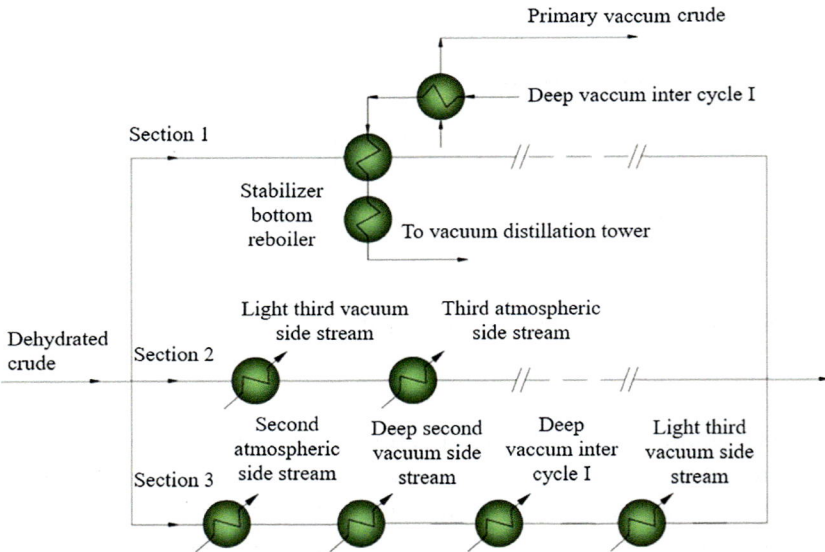

Fig. 3.6 Heat exchanger network after modification.

3.1.4 Heat transfer enhancement technology using elements

The heat exchanger network of crude oil distillation unit is complex which has many heat exchangers, and the differences between the streams' physical properties are large. According to the characteristics of process flow, different heat transfer enhancement techniques can effectively reduce the equipment size and the number of equipment. Heat transfer enhancement techniques such as corrugated pipe heat exchanger, T-shaped grooved tube reboiler, spiral-twisted tube, double-segmental baffle shell heat exchanger, plate shell heat exchanger, fully welded plate heat exchanger, coil-wound heat exchanger, plate air cooler, and surface evaporation air cooler have been successfully applied in crude oil distillation unit [5–12]. The compound efficient air cooler is used for column overhead vapor cooling, which can enhance heat transfer and cancel the water cooler [13]. Taking a 10.0 Mt/a crude oil distillation unit as an example, the analysis is as follows.

3.1.4.1 Heat exchange system of flash column

1. Heat exchange system of crude oil before desalting and demineralization
   Crude oil absorbs about 76.0 MW of heat before entering the electric demineralizing vessel. The heat exchanger network is more complex, as shown in Fig. 3.7.

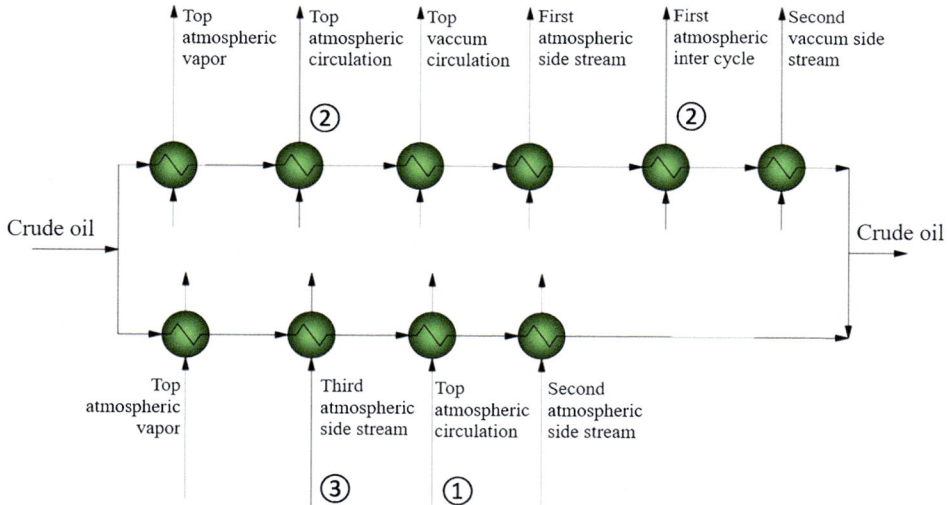

Fig. 3.7 Heat exchanger network of crude oil before the electric demineralizing vessel.

In this heat exchange system, the crude oil exchanged heat with each heat flow in the unit. The temperature increased from 25°C to about 132°C. However, the viscosity was still high. In order to improve its flow ability, it usually flows in the shell side of the heat exchanger. The differences in the physical property and flow rate of the hot streams on tube side are huge and are different upon heat exchangers.

The heat duty of fresh crude/ATM top vapor heat exchanger is as large as 23 MW. The film heat transfer coefficient of tube side and shell side is close. The overall heat transfer coefficient of equipment is about 250 W/(m^2 K). The allowable pressure drop of ATM top vapor is as small as 10 kPa, which does not allow the heat exchanger to increase the flow velocity or use other methods to enhance the film heat transfer coefficient of the inner tube. In recent years, four-sided detachable all-welded plate heat exchangers have been used to condense ATM top vapor, and good results have been obtained [10,11]. As reported in the literature [12], three types of heat exchangers, shell and tube type, welded plate type, and winding tube type, are used in the same crude oil distillation unit. The results of industrial operation indicated that the enhanced heat exchange structure had better results.

The total heat transfer between atmospheric top pump around (ATM top PA) and crude oil is 20 MW. The film heat transfer coefficient of shell side is about 900 W/(m^2 K) with regular heat exchanger. It is only half of the tube side film heat transfer coefficient, accounting for 46% of the heat resistance of the whole heat exchanger. The enhancement of shell side heat transfer can effectively improve the overall heat transfer coefficient. The helical baffle can replace the ordinary segmental baffle, which can eliminate the dead zone on the shell side. It has been successfully applied in the fresh crude/ATM top PA heat exchanger and fresh crude/atmospheric 1st pump around (ATM 1st PA) heat exchanger of several crude oil distillation units.

As for crude oil/vacuum top pump around (vacuum top PA) heat exchanger, crude oil/kerosene heat exchanger, crude oil/ATM 1st PA secondary heat exchanger and crude oil/light diesel heat exchanger, under the condition of making full use of the allowable pressure drop, the overall heat transfer coefficient is about 400 W/(m^2 K) by using the regular heat exchangers, and the film heat transfer coefficient of tube side and shell side is close. The equipment size can be further reduced by the enhancement of both the tube side and shell side.

Regular heat exchangers are used for the crude oil/heavy diesel tertiary heat exchanger and crude oil/MVGO heat exchanger and the overall heat transfer coefficient is generally less than 400 W/(m^2 K). Because the film heat transfer coefficient of the tube side and the shell side is close, the technology of corrugated tube and spiral-twisted tubes can be used to enhance the tube and shell side at the same time. The calibration data shows that the heat transfer coefficient of the corrugated tube is 1.94 times higher than that of the plain tube heat exchanger under the same pressure drop when applied to the crude oil/heavy diesel heat exchanger [14].

2. Heat exchanger system of crude oil after desalting

After being demineralized and dehydrated through the electric demineralizing vessel, the demineralized crude oil is divided into two series to enter the heat exchange system, as shown in Fig. 3.8. After heat is exchanged, the two series of demineralized crude oil are mixed and enter the flash column at about 233°C.

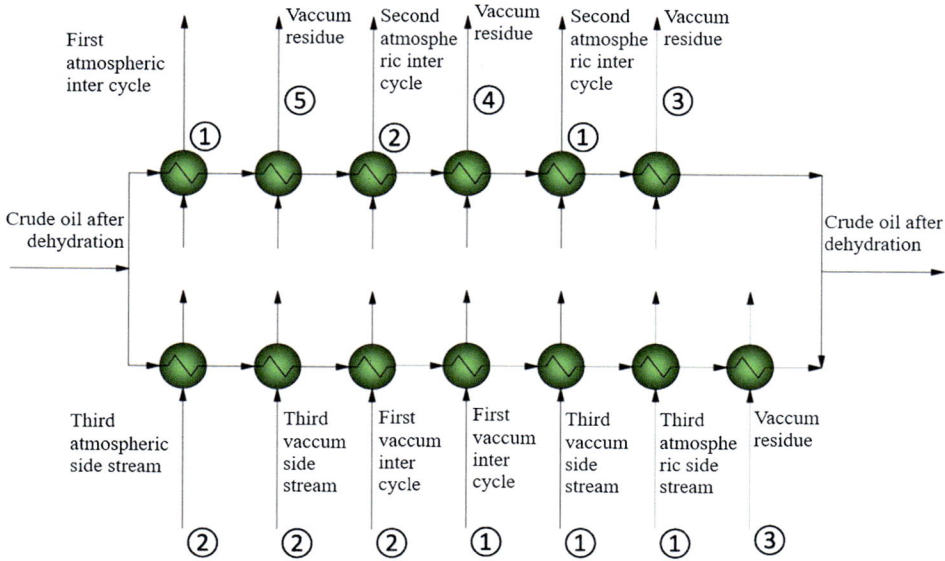

Fig. 3.8 Heat exchanger network of crude oil after demineralized.

The demineralized crude oil was heated by the demineralized oil/ATM 1st PA primary heat exchanger with ATM 1st PA, and the heating duty was nearly 12.0 MW. After the demineralized oil/ATM 2nd PA secondary heat exchanger and the demineralized oil/ATM 2nd PA primary heat exchanger, the total heat exchange of the demineralized crude oil is 15 MW. Under the condition of making full use of the allowable pressure drop, the overall heat transfer coefficient can reach 500 W/(m^2 K) by using the ordinary heat exchanger. The film heat transfer coefficient of tube side and shell side in these heat exchangers are close, and the fouling resistance is large, accounting for about 45% of the total thermal resistance. The heat transfer performance of heat exchangers can be enhanced by reducing scaling.

In the heat exchange system of demineralized crude oil and vacuum residue, the vacuum residue has high viscosity and poor thermal conductivity. Since the stream is more likely to reach turbulent state on the shell side, the residue stream therefore usually flows in the shell side of the heat exchanger, and the film heat transfer coefficient is 300–500 W/(m² K). Even if the flow velocity is increased, the higher film heat transfer coefficient cannot be obtained within a reasonable pressure drop. In the meantime, the fouling resistance is large and accounts for 30% of the total thermal resistance. Therefore the vacuum residue side becomes the thermal resistance control side. To enhance the heat transfer performance of demineralized oil/vacuum residue heat exchanger, the thermal resistance of fouling should be reduced and the degree of turbulence of shell side fluid should be improved. The helical baffle can reduce the dead zone in the shell side and reduce the possibility of fouling deposition. According to the literature [15], the helical-baffle heat exchanger is more effective for heavy oil in crude distillation unit. In addition, the use of corrugated tube is also a suitable enhancing method. For example, the process condition of a demineralized oil/vacuum residue quinary heat exchanger of a crude oil distillation unit is shown in Table 3.2. It can be seen from Table 3.2 that the heat capacity and thermal conductivity of the cold and hot fluid are close, but the viscosity of the hot vacuum residue is still very high. According to the calculation, even if the vacuum residue flows to the shell side, which is more likely to occur turbulent, its film heat transfer coefficient is difficult to reach 300 W/(m² K).

Table 3.2 Process condition of demineralized oil/vacuum residue quinary heat exchanger—Duty: 9.7 MW.

Item	Shell side	Tube side
Fluid name	Vacuum residue	Demineralized oil
Flow rate (kg/h)	283,330	596,300
Temperature (°C)	219 → 164	157 → 180
Liquid density (kg/m³)	941 → 966	770 → 750
Liquid viscosity (mN s/m²)	14.1 → 96.5	0.84 → 0.68
Liquid thermal conductivity (W/(m K))	0.076 → 0.079	0.084 → 0.080
Liquid heat capacity (kJ/(kg K))	2.415 → 2.20	2.40 → 2.48

Note: "→" means from inlet to outlet. The left side of the arrow is the inlet data, and the right side is the outlet data.

The calculation and comparison of high-efficiency heat exchanger with corrugated tubes and double segmental baffle and ordinary heat exchanger are shown in Table 3.3. It can be seen from Table 3.3 that the use of corrugated tubes instead of plain tubes reduces three heat exchangers, and the total weight can be reduced by 30 t. The heat exchange tube can save SS304L stainless steel by 18 t alone. Such a method can save the investment cost of equipment and save the installation cost as well as the operation cost of pipelines.

Table 3.3 Comparison of demineralized oil/vacuum residue quinary heat exchanger.

Option	Plain tube	Corrugated tube
Equipment size (shell diameter × straight tube length) (mm)	1400 × 6500	1800 × 6000
Equipment (set)	6	3
Tube side pressure drop (kPa)	87	38
Shell side pressure drop (kPa)	185	192
Overall heat transfer coefficient (W/(m^2 K))	202	273
Heat transfer area (m^2)	3694	2733
Total weight of equipment (t)	134	104.1

3. Flash column bottom heat exchange system

The flash bottom oil goes through the flash bottom oil/vacuum 2nd pump around (vacuum 2nd PA) secondary heat exchanger, the flash bottom oil/vacuum residue secondary heat exchanger, the flash bottom oil/vacuum 2nd PA primary heat exchanger, and flash bottom oil/vacuum residue primary heat exchanger and is heated to about 319°C, which then enters the ATM furnace to be further heated to 370°C and then enters the atmospheric column.

The heat transfer between flash bottom oil and vacuum 2nd PA was carried out twice, and the duty is about 25 MW. When using ordinary heat exchanger, the tube side and shell side film heat transfer coefficient is greater than 1000 W/(m^2 K). However, due to the relatively large fouling resistance of cold and hot fluid, the overall heat transfer coefficient of heat exchanger is difficult to exceed 300 W/(m^2 K). The fouling resistance of cold and hot fluid accounts for about 40% of the total thermal resistance, which becomes the main factor restricting the heat transfer effect. Heat transfer enhancement technology should take the reduction of fouling resistance into account and adopt heat exchange tubes such as corrugated tubes and spiral-twisted tubes that can enhance the turbulent flow inside and outside the tube. In this way, the scaling trend can be alleviated, the heat transfer effect can be enhanced, and cleaning cycle can be prolonged.

The heat transfer between flash bottom oil and vacuum residue was carried out twice, and the duty was 15 MW. The vacuum residue was cooled from 360°C to 300°C and then further cooled from 300°C to 265°C. Taking the flash bottom oil/vacuum residue primary heat exchanger as an example, the heat transfer enhancement effect is studied and the process condition table is shown in Table 3.4. The comparison of enhancing schemes is shown in Table 3.5. It can be seen from Table 3.4 that the heat capacity and thermal conductivity of the cold and hot fluid are close, while the viscosity of the high-temperature vacuum residue is not high, but the liquid thermal conductivity is good.

Table 3.4 Process condition of flash bottom oil/vacuum residue primary heat exchanger—Duty: 9.0 MW.

Item	Shell side	Tube side
Fluid name	Vacuum residue	Flash bottom oil
Flow rate (kg/h)	188,775	486,156
Temperature (°C)	$360 \rightarrow 297$	$296 \rightarrow 319$
Liquid density (kg/m^3)	$876 \rightarrow 905$	$735 \rightarrow 716$
Liquid viscosity (mN s/m^2)	$1.90 \rightarrow 3.61$	$0.49 \rightarrow 0.42$
Liquid thermal conductivity (W/(m K))	$0.066 \rightarrow 0.071$	$0.067 \rightarrow 0.064$
Liquid heat capacity (kJ/(kg K))	$2.86 \rightarrow 2.68$	$2.79 \rightarrow 2.86$

Table 3.5 Comparison of flash bottom oil/vacuum residue primary heat exchanger.

Scheme	Plain tube	Corrugated tube
Equipment size (shell diameter × tube straight length) (mm)	1400×6500	1400×6000
Equipment (set)	7	4
Tube side pressure drop (kPa)	370	110
Shell side pressure drop (kPa)	84	130
Overall heat transfer coefficient (W/(m^2 K))	198	420
Heat transfer area (m^2)	4631	2180
Total weight of equipment (t)	173.2	94.2

As shown in Table 3.5, it can be seen that the corrugated tube with double-sided heat transfer enhancement effect can effectively enhance the fluid disturbance and replace the plain tube. Such a method can save three sets of heat exchangers: a total weight of 79 t of the heat exchanger can be saved, 52 t of stainless steel materials can be saved, and the equipment investment can be reduced by 45.6%. In addition, it can reduce the floor area, save the cost of civil construction and piping, and also reduce the operation cost. The pressure drop of the tube side of four corrugated tube heat exchangers in series is 110 kPa, which is 260 kPa lower than the pressure drop of the seven shell and tube heat exchangers with plain tube. The hydraulic head of the demineralized oil pump can be reduced and save energy.

3.1.4.2 Atmospheric column heat exchange system

1. Heat transfer system of atmospheric (ATM) column

   The overhead vapor of the ATM column was cooled from 122°C to 78°C by the crude oil/ATM top vapor heat exchanger and then entered the column OH air cooler and water cooler to further cool to 40°C and entered the reflux and product vessel.

   The cooling duty of the ATM column OH air cooler is relatively high. The mass fraction of the gas inlet is about 35% and is fully condensed by the air cooler. In order to ensure the flow rate of the cooled stream, multiple paths should be used for air cooler and the number of tubes in the last path should be reduced as much as possible. In this way, the velocity of stream can be equal as much as possible. For example, for an air cooler bundle with six tube rows, when designed as a two-path structure, the first path can be designed with four tube rows, and the second path can be designed with two tube rows. Compared with equal distribution of tubes, the velocity of condensate in the tube with unequal distribution structure can be increased by 30%, which can effectively improve the heat transfer effect. Alternatively, the surface evaporation air cooler can be used to replace the conventional air cooler and water cooler to directly condense and cool the vapor of ATM top to 40°C. Operation data showed that the ATM top vapor can be cooled to 35°C [6].

The titanium plate and shell heat exchanger are used for the ATM top vapor/hot medium water heat exchanger [16]. The calibration data of all-welded plate heat exchanger showed that the vapor outlet temperature is 5–9°C lower than that of shell and tube heat exchanger [17], which directly reaches the air cooler outlet temperature required by the process. Which means the air cooler can be shut down.

2. Heat exchange system for pump around (PA) and side streams of ATM column
The ATM column usually has three PAs and three side stream extractions. The heat transfer network is shown in Fig. 3.9.

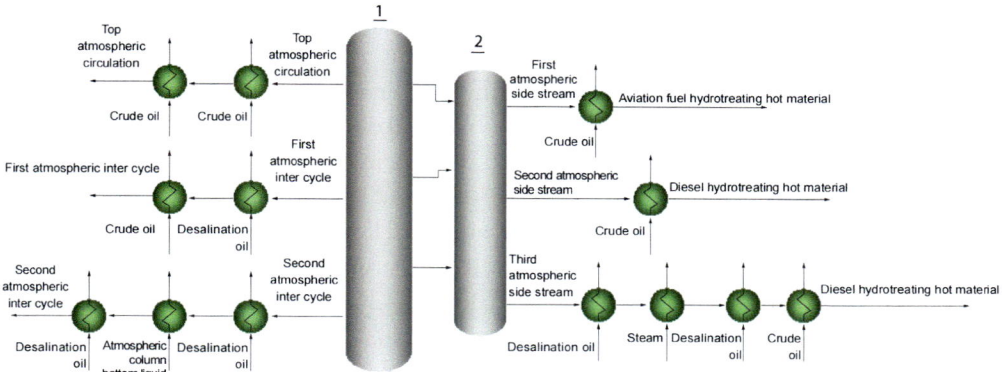

Fig. 3.9 Pump around and side stream effluent heat exchanger network of ATM column. (1) ATM column; (2) stripper.

Heavy diesel is used to generate steam with a duty of 7 MW. The overall heat transfer coefficient of the steam generator is more than 500 W/(m^2 K) by using conventional single-segmental plain tube bundle. However, controlling the thermal resistance is the film heat transfer coefficient of the tube path, which is more than 1500 W/(m^2 K), by making full use of the allowable pressure drop. The heat transfer effect can be enhanced by increasing the inner roughness of the heat exchange tube.

3.1.4.3 Heat exchange system of vacuum column

The vacuum column OH vacuuming system includes ejectors, condensers, liquid separation vessels, and other equipment. Such equipment system consumes relatively more steam and cooling water (or electricity) in the crude oil distillation unit. As far as the vacuum column overhead condenser of the vacuuming system is concerned, the two-stage or three-stage vacuuming system shall be equipped with three "stage" condensers. The first stage vacuum condenser of a 10 Mt/a crude oil distillation unit is required to cool the vacuum column OH vapor, which contains noncondensable gas and water vapor, from 188°C to 36°C, with an allowable pressure drop of 0.667 kPa (5 mmHg), with a cooling duty of 15 MW. The second-stage vacuum condenser cools the vapor from 163°C to 36°C with an allowable pressure drop of 2 kPa (15 mmHg) and a cooling duty of 10 MW. The three-stage condenser needs to cool the vapor from 178°C to 40°C, with an allowable pressure drop of 2.667 kPa (20 mmHg), and the cooling duty is 8 MW. From the above process requirements, it can be seen that the allowable pressure drop of condensers at all levels is very small, and the temperature after cooling is relatively low. In order to meet the process requirements, the structure with heat transfer enhancement characteristics should be properly selected.

If air is used as the cooling medium, ordinary wet air cooler, plate wet air cooler, or surface evaporation wet air cooler can be selected. The ordinary wet air cooler is easy to scale between the fins, which rapidly reduces the heat transfer performance of the air cooler. The heat transfer performance of the plate wet air cooler is better than the ordinary finned air cooler, and the antiscale ability of the plate is better than the fin. A comparison was made with one plate air cooler instead of three ordinary wet air coolers for the cooling system from vacuum column OH of a 5 Mt/a crude oil distillation unit, started up in 2003 [18]. The results showed that the annual electricity consumption is reduced by 648,000 kWh, water consumption by 32,000 tons, and operation cost by 298,000 yuan. It has also been reported that the first, second, and third level air coolers from vacuum column OH of a crude oil distillation unit adopted plate structure. The 48 conventional wet air coolers were replaced by 20 plate bundles, saving 421 m^2 of floor area, 261.76 tons of metal weight, 2.08 million kWh of electricity, and 230,400 tons of tempered water consumption per year. Such method obtained considerable economic benefits [19].

If circulating water is used as the cooling medium, surface condensers are mostly used. In order to reduce the pressure drop as much as possible, the structure of the condenser should be different from the conventional shell and tube heat exchangers. Although the structure of the condenser used in the vacuum system is diverse, its common point is to ensure enough gas space. The condenser inlet and noncondensable gas outlet should have a large tube pass. A steam distribution plate should be installed inside the tube. The condensate leaves the condenser from the bottom, and the noncondensable gas is extracted from the top of the condenser. The structure scheme of surface condenser is shown in Fig. 3.10.

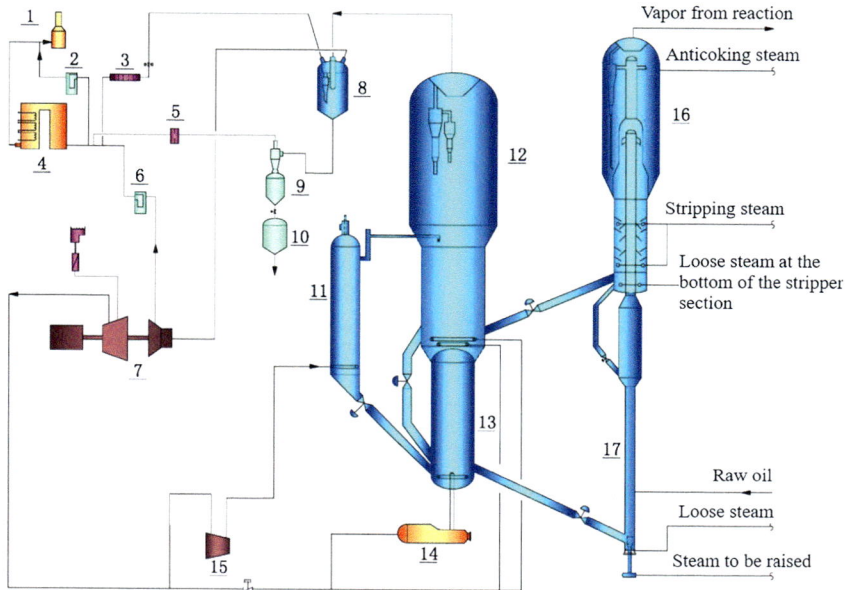

Fig. 3.10 Principle process flow diagram of reaction system. (1) Flue gas desulfurization system; (2) flue gas vent water seal drum; (3) depressurizing orifice chamber; (4) CO incineration boiler; (5) critical flow nozzle; (6) flue gas turbine outlet water seal drum; (7) energy recovery three-machine set; (8) three-stage cyclone separator; (9) fourth-stage cyclone separator and catalyst recovery system; (10) recover catalyst drum; (11) external catalyst cooler; (12) first-stage regenerator; (13) second-stage regenerator; (14) direct fired air heater; (15) regenerator fluffing air compressor system; (16) disengager; (17) riser.

With the gradual improvement of process requirements, more and more research works are focused on the vacuum column overhead condensers. Patent products with different structures have appeared [20–22]. The application of this equipment can help enhance the heat transfer performance of the vacuum column OH vacuum system and save energy.

3.1.4.4 Stabilizer heat exchange system

1. Feed heat exchange system of stabilizer

The 40°C ATM top oil flows to stabilizer as feed and is heated to 148°C by the naphtha/stabilizer feed heat exchanger, then enters the stabilizer. The process conditions of such heat exchangers are shown in Table 3.6. The temperature cross of the cold and hot fluid of the heat exchangers is huge. When ordinary heat exchangers are used, multiple series will be needed.

Table 3.6 Process conditions of naphtha/stabilizer feed heat exchanger—Duty: 15.7 MW.

Item	Shell side	Tube side
Fluid name	Stabilizer feed	Naphtha
Flow rate (kg/h)	105,324	184,646
Temperature (°C)	$40 \rightarrow 148$	$189 \rightarrow 70$
Gas density (kg/m^3)	$19.4 \rightarrow 18.7$	–
Liquid density (kg/m^3)	$561 \rightarrow -$	$527 \rightarrow 665$
Gas viscosity (mN s/m^2)	$0.011 \rightarrow 0.012$	–
Liquid viscosity (mN s/m^2)	$0.153 \rightarrow -$	$0.097 \rightarrow 0.261$
Gas thermal conductivity (W/(m K))	$0.028 \rightarrow 0.033$	–
Liquid thermal conductivity (W/(m K))	$0.12 \rightarrow -$	$0.078 \rightarrow 0.115$
Gas heat capacity (kJ/(kg K))	$1.77 \rightarrow 2.26$	–
Liquid heat capacity (kJ/(kg K))	$2.53 \rightarrow -$	$2.86 \rightarrow 2.33$

Table 3.7 Comparison of naphtha/stabilizer feed heat exchanger.

Scheme	E-shell exchanger	F-shell exchanger
Equipment size (shell diameter × straight tube length) (mm)	1200×6000	1400×7000
Set	4	2
Shell side pressure drop (kPa)	91.3	92.1
Overall heat transfer coefficient (W/(m^2 K))	368	374
Heat transfer area (m^2)	1585	1574
Total weight of equipment (t)	65.9	47.8

Table 3.7 shows the calculation and comparison of ordinary heat exchanger and F-shell heat exchanger schemes. When the ordinary heat exchanger is used, the calculation results show that the film heat transfer coefficients of the tube and shell side are close during the phase change. In the superheated section of gas, the film heat transfer coefficient of shell side decreases significantly, which is lower than tube side, and the overall heat transfer coefficient is not high. Under the same pressure drop condition, the heat transfer performance of two F-shell heat exchangers is basically the same as four E-shell heat exchangers. However, the weight of equipment can be reduced by 27%, the cost of equipment infrastructure construction and pipeline configuration is reduced by 50%, and the floor area is reduced by 50%.

For naphtha/stabilizer feed heat exchanger, all-welded plate heat exchanger can also achieve good results. In a 10 Mt/a crude oil distillation unit, a 287.5 m^2 all-welded plate heat exchanger is used to replace four sets of φ1200 mm×5000 mm conventional shell and tube heat exchanger. The overall heat transfer coefficient was increased from 315 to 1356 W/(m^2 K). The heat transfer area is saved by 1075 m^2, the weight of equipment is reduced by 67.8%, and the capital construction cost and operation cost is saved.

2. Stabilizer OH cooling system

The 64°C stabilizer OH vapor is cooled to 40°C by air cooler and water cooler and enters the stabilizer OH reflux and product vessel. A total duty of 22.0 MW needs to be cooled by stabilizer OH air cooler and water cooler, which is relatively large. See Table 3.8 for the process conditions. Dry air cooler with ordinary water cooler is commonly used. And the calculation and comparison with compound evaporative air cooler are shown in Table 3.9. Scheme I adopts both conventional air cooler and water cooler, scheme II only uses water cooler, and scheme III only adopts compound evaporative air cooler.

Table 3.8 Process conditions of stabilizer OH air cooler and water cooler—Duty: 22.0 MW.

Item	Hot fluid	Cold fluid	
		Air cooler	Water cooler
Fluid name	Naphtha	Air	Cooling water
Flow rate (kg/h)	215,376		
Allowable pressure drop (kPa)	20		<100
Temperature (°C)	$64 \rightarrow 40$	32 (outlet)	$33 \rightarrow 40$
Gas density (kg/m^3)	$25.4 \rightarrow -$		
Liquid density (kg/m^3)	$- \rightarrow 515$		
Gas viscosity (mN s/m^2)	$0.0096 \rightarrow -$		
Liquid viscosity (mN s/m^2)	$- \rightarrow 0.109$		
Gas thermal conductivity (W/(m K))	$0.023 \rightarrow -$		
Liquid thermal conductivity (W/(m K))	$- \rightarrow 0.115$		
Gas heat capacity (kJ/(kg K))	$2.21 \rightarrow -$		
Liquid heat capacity (kJ/(kg K))	$- \rightarrow 1.19$		

Table 3.9 Comparison of stabilizer OH air cooler and water cooler schemes.

Item	Scheme I		Scheme II	Scheme III (Enhanced) Compound evaporative air cooler
	Dry air cooling	Water cooler	Water cooler	
Equipment size (mm)	$12,000 \times 3000^a$	1300×6000^b	1500×6500	9000×3000
Equipment (set)	6	4	4	4
Consumption	Electricity: 180 kW	Cooling water: 1298 t/h	Cooling water: 2671 t/h	Electricity: 236 kW Tempered water: 16 t/h
Energy consumption (kgoe/(t feed))	0.0261	0.0623	0.128	0.0172
Heat transfer area (m^2)	1560	1920	2820	2800
Total weight of equipment (t)	137	70	100	145

^{a}Air cooler size: tube bundle length × width.
^{b}Water cooler size: shell diameter × straight tube length.

It can be seen from Table 3.9 that scheme I needs 10 sets of equipment while scheme II needs 4 sets of equipment. Scheme III only needs four sets of equipment with only two frames. The use of scheme III can save the cost of civil construction, pipeline configuration, and operation, as well as saving valuable floor spaces.

In scheme I, the total equipment weight is 207 tons, and the cooling water is about 1298 t/h. The total equipment weight of scheme II is 100 tons and 52.7% of that in scheme I, but the consumption of cooling water is doubled. Scheme III reduces the total equipment weight by 60 tons compared with scheme I. The energy consumption is the lowest, 13.4% lower compared with scheme II. Moreover, scheme III can save 9.04 Mt/a cooling water based on 8000 h of annual operation time and can save the investment of cooling water plant operation system.

The above analysis shows that the compound evaporative air cooler has significant advantages in various areas, such as one-time investment in equipment, long-term operation cost, energy saving, and consumption reduction. Therefore, the enhanced heat transfer technology can save considerable equipment investments [13].

3. Heat exchange system at stabilizer bottom

The stabilizer bottom reboiler is usually installed horizontally and use the 245°C ATM 2nd PA as heat source. For example, the process conditions of a typical stabilizer bottom reboiler from a domestic unit are shown in Table 3.10. It can be seen from the table that the heat transfer characteristics of cold and hot streams are similar, and the viscosity is not high. Table 3.11 shows the calculation and comparison of the plain-tube reboiler and T-shaped grooved tube reinforced reboiler. It can be seen from Table 3.11 that the heat transfer performance of the reboiler is relatively ideal. The film heat transfer coefficient of the shell side of the reboiler is greatly improved after the T-shaped grooved tube is used to strengthen the boiling process of the shell side. The fouling resistance is reduced as the T-shaped grooved tube increases the vaporization core. Table 3.11 also shows that the diameter of the reboiler is reduced from 1600 to 1400 mm, which can save 25% of the heat exchange area and reduce the weight of the equipment by 25%.

In the naphtha water cooler, a spiral-twisted tube is used instead of a plain tube. The industrial application shows [7] that the shell side pressure drop is reduced by 21.7% and the heat transfer coefficient is increased by 20%. This will solve the problem that the temperature of cooled light naphtha cannot reach the design requirement of 35°C and ensures the stable operation of the light hydrocarbon system.

Table 3.10 Process conditions of reboiler at the bottom of stabilizer—Duty: 11.5 MW.

Item	Shell side	Tube side
Fluid name	Naphtha	Second atmospheric intercycle
Flow rate (kg/h)	462,937	570,152
Allowable pressure drop (kPa)	Thermal siphon	50
Temperature (°C)	$178 \rightarrow 197$	$245 \rightarrow 218$
Gas density (kg/m^3)	$- \rightarrow 34$	–
Liquid density (kg/m^3)	$543 \rightarrow 524$	$672 \rightarrow 694$
Gas viscosity (mN s/m^2)	$- \rightarrow 0.011$	–
Liquid viscosity (mN s/m^2)	$0.0107 \rightarrow 0.094$	$0.244 \rightarrow 0.293$
Gas thermal conductivity (W/(m K))	$- \rightarrow 0.033$	–
Liquid thermal conductivity (W/(m K))	$0.0812 \rightarrow 0.076$	$0.078 \rightarrow 0.083$
Gas heat capacity (kJ/(kg K))	$- \rightarrow 2.57$	–
Liquid heat capacity (kJ/(kg K))	$2.79 \rightarrow 2.89$	$2.73 \rightarrow 2.65$

Table 3.11 Comparison of stabilizer bottom reboiler schemes.

Scheme	Plain tube	T-shaped grooved tube
Equipment size (shell diameter × straight tube length) (mm)	1600×6000	1400×6000
Equipment (set)	1	1
Film heat transfer coefficient of shell side (W/(m^2 K))	1850	4190
Overall heat transfer coefficient (W/(m^2 K))	510	690
Heat transfer area (m^2)	727	540
Total weight of equipment (t)	25.5	19.3

3.1.5 Summary

The energy consumption of crude oil distillation unit occupies a large proportion of oil refinery. The key point of energy saving is to optimize the energy recovery. The pinch point technology is adopted to optimize the heat exchanger network and heat transfer, improve the final temperature of crude oil, and achieve the purpose of reducing the energy consumption of crude oil distillation. In crude oil distillation unit, the heat exchanger network is complex with many sets of heat exchangers, which accounts for 30%–40% of the equipment investment of the crude oil distillation unit. The use of appropriate heat transfer enhancement technology can effectively reduce the metal usage of the equipment. It will save the equipment investment for the owner, achieve the purpose of energy saving, and save space area.

3.2 Catalytic cracking unit

3.2.1 Brief description of process

Catalytic cracking is one of the core units in refinery to process heavy oil into light oil. Compared with other secondary processing units for heavy oil processing, one of the advantages of catalytic cracking unit is the adverse feed sources, including not only vacuum gas oil, coking gas oil, and solvent deasphalting oil (DAO), which can be directly used as feed, but can also be used to process atmospheric heavy oil, vacuum residue, and hydrotreated heavy oil. The products mainly include high-octane gasoline and LPG (rich in propylene), as well as by-products such as diesel blending component, fuel gas, and slurry. Due to the high yield of FCC liquid products, the yield of light oil can reach 60%–70% (mass fraction), and the yield of liquid can reach 70%–90% (mass fraction), which improves the yield of light oil products in the whole plant and makes a great contribution to the efficiency of the refinery [23].

A typical FCC unit usually includes a reaction regeneration system, fractionation system, absorption and stabilization system, and energy recovery system.

The reaction regeneration system is the core system of FCC unit, which is composed of riser, disengager, regenerator, and catalyst cooler lift riser. Taking the reactor riser and the staked two-stage incomplete regeneration technology as an example, the principle flow of the reaction system is shown in Fig. 3.10. After heating up through heat exchange, the feed oil enters the feed nozzle at the lower section of the riser. After the feed and atomized steam are mixed in the feed nozzle, it is sprayed through the feed nozzle to contact with the high-temperature regeneration catalyst from the regenerator and is immediately vaporized in the reaction section of the riser. Under a relatively high reaction temperature and large catalyst-to-oil ratio, the feed is cracked into light products (dry gas, liquefied gas, gasoline and light diesel). The oil and gas vapor generated by the reaction flows upward through the riser with catalyst, and the catalyst is separated through the vortex quick separator at the outlet of the riser and the cyclone separator at the riser overhead and then enters the fractionator for preliminary separation of each product. After the reaction, the coked catalyst enters the regenerator and burns the coke to restore the catalyst activity. The hot regenerated catalyst returns to the bottom of the riser to realize the continuous cycle of the catalyst. The high-temperature flue gas is discharged from the top of the regenerator and enters the flue gas turbine to expand and recover the pressure energy in the flue gas, which then enters the boiler to recover the chemical energy and heat energy in the flue gas. It is then discharged into atmosphere after being treated by the desulfurization and denigration system.

The function of fractionation system is to cut and separate the reaction products to obtain rich gas, crude gasoline, light circulating oil, recycle oil, and slurry fractions.

A typical fractionation system process flow diagram of a catalytic cracking unit is shown in Fig. 3.11. The high-temperature oil and vapor from the reaction system enter the desuperheating section at the fractionator bottom, and the circulating oil slurry is extracted from the column bottom which returns to column after cooling through heat exchange. The cooled circulating oil slurry will quickly cool the catalyst containing superheated oil and gas to a saturation state, and the entrained catalyst dust will be washed down at the same time. The crude gasoline and rich gas are separated from the fractionator OH, which is pressurized and sent to the absorption and stabilization system. Light circulating oil is extracted from the middle of the fractionator as a product and sent OSBL. The catalyst containing slurry from the column bottom can send OSBL as product or can return to the reaction system according to the burning duty and heat balance of the unit.

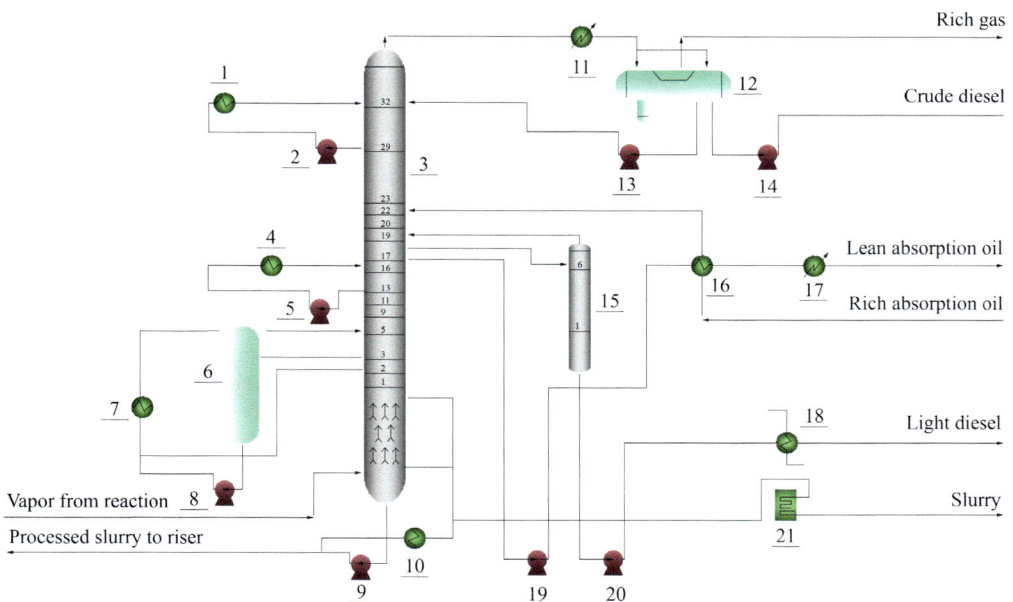

Fig. 3.11 Principle process flow diagram of fractionation system. (1) OH recycle cooler; (2) fractionator OH reflux pump; (3) catalytic fractionator; (4) fractionator 1st pump around (PA) cooler; (5) fractionator 1st PA reflux pump; (6) recycle oil drum; (7) fractionator 2nd PA cooler; (8) recycle oil pump; (9) slurry pump; (10) slurry heat exchanger; (11) fractionator OH cooling system; (12) fractionator OH vapor separator; (13) fractionator OH cold reflux pump; (14) crude gasoline pump; (15) light diesel stripper; (16) lean absorption oil/rich absorption oil heat exchanger; (17) lean absorption oil cooler; (18) light diesel cooler; (19) lean absorption oil pump; (20) light diesel pump; (21) slurry cooler.

The function of the absorption and stabilization system is to separate rich gas and crude gasoline into dry gas ($\leq C_2$), liquefied gas (C_3, C_4), and stable gasoline with qualified vapor pressure by means of absorption, desorption, and distillation. The absorption and stabilization system is mainly composed of absorber, desorption column, stabilizer, reabsorption column, and the corresponding auxiliary equipment. The typical process flow diagram is shown in Fig. 3.12.

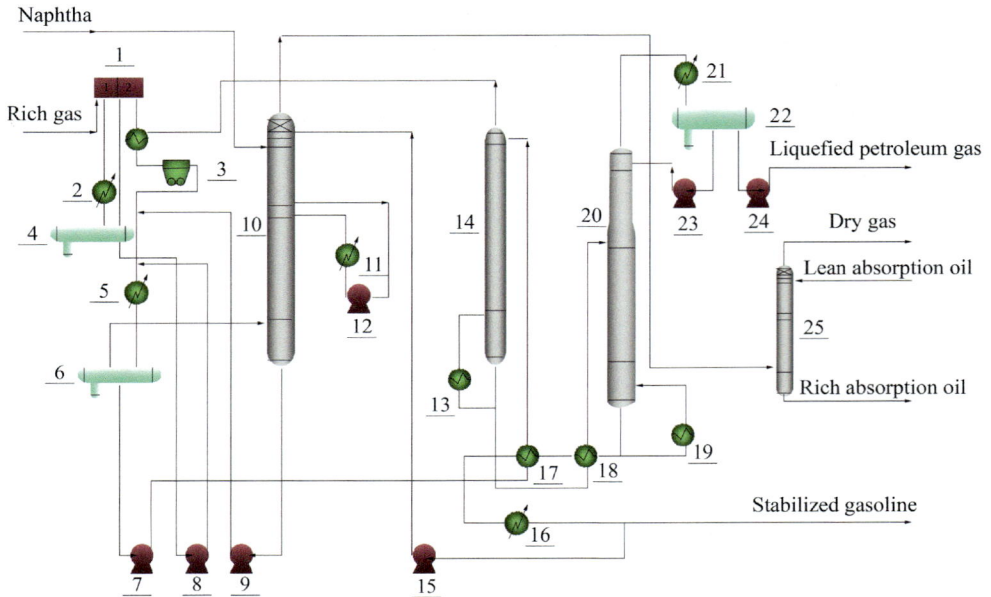

Fig. 3.12 Principle process flow diagram for absorption and stabilization system. (1) Rich gas compressor; (2) air compressor interstage cooler; (3) air compressor outlet air cooler; (4) air compressor interstage condensate drum; (5) air compressor outlet after cooler; (6) air compressor outlet vapor separator; (7) desorption column feed pump; (8) air compressor interstage condensate pump; (9) absorber bottom pump; (10) absorber; (11) absorber PA cooler; (12) absorber PA reflux pump; (13) desorption column bottom reboiler; (14) desorption column; (15) make-up absorbent pump; (16) stabilized gasoline cooler; (17) desorption column feed heat exchanger; (18) stabilizer feed heat exchanger; (19) stabilizer bottom reboiler; (20) stabilizer; (21) stabilizer OH condensing cooler; (22) stabilizer OH reflux drum; (23) stabilizer OH reflux pump; (24) liquefied gas product pump; (25) reabsorption column.

3.2.2 Analysis of heat energy characteristics

The energy consumed by the catalytic cracking unit in the process is mainly provided by the heat from the combustion of the coke generated by the reaction in the regenerator. The heat, from the whole process point of view, is divided into several main sections:

① Supply the heat needed for heating, gasification, and reaction of feed in the reactor.

② The high temperature level heat carried by the reaction vapor is used in the separation of products in fractionation section and absorption and desorption section.

③   In order to maintain the thermal balance of the reaction and regeneration system, the excess heat in the regenerator is recovered by the heat extraction facility to generate saturated steam.

④   The high-temperature flue gas discharged from the regenerator is recovered in the waste heat recovery system (including the pressure energy recovery unit of the flue gas turbine). The setting of the flue gas turbine only improves the grade of the recovered flue gas energy and reduces the overall energy loss of the unit but essentially does not change the total amount of heat exergy recovered from flue gas energy.

⑤   Heat from flue gas and products (flow OSBL) after waste heat recovery.

⑥   Heat loss of equipment and pipeline in the unit.

In addition, the energy (including electric energy, heat energy, etc.) required by the pressurization and heating of the OSBL utility and product heat exchange, or cooling during the process consists of another part of the energy consumption of the unit.

The energy use analysis of each section in FCC is listed in Fig. 3.13.

Heat removal from the regeneration process, flue gas waste heat recovery, and the heat of reaction vapor for fractionation section and absorption and desorption section are part of the energy recovery. The recovered energy is part of the coke burning heat. In the process of the unit's energy consumption statistics, apart from the different energy grades, all the heat comes from the coke burning section. The low-level temperature heat recovered from steam generating and fractionation section of the slurry is part of the reaction heat, which is sent OSBL. Such reaction heat comes from the excess heat of the feed vaporization and reaction [24].

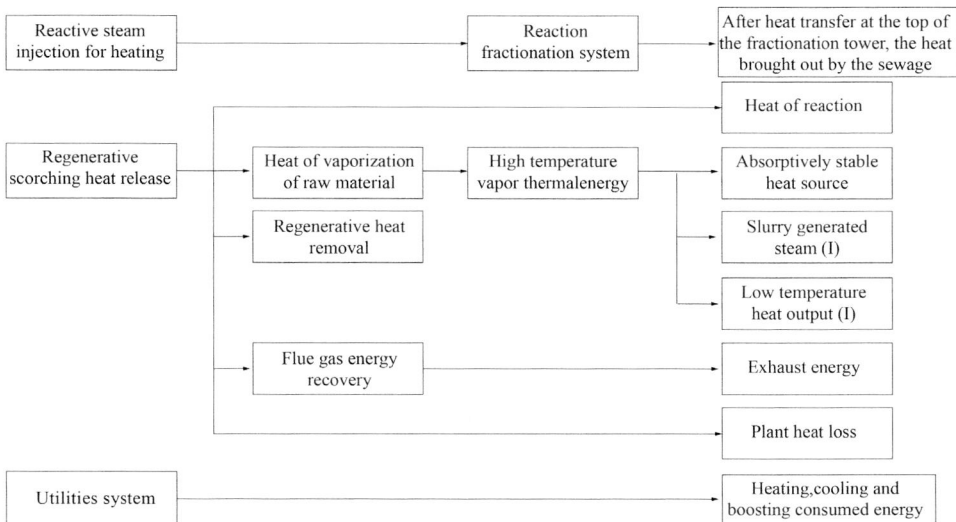

Fig. 3.13 Energy analysis of each section in catalytic cracking unit.

3.2.3 Process enhancement and integration on heat transfer

Through the adoption of technical measures, enhancing the energy recovery of each grade of the process [24], is an important guarantee for energy saving and consumption reduction in catalytic cracking unit. To maximize the utilization of exergy, high-grade energy should be recovered by power generation, medium- or medium–high-pressure steam and high-pressure steam as much as possible. Medium- and low-level temperature heat should exchange heat or undergo heat integration with ISBL or OSBL units, to reduce fuel and steam consumption for the heating of process medium. The consumption of cooling water, air cooling power, and other utilities for cooling process medium can also be reduced to achieve double energy-saving effect, thus further reducing the energy consumption of the unit. In addition, it is necessary to reasonably set the operation variability of the unit and equipment so that the efficiency of the equipment can reach the optimal condition and avoid the increase of energy consumption caused by the actual operation condition and the designed condition.

3.2.3.1 High-temperature-level heat recovery and utilization

The high-temperature-level heat source of the heavy oil catalytic cracking unit mainly comes from two aspects. One is the chemical energy released by the combustion of coke in the regenerator (exceeds the heat required by the reaction), and the other is the heat of the recycled slurry from fractionation system. The temperature of the regenerator is as high as 700°C under normal operating conditions. From the perspective of maximizing the utilization of exergy, high-quality energy such as medium-pressure steam, medium–high pressure, and even high-pressure steam should be generated as much as possible. Usually, the heat removal methods of regenerator can be divided into internal heat removal and external heat removal. Although the internal heat removal technology has little investment and simple structure, it is gradually replaced by external heat removal technology and only becomes the supplement of external heat removal due to disadvantages such as adjustable heat removal duty, low heat removal efficiency, and easy damage of heat removal tube. External catalyst cooler is the main cooling and waste heat recovery equipment in catalytic cracking unit [25–27]. The temperature range of recycled slurry from fractionation system that can be used to generate medium-pressure or medium–high-pressure steam is 330–275°C.

3.2.3.2 Medium-temperature heat recovery and utilization

The superheated high temperature reaction vapor from the reaction section enters the lower section of the fractionator. Under the condition of meeting the gas–liquid separation requirement of the whole column, there is still a lot of excess heat, and it is usually necessary to set up 4–5 pump around (PA) to remove heat. As the tray temperature of the fractionator varies greatly with different tray numbers, the OH reflux, absorbed oil recirculation, 1st PA, 2nd PA, and the oil slurry recirculation are usually set for heat removal.

The temperature utilization range of the 1st PA is usually 268–160°C, and the temperature utilization range of the 2nd PA is usually 310–250°C. These two heat sources belong to the medium-level temperature heat source. They are usually used as the heat source of the bottom reboiler of the column in absorption and stabilization system, and therefore the heat of the fractionation section can be transferred to the absorption and stabilization system to meet the requirements of gas-liquid separation.

3.2.3.3 Low-level temperature heat recovery and utilization

In addition to the high- and medium-level temperature heat sources, the other oils in the fractionation section and absorption and stabilization system are generally regarded as low-level temperature oil. Taking a 3.5 Mt/a heavy oil catalytic cracking unit as an example, the cooling duty of each low-temperature oil is shown in Table 3.12. The total amount of low-level temperature heat of the unit is 147,195 kW. Due to the low-temperature heat recovery measures of the unit, the heat of the left part in Table 3.12 has been recovered and utilized, which is about 78,645 kW, accounting for 53.4% of the cooling duty of the whole unit. As this part of energy is recovered and utilized, according to the "Calculation Method of Energy Consumption in Petrochemical Design," the unit not only outputs heat but also saves cooling water (or electric energy), which actually achieves a "win-win" effect on reducing the energy consumption of the unit. It can also be seen from Table 3.12 that the low-level temperature heat of the FCC unit mainly comes from the condensation and cooling of fractionator OH vapor. The recovered energy from fractionator OH vapor accounts for 42.6% of the total recovered energy [28]. The second largest proportion of recycled energy is the heat from the fractionator OH circulating oil, accounting for about 32.9%. The temperature range of the fractionator OH circulating oil is usually 143–80°C. The remaining low-level temperature heat accounts for 24.5% of the recycled energy.

Table 3.12 Cooling duty statistics of low-temperature oil products.

Name	Recycled		Heat lost	
	Temperature change (°C)	Heat duty (kW)	Temperature change (°C)	Heat duty (kW)
Fractionator OH vapor cooler	120–91	33,525	91–55 (air cooled)	29,181
			55–40 (water cooled)	7930
OH circulation oil cooler	143–113	12,336		
	113–80	13,570		
Light diesel cooler	195–90	5941	90–50 (air cooled)	1967
Lean absorbed oil cooler	85–5	1864	55–40 (water cooled)	885
1st PA cooler	191–160	5860		

Continued

Table 3.12 Cooling duty statistics of low-temperature oil products—cont'd

Name	Recycled		Heat lost	
	Temperature change (°C)	Heat duty (kW)	Temperature change (°C)	Heat duty (kW)
Air compressor outlet front cooler			90–55 (air cooled)	2932
Air compressor exit after cooler			55–40 (water cooled)	8518
Absorber PA cooler			47–40 (water cooled)	2678
Stabilizer OH condenser			70–50 (air cooled)	3060
			50–40 (water cooled)	5152
Stabilized gasoline cooler	96–70	5549	70–55 (air cooled)	3060
			55–40 (water cooled)	3187
Total		78,645		68,550

3.2.4 Heat transfer enhancement technology using elements

3.2.4.1 Heat transfer enhancement of the reaction section

The medium outside the regenerator external heat remover tube is the catalyst particle, and the medium inside the tube is the steam–water mixture. The heat transfer coefficient in the tube includes two parts: the convection heat transfer coefficient of water and the boiling heat transfer coefficient of water. It can be different according to the different recirculating ratio, but generally over $6000 \, \text{W}/(\text{m}^2 \, \text{K})$. For dense phase heat removal, according to the heat transfer experiment and calibration results of industrial units, the heat transfer coefficient outside the tube is generally $600–940 \, \text{W}/(\text{m}^2 \, \text{K})$, so the heat transfer coefficient inside the tube is 6–10 times larger than outside the tube. The control thermal resistance of the external catalyst cooler is on the catalyst side. Improving the heat transfer coefficient of the shell side (catalyst side) can effectively improve the overall heat transfer coefficient.

There are many types of external catalyst coolers. One type is similar to shell and tube heat exchanger. The catalyst flows to the shell side, along the longitudinal of the tube wall, and the lower section fluidizing air will fluidize the catalyst. The structure of the external catalyst cooler is compact, the heat transfer area is large, the water supply flows from bottom to up, and the steam–water mixture flows down, resulting in the bubble lifting trend being opposite to the direction of the water flow, easy to cause local air resistance, and cause the tube local overheating and heat transfer deterioration. In addition, as the wall of the heat transfer tube is thin, it is easy to be worn through by the catalyst. Once a heat transfer tube is worn through, the whole external catalyst cooler fails, and it is very difficult to replace [29–32].

Unit tube group combined box type external catalyst cooler is composed of a shell and a number of heat transfer unit tube groups. Each unit tube group, consisting of multiple heat transfer tubes, is combined with the upper combined box and the lower combined box. The characteristics of this external catalyst cooler is that the catalyst fluidization state is good, and each unit tube group is controlled by valves, which can be opened and closed automatically. When a unit tube group of heat transfer tube fails, it will not affect the use of the whole heat remover. As the heat transfer tube between the upper and lower combined box is in a high-temperature environment, and the diameter is large with many welding seams, the requirement for processing and manufacturing is much higher [31,33].

The diameter of the heat transfer tube of the finned tube-type external catalyst cooler is large, and the surface is covered with longitudinal fins. Each finned tube is also an independent heat transfer component. Since the heat transfer coefficient outside the tube is much smaller than inside the tube, increasing the heat transfer area outside the tube can significantly improve the heat transfer efficiency. When the catalyst is in the sock phase flow, the application effect of this external catalyst cooler is good. However, when the catalyst is in the dense phase fluidized bed, the dense phase catalyst between the fins is not easy to flow, and the "dead interlayer" is easy to form, which reduces the heat transfer effect [31,34–36].

One kind of external catalyst cooler includes stud fin tube bundle. Each stud fin tube is of closed structure, which forms an independent water-steam circuit heat transfer unit. Straight- or L-type studs are staggered and uniformly distributed on the surface of base tube and are perpendicular to the tube surface. Less than 25% of the surface area of base tube is covered by the cross section of studs, retaining the characteristics of plain tube and not affecting the flow of catalysts. The stud can not only increase the heat transfer area outside the tube but also produce local turbulent flow as the stud is perpendicular to the flow direction of the catalyst, which enhances the heat transfer effect, and has a high heat removal efficiency. The results of industrial operation showed that the heat transfer coefficient of the stud fin heat transfer tube can be increased by more than 60% compared with the plain tube. In addition, the stud fin tube has less thermal stress at high temperature, with a better force distribution; therefore it is not prone to rupture and distortion. The nail-head heat transfer tube has a long life cycle and can meet the requirements of long life cycle safe operation of the catalytic unit. It is an ideal heat transfer enhancing component [31,37]. Similar to finned tube, the heat transfer component of nail-head tube can better play its advantage of enhancing heat transfer outside the tube in sock phase heat removal occasions.

In heavy oil catalytic cracking unit, the slurry oil feed is preheated by feed/fractionator bottom slurry heat exchanger and flows to the reactor riser. The physical properties of shell side feed are high viscosity and high fouling thermal resistance. When using shell and tube heat exchanger with single segmental baffle, flow velocity and Reynolds number are low, with a high pressure drop and low film heat transfer coefficient. The helical baffle can significantly increase the flow rate and Reynolds number and reduce dirt deposition, which has an obvious effect on heat transfer optimization.

3.2.4.2 Heat transfer enhancement of the fractionation section

Fractionator bottom slurry contains catalyst solid particles and usually flows in the tube side of shell and tube heat exchanger, to reduce the congestion of the heat exchange tube. Large-diameter heat exchange tubes are usually used. A certain flow velocity of slurry should be maintained to keep the catalyst particles in a suspended state during the design, and the flow velocity should not be too high to avoid heat transfer tube abrasion damage leakage. Spiral plate heat exchanger has been successfully applied for fractionator bottom slurry heat exchanger; for its spiral flow, suspended particles are not easy to deposit and have a certain self-cleaning ability [38].

The fractionator bottom slurry has a large flowrate and a high temperature, so multiple steam generators are needed to recover heat before sending OSBL. As the slurry steam generators often are in series operation, the slurry temperature level of the first equipment is the highest, and the slurry temperature level and heat exchange temperature difference of the other equipment are significantly reduced. High flux tube is suitable to enhance boiling heat transfer outside the tube [39]. The characteristics of the slurry in the tube are large viscosity and low film heat transfer coefficient. The use of tube side enhancing measures such as plugins has a significant effect on improving film heat transfer coefficient and reducing equipment floor area.

The cooling process conditions of catalytic fractionator OH are shown in Table 3.13. The heat duty is large, the oil temperature level is low, and the heat exchange temperature difference is small, so the heat exchanger area is large.

Table 3.13 Process condition of fractionator OH vapor/hot water heat exchanger—Duty: 31.9 MW.

Item	Shell side	Tube side
Fluid name	Fractionator OH vapor	Heat exchanger water
Flow rate (kg/h)	272,142	1,000,000
Allowable pressure drop (kPa)	10	50
Temperature (°C)	$116 \rightarrow 92$	$65 \rightarrow 91$
Gas density (kg/m^3)	$4.00 \rightarrow 4.68$	
Liquid density (kg/m^3)	$\rightarrow 795.7$	$981.2 \rightarrow 958.3$
Gas viscosity (mN s/m^2)	$0.012 \rightarrow 0.011$	

Table 3.13 Process condition of fractionator OH vapor/hot water heat exchanger—Duty: 31.9 MW—cont'd

Item	Shell side	Tube side
Liquid viscosity (mN s/m^2)	→0.310	0.434 → 0.345
Gas thermal conductivity (W/ (m K))	0.027 → 0.026	
Liquid thermal conductivity (W/ (m K))	→0.357	0.656 → 0.657
Gas heat capacity (kJ/(kg K))	2.02 → 1.95	
Liquid heat capacity (kJ/(kg K))	→3.12	4.18 → 4.24

The comparison results of heat exchangers are shown in Table 3.14. The conventional plain tube heat exchange tube requires six heat exchangers in series, and the film heat transfer coefficient of shell side is about 1000 W/(m^2 K), which is much smaller than the film heat transfer coefficient of the cooling water in the tube, so enhancing shell side condensation can significantly improve the heat transfer efficiency. The low-finned tube has the advantages of thin condensation liquid film formed in the condensation outside the tube and high condensation heat transfer coefficient, and the condensation and enhancing effect of light oil is good. If the low-finned tube is used instead of the plain tube, the diameter of the heat exchanger can be reduced from 1500 to 1300 mm, which significantly reduces the weight and investment of the equipment.

Table 3.14 Comparison of fractionator OH vapor/hot water heat exchanger schemes.

Item	Original option	Enhanced option
Equipment size (shell diameter × straight tube length) (mm)	BJS 1500 × 6000	BJS 1300 × 6000
Type of heat exchange tube	Plain tube	Low-finned tube
Equipment (set)	6	6
Pressure drop (shell side) (kPa)	5.7	6.6
Overall heat transfer coefficient (plain tube) (W/(m^2 K))	491.1	713.2
Heat transfer area (m^2)	3855.2	2680.4
Total weight of equipment (t)	139.8	105.1

3.2.4.3 Heat transfer enhancement of absorption-stabilization section

The absorber operating temperature in absorption-stabilization section is low. The absorber has multiple PA coolers. The process medium temperature is low, and heat transfer temperature difference with cooling water is small, which needs more heat transfer area. Under the requirements of design pressure, using compact structure of plate heat exchanger can effectively increase the coefficient of heat transfer and reduce the heat exchange area and equipment floor area.

The desorption column is equipped with a bottom reboiler and multiple intermediate reboilers. The reboiler with low heat source temperature and small heat transfer temperature difference needs a larger heat transfer area. Compared with the plain tube, the T-shaped grooved tube has significantly reduced boiling temperature difference and significantly increased boiling heat transfer coefficient and heat flux, which has a good enhancing effect on shell side boiling and is suitable for application in the reboiler of desorption column and stabilizer [40].

The stabilizer bottom stabilized gasoline passes through the tube side of desorption column feed heat exchanger, the intermediate reboiler of the desorption column, and the stabilizer feed heat exchanger, which then continues to cool to 40°C through the demineralized water heat exchanger, the air cooler, and the water cooler. The allowable pressure drop in this process is relatively low. In the stabilized gasoline/demineralized water heat exchanger and water cooler, the heat transfer resistance is mainly on the shell-side stabilized gasoline. When the double-segmental baffle is used, the film heat transfer coefficient provided per unit pressure drop is significantly higher than that of the single-segmental baffle. Similarly, the stabilizer feed is also a shell-side flow which is limited by allowable pressure drop. Its physical property is similar to stabilized gasoline, which has a small amount of gasification in the process of heating. Double segmental is obviously better than single segmental for such heat exchangers. Take a stabilizer feed heat exchanger of a refinery as an example; the process conditions are shown in Table 3.15.

Table 3.15 Process condition of feed heat exchanger in stabilizer—Duty: 4.8 MW.

Item	Shell side	Tube side
Fluid name	Deethane gasoline	Stabilized gasoline
Flow rate (kg/h)	291,611	240,833
Allowable pressure drop (kPa)	20	50
Temperature (°C)	$133 \rightarrow 145$	$174 \rightarrow 149$
Gas density (kg/m^3)	$34.19 \rightarrow 35.12$	
Liquid density (kg/m^3)	$480.4 \rightarrow 480.3$	$497.8 \rightarrow 567.0$
Gas viscosity (mN s/m^2)	$0.011 \rightarrow 0.012$	
Liquid viscosity (mN s/m^2)	$0.135 \rightarrow 0.131$	$0.121 \rightarrow 0.156$
Gas thermal conductivity (W/(m K))	$0.027 \rightarrow 0.028$	
Liquid thermal conductivity (W/(m K))	$0.088 \rightarrow 0.086$	$0.079 \rightarrow 0.090$
Gas heat capacity (kJ/(kg K))	$2.35 \rightarrow 2.41$	
Liquid heat capacity (kJ/(kg K))	$2.68 \rightarrow 2.72$	$2.86 \rightarrow 2.70$

The comparison results of heat transfer enhancement with double-segmental baffles are shown in Table 3.16. In order to meet the allowable pressure drop, two sets of $\varphi 1300$ mm $\times$ 6000 mm (diameter $\times$ tube length) are required for BES (TEMA type) with single-segmental baffles heat exchanger. With the double-segmental baffle, only one set φ1500 mm $\times$ 6000 mm is needed to meet the allowable pressure drop and heat transfer requirements. The heat transfer area and the total weight of the equipment can be reduced by more than 30%, which has obvious advantages.

Table 3.16 Comparison of feed heat exchanger schemes for stabilizer.

Item	Original option	Enhanced option
Equipment size (shell diameter $\times$ straight tube length) (mm)	BES 1300 $\times$ 6000	BES 1500 $\times$ 6000
Baffle type	Single segmental	Double segmental
Equipment, set	2	1
Pressure drop (shell side) (kPa)	13.7	13.9
Overall heat transfer coefficient (W/(m^2 K))	298.7	427.3
Heat transfer area (m^2)	952.9	639.0
Total weight of equipment (t)	36.2	24.6

Lean absorption oil is extracted from the middle section of catalytic fractionator and used for the heat source of desorption column bottom reboiler and rich absorption oil heat exchange. It is further cooled in desalting water heat exchanger and water cooler and sent to the reabsorption column. Although lean absorption oil is transferred by lean absorption oil pump, there are many heat exchangers during the process, which is also stream limited by allowable pressure drop. As the shell side fluid, with double segmental baffle instead of a single segmental baffle, the heat transfer coefficient per unit pressure drop can be increased by more than 50%, which has an obvious effect on reducing the pressure drop and making it easier to meet the design requirements.

The light diesel column bottom reboiler in the fractionation section is similar to other reboilers in the absorption-stabilization section, which is also suitable for the use of T-shaped grooved tube to enhance boiling [41]. Before the light diesel is sent OSBL, the low-level temperature is recovered by desalting water heat exchanger. Light diesel flows in the shell side, with high viscosity and low thermal conductivity. In the cases related to this heat exchanger, corrugated tubes have been applied to enhance heat transfer. During plant modification projects, the plain tube is replaced by corrugated tube, and the heat transfer enhancement effect is obvious [42].

3.2.5 Summary

FCC unit can produce excess amount of heat. Through the different recovery and utilization methods of high, medium, and low temperature potential heat in FCC units, especially the heat integration through units, the overall energy consumption of catalytic process can be optimized, and the energy consumption of the whole plant can be reduced.

The catalyst regenerator and the external catalyst cooler in the catalytic cracking unit are important energy recovery equipment. Optimizing the flow state and heat transfer coefficient of the gas-solid mixture of catalyst is the key to enhancing heat transfer of the equipment. There are many applications of the remover tube in industry. The fractionation section of the slurry medium is special; to enhance the heat transfer of such slurry, we will need to consider ensuring the long life cycle of the equipment. Light oil and absorption-stabilization section is suitable for the use of regular light oil heat transfer enhancement technology.

3.3 Hydrocracking unit

3.3.1 Brief description of process

Hydrocracking is one of the main processes for heavy fractionated oil deep processing. According to the world's oil reserves, light crude oil with low sulfur content from shallow and medium-depth zones is becoming harder to explore. Heavy crude oil from deep zone, even though hard to explore, is increasingly explored. Nowadays, countries pay more attention to environmental protection around the world, which means higher-quality light fuel such as gasoline, coal, and diesel is required. Hydrocracking process is becoming more and more important and the production capacity is rising year by year [43]. Hydrocracking refers to the hydrogenation technology that convert more than 10% of the large molecules in the feed oil to smaller molecules by hydrogenation reaction [44]. The main purpose is to convert heavy oil into light oil products while removing impurities such as sulfur, nitrogen, and oxygen, therefore producing high-quality products such as gasoline, kerosene, and diesel.

There are many types of hydrocracking processes, such as single-stage process, two-stage process, once-through process, full-cycle process, and a combination thereof, etc. According to the pressure level, it can be divided into high-pressure hydrocracking, medium-pressure hydrocracking, moderate hydrocracking, etc. According to the feed, it can be divided into diesel hydrocracking, gas oil hydrocracking, residual oil hydrocracking, etc. According to the reactor type, it can be divided into fixed bed hydrocracking, bubbling bed hydrocracking, slurry bed hydrocracking, etc. Therefore, the energy consumption of different hydrocracking units is quite different. According to the total processing volume and comprehensive energy consumption index of hydrocracking unit of Sinopec, it can be seen that the processing volume increases by years and the energy consumption decreases by years.

Although there are many types of hydrocracking processes, they can be divided into reaction section (high pressure) and product fractionation section (low pressure) according to the basic process principle. Taking a typical once-through hydrocracking process in China as an example, the process flow is shown in Fig. 3.14. The reaction section generally includes the heating and pressurization system of the feed. As well as the furnace system, the hydrogenation reactor system, the reaction product cooling system, the high–pressure flash system, the low–pressure flash system, the recycle gas, the make–up hydrogen pressurization system, etc. The reaction section completes the hydrogenation reaction and the preliminary gas–liquid separation. The fractionation section generally includes H_2S stripping, product fractionator, naphtha fractionator, absorption and desorption column, and stabilizer. Fractionation section mainly separates the reaction products into light hydrocarbons, liquefied gas (LPG), light naphtha, heavy naphtha, kerosene, diesel, unconverted oil (UCO), and other products.

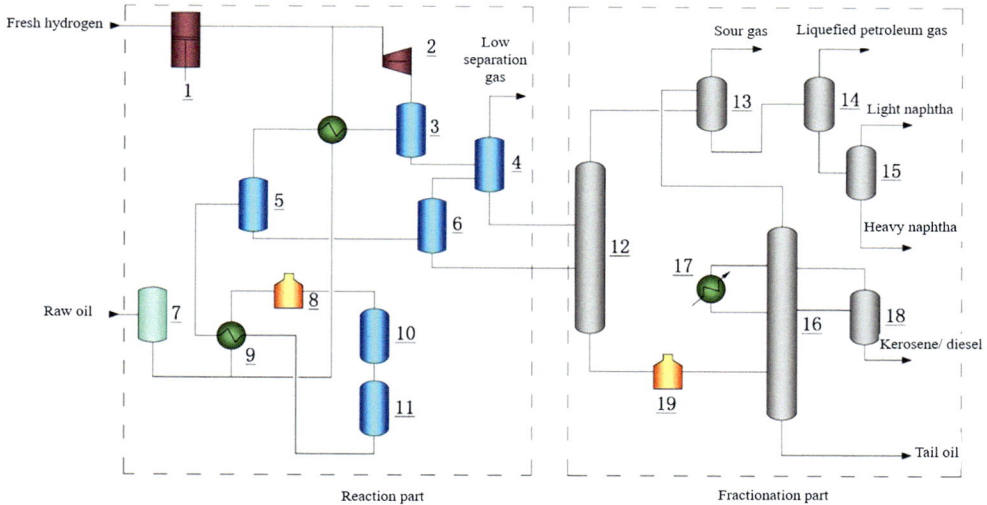

Fig. 3.14 Schematic diagram of typical single section once-through hydrocracking process. (1) Make-up hydrogen compressor; (2) recycle gas compressor; (3) cold high-pressure separator; (4) cold low-pressure separator; (5) hot high-pressure separator; (6) hot low-pressure separator; (7) feed surge drum; (8) reaction feed furnace; (9) effluent/mix feed heat exchanger; (10) hydrotreating reactor; (11) hydrocracking reactor; (12) H_2S stripper; (13) absorption/desorption column; (14) stabilizer; (15) naphtha splitter; (16) product fractionator; (17) pump around cooler; (18) kerosene/diesel side-draw stripper; (19) fractionator feed furnace.

3.3.2 Analysis of heat energy characteristics

The heat energy characteristic of hydrocracking process is the exothermic reaction which can release huge amount of heat. According to the experience, the total reaction temperature rise is generally more than 70°C. Therefore, rational utilization of reaction heat through process optimization is one of the directions of process energy enhancing. Due to the high reaction temperature required by hydrocracking (about 370°C) and the fact that the acid gas, liquefied gas, light naphtha, heavy naphtha, kerosene, diesel, hydrotreating tail oil, and other products need to be separated from fractionation section, the number of furnaces and fractionators or strippers are large. The second heat energy characteristic is that the unit requires large amount of heat. The third heat energy characteristic is the various low–level temperature heat sources. These are due to the lack of cold sources in the unit while multiple products need to be cooled. Therefore, reasonable optimization of heat exchange process and low-temperature heat utilization is the second direction of process energy enhancement. According to the average energy consumption index distribution of 21 hydrocracking units in Sinopec, it can be seen that the main energy consumption is electricity, medium–pressure steam, and fuel. Firstly, due to the high reaction pressure (about 17.0 MPa (gauge pressure)), it is necessary to increase the pressure of the feed, which leads to high power consumption. Secondly, the reaction needs a lot of recycle gas and cold hydrogen, so it needs to consume a lot of medium-pressure steam to drive the recycle gas compressor. Thirdly, the unit needs a large amount of heat, so the fuel consumption is large. To reduce the power and steam consumption, it is necessary to add energy conversion equipment (such as hydraulic turbine) or improve the process technology (such as improved catalyst performance, catalyst arrangement, liquid phase hydrogenation technology, etc.). However, to reduce fuel gas consumption, it is necessary to enhance heat transfer in the process to reduce heat energy loss and finally achieve the goals of low carbon, energy saving, and low investment.

Due to the great difference in operating pressure between the reaction section (17 MPa (gauge pressure)) and the fractionation section (less than 1 MPa (gauge pressure)), the heat exchanger network is generally divided into high–pressure and low-pressure heat exchanger networks. In the case of excess high-pressure heat source, the low-pressure cold source can be considered to exchange heat. The network of high-pressure section heat transfer from the beginning of the reaction effluent to the cold high-pressure separator is shown in Fig. 3.15, and the excess heat load can be transferred to the stripper feed.

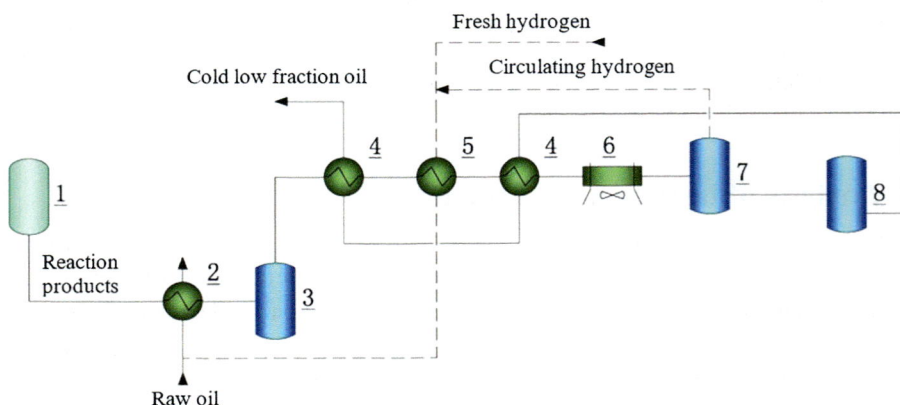

Fig. 3.15 Heat exchanger network of high pressure section. (1) Reactor; (2) effluent/mixed feed heat exchanger; (3) hot high-pressure separator; (4) hot high-pressure separator gas/stripper feed heat exchanger; (5) hot high-pressure separator gas/mixed hydrogen heat exchanger; (6) hot high-pressure separator gas air cooler; (7) cold high-pressure separator; (8) cold low-pressure separator.

    The high-pressure heat exchange section of hydrocracking unit has the following constraints due to safety and other factors:

① The temperature difference between the inlet and outlet of the reactor furnace must be no less than 28°C. Therefore the maximum temperature of feed heat exchange is limited. Taking 1.8 Mt/a hydrocracking unit as an example, the heating duty is about 8.4 MW, which is 3.38 kgoe/(t feed), accounting for about 11% of the total energy consumption of the unit.

② In order to prevent the ammonium salt crystallization from blocking the pipeline, the hot high-pressure separator gas needs continuous water injection at about 160°C. After the injection, the temperature decreases by 30–40°C, but the medium is highly corrosive. Therefore, the air cooler can only be used to cool the gas to 50°C, and the heat cannot be recovered. Taking a 1.8 Mt/a hydrocracking unit as an example, this part of heat is about 21.6 MW. The energy consumption is 0.21 kgoe/(t feed), accounting for about 0.7% of the total energy consumption of the unit.

③ If there is no safety valve on the top of the hot low–pressure separator, the hot low–pressure separator gas pipeline must consider the upstream hot high-pressure separator pressure breakthrough accident. The venting gas is cooled by the air cooler or water cooler and then vented through the cold low–pressure separator safety valve. Therefore the heat of the hot low–pressure separator gas cannot be used. Taking a 1.8 Mt/a hydrocracking unit as an example, this part of the heat is only about 0.83 MW.

④ Hydrocracking unit feed is generally the reduced pressure gas oil in the range of 350–540°C. The fraction is dirty and more likely to scale and block, which needs regular cleaning. Therefore, plate heat exchangers, grooved tubes, high–flux tubes, and other efficient heat exchangers that are not easy to clean cannot be used.

⑤ The catalyst reaction is divided into Start of Run (SOR) and End of Run (EOR), and the reaction temperature and reaction heat release are different during SOR and EOR. When SOR heat transfer optimization is met, it is difficult to meet the EOR heat transfer optimization, and vice versa. Therefore, it is necessary to find the most stringent operating conditions of SOR and EOR. However, the optimal condition under most stringent operating conditions is not necessarily optimal in all operating conditions.

The heat transfer process of the low-pressure section of the hydrocracking unit includes the preheating of the feed and the heat transfer of the whole fractionation section, as shown in Fig. 3.16. Fractionation section is to separate sour gas, liquefied gas, light naphtha, heavy naphtha, kerosene, diesel, and hydrotreating tail oil. There are many columns for a variety of products. Although the process is complex, the correlation between different cold and hot streams is less. Therefore, it is important to consider reasonable temperature matching, using logarithmic temperature driving force, reducing the heat transfer area and equipment investment in the optimization of heat exchanger network. In addition, increasing the utilization of low temperature heat and heat discharge measures to reduce the waste energy is very important for the heat transfer enhancement of the process and the optimization design of energy saving of the unit.

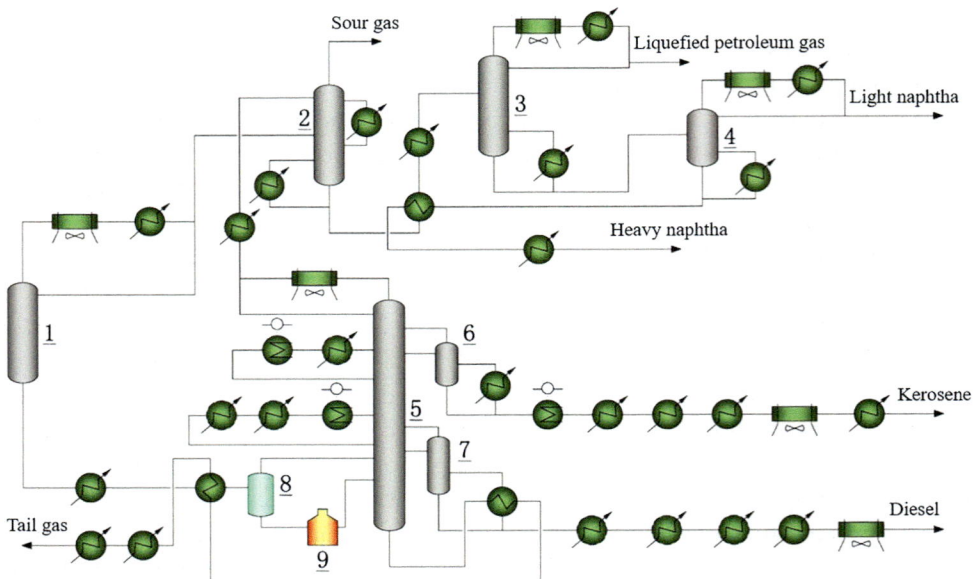

Fig. 3.16 Heat exchanger network of low pressure section. (1) Stripper; (2) absorption/desorption column; (3) stabilizer; (4) naphtha splitter; (5) product fractionator; (6) kerosene stripper; (7) diesel stripper; (8) flash drum; (9) furnace.

3.3.3 Process enhancement and integration on heat transfer

3.3.3.1 Optimization of heat transfer process by pinch point analysis technology

Pinch point analysis is one of the most successful integrated optimization methods for energy systems applied in engineering so far. According to the duty of hot and cold fluid, the cold and hot duty curves can be drawn, respectively. And then determine the minimum heat exchange temperature heat difference of heat exchanger network, by moving the cold and hot duty curves. When the curves reach the smallest vertical distance, the minimum heat exchange temperature difference can be found. This point is therefore the pinch point of heat exchanger network. Then the heat exchanger network can be optimized according to the three principles of the pinch point design.

The temperature-enthalpy diagram of the optimized high-pressure section heat exchanger network of a 1.8 Mt/a hydrocracking unit is shown in Fig. 3.17. The characterization of this heat exchanger network is one hot fluid and multiple cold fluids. Unlike the regular heat exchanger network of multiple hot fluids and multiple cold fluids, it does not have cross–pinch point heat transfer. The heat of reactor effluent can be fully utilized by matching with feed, recycle gas, stripper feed, etc. When the feed is heated to 348°C, the temperature difference of the furnace reaches the limit requirements. When the reactor effluent is cooled to 166°C, water injection begins. Both feed and reactor effluent meet the process requirements.

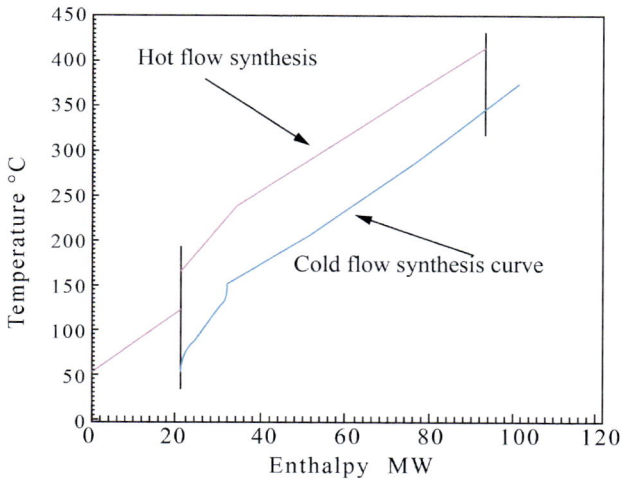

Fig. 3.17 The *T-H* diagram of high-pressure heat exchanger network.

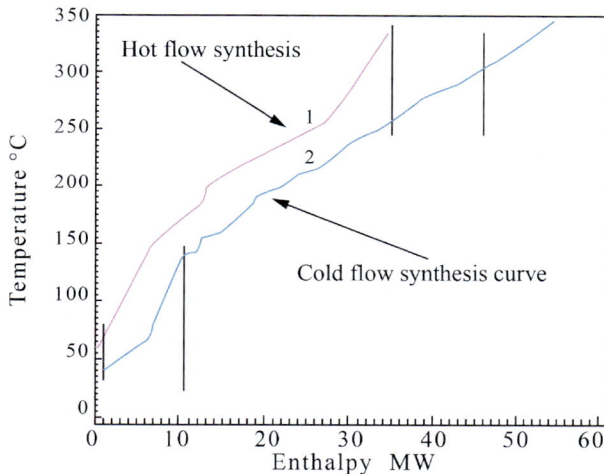

Fig. 3.18 The *T-H* diagram of low pressure heat exchanger network.

The temperature-enthalpy diagram of the low–pressure heat exchanger network is shown in Fig. 3.18. Although there are many products and complex processes in the low-pressure section, different hot and cold fluids have little correlation with each other. The stream that needs to be heated not only is at high temperature but also needs a large heat capacity flow rate. Therefore, a reasonable temperature and heat capacity flow rate match is very important. The feed temperature of the product fractionator needs to be higher than the heat source temperature of the fractionation section, so the fractionation feed furnace is essential. The full utilization of product waste heat is the key to energy saving in fractionation section. As can be seen from Fig. 3.18, the fractionation section of the hydrocracking unit makes full use of the waste heat of the products, which is used to generate steam and heat the demineralized water. Therefore, air coolers and water coolers are no longer used during normal operation before sending OSBL.

3.3.3.2 Heat transfer enhancement of the process optimization
In addition to optimizing the heat transfer process of the whole unit by pinch point analysis, the influence of different process schemes on heat transfer enhancement through process simulation and heat exchanger simulation software was also compared and analyzed. This will help select the process scheme which saves energy and has economic benefit.

1. Cold–side heat transfer enhancement

   There are two kinds of heating processes of feed and hydrogen: "mixing hydrogen with feed at inlet of heater" and "mixing feed with hydrogen at outlet of hydrogen heater". For "mixing hydrogen with feed at inlet of heater" process, feed and hydrogen are mixed and become a two-phase stream, then exchange heat with reactor effluent, and enter the reaction feed furnace. For "mixing feed with hydrogen at outlet of hydrogen heater", feed and hydrogen exchange heat with reactor effluent in single phase, respectively. Heated feed is then mixed with heated hydrogen at the outlet of the hydrogen furnace and enters the reactor furnace. The heat transfer network of these two methods is shown in Fig. 3.19. Taking the 1.8 Mt/a hydrocracking unit as an example, the two heat exchange methods are calculated with the conventional shell and tube heat exchangers as the benchmark under the condition of zero fouling coefficient and similar total pressure drop. The comparison of the calculation results is shown in Table 3.17. It can be seen from the table that mixing hydrogen with feed at inlet of heater scheme enhances the heat transfer coefficient of cold side and reduces the heat transfer area by 1.5 times. The heat exchange process is simple and the economic benefit is significantly improved. Therefore, mixing hydrogen with feed at inlet of heater scheme is preferred.

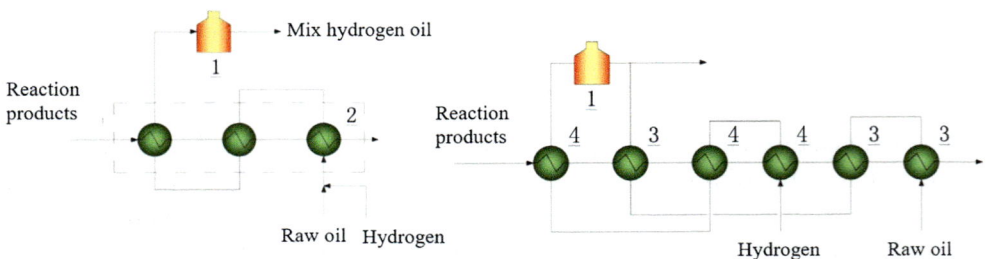

Fig. 3.19 Heat exchanger network of mixing hydrogen before heater and mixing feed after heater. (1) Reaction feed furnace; (2) reactor effluent/mixed feed heat exchanger; (3) reactor effluent/feed heat exchanger; (4) reaction product/hydrogen heat exchanger.

Table 3.17 Comparison of heat transfer between mixing hydrogen before heater and mixing feed after heater.

	Cold side	Average heat transfer coefficient at cold side	Heat transfer area
Mixing hydrogen with feed at inlet of heater	Mixed feed (two phases)	Standard	Standard
Mixing feed with hydrogen at outlet of hydrogen heater	Feed (single phase) Hydrogen (single phase)	0.2 × Standard 0.8 × Standard	2.5 × Standard

2. Hot-side heat transfer enhancement

There are two processes for cooling and separation of reactor effluent in hydro-cracking unit: "hot high-pressure separator" and "cold high-pressure separator" process. "Hot high-pressure separator" refers to the reactor effluent cooling to 240°C or higher temperature, which then enters the hot high-pressure separator for gas-liquid separation. The separated liquid is not cooled but flows to fractionation section after depressurization. The separated gas continues to be cooled to 50°C and then enters the cold high-pressure separator. "Cold high-pressure separator process" refers to the reactor effluent cooling to 50°C and then entering the cold high-pressure separator for gas-liquid separation. After the separated liquid is depressurized, it exchange heat with the reactor effluent, reaches 230°C and enters the fractionation section. The schematic diagram of the heat transfer network of these two methods is shown in Fig. 3.20. The advantage of hot high-pressure separator process is that part of the reactor effluent does not need to be cooled and then heated by itself, which reduces the energy loss caused by repetitive heating. Taking a 1.8 Mt/a hydrocracking unit as an example, under the condition of zero fouling resistance and similar total pressure drop, the conventional shell and tube heat exchanger is taken as the calculation benchmark. The calculation results of hot high-pressure separator and cold high-pressure separator are shown in Table 3.18. It can be seen that the hot high-pressure separator process enhances heat transfer of hot stream (reactor effluent). The total heat transfer area required is reduced by 40% compared with the cold high-pressure separator process, and the fuel energy consumption is saved by about 1.38 kgoe/(t feed).

Heat transfer process of thermal high separation scheme

Heat transfer process of cold high separation scheme

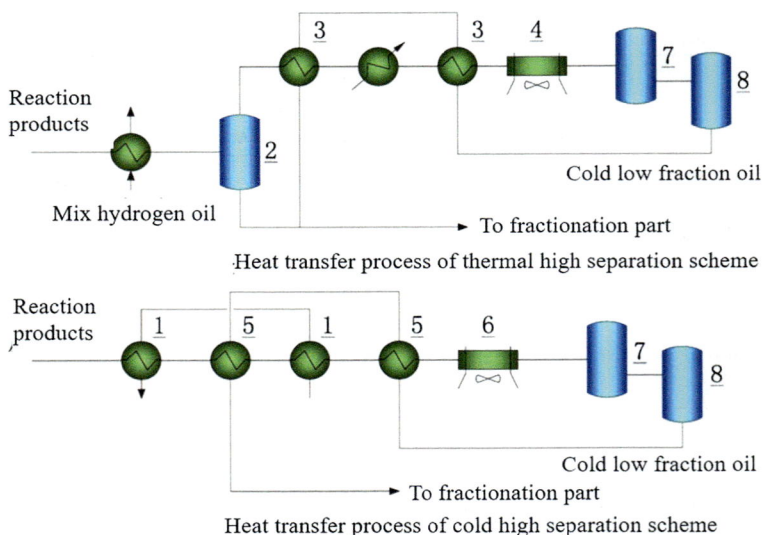

Fig. 3.20 Heat exchanger network of hot high and cold high-pressure separator schemes. (1) Reaction product/mix hydrogen oil heat exchanger; (2) hot high-pressure separator; (3) hot high-pressure separation gas/cold low fraction oil heat exchanger; (4) hot high-pressure separation gas air cooler; (5) reaction products/cold low fraction oil heat exchanger; (6) reaction product air cooler; (7) cold high-pressure separator; (8) cold low-pressure separator.

Table 3.18 Comparison of heat transfer.

	Furnace duty	Heat exchanger area	Air cooler area
Hot high-pressure separator process	Standard	Standard	Standard
Cold high-pressure separator process	1.4 × Standard	1.9 × Standard	1.4 × Standard

3. Other processes to enhance heat transfer

① Addition of a gas–liquid flash drum at the inlet of fractionator feed furnace. After preheating, fractionator feed usually enters the furnace as two-phase stream before entering fractionator. This process does not have flash drum at the inlet of furnace. The process of adding flash drum is the preheated fractionator feed entering the flash drum for gas–liquid separation. The liquid at the bottom of the drum is separated equally through control valves into multiseries and heated by fractionator feed furnace. The gas directly enters the fractionator. Taking a 1.8 Mt/a hydrocracking unit as an example, under the condition of similar product cutting accuracy such as the consistent temperature difference between 5% tail oil fraction and 95% diesel oil fraction, the simulation results show that the load of furnace with flash drum at the inlet of fractionator furnace is reduced by 8% compared with that without flash drum, saving fuel consumption. Gas and liquid load on tray are compared and are shown in Figs. 3.21 and 3.22, respectively. It can be seen that with a flash drum, the gas enters the upper section trays of fractionator. This can avoid part of light component backmixing with heavy component in distillation section, which would need further separation. Therefore, the operation duty of the trays decreases, the gas duty decreases by around 10%, and liquid duty decreases by 20%. In addition, the operational flexibility is better. Therefore, adding a flash drum at the inlet of fractionator feed furnace not only solves the problem of single–phase (liquid) multiseries distribution into fractionator feed furnace under low pressure through control valves but also enhances heat transfer (reducing energy consumption) and mass transfer (reducing tray operating duty).

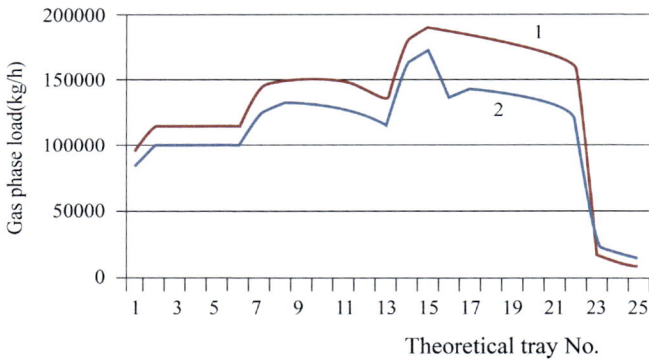

Fig. 3.21 Column tray gas phase load. (1) With flash drum for fractionator feed; (2) without flash drum for fractionator feed.

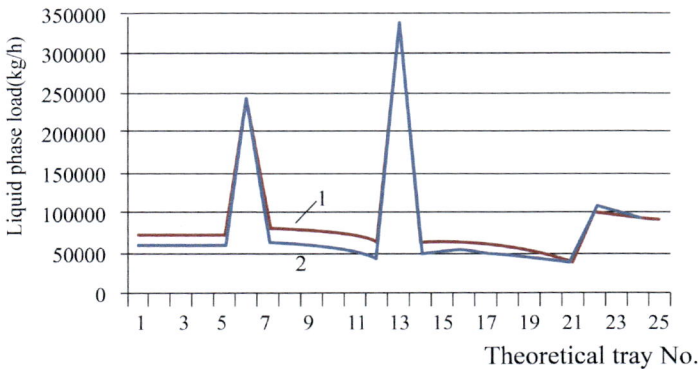

Fig. 3.22 Column tray liquid phase load. (1) With flash drum for fractionator feed; (2) without flash drum for fractionator feed.

② Low-temperature heat utilization of fractionator overhead (OH) gas. Take a 1.8 Mt/a hydrocracking unit as an example; the fractionator OH gas needs to be cooled from 150°C to 55°C. If there is no suitable cold source, the 22 MW heat duty of fractionator OH gas can only be completely cooled by air cooler. In the meantime, 75 t/h of cold demineralized water needs to be heated and reused in a thermal power plant. After adding such heat exchanger, the total heat exchange area is reduced by 5%, and the energy consumption of the unit is reduced by about 1.2 kgoe/(t feed). It can be seen that the utilization of low-temperature heat and even the joint utilization of low-temperature heat and heat discharge between different units can enhance the utilization and recovery of low-temperature potential heat and achieve the purpose of reducing the energy consumption of the unit.

③ Optimization of fractionator feed heat transfer enhancement application. Optimization of feed temperature and feed tray position has a great impact on the design of condenser and bottom reboiler. These directly affect the energy consumption, equipment investment, and operating costs of the unit. The feed optimization mainly adopts the sensitivity analysis method. Under the same condition of product yield and purity, the duty changes of the overhead condenser and OH reboiler are investigated by changing the feed temperature or the position of the feed tray. In this way, the maximum economic benefit can be found. For example, the light and heavy naphtha fractionator of a 1.8 Mt/a hydrocracking unit is equipped with an overhead condenser and a bottom reboiler. The main purpose is to separate the light and heavy naphtha fractions. Through sensitivity analysis method, under the same condition of product yield and purity, the feed tray position is changed and the duty change of OH condenser and bottom reboiler is shown in Fig. 3.23. The results show that when feed location changes from top to bottom, the duty of overhead condenser and bottom reboiler reduces hugely. When the feed position is at theoretical tray 9–12, the duty of OH condenser and bottom reboiler changes very little and is basically at the bottom of the curve. When the feed position continues to fall to the 13th theoretical tray, the duty of overhead condenser and bottom reboiler begins to increase again. Therefore, the feed position can be selected between 9th and 12th theoretical trays.

Take the light and heavy naphtha fractionator above as an example. Under the condition of same product yield and purity, the feed temperature is changed, and the stream is therefore changed from liquid phase supercooling → bubble point temperature → gas-liquid phase → dew point temperature → full gas phase superheating feed. The figure of the relationships between the duty change of OH condenser and bottom reboiler with feed temperature is shown in Fig. 3.24. The duty of OH condenser increases with the increase in feed temperature, while the duty of bottom reboiler decreases with the increase in feed temperature. When choosing appropriate feed temperature, the cold source of OH condenser and hot source of bottom reboiler should be analyzed. When using low-temperature heat as the heat source of bottom reboiler and cooling water as the cold source of overhead condenser, the feed temperature with a lower duty of condenser and higher duty of reboiler is preferred in order to decrease the amount of cooling water usage and lower the energy consumption of the unit. When using steam as the heat source of bottom reboiler and cooling water as the cold source of overhead condenser, the feed temperature with a higher duty of condenser and lower duty of reboiler is preferred due to the fact that the unit energy consumption of steam is much higher than that of cooling water. Therefore, the selection of the optimal feed temperature should not be judged simply by the duty of condenser or reboiler but from the perspective of maximizing economic benefits.

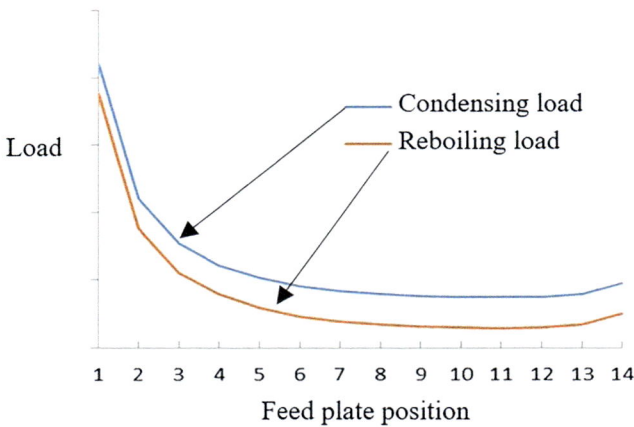

Fig. 3.23 Diagram of OH condenser and bottom reboiler duty change with various feed tray locations.

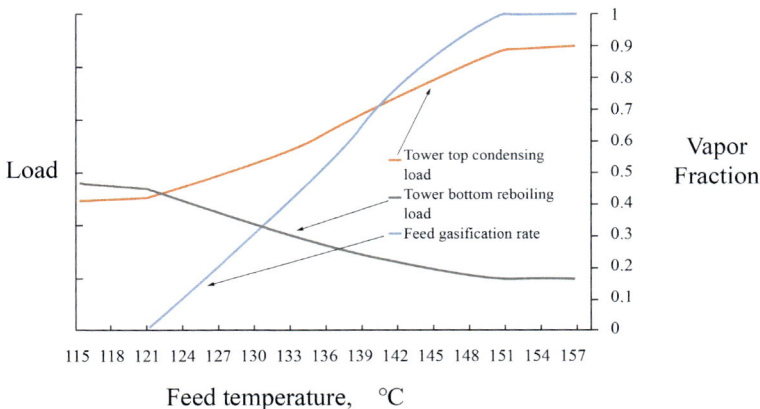

Fig. 3.24 Diagram of OH condenser and bottom reboiler duty change with various feed temperatures.

3.3.4 Heat transfer enhancement technology using elements

3.3.4.1 Enhancing of heat transfer component in reaction section

The reactor effluent and mixed feed heat exchanger is an important heat exchange equipment in the reaction section. The process conditions are shown in Table 3.19.

Table 3.19 Process condition of reactor effluent/mixed feed heat exchanger—Duty: 57.7 MW.

Item	Shell side	Tube side
Fluid name	Mixed feed	Reactor effluent
Flow rate (kg/h)	260,824	274,855
Allowable pressure drop (kPa)	200	200
Temperature (°C)	$150 \rightarrow 348$	$415 \rightarrow 240$
Gas density (kg/m^3)	$15.4 \rightarrow 12.4$	$32.5 \rightarrow 24.6$
Liquid density (kg/m^3)	$802.8 \rightarrow 666.5$	$658.4 \rightarrow 663.9$
Gas viscosity (mN s/m^2)	$0.013 \rightarrow 0.017$	$0.028 \rightarrow 0.024$
Liquid viscosity (mN s/m^2)	$1.694 \rightarrow 0.112$	$0.039 \rightarrow 0.167$
Gas phase thermal conductivity (W/(m K))	$0.210 \rightarrow 0.281$	$0.255 \rightarrow 0.219$
Liquid phase thermal conductivity (W/(m K))	$0.075 \rightarrow 0.056$	$0.044 \rightarrow 0.062$
Gas heat capacity (kJ/(kg K))	$9.04 \rightarrow 8.40$	$4.52 \rightarrow 5.39$
Liquid heat capacity (kJ/(kg K))	$2.42 \rightarrow 2.99$	$3.19 \rightarrow 2.76$

This heat exchanger is characterized by its high duty, cold and hot fluid temperature deep cross, high temperature, and high-pressure medium. A threaded locking ring or diaphragm sealed is usually used for these types of high-pressure heat exchangers. The F-shell can achieve pure reverse flow of the tube and shell streams, avoid temperature cross, and improve the effective average temperature differences. See Table 3.20 for the comparison scheme.

Table 3.20 Comparison of reactor effluent/feed heat exchanger scheme.

Item	Original option	Enhanced option
Equipment size (shell diameter × straight tube length) (mm)	DEU 1450 × 7500	DFU 1400 × 6500
Baffle type	Single segmental	Double segmental
Equipment (set)	3	3
Pressure drop (shell side) (kPa)	182.3	165.2
Overall heat transfer coefficient (W/(m^2 K))	498.6	520.4
Heat transfer area (m^2)	1901.6	1629.5
Total weight of equipment (t)	259.6	234.9

Three 1450 mm × 7500 mm (diameter × tube length) DEU (TEMA type) exchangers with single-segmental baffle will be needed for regular design. Compared with the DEU type exchanger, the effective average temperature difference of DFU type is increased from 67.0°C to 75.7°C. This will increase the shell flow rate, and the film heat transfer coefficient can be increased from 2467 to 2767 W/(m^2 K). The heat transfer area is reduced by 14.3% and equipment weight is reduced by nearly 10%.

The coil-wound heat exchanger has a compact structure, which can save space and investment. It has been applied in the reactor effluent/mixed feed heat exchanger, reactor effluent, and stripper feed heat exchanger of hydrocracking unit. Taking a 1.5 Mt/a hydrocracking unit as an example, compared with the ordinary heat exchanger, the steel weight can be saved by about 50%, and the designed total weight is reduced by about 130 t, in the meantime significantly reducing the investment [45,46].

3.3.4.2 Enhanced heat transfer component in the fractionation section

The duty of product fractionator OH air cooler is large, which means that using conventional air cooler will cover large space area. When the climate and utility conditions are right, surface evaporation air cooler can be considered [13]. The reboilers of kerosene and diesel stripper or naphtha splitter have lower boiling heat transfer coefficient, which means it is more suitable to use T-shaped grooved tube to improve the shell side boiling heat transfer coefficient and reduce the heat transfer area.

The tail oil at the bottom of the product fractionator is an important high-temperature heat source for the fractionation section. Before sending OSBL, it is used as the heat source for side stream stripper reboilers, fractionator feed heat exchanger, feed heat exchanger and demineralized water heat exchanger. Since the tail oil contains heavy components, the viscosity of the tail oil increases significantly with the decrease in temperature. When the bottom temperature is 350°C, the viscosity is less than 0.4 mN s/m^2. And the viscosity is greater than 7.5 mN s/m^2 when the outlet temperature of demineralized water heat exchanger is 90°C. Although the allowable pressure drop is increased by increasing the static head of the centrifugal pump, the heat transfer coefficient decreases significantly at low temperature and with high viscosity. Therefore, the heat exchanger at low temperature is suitable for using internal plug-in or spiral-twisted tube to enhance the heat transfer inside the tube. Feed oil before entering the feed filters is heated through the heat exchanger with product fractionator tail oil. Low-temperature feed has similar viscosity to tail oil. Low-temperature and high–viscosity streams are in both the shell and tube side of the heat exchanger, which means the film heat transfer coefficients are both low. Spiral-twisted tube is more suitable to enhance heat transfer on both sides and has been successfully applied in other units [47].

The research group led by Sinopec Engineering Incorporation studied the thermodynamic characteristics of the spiral twist tube, and the results are applied to the vacuum bottom oil and feed heat exchanger in a 250,000 t/a hydrocracking tail oil vacuum distillation unit. The process conditions are shown in Table 3.21.

Table 3.21 Process condition of vacuum bottom oil/feed heat exchanger—Duty: 1.34 MW.

Item	Shell side	Tube side
Fluid name	Feed	Vacuum bottom oil
Flow rate (kg/h)	26,400	20,000
Allowed pressure drop (kPa)	50	50
Temperature (°C)	177 → 240	286 → 204
Liquid density (kg/m^3)	765.4 → 730.0	716.0 → 761.4
Liquid viscosity (mN s/m^2)	3.425 → 1.326	0.920 → 2.804
Liquid thermal conductivity (W/(m K))	0.053 → 0.049	0.045 → 0.051
Liquid heat capacity (kJ/(kg K))	2.700 → 2.977	3.165 → 2.801

The characteristic of this heat exchanger is that the viscosity of the tube and shell side fluid increases with the decrease in temperature. And the film heat transfer coefficient on both sides is low. Spiral-twisted tube is suitable to enhance heat transfer. The design scheme of the heat exchanger with plain tube and spiral-twisted tube is shown in Table 3.22.

Table 3.22 Comparison of vacuum bottom oil/feed heat exchanger.

Item	Original option	Enhanced option
Equipment size (shell diameter × straight tube length) (mm)	BEU 500 × 6000	BFU 500 × 6000
Type of heat exchange tube	Plain tube	Spiral-twisted tube
Equipment (set)	4	3
Pressure drop (tube side) (kPa)	63.3	21.3
Overall heat transfer coefficient (W/(m^2 K))	175.1	238.6
Heat transfer area (m^2)	223.5	168.9
Total weight of equipment (t)	11.0	8.3

The use of spiral-twisted tube can save one set of heat exchanger. The overall heat transfer coefficient is increased by more than 30%. The heat transfer area and the total weight of the equipment are decreased by more than 20%. The pressure drop of the tube side is reduced by 66% compared with the plain tube. The calibration data of industrial applications shows that the heat transfer requirements of process flow can be completed by using the spiral-twisted tube heat exchanger, and the heat transfer enhancement effect is significant.

3.3.5 Summary

From the above examples of heat transfer enhancement optimization schemes, it can be seen that heat transfer enhancement technology in process plays an important role in reducing energy consumption. Through pinch point analysis, the hot and cold sources can be matched more reasonably and the heat loss can be reduced. Taking a 1.8 Mt/a high-pressure hydrocracking unit as an example, the film heat transfer coefficient on the cold side is enhanced by mixing hydrogen at the inlet of feed furnace. In this way, the total heat transfer area is only 40% of that of mixing feed at the outlet of hydrogen furnace, which saves tens of millions of dollars in equipment investment. The heat transfer on the hot side is enhanced by using the hot high-pressure separator process compared with the cold high-pressure separator process. The total heat transfer area is reduced by 40%, and the fuel consumption of the furnace is saved by 1.38 kgoe/(t feed). Adding flash drum at the inlet of fractionation furnace not only enhances heat transfer (saving 8% fuel) but also enhances mass transfer. The utilization of low-temperature heat reduces the duty of air cooler and the consumption of cooling water, which is of great significance to reduce the energy consumption of the unit. The optimization of fractionator feed temperature and feed tray location plays an important role in reducing the duty of overhead condenser and bottom reboiler, which can maximize the economic benefit of the fractionator. With the development of science and technology, heat transfer enhancement technology from engineering prospects will constantly be bringing forth new, advanced heat transfer enhancement technology in the future.

Most of the streams in reaction section have the characteristics of high temperature, high pressure, and containing hydrogen. The requirements for the selection of heat transfer equipment, material, and manufacturing are high. The application of heat transfer component enhancement technology is a very obvious advantage for high-pressure equipment cost. The heat transfer enhancement technology should make sure the whole system is sealed. The coil-wound heat exchanger can stand high pressure and save space, which has been used in industry. The ammonium salt crystallization of the heat exchanger for hot high-pressure separator gas should be considered. The temperature and pressure in fractionation section are lower than in reaction section; the duty is mainly on overhead condenser, product waste heat recycle, and cooling. The heat transfer enhancement of fractionation section requires special consideration of the viscosity increase of fractionator bottom oil and feed at low temperature.

3.4 Hydrotreating unit

3.4.1 Brief description of process

Hydrotreating catalytic hydrogenation reaction in hydrogen atmosphere, the conversion of cracked feed is less than 10%. The main purpose is to remove most of the impurities such as sulfur or nitrogen from the feed, to improve the quality of the product oil.

Especially for secondary processing products such as coking gasoline and diesel, unstable impurities like olefins and aromatics can be saturated by hydrotreating. The color, storage safety, and combustibility of secondary processing products can be improved. Therefore high-quality products with good stability and high quality can be obtained.

Hydrotreating process, according to the feed, includes liquefied gas hydrotreating, naphtha hydrotreating, kerosene hydrotreating, diesel hydrotreating, gas oil hydrotreating, lubrication base oil hydrofinishing, special oil deep hydrotreating, to remove aromatics, reformate selective hydrogenation to remove olefins, etc. The diesel fixed-bed hydrotreating unit is widely used in refineries around the world, which is an important process for producing clean diesel products. The typical process flow of diesel hydrotreating is shown in Fig. 3.25.

The diesel hydrotreating process can be generally divided into reaction section and product fractionation section. The reaction section includes feed pressure boost, feed heating, reaction feed furnace, reactors, reactor effluent cooling, reactor effluent separation, recycle gas compressor system, make-up hydrogen compressor system, etc. In fractionation section, a two-column process includes stripper and product fractionator to ensure the water content of refined diesel products. Nowadays, the product fractionator usually uses steam stripping furnace as reboiler instead of using feed furnace and steam stripping.

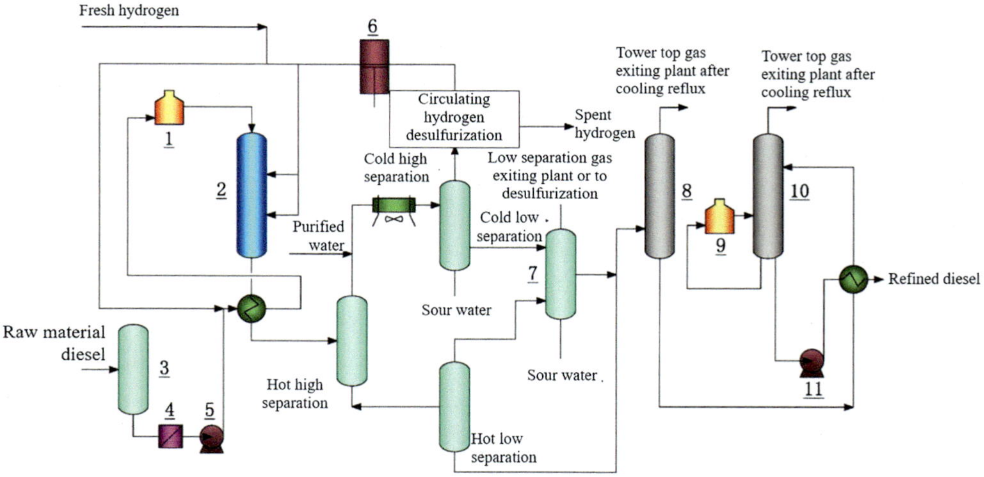

Fig. 3.25 Typical process flow diagram of diesel hydrotreating. (1) Furnace; (2) reactor; (3) feed surge drum; (4) filter; (5) feed pump; (6) recycle gas compressor; (7) cold low-pressure separator; (8) stripper; (9) furnace; (10) product fractionator; (11) refined diesel pump.

3.4.2 Analysis of heat energy characteristics

Hydrotreating reactions are exothermic reactions. With the increase of sulfur and other impurities in the feed and the proportion of secondary processing oil, especially the increase of hydrocracking degree, the heat release of hydrotreating reactions gradually increases. Although the heat release of diesel hydrotreating unit is less than that of hydro-cracking and residue hydrotreating units, it still needs to be effectively utilized. The heat input for the feed in the hydrotreating unit mainly comes from the conversion of the heating equipment ISBL or the heat input from OSBL, such as the reaction feed furnace, fractionator feed furnace, reboiler furnace, steam, etc. The excess heat energy in the hydrotreating unit is firstly recovered by steam and low-temperature thermal power generation. However, the heat is difficult to recover completely and will be discarded by air cooling and water cooling. Taking a 2.4 Mt/a diesel hydrotreating unit as an example, the application of heat energy is briefly analyzed as follows.

If the heat input process for streams only comes from heating equipment ISBL and the heat is only removed by cooling equipment AISBL, it means there will be no heat exchange inside the unit. That is, feed can only be heated by furnace to reach the required temperature, and the separation accuracy of hydrotreated products can only be done by fractionator feed furnace or reboiler furnace. The heat energy flow of the whole process is shown in Fig. 3.26.

It can be seen from Fig. 3.26 that the heat required by hydrotreating reaction can be provided by feed furnace, fractionator feed furnace, and reboiler furnace, which can maintain the separation accuracy of the products. The total heat requirement is about 112.9 MW. In addition, the hydrotreating reaction can provide 14.7 MW reaction heat, so the total heat input of the unit is 127.6 MW. In order to meet the process requirements of the stream temperatures, heat removal should be considered at the key point of the process. At the same time, the temperature of hydrotreated products needs to be cooled lower than the flash point before it can be sent to the storage tank. Therefore, the heat needs to be removed from the unit. The total heat that needs to be converted and removed in the unit is 127.6 MW.

It can be seen from Fig. 3.26 and Table 3.23 that both the heating energy and cooling energy of the diesel hydrotreating unit are large. Therefore, it is necessary to rationally utilize the heat energy of the unit and achieve the purpose of energy saving through the integration and enhancement of the heat exchange process. Considering the characteristics of hydrorefining process, the enhancement and integration method of process should follow the following principles:

① From the perspective of process and engineering, the upstream and downstream heat exchange equipments are interrelated and should be considered a whole unit. Each exchanger cannot be analyzed individually. And the effect of each exchanger is to serve the whole unit. For example, steam generator heat exchange area is easily affected by logarithmic mean temperature difference between hot and cold streams. For hot stream, even 1.0°C difference at the outlet will significantly affect the steam generator heat exchange area. But such difference only has a minor effect on process design and energy index for the whole unit. Equipment manufacturing, investment, floor area, and other factors also need to be considered instead of focusing on the amount of generated steam.

② The gradient utilization of medium temperature in the unit should be fully considered.

③ The matching of the medium operating pressure in the unit should be fully considered. The heat exchange between high- and low-pressure streams should be avoided as much as possible for the essential safety concept of equipment manufacturing cost and engineering design. This is to avoid the risk of high-pressure breakthrough caused by the rupture of heat exchange tube.

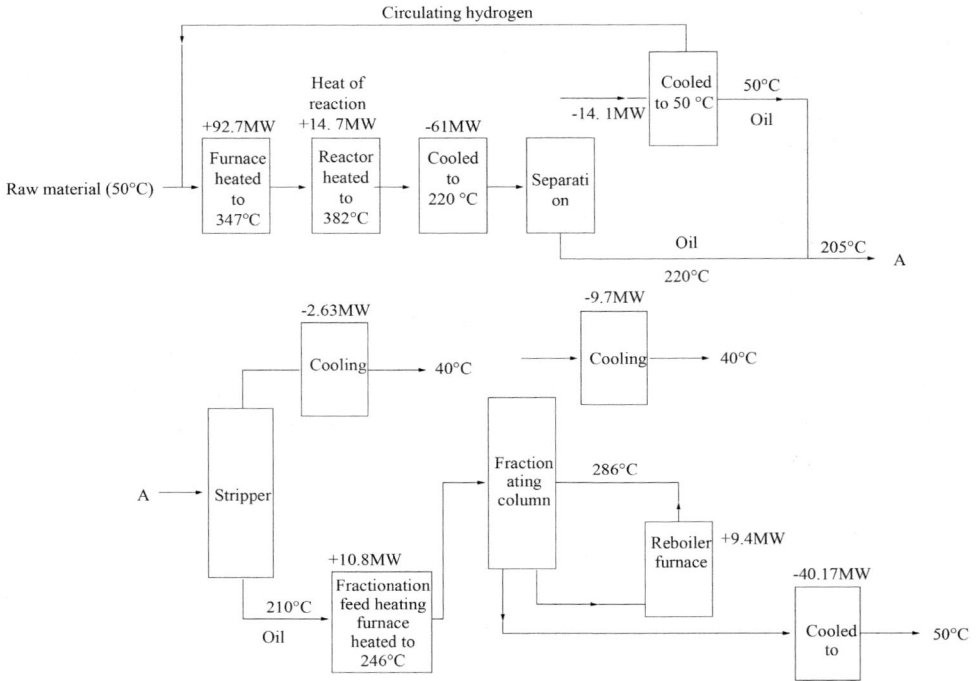

Fig. 3.26 Energy flow diagram of the unit.

Table 3.23 Heat energy flow statistics.

Source	Corresponding equipment	Heat (MW)
External heat	Reaction furnace, fractionation feed furnace, reboiling furnace, stripping steam	112.9
Reaction heat	Reactor	14.7
Required cooling duty	Heat exchange equipment	−127.6

3.4.3 Process enhancement and integration on heat transfer

3.4.3.1 Heat transfer enhancement of process optimization

E-1, E-2, E-3, E-4, and other heat exchange equipments are used to enhance and integrate the heat energy process of the unit, as shown in Fig. 3.27. In the figure, "+" represents the heat energy needed from external supply, and "−" represents the heat energy recovered and discarded by the unit. Duty without "+" or "−" represents the heat exchange between cold and hot streams.

E-1 and E-2 are optimized according to the operating pressure matching of cold and hot streams. The medium pressure difference between the tube and shell side is less than 2.0 MPa. The products at the bottom of the fractionator conform to the principle of gradient utilization of temperature between medium. The high-temperature stream is used to exchange heat with the fractionator feed first, such as E-3, then generate steam through E-4, and exchange heat with 50°C cold feed from the storage tank.

After the enhancement of the heat exchange process, the external heat required for the unit is about 18.55 MW. The heat taken from the unit is about 28.81 MW. The heat released from the reaction is 14.7 MW, and the heat recovered is about 4.44 MW. The statistical data is shown in Table 3.24.

The comparison of Figs. 3.26 and 3.27 is shown in Fig. 3.28. After enhancing the process, E-3 can replace the fractionator feed furnace. The duty of reaction feed furnace is greatly reduced, and the reaction heat has been effectively utilized. In addition, the duty of cooler J and cooler H are greatly reduced. It can also be seen that the duty column overhead condenser (E, F) is large (10.8 MW), and the duty of product cooling (H) is also large. These are low-temperature heat, which needs to be utilized through the unit heat combination or the whole unit low-temperature heat utilization (such as heating demineralized water) to further recover low–temperature heat.

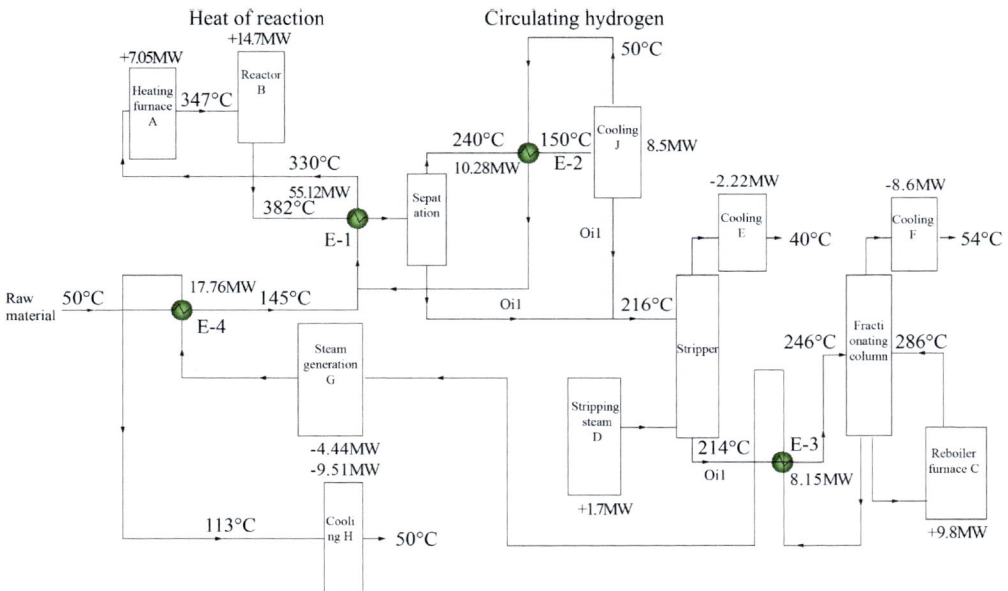

Fig. 3.27 Diagram of enhanced heat (cold) energy in the heat exchange flow.

Table 3.24 Statistical of enhanced heat (cold) energy flow during heat exchange of the unit.

Source	Corresponding equipment	Heat (MW)
External heat	Reaction furnace, reboiling furnace, stripping steam	18.55
Heat of reaction	Reactor	14.7
Heat removal	Water cooler, air cooler	−28.81
Recycling	Steam generator	−4.44

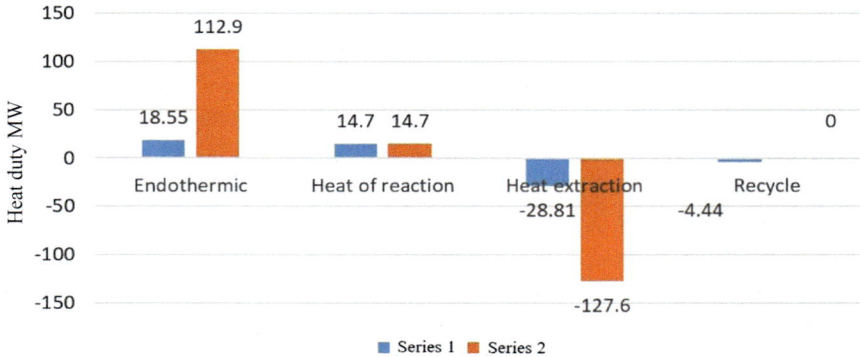

Fig. 3.28 Comparison of enhancement and integration effect of heat transfer.

3.4.3.2 Enhancing process by making full use of reaction heat

The hydrogenation reaction process is complex, and the reaction is an exothermic reaction. Therefore, making full use of reaction heat of hydrotreating unit is crucial to energy saving. For example, the heat transfer process optimization of a 260 Mt/a diesel hydrodesulfurization unit [48] is shown in Fig. 3.29, according to the characteristics of exothermic reaction and reboiler furnace. Reactor effluent exchange heat with feed as much as possible and use efficient heat exchangers in the reaction section having only 5°C temperature differences. Through the above adjustments, in normal operation, the reaction feed furnace can be shut down. The fuel gas consumption of the unit can be reduced by 930 kg/h, and the energy consumption can be reduced by 1.54 kgoe/(t feed).

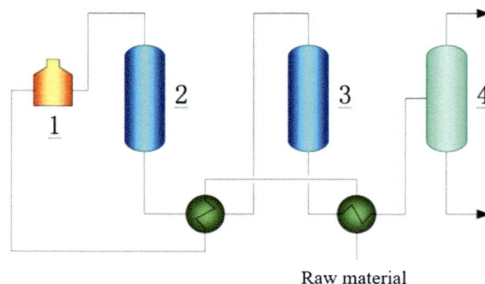

Fig. 3.29 Heat transfer process diagram of reaction heat fully used in diesel hydrodesulfurization unit. (1) Furnace; (2) reactor 1#; (3) reactor 2#; (4) hot high-pressure separator.

3.4.4 Heat transfer enhancement technology using elements

From light to heavy, hydrotreating unit feed can be divided into gasoline, kerosene, diesel, gas oil, and residue. The properties of these feed gradually deteriorate, the heat transfer coefficient gradually decreases, and the scaling trend increases. The need for mechanical cleaning may become shorter. Although ordinary shell and tube heat exchangers have strong adaptability and are widely used, the study of heat transfer enhancement technology will help to improve the efficiency of energy use, which is of great significance for reducing energy consumption. According to the characteristics of streams, heat transfer component enhancement can focus on tube inside, tube outside, or both sides. This method can increase theoretic heat transfer area per unit volume and increase the degree of turbulence in the single-phase flow, as well as increase the boiling or condensation of film heat transfer coefficient and reduce the fouling thermal resistance. Therefore, the equipment size can be reduced in practice and decrease heat exchange temperature difference, to achieve the purpose of low-carbon, energy saving, and environmental protection.

3.4.4.1 Application of coil-wound heat exchanger in hydrotreating unit

Spiral sounded heat exchanger is a compact, high-efficiency, and energy-saving heat exchange equipment. It is a recuperative tube and shell heat exchanger. The advantages of such heat exchangers are that they have high-pressure resistance like tube and shell heat exchangers, compact structure, and high heat transfer efficiency. It can be considered the kind of heat exchanger that can increase heat transfer area per unit volume and enhance tube outside turbulence and is a comprehensive enhanced heat transfer equipment with reverse flow. In recent years, the application of coil-wound heat exchangers has become more and more widespread [49]. For reactor effluent/mixed feed heat exchangers, coil-wound heat exchangers are used to replace the traditional threaded lock ring tube and shell heat exchangers in many hydrotreating units in China which has a long life cycle.

Reactor effluent/mixed feed heat exchanger in diesel hydro-treating unit is the key equipment in the reaction section. After the cold feed is heated by reactor effluent heat exchanger, the stream is further heated to the required tem-perature by the furnace and enters the reactor. Take a 2.0 Mt/a diesel hydrotreating unit as example; the hot and cold fluids of heat exchanger are shown in Table 3.25. As the hot and cold fluids are two-phase flows, although the film heat transfer coefficients of the conventional shell and tube heat exchanger on both sides have been increased (average 2000 W/$(m^2$ K)) by the process enhancing technology, in order to continue improving its energy efficiency, coil-wound heat exchanger is used to enhance the heat transfer component. On the premise of meeting the allowable pressure drop, the calculation and comparison of before and after heat transfer enhancement is shown in Table 3.26.

As can be seen from Table 3.26, the heat transfer design of the reactor effluent/mixed feed heat exchanger is enhanced by the use of coil-wound heat exchanger. The number

Table 3.25 Process condition of reactor effluent/mixed feed heat exchanger—Duty: 52 MW.

Item	Cold fluid	Hot fluid
Fluid name	Mix hydrogen oil	Reaction products
Flow rate (kg/h)	264,058	269,087
Allowable pressure drop (kPa)	150	110
Temperature (°C)	$136 \rightarrow 339$	$396 \rightarrow 220$
Gas density (kg/m^3)	$12.9 \rightarrow 38$	$50 \rightarrow 18$
Liquid density (kg/m^3)	$707 \rightarrow 587$	$- \rightarrow 625.8$
Gas viscosity (mN s/m^2)	$1.3 \rightarrow 1.7$	$1.8 \rightarrow 1.6$
Liquid viscosity (mN s/m^2)	$0.17 \rightarrow 0.03$	$- \rightarrow 0.08$
Gas thermal conductivity (W/(m K))	$0.19 \rightarrow 0.16$	$0.13 \rightarrow 0.18$
Liquid thermal conductivity (W/(m K))	$0.10 \rightarrow 0.05$	$- \rightarrow 0.08$
Gas heat capacity (kJ/(kg K))	$6.4 \rightarrow 3.6$	$3.3 \rightarrow 4.3$
Liquid heat capacity (kJ/(kg K))	$2.4 \rightarrow 3.0$	$- \rightarrow 2.7$

Table 3.26 Comparison of reactor effluent/mixed feed heat exchanger scheme.

Item	Original option	Enhanced option
Equipment size (shell diameter × straight tube length) (mm)	1500 × 6000	1700 × 11,880
Equipment (set)	3	1
Overall heat transfer coefficient (W/(m^2 K))	530	–
Heat transfer area (m^2)	1871	–
Total weight of equipment (t)	238	114.4
Equipment investment, 10,000 yuan	Benchmark	0.78 × benchmark

of equipment is reduced from three to only one set, the weight of equipment is reduced by more than 50%, and the investment saved is more than 20%. In addition, as high pressure heat exchangers should not be arranged overlapping in series, the three shell and tube heat exchangers can only be juxtaposed. The use of coil-wound heat exchangers can save more than 40% of the floor area.

3.4.4.2 Application of shell and plate heat exchanger in hydrotreating unit

Shell and plate heat exchanger is similar to plate heat exchanger and shell heat exchanger. Similar to plate heat exchanger, shell and plate heat exchanger has high heat transfer efficiency and compact structure and its shell can withstand high temperature and high pressure. Shell and plate heat exchangers are widely used in oil refining and petrochemical industry around the world [50]. In particular, it solves the problems of large floor area, large weight, and high energy consumption of conventional heat exchanger when the units become larger.

1. Research and application of diesel hydrotreating unit

The use of shell and plate heat exchangers in hydrotreating units has been studied abroad since the mid-1990s. According to the research [51], for a 4.5 Mt/a diesel hydrotreating unit, the use of shell and plate heat exchangers instead of shell and tube heat exchangers can save US$5–8 million per year in operating costs. In 1996, Taiwan FPC refinery planned to build two sets of 3.0 Mt/a diesel hydrotreating units. The licensor compared the economic benefits of replacing traditional shell and tube heat exchangers with shell and plate heat exchangers for the effluent/mixed feed heat exchanger and the product/stripper feed heat exchanger in cold high-pressure separator process as well as the effluent/mixed feed heat exchanger and the product/feed heat exchanger in hot high-pressure separator process, respectively. The positions of the two heat exchangers (equipment 1 and 5) are shown in Fig. 3.30 (cold high-pressure separation process) and Fig. 3.31 (hot high-pressure separation process). The calculation and comparison of the cold high-pressure separator process are shown in Table 3.27.

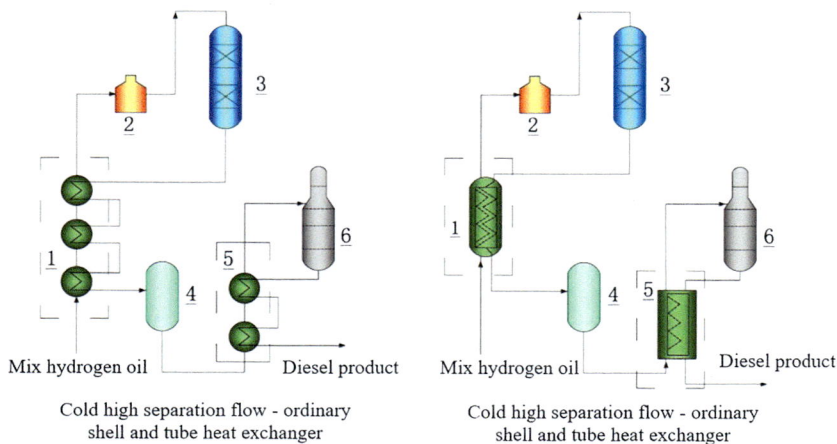

Fig. 3.30 Equipment location diagram of cold high-pressure separation process. (1) Reactor effluent/mixed feed heat exchanger; (2) reaction feed furnace; (3) hydrotreating reactor; (4) cold high-pressure separator; (5) diesel/stripper feed heat exchanger; (6) product fractionator.

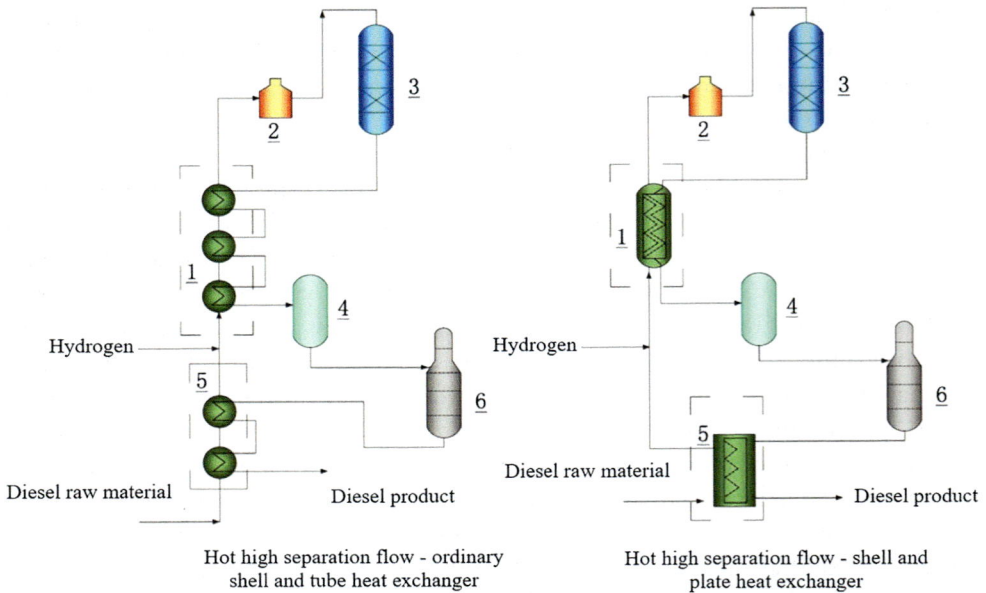

Hot high separation flow - ordinary shell and tube heat exchanger

Hot high separation flow - shell and plate heat exchanger

Fig. 3.31 Equipment location diagram of hot high-pressure separation process. (1) Reactor effluent/mixed feed heat exchanger; (2) reaction feed furnace; (3) hydrotreating reactor; (4) hot high-pressure separator; (5) diesel/feed heat exchanger; (6) product fractionator.

Table 3.27 Comparison of shell and tube heat exchangers with shell and plate heat exchangers scheme.

Item	Original option	Enhanced option
Equipment (set)	8	2
Equipment 1 heat flow heat exchange temperature difference (°C)	58	23
Equipment 5 heat flow heat exchange temperature difference (°C)	53	15
Reaction system pressure drop (MPa)	1.9	1.1
Furnace duty (MW)	23	5
Product cooling duty (MW)	8	4
Floor area (m^2)	920	820
Equipment investment savings, $10,000	–	800
Saved operating costs, $10,000/a	–	480

From Table 3.27, it can be seen that the application of shell and plate-type high-efficiency heat exchanger in the cold high-pressure separator process has great advantages. The number of equipment is reduced, the floor area is less, and the construction investment is reduced. In addition, lower heat exchange temperature difference can be achieved, so the fuel consumption is reduced by more than 80%, and the product cooling energy consumption is reduced by more than 50%. The pressure drop of the reaction system is reduced by 40%, and the energy consumption of the compressor is reduced. The operating costs can save $4.8 million per year. Similarly, through the application of shell and plate high-efficiency heat exchanger instead of the conventional shell and tube heat exchanger in the hot high-pressure separator process, the licensor found that the pressure drop of the reaction system can be reduced by 30%, the temperature difference of heat exchange is reduced, the furnace can be shut down, the cooling duty of the product can be reduced by 60%, and the annual operating cost can be saved by US$3 million. After years of operation practice proving, with the special structural characteristics of shell and plate heat exchanger, the scaling time is longer, the blockage rate of ammonium salt is low, the corrosion rate is low, and it is suitable for long life cycle operation. It can be seen that the application of high-efficiency shell and plate heat exchanger in hydrotreating unit brings great economic benefits.

2. Optimization design of jet fuel hydrotreating unit

   The heat exchangers of reactor effluent and mixed feed in traditional jet fuel hydrotreating units in China usually use horizontal layout shell and tube heat exchangers. Take a 0.6 Mt/a jet fuel hydrotreating unit as an example, the process conditions of hot and cold fluid are shown in Table 3.28. This exchanger is similar to diesel hydrotreating unit as described before. In order to further effectively use the reaction heat and reduce the consumption of the furnace, the process is enhanced to use shell and plate high-efficiency heat exchanger. High-efficiency heat exchanger and conventional heat exchanger are calculated. The technical parameters and economic benefits [52] are shown in Table 3.29.

Table 3.28 Process condition of reactor effluent/mixed feed heat exchanger—Duty: 10.9 MW.

Item	Cold fluid	Heat flow
Fluid name	Mixed feed	Reactor effluent
Flow rate (kg/h)	72,641	72,641
Allowable pressure drop (kPa)	100	220
Temperature (°C)	$80 \rightarrow 270$	$282 \rightarrow 115$

Table 3.29 Comparison of reactor effluent/mixed feed heat exchanger scheme.

Item	Shell and tube heat exchanger	Shell and plate heat exchanger
Equipment size (shell diameter × straight tube length) (mm)	800 × 6000 (T/T)	1100 × 10,000 (T/T)
Equipment (set)	4	1
Pressure drop (kPa) (tube/shell)	219/99	49/40
Overall heat transfer coefficient $(W/(m^2 K))$	384	982

Continued

Table 3.29 Comparison of reactor effluent/mixed feed heat exchanger scheme—cont'd

Item	Shell and tube heat exchanger	Shell and plate heat exchanger
Heat transfer area (m²)	1020	600
Total weight of equipment (t)	40	24.9
Equipment investment (Yuan)	2,200,000	2,100,000
Floor area (m²)	54	25
Achievable heat duty (MW)	10.9	11.4
Achievable heat exchange temperature difference (°C)	27.8	19.3
Saved fuel cost (Yuan/a)	–	977500
Saved electricity (Yuan/a)	–	84000

According to the comparison of economic benefits between shell and plate heat exchangers and shell and tube heat exchangers in Table 3.29, the advantages of shell and plate heat exchangers are significant. The heat transfer area is reduced by 41%, and the heat transfer coefficient is increased by 156%. In addition, the temperature difference at the hot end is reduced by 6°C and the temperature difference at the cold end is reduced by 7°C. The heat recovery is therefore increased by 4.6%, and the system pressure drop is reduced by 2.29 MPa. The equipment weight is reduced by 38%, the floor area is saved by 54%, and the equipment investment is saved by 4.5%. According to the annual operation time of 8400 h, low calorific value of 41,680 kJ/kg for fuel gas, and furnace efficiency of 92%, fuel gas can be reduced by 391 t/a. The fuel cost can be reduced by 977,500 yuan/a at the price of 2500 yuan/t, which means a total saving of 19.55 million yuan for 20 years. The economic benefits are very considerable. In addition, as the pressure drop of shell and plate heat exchanger is smaller than tube and shell heat exchanger, the compressor shaft power is reduced by 20 kW. The electricity price electricity cost is therefore reduced by 84,000 yuan/a at the price of 0.5 yuan/kWh.

3.4.4.3 Other applicable enhancement components

Apart from the heat exchanger mentioned above, other main heat exchange equipment of hydrotreating unit includes bottom reboiler, product steam generator, product cooler, OH air cooler, etc. The reboiler is generally used in the kerosene and diesel hydrotreating units with small processing capacity. In current kerosene and diesel hydrotreating units with large processing capacity, the reboiler furnace is used instead of the reboiler.

For the bottom reboiler, the horizontal thermosiphon reboiler is used. The hot stream flows through the tube side and the cold stream flows through the shell side, which reaches boiling state outside the tube. The shell side stream can be gasoline, kerosene, or diesel, and the film heat transfer coefficient is 1500 W/(m² K). Due to the large boiling range difference, the superheating temperature is high, so it is difficult to strictly control near the nucleated boiling area. In order to increase the shell side film heat transfer coefficient, it is necessary to reduce the thermal resistance of the gas film formed outside the tube. T–shaped grooved tube and other enhancing components can be selected to solve these problems.

When the product heat excess, product steam generator is usually used to recover heat. The steam side film heat transfer coefficient of the outer tube is high ($>3500 \, W/(m^2 \, K)$). Therefore, enhanced components can be used in the hot side tube to increase the turbulence of single–phase flow and reduce the negative effect of boundary layer, such as using internal plug-ins to enhance heat transfer technology.

For product cooling equipment, apart from the plate heat exchanger mentioned above, shell and tube heat exchangers with helical baffle can be used to enhance the shell side fluid heat transfer performance. Spiral-twisted tube can also be used for double-side heat transfer enhancement. These heat transfer enhancement equipments in the petro-chemical industry have been applied in recent years.

Air cooler is usually used for column OH gas condensing. The film heat transfer coefficient of the inner tube is generally low ($<2000 \, W/(m^2 \, K)$), but due to the limitation of the column OH pressure drop, it is not recommended to use the heat transfer enhancement by increasing the pressure drop inside the tube. A new type of heat transfer enhancement finned tube such as elliptic finned tube can be used, according to the research [53]. The heat transfer performance of elliptic finned tube is 11% higher than that of conventional circular finned tube. The enhancement technology can also be applied to the product air cooler before the product is sent OSBL.

3.4.5 Summary

Hydrotreating is an exothermic reaction. Nowadays, feed with more sulfur and other impurities and secondary processing oil ratio is increasing. Especially when the feed is more and more cracked, the heat released from hydrotreating reaction will gradually increase. The large amount of heat should be effectively used. It is an important approach to reduce unit energy consumption and is an important method of heat transfer process intensification and integration. In the process of heat transfer enhancement and integration, it is also necessary to consider the technical characteristics of the process and the constraints of intrinsically safe. The principle of gradient utilization and pressure matching should be followed. Low-temperature heat utilization is the potential direction of enhancing heat transfer process in hydrotreating units.

The comparison of optimization schemes and the application of heat transfer enhancement technology using elements show that a combination of process enhancement with enhancement using elements can further reduce the energy consumption, floor area, investment and operation cost of process units. It will have significant economic and social benefits. With the upsizing of the unit and the improvement of social environmental protection requirements, there are more and more demands for the upsizing and high reliability of the efficient heat transfer components, which requires more research and development.

3.5 Heavy oil hydrotreating unit

3.5.1 Brief introduction of process

With the world's crude oil becoming heavier and worse, there comes more and more heavy oil feed. In order to meet the increasing demand for light oil and middle distillate in the market, refineries need to convert heavy oil into light oil as much as possible. Heavy oil hydrotreating technology has high liquid yield without producing low-value by-products and is more environmentally friendly. Therefore, heavy oil hydrotreating units are playing an increasingly important role in refineries.

Heavy oil in heavy oil hydrotreating process usually refers to atmospheric residual oil (AR) and vacuum residual oil (VR). In order to reduce the difficulty of reaction or adjust the material balance of the whole refinery, sometimes straight run gas oil, coking gas oil, catalytic cracking circulating oil, and catalytic diesel are added as dilution. Nowadays, the most common technology for heavy oil hydrotreating is the fixed-bed process, which consists of two sections: reaction and fractionation section. The reaction section includes the feed oil preheating and filtration system, feed pressure increasing system, reactor feed heat exchange and heating system, reactor system, reactor effluent separation system, water injection system, recycle gas desulfurization system, recycle gas compressor system, and make-up hydrogen compressor system. The fractionation section is mainly divided into stripper system and fractionator system. The schematic flow diagram is shown in Fig. 3.32.

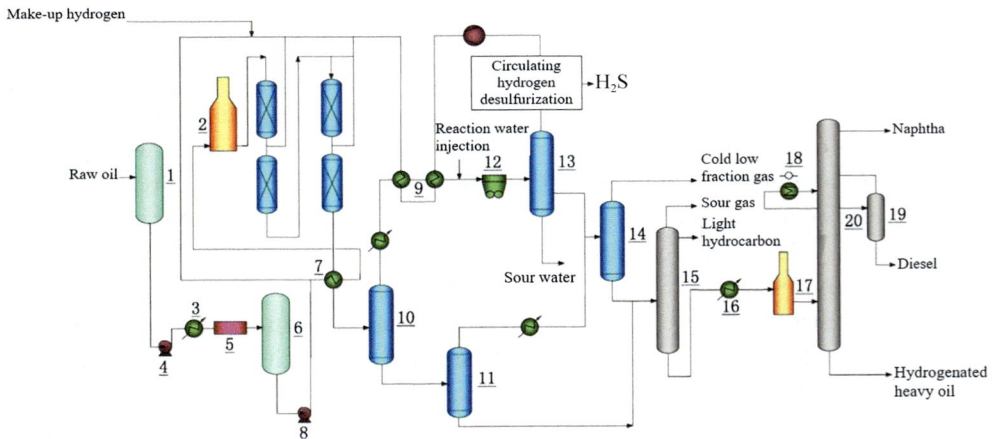

Fig. 3.32 Flow diagram of typical fixed-bed heavy oil hydrotreating unit in China. (1) Feed surge drum; (2) reaction feed furnace; (3) feed/hydrotreated heavy oil heat exchanger; (4) feed booster; (5) automatic backwash filter; (6) filtered feed surge drum; (7) reactor effluent/mixed feed heat exchanger; (8) reaction feed pump; (9) hot high-pressure separator gas/recycle gas heat exchanger; (10) hot high-pressure separator; (11) hot low-pressure separator; (12) hot high-pressure separator gas air cooler; (13) cold high-pressure separator; (14) cold low-pressure separator; (15) stripper; (16) fractionator feed/hydrotreated heavy oil heat exchanger; (17) fractionator feed furnace; (18) pump around steam generator; (19) diesel stripper; (20) fractionator.

3.5.2 Analysis of heat energy characteristics

The heat in the heavy oil hydrotreating unit comes from the following ways. ① Reaction heat; ② heat provided by the furnace; ③ the hot feed from OSBL and the hot product going OSBL; and ④ condensing heat provided by steam and sensible heat provided by temperature difference when steam heater is available. The above points ① and ② are the main heat sources of the unit.

3.5.2.1 Heat released from reaction

In the process of heavy oil hydrotreating, the main reactions are hydrogenation of desulfurization, denitrification, demetalization, removal of residual carbon, and hydrocracking. Hydrodesulfurization is an exothermic reaction, and the total reaction heat is about $2365\,kJ/m^3$ hydrogen (standard state, same for later figures unless otherwise specified). Due to the high sulfur content in heavy oil, hydrodesulfurization reaction is the main reaction in the hydrotreating process of heavy oil, which contributes a large amount of heat to the total reaction heat. Hydrogenation of denitrification is also an exothermic reaction, and the reaction heat is about $2638-2952\,kJ/m^3$ hydrogen. Although the nitrogen content in heavy oil is high, the denitrification rate is only 50%–60%, so the denitrification reaction contributes less to the total reaction heat than the desulfurization reaction. The hydrogenation for residual carbon removal is closely related to the hydrogenation of polycyclic aromatic hydrocarbons and heterocyclic aromatic hydrocarbons. The hydrogenation saturation reaction of aromatic hydrocarbons is also an exothermic reaction, and the reaction heat is about $1570-3140\,kJ/m^3$ hydrogen [54]. The cracking reaction that occurs in the process of heavy oil hydrogenation is an endothermic reaction, but in the process of heavy oil hydrogenation, the heat absorption of cracking reaction is significantly lower than the heat release of hydrogenation reaction. Therefore the hydrotreating process is essentially an exothermic process.

3.5.2.2 Heat provided by the furnace

Heavy oil hydrogenation needs to heat the feed and hydrogen to a certain temperature before it can react under the action of the catalyst. At the same time, in the fractionation section, acid gas, naphtha, diesel, and sometimes gas oil from the hydrotreated heavy oil need to be separated. In that case, the unit must have enough heat to complete the above reaction and separation processes. Therefore, apart from the reaction heat, it is also necessary to set up a furnace to provide enough heat for the unit. The heat provided by the furnace can be reflected by the fuel gas consumption. The average fuel gas consumption of heavy oil hydrotreating units in Sinopec accounts for 30%–50% of energy consumption in the unit. In order to reduce energy consumption, it is necessary to enhance heat transfer, reasonably distribute heat sources, and reduce heat exchange losses.

3.5.2.3 Analysis of heat transfer process characteristics

The heavy oil hydrotreating unit includes two sections: reaction and fractionation sections. The operating pressures of the two sections are very different. The operating pressure of the high-pressure section is about 14–18 MPa (gauge pressure), and the low-pressure section is about 0.05–1.0 MPa (gauge pressure). According to the different operating pressures, the heat transfer/heating process can be divided into two categories: high-pressure heat transfer process and reaction feed furnace, as well as low-pressure heat transfer process and fractionator feed furnace. Fig. 3.33 shows a typical heat transfer and heating process in high-pressure section, while Fig. 3.34 shows a typical heat transfer and heating process in low-pressure section. As can be seen from Fig. 3.33, a typical heat transfer process includes the heating of high-pressure feed and hydrogen as well as the cooling of the reactor effluent. Fig. 3.34 shows that a typical heat transfer process includes low-pressure preheating of the feed, the cooling of each reaction effluent, and the heat supply and utilization.

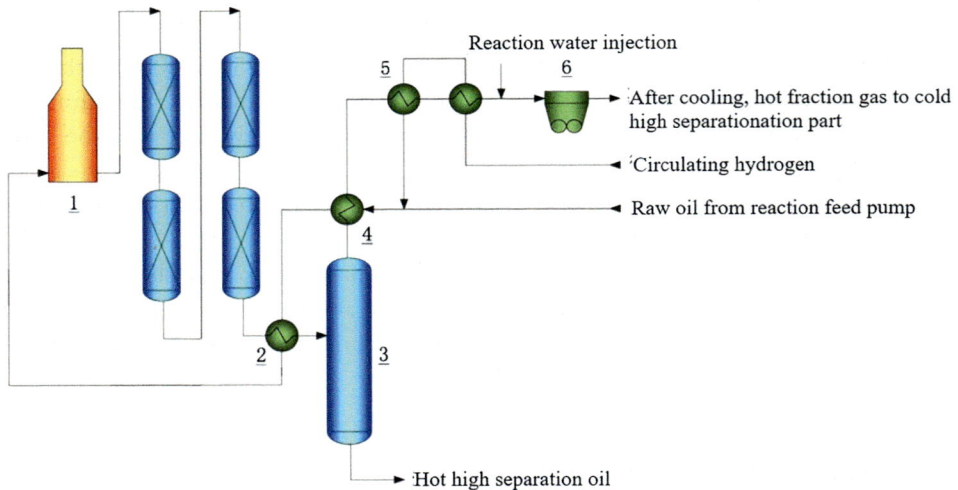

Fig. 3.33 Typical heat transfer/heating flow in high-pressure section. (1) Reaction feed furnace; (2) reactor effluent/mixed feed heat exchanger; (3) hot high-pressure separator; (4) hot high-pressure separator gas/mixed feed heat exchanger; (5) hot high-pressure separator gas/recycle gas heat exchanger; (6) hot high-pressure separator gas air cooler.

In the two heat exchanger networks, the heat from reaction section and fractionation section can transfer between each other. The two parts of heat transfer are a hot high-pressure separator and filtered feed surge drum. By changing and controlling the temperature of these two points, the heat distribution in reaction and fractionation sections can be adjusted. For example, when the operating temperature of the hot high-pressure separator rises, hot low–pressure separator oil temperature will also increase. The heat of the reaction section (high–pressure thermal network) can therefore transfer to the fractionation section (low–pressure thermal network). When the operating temperature of the hot high–pressure separator decreases, more heat is left in the reaction section (high–pressure thermal network), which can be used to heat the reaction feed. When feed is more heated in the low–pressure heat exchanger network, more of the heat from the fractionation section is transferred to the reaction section, thus reducing the high–pressure heat transfer, and vice versa. The total heat required by the unit can therefore be balanced between the reaction feed furnace and the fractionator feed furnace.

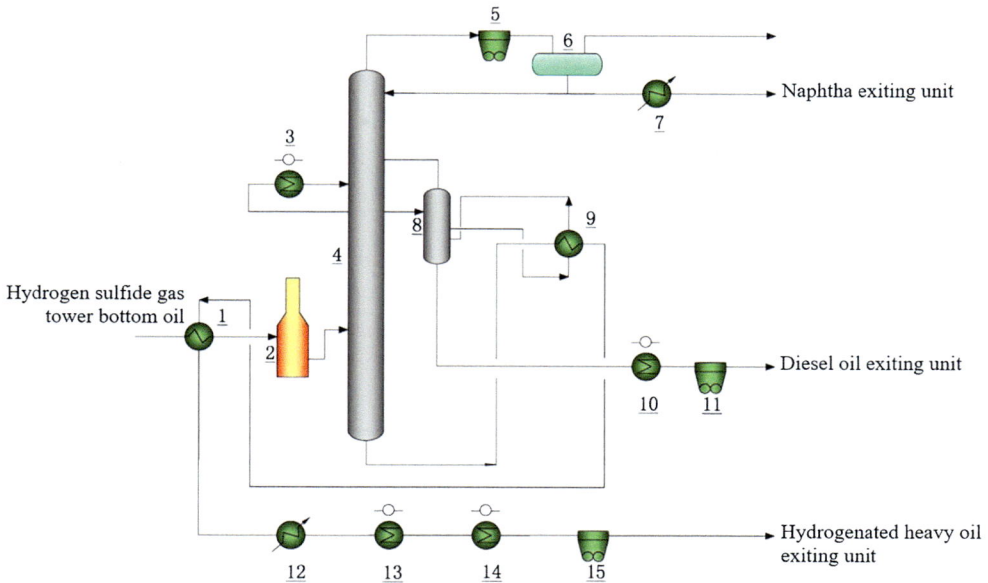

Fig. 3.34 Typical heat transfer/heating flow in low-pressure section. (1) Fractionator feed/hydrogenated heavy oil heat exchanger; (2) fractionator feed furnace; (3) pump around steam generator; (4) fractionator; (5) fractionator overhead air cooler; (6) fractionator overhead reflux drum; (7) naphtha water cooler; (8) diesel stripper; (9) diesel stripper reboiler; (10) diesel low-pressure steam generator; (11) diesel air cooler; (12) hydrogenated heavy oil/feed heat exchanger; (13) hydrogenated heavy oil low-pressure steam generator; (14) hydrogenated heavy oil low-pressure steam generator; (15) hydrogenated heavy oil air cooler.

Due to the unique properties of feed, such as high density, high viscosity, high content of impurities such as sulfur and nitrogen, and high content of colloid and asphaltene, the special requirements and restrictions in the design of heat exchange process for heavy oil hydrotreating unit are as follows.
① Heavy oil viscosity is large, and the viscosity decreases very fast with the increase in temperature. In order to ensure the filtration efficiency of the feed, the operating

temperature is usually between 240°C and 290°C, which is the target temperature of the feed heat transfer in the low-pressure preheating section. This part of the heat comes from the fractionation section.

② Selection of hot high-pressure separator temperature: reducing the temperature of the hot high-pressure separator can reduce the duty of the reaction feed furnace to a certain extent. Increasing the utilization of reaction heat can reduce the consumption of fuel gas. However, for the heavy oil hydrotreating unit, when the hot high-pressure separator temperature is too low, the residue fluidity becomes poor, and it is easy to block the instrument. Therefore, according to the different properties of feed, the hot high-pressure separator temperature is usually selected between 340°C and 380°C, which limits the final temperature of high-pressure heat transfer on the side of reactor effluent to a certain extent.

③ Considering the safety factors, in order to effectively control the reactor temperature and prevent the occurrence of temperature runaway and other hazards, the temperature difference between the inlet and outlet of the furnace of the reaction feed should not be less than 20°C. This requirement limits the final temperature of heat transfer on the side of the reaction feed.

④ In heavy oil hydrotreating process, the operating pressure is high, with high sulfur and nitrogen content in the feed and hydrogen sulfide and ammonia concentration in the reactor effluent. In order to prevent the ammonium salt crystallization from blocking heat exchanger and air cooler, water injection starts when the hot high-pressure separator gas end temperature reaches 190–220°C. The temperature will cool the hot high-pressure separator gas by around 50–60°C after water injection, which is further cooled to 50–55°C with high-pressure air cooler. Due to the highly corrosive stream, the low-temperature heat can only be cooled by air cooler and is not suitable for low-temperature heat recovery. The heat that can be effectively recycled from hot high-pressure separator gas is greatly limited.

⑤ Heavy oil has large viscosity and high freezing point. When going ISBL or OSBL, unit will usually provide hot feed and hot product at the temperature of 90–180°C. The hot feed and product can also reduce the viscosity of feed heavy oil and hydrogenated heavy oil, so as to reduce the electrical energy consumed by pumping. On the other hand, it can form a thermal combination between units, avoid repeated cooling and heating of oil products, and reduce the fuel consumption of furnace.

3.5.3 Process enhancement and integration on heat transfer

The enhancement of heat exchange process in heavy oil hydrotreating unit includes two parts: one is the effective transfer of reaction heat in the reactor, and the other is the full utilization of reaction heat and the optimization of heat exchange process to reduce the fuel gas consumption of the unit, thus reducing energy consumption.

3.5.3.1 Selection and optimization of high-pressure heat exchanger network

The high-pressure heat transfer process of the heavy oil hydrotreating unit is relatively fixed. Fig. 3.35 shows two typical heat transfer modes. Both of the two heat transfer processes adopt hydrogen mixing at the inlet of the furnace, and the difference is the positions of hydrogen and feed mixing point.

Fig. 3.35A shows that the feed and the mixed hydrogen heat exchanged, respectively, and then mixed at the inlet of the furnace. Fig. 3.35B shows that the feed and mixed hydrogen are mixed first, exchange heat, and are then heated by furnace. The process of separate heat exchange of two streams is convenient to adjust the flow rate of feed to multiple furnace tubes through control valves. However, for residue feed, the heat transfer coefficient is small and the heat transfer area required is large. Coking is more likely to occur at the residue side of the heat exchanger. At the end of run (EOR), with the increase of coking possibility, the pressure drop increases, and the heat transfer coefficient decreases significantly. The mixed two-phase heat transfer process has the advantages of high heat transfer coefficient, less coking on the residue side of the heat exchanger, and requirement of less heat transfer area. Therefore, from heat transfer enhancement point of view, it is recommended to adopt the mixed two-phase heat transfer process as shown in Fig. 3.35B.

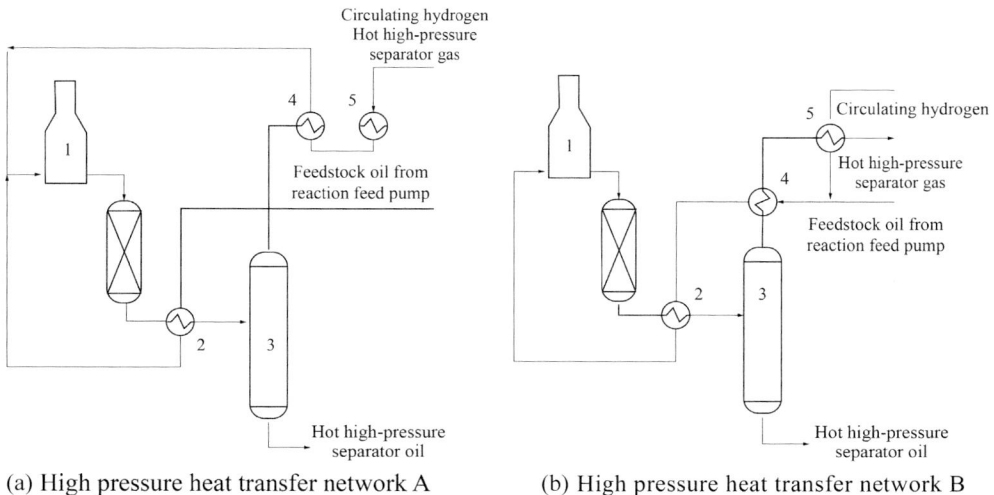

(a) High pressure heat transfer network A   (b) High pressure heat transfer network B

Fig. 3.35 Schematic diagram of heat exchanger network in high-pressure section. (1) Reaction feed furnace; (2) reactor effluent/feed heat exchanger (reactor effluent/mixed feed heat exchanger); (3) hot high-pressure separator; (4) hot high-pressure separator gas/mixed feed heat exchanger (hot high-pressure separator gas/recycle gas heat exchanger); (5) hot high-pressure separator gas/recycle gas heat exchanger.

3.5.3.2 Optimization of low-pressure heat exchanger network

In the low-pressure heat transfer section, the heating of feed heavy oil and the cooling of hydrogenated heavy oil are single-phase heat transfer with small heat transfer coefficient. Therefore, the low-pressure heat exchanger network mainly optimizes the heat transfer sequence and heat exchanger structure to realize the cascade utilization of heat, so as to improve the heat transfer efficiency. Fig. 3.36 shows an example of optimization.

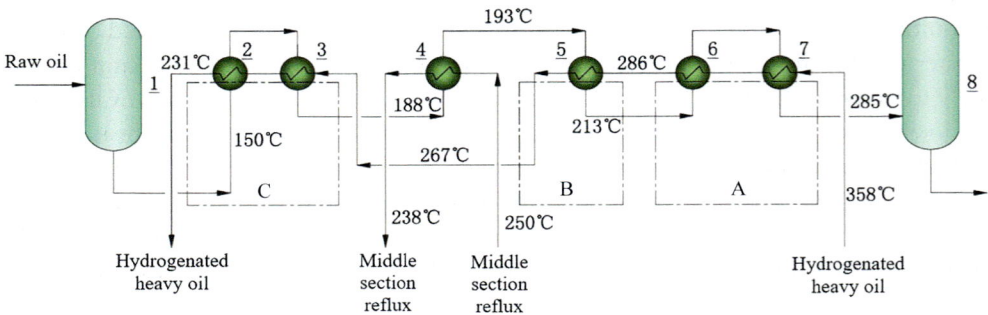

Fig. 3.36 Schematic diagram of heat exchanger network in low-pressure section. (1) Feed surge drum; (2, 3) feed/hydrogenated heavy oil heat exchanger; (4) feed/fractionator PA heat exchanger; (5) hot feed/hydrogenated heavy oil heat exchanger 1; (6, 7) hot feed/hydrogenated heavy oil heat exchanger 2; (8) filtered feed surge drum.

After the hydrogenated heavy oil is extracted from the fractionator bottom, the highest temperature of 365–358°C is used as the heat source of the diesel stripper reboiler, which is then used to preheat feed. The remaining heat is used to generate steam. The product is therefore sent to the downstream catalytic cracking unit as hot product. When the catalytic cracking unit does not accept the feed, it is sent to the storage tank area after cooling by the air cooler. In feed preheating process, according to the gradual decrease of the temperature level, it is divided into three sections (A, B, and C) for heat transfer. In the A–B section, the boundary temperature of the hydrotreated heavy oil is 286°C, which matches the final temperature of the feed preheating and is used as the backwash oil of the feed filter. The temperature of the feed preheating in the C section is 188°C, and in the middle of the B–C section, this temperature can be matched with the PA temperature. This will make full use of the waste heat of the fractionator.

3.5.3.3 Heat analysis and utilization of fractionator

The heat source of the fractionator of the heavy oil hydrotreating unit comes from reaction section (hot low-pressure separator oil) and fractionation section (fractionator feed furnace). A typical fractionator heat distribution diagram is shown in Fig. 3.37.

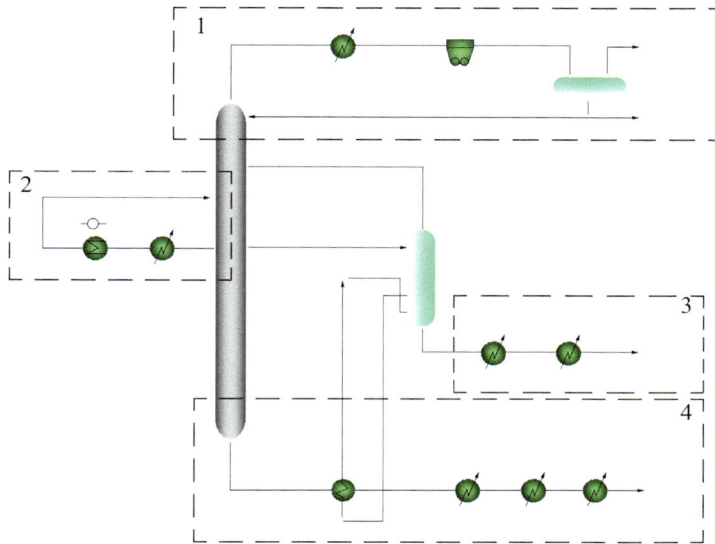

Fig. 3.37 Schematic diagram of heat distribution in fractionator. (1) Fractionator OH heat; (2) middle section heat; (3) diesel product heat; (4) fractionator bottom heat.

1. Fractionator OH heat

   Typical operating temperature at the fractionator OH is 120–130°C. Taking a 4.0 Mt/a heavy oil hydrotreating unit as an example, if the temperature reaches 55°C by air cooling, the duty is about 14 MW, which is difficult to utilize such low-temperature heat. However, it can be used to preheat demineralized water, heating water, warm water, and condensate water from OSBL. Therefore, considering the recovery of low-temperature heat is an important way to enhance and integrate heat transfer for the whole plant.

2. Pump around (PA) heat

   The PA can effectively reduce the column OH cooling duty with a higher temperature level. Taking a 4.0 Mt/a heavy oil hydrotreating unit as an example, the feed can be heated by 5–12°C with the PA, and the remaining heat is used to generate low-low-pressure steam.

3. Heat of diesel product

   In recent years, in order to avoid water in the diesel product in newly built heavy oil hydrotreating units, the diesel stripper uses reboiler to provide heat required for stripping. Therefore, the bottom temperature of the diesel stripper is high. However, the yield of diesel is only 7%–15% (mass fraction), which means the heat flow is small. It can only be used for the heating of cold streams with small flow rate, generating steam, or sending OSBL as a low-temperature heat source.

4. Heat of fractionator bottom

   The flow rate of high-temperature stream at fractionator bottom is large and stable. It is a very high-quality heat source which can be used as the heat source of diesel stripper reboiler, preheating fractionator feed, heating feed, or other important uses. The remaining heat can be used to generate steam. The product then can be sent to the downstream catalytic cracking unit as hot product. When the catalytic cracking unit is shut down, the stream is cooled to the appropriate temperature by air cooler and sent to storage tank area.

3.5.3.4 Optimization of units heat combination and integration

The optimization of heat exchanger network in a single unit is limited due to the limited distributable flow. Heat combination, integration, and distribution among units can utilize heat more fully and improve the efficiency of heat utilization. For example, direct hot feeding and discharging between units can reduce the repeated cooling and heating of streams and form an overall optimized heat utilization network.

3.5.3.5 Other process optimization for heat transfer enhancement

Heavy aromatic hydrocarbons, colloids, and asphaltene are high in heavy oil. In the heat transfer process of feed, it is more likely to form coke and scale in heat exchangers due to the high temperature. This may cause a decrease in heat transfer efficiency and even affect the operation cycle of the unit. For example, the reactor effluent/feed heat exchanger is severely scaled in a heavy oil hydrotreating unit, and the heat exchange efficiency continues to decrease. The final temperature of the heat exchanged feed decreases by 2–3°C every month, which causes the outlet temperature of the furnace fail to reach the required reaction temperature. The unit has to reduce the feed input and shut down for maintenance before a full production cycle is completed. In order to enhance the heat exchange effect of heat exchanger, the following methods are used. Make sure the feed storage tank is protected by inert gas to avoid air contact, which can reduce the feed scaling tendency. Eliminate the dead zone in the shell design of high-pressure heat exchanger. The scale inhibitor is injected into the feed to reduce the scaling tendency in the heat exchanger.

3.5.4 Heat transfer enhancement technology using elements

The feed of heavy oil hydrotreating unit has high viscosity, easy scaling, and poor heat transfer performance. Different heavy oil hydrotreating processes have their own unique heat transfer processes and characteristics. Taking a typical unit as an example, the characteristics of cold exchange equipment in heavy oil hydrotreating unit were analyzed, and the application prospect of heat transfer enhancement technology was explored.

3.5.4.1 Feed heat exchange system

The feed of the heavy oil hydrotreating unit comes from the storage tank or upstream unit, and the viscosity of the feed is high. If the hot feed comes from the upstream unit, the temperature is usually more than 200°C, while the temperature from storage tank is usually about 150°C. Taking a 3.9 Mt/a residue hydrotreating unit as an example, the viscosity of 150°C residue is 46.3 mN s/m^2, which must be heated to about 260°C by the feed/hydrogenated heavy oil heat exchanger before flowing to feed surge drum.

Due to the high viscosity and density of the feed, it flows in the shell side in order to enter the turbulent flow area at a low Reynolds number and obtain a high film heat transfer coefficient. The physical properties of the cold and hot fluid in the heat exchanger are shown in Table 3.30. When ordinary heat exchanger is used during design, it is found that heat transfer thermal resistance and fouling resistance account for 75% of the total thermal resistance in the shell side. The fouling resistance of hydrogenated heavy oil and feed is 0.00069 m^2 K/W and 0.00118 m^2 K/W, respectively. The film heat transfer coefficient of shell side is only 50% of tube side. Therefore, it is necessary to take measures to improve the film heat transfer coefficient of shell side and reduce the possibility of scaling. Corrugated tubes and spiral-twisted tubes have the effect of enhancing heat transfer on both sides. The calculation and comparison results of ordinary heat exchangers and enhanced heat exchangers are shown in Table 3.31.

Table 3.30 Process condition of feed/hydrogenated heavy oil heat exchanger—Duty: 33.5 MW.

Item	Shell side	Tube side
Fluid name	Feed	Hydrogenation of heavy oil
Flow rate (kg/h)	464,286	388,004
Allowable pressure drop (kPa)	200	200
Temperature (°C)	150 → 255	350 → 241
Liquid density (kg/m^3)	919 → 855	741 → 813
Liquid viscosity (mN s/m^2)	46.3 → 7.6	0.95 → 2.57
Liquid thermal conductivity (W/(m K))	0.094 → 0.084	0.066 → 0.076
Liquid heat capacity (kJ/(kg K))	2.267 → 2.659	2.998 → 2.658

Table 3.31 Comparison of feed/hydrogenated heavy oil heat exchanger scheme.

Scheme	Plain tube	Spiral-twisted tube
Equipment size (shell diameter × straight tube length) (mm)	1400 × 7500	1400 × 7500
Equipment (set)	3	2
Tube side pressure drop (kPa)	146	166
Shell side pressure drop (kPa)	164	126
Overall heat transfer coefficient (W/(m^2 K))	190	288
Heat transfer area (m^2)	2162	1442
Total weight of equipment (t)	80	51

It can be seen from the calculation results in Table 3.31 that using corrugated tube bundle instead of plain tube can reduce the required heat transfer area by 33% and the equipment weight by 36%. At the same time, the construction cost, piping cost, and later operation cost are reduced due to the reduction of one heat exchanger. And because of the particularity of the medium flow, it can slow down the possibility of scale and keep the equipment long-term working at best performance with a prolonged cleaning cycle. The composite heat transfer enhancement technology of corrugated tubes and helical baffles can also be used to reduce the dead zone.

The literature [55] reported that the overall heat transfer coefficient can be increased by 85% compared with the same type of plain tube heat exchanger in parallel operation with same pressure drop. After 2 years of operation, it is found that there is little sump oil, and the trough peak and trough of the transverse groove are in good shape. In addition, no corrosion is found on the surface of the transverse groove tube, which shows that the corrugated low-finned tube has good antifouling performance.

3.5.4.2 High-pressure heat exchange system

The D-type special high-pressure tube box of high-pressure heat exchanger in hydrotreating unit is processed by forging. The sealing structure is complex, which accounts for a large proportion of the equipment cost. The number of equipment means that the cost of increasing heat transfer area will be less. Therefore, the heat transfer enhancement technology should focus on reducing the number and the diameter of equipment. The use of F-shell exchanger or coil-wound heat exchanger structure can reduce the hot approach temperature [56]. The duty of furnace can be reduced when used for reactor effluent/mixed feed heat exchanger furnace.

The process condition and physical properties of a hot high-pressure separation gas/mixed hydrogen heat exchanger are shown in Table 3.32, and the thermodynamic calculation results are shown in Table 3.33.

Table 3.32 Process condition of hot high-pressure separator gas/mixed hydrogen heat exchanger—Duty: 11.8 MW.

Item	Shell side	Tube side
Fluid name	Hydrogen	Hot high-pressure separator gas
Flow rate (kg/h)	47,916	97,930
Allowable pressure drop (kPa)	50	50
Temperature (°C)	$89 \rightarrow 211$	$270 \rightarrow 190$
Gas density (kg/m^3)	$25.8 \rightarrow 19.6$	$23.6 \rightarrow 25.9$
Liquid density (kg/m^3)	–	$650 \rightarrow 677$
Gas viscosity (mN s/m^2)	$0.0124 \rightarrow 0.0148$	$0.0172 \rightarrow 0.0154$
Liquid viscosity (mN s/m^2)	–	$0.0158 \rightarrow 0.0402$

Table 3.32 Process condition of hot high-pressure separator gas/mixed hydrogen heat exchanger—Duty: 11.8 MW—cont'd

Item	Shell side	Tube side
Gas thermal conductivity (W/(m K))	$0.1696 \rightarrow 0.2167$	$0.2104 \rightarrow 0.1851$
Liquid thermal conductivity (W/(m K))	–	$0.0749 \rightarrow 0.0875$
Gas heat capacity (kJ/(kg K))	$7.167 \rightarrow 7.456$	$5.466 \rightarrow 5.5$
Liquid heat capacity (kJ/(kg K))	–	$2.773 \rightarrow 2.518$

It can be seen from Table 3.33 that even with E-shell shell and tube heat exchangers, the overall heat transfer coefficient of single-phase heat transfer area exceeds 500 W/(m² K). However, due to the deep approach temperature between the hot high–pressure separator gas and the mixed hydrogen, it is necessary to connect multiple sets in series to avoid temperature crossing. The F-shell structure can realize countercurrent heat transfer, reduce the number of heat exchangers, improve the effective temperature difference, and reduce the required heat transfer area. After using F-shell in later design, the weight of the equipment can be reduced by 46% and it can reduce one set of equipment. The operation and maintenance cost can be therefore reduced. In addition, the heat transfer enhancement effect is remarkable and is an effective heat transfer enhancement structure suitable for the process condition of this heat exchanger.

Table 3.33 Comparison of hot high-pressure separator gas/mixed hydrogen heat exchanger scheme.

Scheme	E-shell exchanger	F-shell exchanger
Equipment size (shell diameter × straight tube length) (mm)	1100×2800	1000×4800
Equipment (set)	2	1
Tube side pressure drop (kPa)	32	25
Shell side pressure drop (kPa)	23	32
Overall heat transfer coefficient (W/(m² K))	612	668
Heat transfer area (m²)	289	257
Total weight of equipment (t)	89.6	48.5

3.5.4.3 Fractionation heat transfer system

① Stripper overhead (OH) cooling system. The OH heat is mainly removed by air cooler. Due to the high temperature of process medium, when the effective temperature difference of the air cooler is more than 50°C, the heat transfer area is not large. If the equipment size is to be further reduced, the elliptical tube high fin can be used to replace the commonly used round tube high fin to enhance the heat transfer in and outside the tube.

② Fractionator bottom oil heat exchange system. The fractionator bottom is cooled through the stripper bottom/hydrotreated residue heat exchanger, the feed/hydrotreated residue heat exchanger and the residue steam generator successively. When the temperature drops to 182°C, it is sent to the heavy oil catalytic cracking unit.

   The tube side and shell side film heat transfer coefficient of the stripper bottom/hydrotreated residue heat exchanger are close, and the film heat transfer coefficient is more than $1500 \, W/(m^2 \, K)$ when the ordinary plain tube and shell heat exchanger is used. The control thermal resistance is fouling thermal resistance, which accounts for about 45% of the total thermal resistance. Spiral-twisted tube and corrugated tube are more suitable heat transfer enhancement components.

   In the residue steam generator, the film heat transfer coefficient is less than $500 \, W/(m^2 \, K)$ when the flow rate of the residue in the tube reaches $1.5 \, m/s$, which is only 10% of the shell side boiling heat transfer. It is used as control thermal resistance of the equipment. If the spiral-twisted tube is used instead of the plain tube, the heat transfer of the tube side can be effectively enhanced and the total heat transfer performance of the steam generator can be improved.

③ Fractionator OH and product cooling system. Fractionator OH gas is about 127°C, which is condensed and cooled by the OH cooler to 70°C and enters the OH reflux drum. Although the duty of OH air cooler is large, the heat transfer area required is not much as the effective temperature difference is large. To further reduce the equipment size, the elliptical tube can be used to replace round tube as the base tube. After the diesel product extracted from the fractionator is stripped by diesel stripper, the diesel product is further cooled to 170°C after generating steam. The film heat transfer coefficient of diesel in the diesel steam generator is about 20% of boiling heat transfer in the shell side, accounting for 40% of the thermal resistance of the equipment. The literature reported that [57] spiral-twisted tube in hydrocracking tail oil (HVGO) product cooler is used to reduce the heat transfer area by 50%. The heat transfer enhancement technology can also be applied to diesel steam generator, which can effectively enhance the heat transfer in the tube and improve the boiling heat transfer coefficient in the shell.

3.5.5 Summary

To sum up, in the design of heavy oil hydrotreating unit, the unique properties of heavy oil and chemical reaction process need to be fully understood. On this basis, the efficient reaction heat removal and utilization, heat exchanger network optimization, and heat integration between units through process and equipment improvement. Therefore, safety and smooth operation can be ensured to save energy.

   Due to the high viscosity of the feed at less than 200°C in the heavy oil hydrotreating unit, the heaters of the feed are the key point to enhance heat transfer. The heat exchange equipment of hot high-pressure separator gas has high pressure and complex structure, so it is necessary to improve the heat exchange performance and reduce the equipment size as much as possible to reduce the equipment investment.

3.6 Hydrogen production unit

3.6.1 Brief introduction of process

There are many methods for industrial hydrogen production, such as partial oxidation of residual, petroleum coke, or coal. Light hydrocarbon steam reforming, purification and separation of hydrogen–rich gas, methanol steam reforming, ammonia cracking, and water electrolysis can also be used as well. In large-scale industrial hydrogen production unit, the process of light hydrocarbon steam reforming is more reliable with low investment and convenient operation.

The light-hydrocarbon steam reforming unit consists of feed refining, steam reforming, shift, and product refining. There are several chemical reactions during the processes, including feed hydrogenation reaction, desulfurization reaction, steam reforming reaction, prereforming reaction, and reforming reaction. The typical domestic light-hydrocarbon steam reforming process is shown in Fig. 3.38.

Feed gas from OSBL is pressurized to 3.2 MPa by feed compressor and heated to 220°C by heat exchange with the medium-temperature shift gas. The heated feed then enters the hydrotreating reactor. The organic sulfur converts to hydrogen sulfide and olefins are saturated in the reactor. After the reaction, reactor effluent reaches 360°C and flows to desulfurization reactor. The total sulfur content is less than 0.2 μL/L, and hydrogen chloride is less than 0.3 μL/L, after the adsorption of chloride and hydrogen sulfide in the desulfurization reactor. After purification, the feed gas is mixed with the self-produced steam of the unit (the molar ratio of H_2O/C is 3.2) and heated to 500–520°C in feed preheating section of the steam reformer and exchanges heat with high-temperature flue gas, which then enters the steam reformer tube. The feed gas and steam react to generate H_2, CO, CO_2, and methane under the action of catalyst. The outlet temperature of the steam reformer is 840–860°C, the pressure is 2.7 MPa, and the residual methane is about 5.0% (dry basis volume fraction).

The 850°C reforming gas generates 3.5 MPa medium-pressure steam in the reformer gas waste heat boiler and is cooled to about 340°C before sending to the medium temperature shift reactor. The CO in the reformer gas and water vapor undergoes shift reaction to generate H_2 and CO_2, and the CO at the outlet is less than 3% (dry basis volume fraction). The medium temperature shift gas at the outlet of the reactor preheats feed by the medium temperature shift gas/feed heat exchanger, then passes through the medium temperature variant gas/deoxygenated water heat exchanger and enters medium temperature shift gas #1. After the liquid is divided, the medium-temperature variant gas exchanges heat with the demineralized water in the medium-temperature shift gas-demineralized water heat exchanger and separated in the medium-temperature shift gas #2. After being cooled to 40°C by air cooler and water cooler, it is sent to medium–temperature shift gas #3. The medium–temperature shift gas is therefore sent to PSA section after liquid separation.

The medium–temperature shift gas is physically adsorbed in the PSA section. The product hydrogen is sent OSBL from adsorption drum overhead. The low–pressure desorption gas during adsorbent regeneration is used as furnace fuel after the pressure and component are stabilized by surge drum.

Sour condensed water is separated from the medium–temperature shift gas separator. It is then mixed with demineralized water after stripping by sour water stripper. The sour condensed water therefore exchanges heat with medium–temperature shift gas in the medium–temperature shift gas/demineralized water exchanger and flows to desecrator at 90°C. The 104°C deoxygenated water is boosted by the boiler feed water pump and enters the steam drum for steam generation after being heated to 220°C by medium–temperature shift gas/deoxygenated water heat exchanger. Most of the steam is mixed with the purified feed gas for steam reformer, and the remaining is sent to the 3.5 MPa steam network.

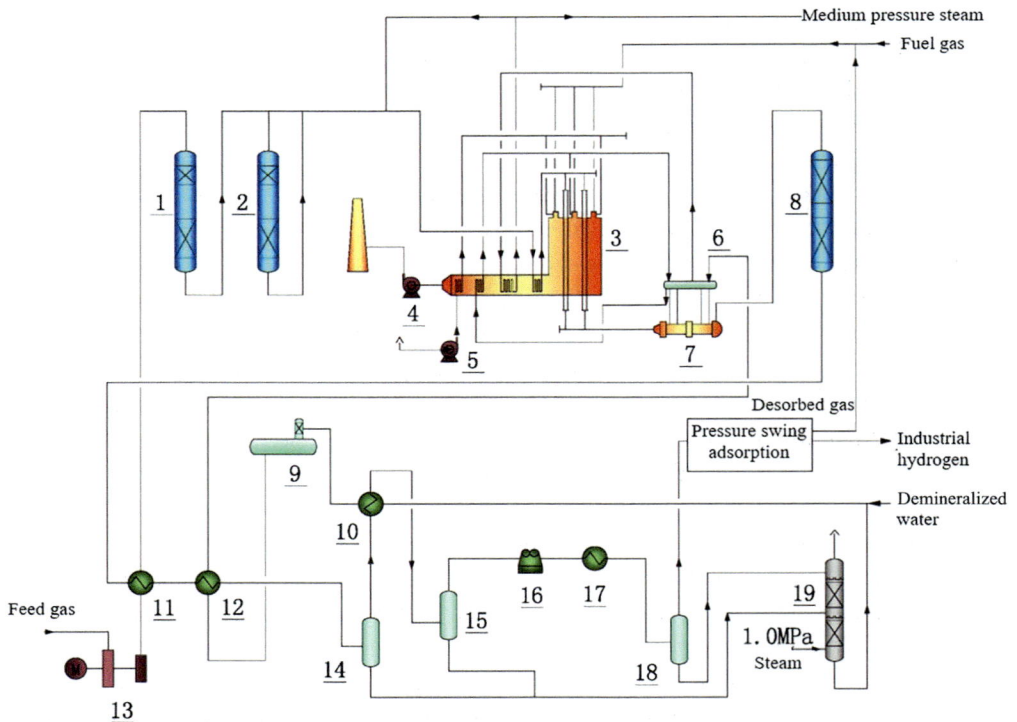

Fig. 3.38 Schematic diagram of light-hydrocarbon steam reforming process. (1) Hydrotreating reactor; (2) desulfurization reactor; (3) steam reformer; (4) induced draft fan; (5) air blower; (6) team drum; (7) reforming gas waste heat boiler; (8) medium-temperature shift reactor; (9) deoxygenation drum; (10) medium-temperature (MT) shift gas/demineralized water heat exchanger; (11) MT shift gas/feed heat exchanger; (12) MT shift gas/deoxygenated water heat exchanger; (13) feed compressor; (14) MT shift gas knockout drum #1; (15) MT shift gas knockout drum #2; (16) MSG air cooler; (17) MT shift gas water cooler; (18) MT shift gas knockout drum #3; (19) sour water stripper.

3.6.2 Analysis of heat energy characteristics

There are high-, medium-, and low-level temperatures in hydrogen production unit for the multiple chemical reactions during the process. Steam reformer is the main heat source of high-temperature level, which includes high-temperature flue gas and effluent. The reforming reaction is a strong endothermic reaction at 850–870°C. The required reaction heat is provided by the combustion of PSA off gas and makeup fuel gas through furnace tubes. Therefore, the high-temperature flue gas after combustion (about 1000°C) and high-temperature reforming gas (800–900°C) contain a large amount of high-temperature level waste heat. The 400–420°C shit gas from shift reactor is the source of the medium-level temperature heat in the unit. CO shift reaction is an exothermic reaction. The reaction temperature is 250–420°C. After the reaction, whether in high-, medium-, or low-temperature shift reaction, the shift gas contains a large amount of medium- or low-level temperature waste heat. After the shift gas exchanged heat and generated steam (when needed), the 200°C shift gas still contains big amount of heat. Especially when it is cooled to 180°C, the shift gas vapor begins to condense, which can release huge amount of heat.

The energy consumption of hydrogen production units is mainly from fuel and electricity, which accounts for more than 95% according to the average energy consumption index of hydrogen production units of Sinopec from 2020 to 2022. The strong endothermic characteristic of hydrogen reforming reaction determines the large fuel consumption. The power consumption of the unit is mainly on key rotating equipment, such as on feed pressure boost, induced draft fan, air blower, and boiler feed water pump. The medium-pressure steam used in the unit mainly includes the required steam for the process and the produced medium-pressure steam which can send OSBL (minus the required process steam). Therefore, reducing the energy consumption of the unit should focus on reducing the consumption of fuel gas, steam, and electricity.

To reduce fuel consumption, it is crucial to enhance heat transfer; reasonably distribute heat source; preheat feed, fuel gas, or air by different level temperatures; and reduce heat exchange losses. The improvement of process technology (such as improving the performance of the reforming catalyst, using higher conversion temperature and lower water-carbon ratio, etc.) to reduce the amount of steam and reasonable use of waste heat at different temperatures for generating steam will reduce the steam energy consumption. For an $80,000 \, m^3/h$ of a typical hydrogen production unit, the waste heat of the whole unit is as high as 212.0 MW. Among these, there is 90.0 MW high-temperature-level heat higher than 360°C, which accounts for about 42% of the total waste heat. Therefore, it is crucial to reasonably utilize the high-, medium-, and low-level temperature heat of the hydrogen production unit. In addition, the reduction of fuel gas and steam will also contribute to the energy saving of the unit.

3.6.3 Heat optimal utilization of different temperature levels

3.6.3.1 Heat utilization of high temperature level

1. Superhigh-pressure steam generated

   The grade use of steam will improve the utilization rate of exergy. The high-temperature heat of the reforming furnace flue gas and the reforming gas at the outlet of the steam reformer tube of hydrogen production unit are usually used for generating high-pressure steam. The steam can be directly used as reaction feed. The superheated steam can also be used as steam reformer tube feed and the rest is sent OSBL as by-product. The steam pressure is 3.5 MPa.

   However, this part of the high-temperature heat can also produce 10.0 MPa high-pressure steam. The superheated 520°C 10.0 MPa steam can generate power or as driving power of rotating machinery. The backpressure steam (3.5 MPa) can therefore be used for the process and the rest is sent to steam network. By improving the grade of steam generated and adopting step-by-step utilization of steam, the utilization rate of exergy can be greatly improved. For a 40,000 m^3/h hydrogen production unit, 10 MPa high-pressure steam backpressure to 3.5 MPa can generate 4.0 MW electricity or duty. Under the same heat efficiency, this part of the duty cannot be effectively recovered when directly generating 3.5 MPa steam.

   The generation and grade use of superhigh-pressure steam will increase the construction investment. Therefore, only in the large-scale hydrogen production unit will such a scheme be used to obtain a better economy. Such a scheme has been used in a few units in China. Therefore, for larger-scale hydrogen production units, planning and grade use of steam will undoubtedly have more obvious energy-saving effects and economic benefits.

2. Gas reformed to provide part of heat energy of reforming reaction

   The high-temperature-level reforming gas (850–870°C) from steam reformer tube is used to provide the heat needed for reforming reaction for the mixed feed without going through the steam reformer in the heat exchange reformer (the heat exchange tube is equipped with catalyst). In this way, the high-temperature-level heat of the reforming gas at the outlet of the steam reformer tube is better used. The amount of supplementary fuel and the scale of the steam reformer therefore can be reduced, and the energy utilization efficiency of the unit can be effectively improved. There are two kinds of heat exchange reformer in the hydrogen production unit which have been applied in industry. One is the post reformer with two inlets and one outlet, and the other is a gas heating reformer with two inlets and two outlets.

   Table 3.34 shows a comparison of the designed performance of a conventional steam reformer and a postreformer. Postreformers can be used in newly built and existing units. For unit modification, the transforming ability can be increased by more than 30%.

Table 3.34 Comparison of 25,000 m^3/h scale hydrogen production unit with and without postreformer.

	Steam reformer +PSA	Steam reformer +postreformer+PSA
Feed and fuel gas (m^3/h)	10,960	9384
Steam reformer size (%)	100	71
One-stage furnace radiates thermal efficiency (%)	40–50	55–60
Investment (%)	100	95

Gas–heated reformers are developed in synthetic ammonia process and can be used in hydrogen production units. The gas heating reformer and steam reformer can be used in series or parallel. The main advantages include reducing the size of steam reformer and fuel consumption, NO_x and CO_2 in flue gas emissions, external steam transport, and investment and improving the utilization rate of feed gas.

The technical evaluation shows that the operation cost can be reduced by 10% and the investment can be saved by about 5%–8% after the gas heating reformer is used.

3. High-temperature-level flue gas to provide part of the reaction heat of the reforming reaction

Compared with the process of conventional tube steam reformer, the process of convective steam reformer uses less fuel gas during reforming reaction. This is because a part of the heat energy is provided through the heat exchange of feed gas and steam reformer gas as well as flue gas. The process of convective steam reformer also has other characteristics. Firstly, the convective steam reformer occupies a small floor area and has a compact structure. This is because it combines the radiation chamber (including tube and burner) and the high-temperature part of the convective section. Secondly, the heat transfer efficiency of convective steam reformer is higher than the radiation heat transfer efficiency of conventional tube steam reformer. Thirdly, less excess steam is sent OSBL.

3.6.3.2 Utilization of waste heat at medium temperature level

The medium-temperature (MT) waste heat mainly comes from MT shift gas. For the process of preheating feed by feed furnace, the 410°C MT shift gas is mainly used to generate medium-pressure steam. The steam therefore exchanges heat with deoxidized water and demineralized water, respectively, and is sent to PSA after air cooling and water cooling, as shown in scheme III. In scheme I, 1.0 MPa steam is generated by using MT reforming gas. Scheme II is to make full use of low-temperature heat to produce 0.3 MPa steam.

Scheme I:

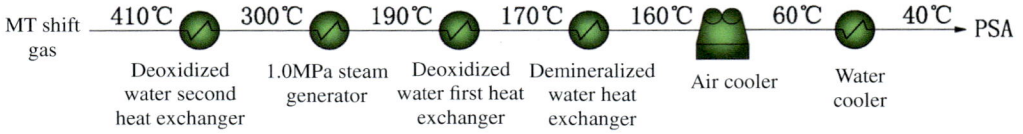

Scheme II:

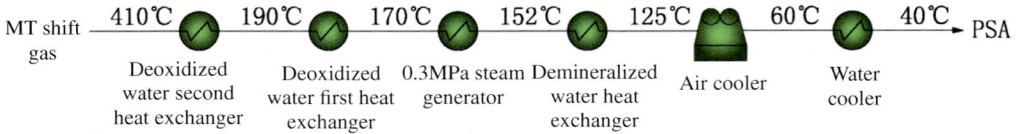

Scheme III:

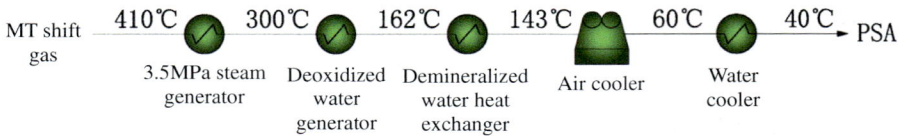

For scheme I, although there is little difference in thermal efficiency between generating 1.0 MPa steam and preheating boiler feed water, for the whole system, preheating deoxidized water can increase the production of medium-pressure steam and improve the overall efficiency of exergy recovery. Therefore, scheme I is not ideal. Scheme II makes full use of the low-temperature heat source to generate 0.3 MPa steam and improves the heat utilization efficiency, which is suitable for northeast China. However, it is still not ideal for most refineries with excess low-pressure steam. The difference between scheme III and the other two schemes is that although the waste heat of MT is fully utilized to produce 3.5 MPa steam, and the utilization of exergy is reasonable, the utilization of low-temperature heat is not sufficient.

The above three schemes all have furnaces. Although feed furnace has flexible operation, it is not reasonable in terms of exergy utilization. For a 30,000 m^3/h hydrogen production unit, the duty for feed preheating is 2.33 MW, which means the process includes a low-efficiency small furnace (50%) and a large amount of the MT shift gas waste heat needs to be recovered at the same time. The low thermal efficiency and exergy efficiency are obvious. Recently, many hydrogen production units have used MT shift gas-feed preheater instead of furnace during normal operation, which is shown in scheme IV. Heat recovery and economic benefits are significant. After the heat exchange between the MT shift gas and feed, the temperature drops to about 340°C. The temperature makes it harder to produce high-quality 3.5 MPa steam. After pinch point analysis, the process is further optimized, shown in scheme V. In scheme V, the MT shift gas is divided, and the low-level temperature heat energy of the MT shift gas is better used. For a 30,000 m^3/h hydrogen production unit, the heat below 300°C is utilized, and the 1.16 MW duty can produce 3.0 t/h more steam than scheme IV and reduce 1.16 MW cooling duty. This will improve both heat exchange efficiency and exoergic efficiency.

Scheme IV:

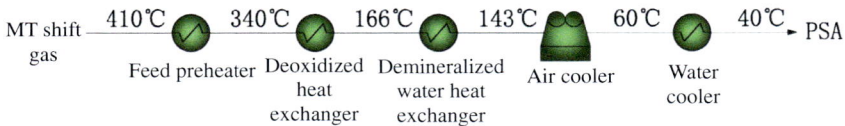

Scheme V:

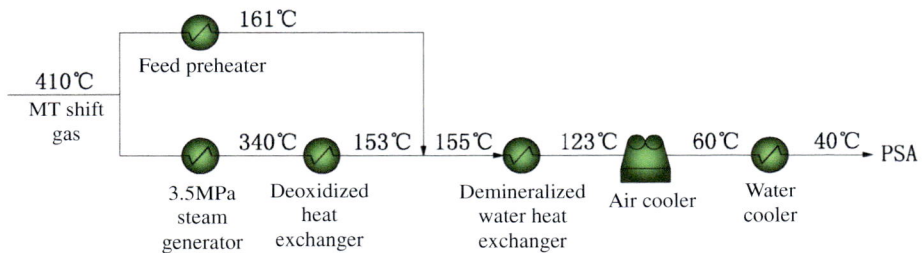

3.6.3.3 Low-level temperature potential waste heat utilization

The dew point of the MT shift gas is usually about 160–170°C. At such temperature, the steam in the MT shift gas will condense with the decrease of temperature. The condensation process will have a large amount of low-temperature heat release. The heat, apart from preheating desalted water, is released into the environment through the air cooler. In some refineries, the heat can be used to heat the heat-tracing water or preheat desalted water. It can also produce low-pressure steam (when needed). Therefore, the temperature of MT shift gas at air cooler inlet will be lower which will improve the energy efficiency.

3.6.4 Heat transfer enhancement technology using elements

The heat exchanger network of hydrogen production unit is relatively simple, which includes MT shift gas heat exchange and steam generation.

3.6.4.1 MT shift gas heat exchange system

The temperature of the MT shift reactor is generally about 410°C, which is cooled to 40°C by a series of heat exchange equipment and then enters the PSA unit for hydrogen purification. The MT shift gas heat exchange system has two characteristics:

① The molecular weight of the MT shift gas is low; the average molecular weight of the dry base is about 10.8, and the molecular weight of the MT shift gas mixed with water vapor is only about 12.6. In the heat exchanger, the flow rate of the gas side is high and the pressure loss is large. Therefore, the top priority is to ensure the pressure drop when selecting the heat exchanger.

② The gas contains a large amount of carbon monoxide and water vapor, and the dew point is usually between 160 and 170°C. The heat transfer coefficient difference around the dew point temperature is huge; the MT shift gas temperature above the dew point is high and the heat is less; and the MT shift gas temperature below the dew point is low, but the heat is high.

Taking an 80,000 m³/h hydrogen production unit as an example, the MT shift gas is cooled through feed gas preheater II, process condensing water steam generator, process condensing water heat exchanger, boiler feed water heat exchanger, MT shift gas low-pressure steam generator, feed preheater I, demineralized water heat exchanger, air cooler, and water cooler. The whole heat exchanger network has no external power, so it is necessary to strictly control the pressure drop of the MT shift gas in the heat exchange equipment along the way. The heat exchanger network is shown in Fig. 3.39.

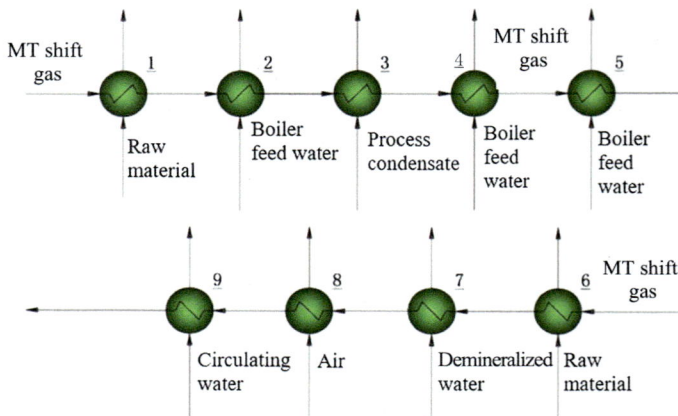

Fig. 3.39 MT shift gas heat exchanger network. (1) Feed preheater II; (2) process condensing water evaporator of MT shift gas; (3) process condensing water heat exchanger of MT shift gas; (4) MT shift gas/boiler feed water heat exchanger; (5) MT shift gas low-pressure steam generator; (6) MT shift gas/feed preheater I; (7) MT shift gas/demineralized water heat exchanger; (8) air cooler; (9) MT shift gas water cooler.

In order to reduce the pressure loss, the MT shift gas is arranged to flow in the tube side of the heat exchanger. The U-shaped tube heat exchanger structure is adopted, and the heat exchange tube is made of stainless steel such as 00Cr19Ni.

Because there is no condensation of water vapor, the heat transfer effect of high-temperature section heat exchanger is poor than that of low-temperature section heat exchanger, and the film heat transfer coefficient of tube side is low. For example, the film heat transfer coefficient of the shell side of the process condensing water steam generator and process condensing water heat exchanger is close to $5000\,W/(m^2\,K)$, and the film heat transfer coefficient of the tube side is no more than $2000\,W/(m^2\,K)$. Due to the pressure drop limitation, the pure gas phase cooling process of the tube side is not easy to enhance. However, the shell side of feed-gas preheater II is heated by pure gas phase, and the film heat transfer coefficient of shell side is less than $500\,W/(m^2\,K)$. Although the film heat transfer coefficient of tube side does not exceed $2000\,W/(m^2\,K)$, control of the thermal resistance is still on the shell side, and measures can be taken to enhance the heat transfer effect of the shell side.

Taking a hydrogen production unit as an example, the process conditions of the feed gas preheater are shown in Table 3.35.

Table 3.35 Process condition of the feed preheater—Duty: 4.4 MW.

Item	Hot fluid	Cold fluid
Fluid name	MT shift gas	Feed
Flow rate (kg/h)	95,076	22,667
Allowable pressure drop (kPa)	25	50
Temperature (°C)	$404 \rightarrow 350$	$135 \rightarrow 348$
Gas density (kg/m³)	$6.18 \rightarrow 6.73$	$17.24 \rightarrow 11.55$
Gas viscosity (mN s/m²)	$0.0187 \rightarrow 0.0176$	$0.0151 \rightarrow 0.02$
Gas thermal conductivity (W/(m K))	$0.129 \rightarrow 0.121$	$0.058 \rightarrow 0.094$
Gas heat capacity (kJ/(kg K))	$2.806 \rightarrow 2.776$	$2.683 \rightarrow 3.303$

Shell-side flow rate is small, with low thermal conductivity. In order to improve the film heat transfer coefficient of the shell side, a relatively simple method is to reduce the shell diameter and increase the flow velocity of the shell side medium. The heat transfer effect can also be enhanced by increasing the secondary heat transfer surface of the shell side, such as low-finned tube and transversally corrugated tube. The comparison results are shown in Table 3.36.

Table 3.36 Comparison of feed preheater schemes.

Scheme	Plain tube	Corrugated tube	Low-finned tube
Equipment size (shell diameter × straight tube length) (mm)	1100×2200	1100×1400	1150×1200
Equipment (set)	1	1	1

Continued

Table 3.36 Comparison of feed preheater schemes—cont'd

Scheme	Plain tube	Corrugated tube	Low-finned tube
Tube side pressure drop (kPa)	25	22	25
Shell side pressure drop (kPa)	5.9	16.3	4.5
Overall heat transfer coefficient $(W/(m^2\,K))$	331	411	536
Heat transfer area (m^2)	168	111	109
Total weight of equipment (t)	10.2	8.5	8.7

As the pressure drop of the tube pass is limited, the number of tubes in a single tube pass cannot be too low. Since the temperature and pressure difference of the streams in heat exchanger is high, it is not suitable to reduce shell diameter size for a single tube pass. F-shell side design can effectively improve the heat transfer coefficient of shell side film while ensuring the pressure drop of pipe side. As can be seen from Table 3.36, in order to ensure the pressure drop of the tube side, the heat transfer enhancement Option I of the F-shell exchanger can shorten the tube length by 35% and reduce the weight of the equipment by 17% under the condition that the shell diameter is unchanged. Enhanced heat generation scheme II with low-finned tube instead of plain tube can reduce the weight of the equipment by 15%. Compared with the F-shell option, although the weight of the equipment is slightly heavier, the overall cost of the scheme is low, and the operation cost is low.

In the low-temperature section, the MT shift gas starts to condense, and the cold and hot stream heat transfer capacity of the heat exchanger is relatively high. Among them, the boiler feed water heat exchanger, the MT shift gas low-pressure steam generator, the demineralized water heat exchanger, and the tube side film heat transfer coefficient of these heat exchangers are more than $8000\,W/(m^2\,K)$. The shell side medium of these heat exchangers is all water, and the film heat transfer coefficient is about 4000–$6500\,W/(m^2\,K)$. The overall heat transfer performance of these heat exchangers is relatively good. When further improving their thermodynamic performance, the heat transfer enhancement technology that can effectively reduce the fouling resistance can be considered.

Although the film heat transfer coefficient of the feed preheater I tube side in the low-temperature section is also more than $8000\,W/(m^2\,K)$, 40% of the duty of shell side medium is used for pure gas phase heating, and the film heat transfer coefficient of shell side is only $600\,W/(m^2\,K)$ or so; it is necessary to adopt technology that can enhance shell side heat transfer.

Different from the conventional water cooler, the cooling water is generally arranged in the shell side of the MT shift gas water cooler based on the consideration of the pressure loss on the gas side and the corrosion of carbon dioxide. The film heat transfer coefficient of the cold and hot fluid in the MT shift gas water cooler is relatively high, and the overall heat transfer coefficient is close to $700\,W/(m^2\,K)$. The fouling resistance accounts for 55% of the total thermal resistance, and the heat transfer enhancement technology that can effectively reduce the scale resistance should be adopted to improve the performance of the equipment. Considering that the scale of cooling water in the shell side is easy to occur, and scale corrosion may lead to equipment failure, the "accordion" effect of low-finned tube can effectively reduce the scale rate, improve the heat transfer performance of equipment, and prolong the cleaning cycle.

Limited by the allowable pressure drop, the MT shift gas cooler can only use relatively short heat exchange tubes. The MT shift gas is condensed in the air cooler tube, the film heat transfer coefficient is about $2000\,W/(m^2\,K)$, and the heat transfer effect is good. When further enhancing the heat transfer, the oval tube can be used instead of the circular base tube of the high-finned tube.

3.6.4.2 Steam heat exchange system

The steam heat exchange system of hydrogen production unit is composed of two parts: steam generation system of reforming gas waste heat boiler and steam generation system of steam reformer waste heat boiler. The process condensate from various MT shift gas knockout drums is deoxygenated, then preheated by MT shift gas/process condensate heat exchanger, and then sent into the condensate evaporator of the MT shift gas and the steam reformer waste heat boiler, respectively.

① Steam generation system of reforming gas waste heat boiler. The reforming gas waste heat boiler is a horizontal natural circulation fire tube waste heat boiler, which is set downstream of the steam reformer. Its function is to generate medium-pressure saturated steam by using deoxygenated water to conduct heat transfer with the reforming gas of the unit. The 870°C reforming gas is cooled to 330–370°C and then enters the MT shift reactor.

The heat exchange furnace tube of the reforming gas waste heat boiler is generally made of seamless steel tube with φ38 mm of 15CrMoG. The temperature gradient of the reforming gas in the tube side is very large. In order to reduce the temperature difference stress of reformer gas waste heat boiler under high temperature and high pressure and improve the reliability of equipment operation, the reformer gas waste heat boiler is designed with flexible thin tube plate structure. The control thermal resistance of the reforming gas waste heat boiler is in the tube side, but because of the high temperature of the reforming gas and the complex structure design of the equipment, the special heat transfer enhancement structure is generally not adopted.

② Steam generation system of steam reformer waste heat boiler. The steam reformer waste heat boiler is a natural circulation water tube flue waste heat boiler. Taking the 80,000 m³/h hydrogen production unit of a petrochemical enterprise as an example, the waste heat boiler of the reformer is composed of two parts:

high-temperature heat exchange section and low-temperature heat exchange section. The high-temperature heat exchange section is composed of feed preheating section II, feed preheating section I, and steam superheating section. The low-temperature heat transfer part is composed of process condensing water steam superheating section and process condensing water evaporation section.

The flue gas temperature of the waste heat boiler of the steam reformer is about 1000°C. In order to enhance heat transfer and make full use of the waste heat of the high-temperature flue gas, the high-temperature flue gas enters the waste heat boiler of the steam reformer successively and exchanges heat with the preheating section II of the feed, the preheating section I of the feed, and the overheating section of the steam. In the feed preheating section II, the high-temperature flue gas preheats the feed to about 630°C and is sent to the steam reformer for steam reforming reaction. In the feed preheating section I, the high-temperature flue gas preheats the feed to about 500°C and enters the preconversion reactor. In the steam superheating section, the saturated steam produced by the unit is overheated to 425°C. Because the temperature of the flue gas outside the tube and the temperature of the heated medium inside the tube is very high, the three vertical heating surfaces of the high-temperature heat exchange part are mostly composed of φ42 mm, φ51 mm, or φ60 mm plain tubes. In order to control the temperature of the metal wall, the heat transfer enhancement structure is rarely used due to the limitation of material properties.

The flue gas temperature after heat exchanging in the superheated section of steam enters the high-temperature air preheater at 630°C, and the flue gas temperature after heat exchange with the air in the high-temperature air preheater enters the waste heat boiler in the low-temperature section at about 510°C.

In the low-temperature section of the waste heat boiler process condensing water steam superheated section and process condensing water evaporation section, the fuel gas exchanges heat, and the 290°C flue gas is sent into the low-temperature air preheating section. It is necessary to confirm whether the heat transfer enhancement structure can be adopted in the superheated section according to the medium conditions. In the process condensing water evaporation section, the purpose of heat transfer enhancement can be achieved by replacing the plain tube with a high-finned tube to expand the secondary heat transfer surface.

3.6.5 Summary

From the above examples, it can be seen that heat transfer enhancement technology in process plays an important role in reducing energy consumption of hydrogen production unit. By improving the grade of generated steam, the backpressure steam after work or power generation can be used as process steam, so the utilization efficiency of exergy of the system can be improved by the grade use of steam. The high-temperature potential heat of the reforming gas or flue gas waste heat is used to provide part of the heat energy for the conversion reaction, which makes better use of the high-temperature level potential heat of the unit, reduces the scale and floor area of the steam reformer, and saves the consumption of fuel gas. The pinch point analysis method is used to reasonably match the heat source and cold source in the medium temperature part of the unit and optimize the heat exchange process, which can produce more steam as by-product and reduce the subsequent air cooler duty, improve the heat exchange efficiency and exoergic efficiency, and reduce the energy consumption of the unit. With the continuous progress of science and technology, more advanced heat transfer enhancement technologies will be applied to hydrogen production units, thus promoting the development of hydrogen production technology.

3.7 Delayed coking unit

3.7.1 Brief description of process

With the increased import of crude oil and the decreased quality, delayed coking unit has become one of the main processing equipment for residue oil. The delayed coking unit can process almost all the low-quality heavy oil residue discharged from the refinery, including atmospheric and vacuum residue, visbreaking residue, FCC slurry, heavy crude oil, heavy fuel oil, coal tar, and refinery oil-containing sludge. Through a series of thermal cracking and condensation reactions, gas, gasoline, diesel, gas oil, and coke are generated. It is one of the main processing units to increase the yield of light oil and produce petroleum coke in refineries. The required reaction temperature is provided by furnace. The feed is heated and leaves the furnace at high velocity and short residence time before cracking under high temperature and is sent to adiabatic coke tower. Delayed coking process means the cracking and condensation of feed is "delayed" [58].

Delayed coking process is an intermittent-continuous process. The process usually uses one furnace providing feed for two coke towers. The furnace is feeding continuously while two parallel coke towers are in parallel intermittent switch operation. There are many delayed coking processes and each has its own characteristics. The energy consumption is also different, which is mainly affected by feed properties, product requirements, operating conditions, and equipment performance.

According to the number of furnace and the coke tower, delayed coking process can be divided into "one furnace with two towers" process and "two furnaces with four towers" process. According to the process flow, it can be divided into coking fractionation section, steaming and venting section, and absorption-stabilization section. In the four-tower process, absorption-stabilization section includes coking-rich gas compressor part, gasoline absorption part, diesel absorption part, gasoline desorption, and stability part. The absorption-stabilization section in some refineries is relatively simple. Some only by one-stage gasoline absorption with only one absorption column. Some by gasoline and diesel two-stage absorption with absorber and reabsorber. There the absorption-stabilization process of the two-tower and four-tower process are illustrated as examples. The schematic diagram of the typical delayed coking unit is shown in Fig. 3.40.

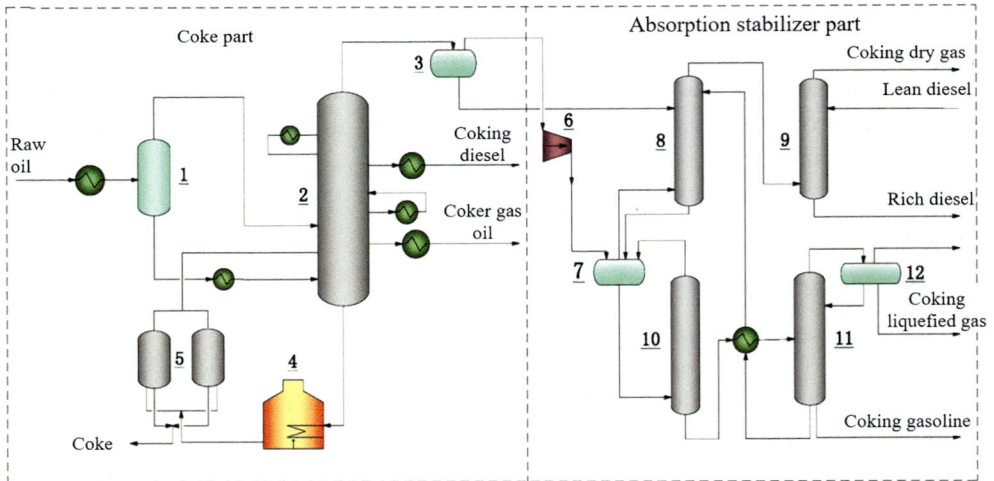

Fig. 3.40 Schematic process flow diagram of typical delayed coking unit. (1) Feed surge drum; (2) fractionator; (3) fractionator OH knockout drum; (4) furnace; (5) coke tower; (6) rich gas compressor; (7) rich gas knockout drum; (8) absorber; (9) reabsorber; (10) desorption tower; (11) stabilizer; (12) reflux drum.

Vacuum residue usually exchanges heat with diesel and is sent into the feed surge drum. The feed is further heated through diesel, pump around (PA), gas oil, and circulating oil to about 300°C and sent into the bottom of the fractionator after mixing with the desuperheating recycle oil of the 420°C high-temperature oil gas from coke tower. Coking oil from fractionator bottom is pressurized by coking furnace feed pump and sent to the furnace. The outlet temperature of the furnace is generally about 500°C. The feed is divided into four streams before sending into the furnace with 1–3 steam injection points on each of the streams. This is to improve the flow rate of coking oil so that it can quickly enter the coke tower without coking and finish cracking and condensation reaction in the coke tower. The generated high-temperature oil gas enters the fractionator after cooling by the distillate and separates the gas, gasoline, diesel, gas oil, heavy gas oil, and circulating oil. Gas and gasoline enter the absorption-stabilization section for further processing to separate dry gas, liquefied gas, and stable gasoline. After heat exchange, part of the diesel, gas oil, and circulating oil return to the fractionator as cold reflux, to ensure the temperature gradient in the tower and vapor-liquid phase balance. The other part of the product is sent to the downstream unit for further refining treatment. The generated high-temperature coke stays in the coke tower and is cooled with steam and water to about 100°C, and the coke is removed to the coke pool with high-pressure water.

3.7.2 Analysis of heat energy characteristics

The average energy consumption and utility consumption of delayed coking units in Sinopec decreased year by year. The energy consumption of delayed coking units is mainly fuel gas, steam, and electricity consumption, which account for 65%–70%, 25%–30%, and 15%–20% of the total energy consumption, respectively. Fuel gas mainly provides energy for the furnace, heats residue from 300°C to 500°C. Heat absorption is required for thermal cracking reaction, and fuel gas consumption accounts for the highest proportion of the total energy consumption of the unit. Among the steam consumed, the steam injection of the furnace accounts for about 3%–5% of the total energy consumption. The compressor is driven by electric motor, and the consumed electricity accounts for 20%–25% of the total energy consumption. If the compressor is driven by steam, the steam consumed will account for about 10%–15% of the total energy consumption. Therefore, reducing the energy consumption of the delayed coking unit should mainly focus on reducing the consumption of fuel gas, steam, and electricity.

The thermal characteristic of delayed coking unit is that cracking reactions need a huge amount of external heat input. Although the feed of delayed coking unit undergoes cracking and condensation reaction in furnace, the proportion of condensation reaction is very small, and the heat release is small. Thermal cracking reaction, as the main reaction, needs a lot of external heat. Fuel gas consumption accounts for the largest proportion. Therefore, optimizing the heat exchange process and improving the temperature of residue in the furnace can save fuel gas consumption and reduce the total energy consumption of the unit.

Another characteristic of the delayed coking unit is that it has a lot of low-temperature heat. Low-temperature heat is divided into two kinds: stable low-temperature heat, which can be fully used, and unstable low-temperature heat, which is not convenient to recycle. The former mainly refers to the remaining low-temperature heat of the fractionator side streams after heat exchange with residue. This part of low-temperature heat can be used to generate self-using steams for the unit according to the temperature which will save steam consumption in the system. The lower level heat can be used to heat cold medium, such as fuel gas or deoxygenated water. The latter kind of unstable heat refers to the heat from periodic operation of the coke tower and the heat of vent from tower OH to fractionator and vent tower, as well as the heat of dumping oil from coke tower to vent tower. When coke is cold, small amount of steam blowing is to blow the vapor from coke tower to the fractionator. When the steam blowing amount is large with small amount of water supply, the generated vapor and steam are discharged to the vent tower. This part of heat can only be cooled by air cooler or cooling water, which cannot be used rationally. Taking a 1.6 Mt/a delayed coking unit of a refinery as an example, the specific analysis is as follows:

① The vapor outlet temperature of the fractionator OH is generally 115–120°C. In order to prevent corrosion of equipment and pipelines by ammonium salt crystallization, corrosion inhibitors and sulfur-containing sewage separated from the fractionator OH drum should be injected before the air cooler. After water injection, the temperature of the OH vapor generally drops 5–10°C, and then cooled by air cooler at 60°C, and further cooled to 40°C before going into fractionator OH drum. This part of the heat cannot be recovered, which is about 11.5 MW, and the energy consumption equals about 0.23 kgoe/(t feed), accounting for about 0.99% of the total energy consumption of the unit.

② The vapor outlet temperature of the vent tower OH is generally 150°C, which is cooled to 50°C by air cooler. This part of the heat is 43.0 MW and equals about 0.21 kgoe/(t feed) energy consumption, accounting for 0.9% of the energy consumption of the whole unit.

③ The dumping oil cooler at the vent tower bottom reduces the dumping oil temperature from 260°C to 90°C and partially returns to the vent tower OH as reflux, and the other part is recycled to the fractionator. Due to intermittent operation, the heat should not be recovered, and the heat is about 5.0 MW.

3.7.3 Process enhancement and integration on heat transfer

3.7.3.1 Enhancement technology of feed heat exchange process

The purpose of feed heating in delayed coking unit is to firstly reduce the viscosity of residue into the feed pump to reduce the power of the feed pump. Secondly, part of the light components can be sent directly into the fractionator after high-temperature residue is flashed in the feed surge drum. Coking feed is heated to 290–320°C by exchanging heat with side-stream diesel, PA, gas oil, heavy gas oil, or circulating oil along the way from feed pump to surge drum. The feed then enters the buffer section of coking fractionator bottom, mixes with the condensed recycling oil from upper desuperheating section, and is then sent to coking furnace by feed pump. The typical heat exchange process of residue is shown in Fig. 3.41.

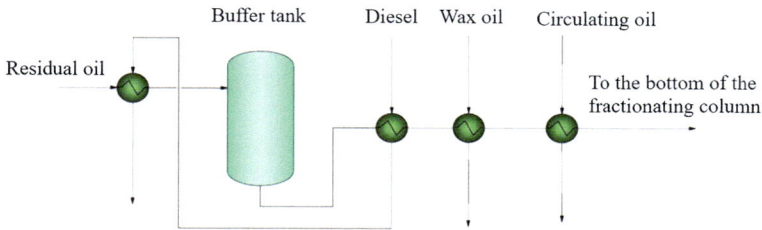

Fig. 3.41 Schematic diagram of typical residue heat transfer process.

Taking the hot feed as an example, the residue is heated from 160°C to 300°C by heat exchange with multiple side streams of the fractionator. This part is a single cold fluid and multihot fluid heat exchange. As long as the temperature level of each side steam is reasonably matched to maximize the final temperature of the residue before flow into the fractionator, there will be no pinch point cross between cold and hot fluid. According to the large fouling coefficient of residue, in order to enhance the heat transfer effect, in the heat transfer process between each side stream and residue, the residue flows through the shell side to improve the heat transfer coefficient of residue. Taking a 1.6 Mt/a delayed coking unit as an example, the heat exchange of the above part is about 16.5 MW, which is equivalent to 6.6 kgoe/(t feed), accounting for about 28% of the total energy consumption of the unit and about 1300 kg/h of fuel gas of the furnace can be saved.

3.7.3.2 Fractionation section of heat transfer enhancement technology

① Enhance the pump around (PA) heat removal process. The high–temperature vapor of coke tower enters the fractionator bottom and exchanges heat with coking gas oil, circulating oil, or feed. According to the different relative volatility of each product, the internal reflux is set up by external heat removal and forms a temperature gradient from fractionator bottom to top on the internal trays of the coking fractionator. Fractionation system heat reflux usually includes all or part of the fractionator OH cold reflux, fractionator OH reflux, diesel reflux, PA reflux, gas oil reflux, and heavy gas oil reflux. The heat removal proportion relates to the product yield and heat exchange process. Usually the heat removal amount above the temperature of diesel extraction accounts for 35%–45% of the total heat removal. And the heat extraction amount below the temperature of diesel extraction accounts for 55%–65% of the total heat extraction [59]. A typical heat removal flow diagram of fractionator is shown in Fig. 3.42.

In order to ensure the flash point of diesel and gas oil, improve the separation resolution of products, and reduce the height of the fractionator, diesel stripper and gas oil stripper can be set up for the fractionator. The typical fractionator with stripper flow diagram is shown in Fig. 3.43.

The two different heat removal processes of fractionator are mainly in the consumption of stripping steam and the power consumption of the pump. Whether to adopt the process with stripping column mainly depends on the requirements of product quality. The main purpose of the stripper is to reduce the overlap of gasoline and diesel and the overlap of diesel and gas oil; improve the flash point of diesel to meet the requirements of safety, stable, and flexible operation; ensure that the vapor-liquid duty is relatively uniform to make the diameter of the fractionator reasonable; and maximize the recovery of heat and reduce power consumption to reduce energy consumption and improve efficiency.

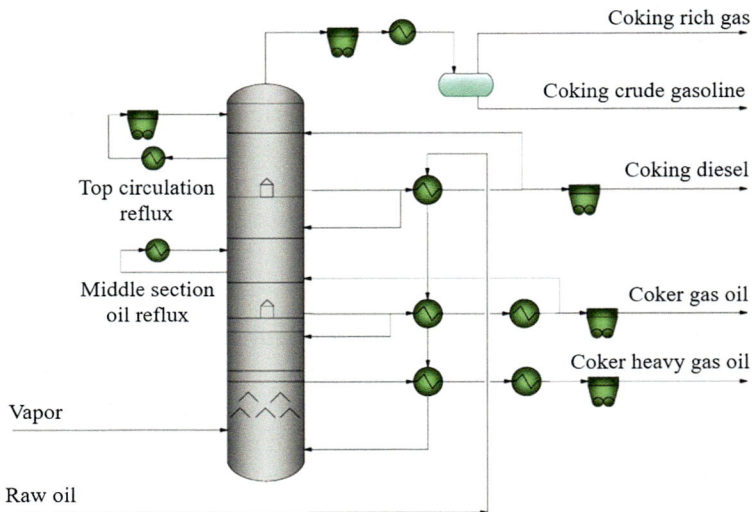

Fig. 3.42 Schematic diagram of typical heat removal flow of fractionator.

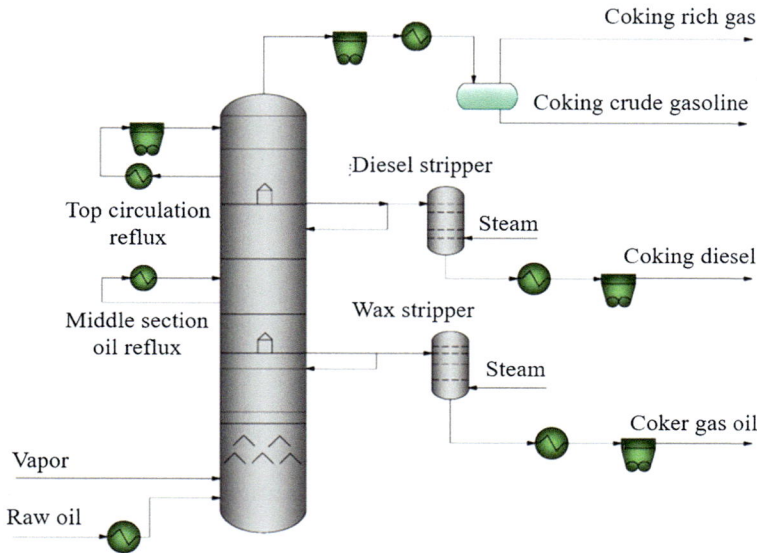

Fig. 3.43 Schematic diagram of typical process flow of fractionator with stripper.

② Application of heat transfer enhancement of fractionator gas oil reflux (return point below extraction point). The strengthening of the inlet distributor of the lower reflux of the fractionator gas oil can reduce the carrying of coke powder into each side stream and the carrying of heavy components into light components and improve the operation cycle of the unit. Taking a delayed coking unit as an example, without reinforcement of lower reflux of the fractionator gas oil, coke powder is carried into the diesel, gas oil, and other side-stream products. At the same time, the coke on the tray, bottom gathered coke, coking, and bottom line export filters need to be dealt with frequently. Dealing with high temperatures and easy spontaneous combustion medium increases the potential danger for operation personnel and will affect the safety, stable operation, and long life cycle of the unit. Through the improvement of the type of gas oil inlet distributor and the improvement of the fractionator bottom structure, the contact area between the lower reflux of the fractionator gas oil and high-temperature oil and gas is increased, and the heat and mass transfer effect of cold and hot streams are strengthened. At the same time, the cooled heavy circulating oil carrying coke powder adopts the full extraction type structure to avoid the coke powder entering the bottom of the fractionator, thus preventing harm to the coking furnace feed pump furnace and greatly improving the safe, stable, and long life cycle performance of the unit.

3.7.3.3 The heat integration of absorption and stability section and the coking section to reduce steam consumption

The main role of the desorption tower is to recover C_3 and C_4 liquefied gas components but also to control the C_2 in liquefied gas. The stabilizer removes the liquefied gas component from saturated gasoline and controls the C_5 in liquefied gas and vapor pressure of gasoline. Taking a 1.6 Mt/a delayed coking unit as an example, the process of heat integration of absorption and stability section and the coking section is adopted. The heat source of desorption tower reboiler can use diesel or PA as the heat source, and the heat source of stabilizer reboiler can use gas oil reflux. The heat of this part is equivalent to 2.65 kgoe/(t feed) and accounts for 11% of the energy consumption of the whole unit. When converted into 3.5 MPa steam, the steam consumption can be reduced by about 24 t/h. At the same time, the design also considers stable gasoline as the heat source of the reboiler in the desorption tower. Converted into 1.0 MPa steam, it can reduce the steam consumption by about 5–6 t/h. Through the integration of the above heat transfer enhancement process, the energy consumption of the unit can be reduced, and the operation cost of the unit can be saved.

3.7.3.4 Application of heat transfer enhancement technology in vent tower

A large amount of high-temperature vapor produced during the cooling of coke on coke tower OH enters the vent tower for cooling and recovers dirty oil and sewage. The dirty oil can be recycled back to the coking section, and the sewage is sent OSBL for purification. After cooling the dirty oil at the bottom of the vent tower, a cold reflux is returned to the vent tower OH. The distributor of the cold reflux into the vent tower adopts the type of spray nozzle to enhance the area of countercurrent contact with high-temperature vapor, improve the rate of mass and heat transfer, and fully wash the coke powder carried in high-temperature vapor. The application of this heat transfer enhancement technology greatly increases the recycling rate of dirty oil, reduces the content of oil in sewage, and improves the operation stability of the downstream sour water stripping unit.

3.7.3.5 Low temperature heat utilization

① Steam generated by gas oil. The temperature level at which the gas oil returns from the stabilizer bottom reboiler is still very high. The heat exchange process is optimized by adopting a low-pressure steam generator to make use of the high-temperature heat of the reflux gas oil. For example, in the 1.6 Mt/a delayed coking unit, the reflux gas oil can produce 4–5 t/h low-pressure steam under normal conditions, which is equivalent to about 1.5–2.0 kgoe/(t feed), which is very beneficial to reduce the energy consumption of the whole unit.

② Heating the low-temperature heating water. In order to reduce the duty of column overhead condenser, the OH reflux heat removal can be used on fractionator OH. However, due to low temperature of the OH recycle reflux, it can only be utilized to heat the fuel gas and low-temperature hot water of the whole plant, and the heat duty is very low. Take a 1.6 Mt/a delayed coking unit as an example; such amount of heat approximately equals to 0.05 kgoe/(t feed).

The gasoline at the bottom of the stabilizer is heated by the desorption tower middle reboiler to 120°C, and the temperature of gas oil product is about 200°C after heat exchange. The excess heat can be used to heat the heating water of the whole plant to 90°C, which not only saves the heat of the whole plant in winter but also reduces the heat duty of the air cooler before sending OSBL. Taking a 1.6Mt/a delayed coking unit as an example, the power consumption saved by the two sections of low-temperature heat utilization is equivalent to about 1.0–1.5kgoe/(t feed).

3.7.4 Heat transfer enhancement technology using elements

During the operation of the unit, as the coke tower needs to switch frequently, the high-temperature vapor preheating of the other tower is required during the switch, which leads to the fluctuation of the flow and heat of the fractionator, which raises higher requirements for the heat exchange equipment of the fractionator system. In addition, the streams in the delayed coking unit contain coke powder, and the fouling resistance is relatively large. The characteristics of process medium and the large-scale unit promote the development and application of heat transfer enhancement of high–efficiency heat exchange equipment. In recent years, heat transfer enhancement components such as low-finned tube and helical baffle have been applied to delayed coking units, and surface evaporative wet air cooler and compound evaporative air cooler have also been successfully applied. According to the characteristics of delayed coking heat transfer streams, the appropriate heat transfer enhancement technology is analyzed according to the process.

3.7.4.1 Feed heat exchange system

The feed of coking unit has two ways: hot feed and cold feed. The hot feed is generally about 160°C, the cold feed is generally no higher than 100°C, and the feed viscosity is high.

The viscosity of feed at 160°C is about 77 mN s/m². Usually after heat exchange with diesel, the feed is sent into the surge drum to reduce the viscosity. The feed viscosity at 190°C can be reduced to 44 mN s/m². The viscosity of diesel for heat exchange with feed is only 0.4 mN s/m². The feed then enters the fractionator after heat exchange with the PA, gas oil, and circulating oil. Before entering the fractionator, the feed is usually heated to about 300°C, and the viscosity is about 9 mN s/m². The viscosity of the PA is about 0.3 mN s/m², the gas oil is about 0.5 mN s/m², and the circulating oil is about 0.25 mN s/m². In order to obtain higher film heat transfer coefficient at lower Reynolds number for feed with higher viscosity, the feed is usually arranged in the shell side of the heat exchanger.

Taking a coking unit as an example, the conventional single-segmental baffle shell and tube heat exchanger are used. Even after the structural optimization design, the film heat transfer coefficient of the hot fluid in the tube side is more than 1500 W/(m² K) in the feed/diesel heat exchanger, the feed/PA heat exchanger, the feed/gas oil heat exchanger, and the feed/circulating oil heat exchanger, the film heat transfer coefficient of the shell side is less than $500 \, W/(m^2 \, K)$. It is obvious that the heat transfer effect of the shell side is much lower than that of the tube side. In order to improve the heat transfer coefficient of the whole heat exchange unit, heat transfer enhancement measures should be applied in the shell side of all feed heat exchangers.

Due to the large viscosity of the feed, the flow pressure loss along the pipe is large. If the pressure drop is reduced by reducing the flow rate, the film heat transfer coefficient will therefore inevitably decrease. At the same time, it will lead to the increase of the dead zone and reduce the effective heat transfer area of the heat exchanger, which will reduce the heat transfer performance of the heat exchanger. Using helical baffle instead of segmental baffle can reduce the pressure drop by changing the flow direction, and the spiral motion reduces the dead zone. The combined effect of the two heat transfer enhancement technologies can reduce the pressure drop in the shell side and improve the overall heat transfer coefficient of the heat exchanger, reduce the required heat transfer area, and reduce the equipment weight and investment of the equipment.

The process condition table of diesel/feed heat exchanger in a coking unit is shown in Table 3.37, and the heat transfer calculation results are shown in Table 3.38. The heat transfer area of the low-finned tube heat exchanger in the table refers to the heat transfer area of the base tube, and the secondary heat transfer area is not considered.

Table 3.37 Process condition of diesel/feed heat exchanger—Duty: 3.1 MW.

Item	Hot fluid	Cold fluid
Fluid name	Raw oil	Diesel
Flow rate (kg/h)	297,620	465,640
Allowable pressure drop (kPa)	50	50
Temperature (°C)	$187 \rightarrow 203$	$213 \rightarrow 204$
Liquid density (kg/m^3)	$942 \rightarrow 933$	$708 \rightarrow 716$
Liquid viscosity $(mN \, s/m^2)$	$44.48 \rightarrow 33.51$	$0.346 \rightarrow 0.371$
Liquid thermal conductivity $(W/(m \, K))$	$0.088 \rightarrow 0.086$	$0.086 \rightarrow 0.088$
Liquid heat capacity $(kJ/(kg \, K))$	$2.304 \rightarrow 2.364$	$2.554 \rightarrow 2.535$

Table 3.38 Diesel/feedstock heat exchanger option comparison.

Item	Original option	Enhanced option
Equipment size (shell diameter × straight tube length) (mm)	1100×6000	1100×6000
Equipment (Set)	4	3
Tube side pressure drop (kPa)	82	92
Shell side pressure drop (kPa)	134	100
Overall heat transfer coefficient $(W/(m^2 K))$	216	375
Heat transfer area (m^2)	335×4	251×3
Total weight of equipment (t)	51	41

As can be seen from Table 3.38, the conventional single-segmental baffle floating head heat exchanger needs four sets of heat exchangers in series, and the effective heat transfer area is $1340\,m^2$. With the same internal structure of the equipment, low-finned tube is used instead of plain tube; only three sets are needed instead of the original four sets with the same specifications. After using low-finned tube instead of plain tube, the metal consumption is reduced by 20% and the equipment investment is reduced. The number of heat exchangers is reduced, and the capital construction investment and operation cost are reduced.

Within the allowable pressure drop, the film heat transfer coefficient of the shell side can also be effectively improved by using the outer corrugated tubes to reduce the boundary layer. The successful applications of heat transfer enhancement in industry include external corrugated tubes, helical baffles, and their combination of enhancement technologies.

3.7.4.2 Heat transfer system of fractionator

1. Fractionator OH vapor cooling system

   The gasoline and rich gas containing H_2S, NH_3, and Cl^- from fractionator OH is about 110°C, which is prone to ammonium salt crystallization and requires water injection and corrosion inhibitor injection. The medium temperature after water and corrosion inhibitor injection is about 75°C. In order to utilize low-temperature heat, the OH gas usually enters the fractionator OH gas/tempered water heat exchanger and uses 40°C tempered water to cool the OH stream to about 55°C. Then the fractionator OH gas is cooled by OH air cooler and OH water cooler to 40°C.

   Although the shell side of the fractionator OH/tempered water heat exchanger and the fractionator water cooler undergo gas condensation, the shell side film heat transfer coefficient of the two heat exchange units is much lower than that of the tube side film heat transfer coefficient. When the flow velocity of the tempered water and cooling water in the heat exchange tube is 0.8 m/s, the film heat transfer coefficient can reach $4100\,W/(m^2\,K)$, and the film heat transfer coefficient of the OH gas is only $500\,W/(m^2\,K)$. It can be seen that the control thermal resistance of the two heat exchangers is in the shell side. The fundamental reason for the low film heat transfer coefficient of the shell side is that the OH gas contains noncondensable gas, which forms a film on the outer wall of the heat exchange tubes that slows the condensation process. Therefore, the main enhancement measure for the fractionator OH tempered water heat exchanger and fractionator OH water cooler is to break the noncondensable gas film close to the outer surface of the heat exchange tube and make the condensable gas flow to the cooling wall. Using spiral-twisted tube bundle instead of baffle plain tube bundle, the light noncondensable gas is centered in the middle of the tube, and the relatively heavy condensable medium is dumped to the heat exchange tube wall through the spiral movement of the medium outside the heat exchange tube, so as to achieve the effect of heat transfer enhancement.

The helical baffle also has the effect of spiral flow, as the fractionator OH noncondensable gas content is high; compared with the spiral-twisted tube heat exchanger, the helical baffle heat exchanger heat transfer enhancement effect is slightly worse.

The heat duty of the water cooler (connected in series after the air cooler) at the fractionator OH is small. However, with 70% (molar fraction) of noncondensable gas at the fractionator OH, the film heat transfer coefficient of the shell side is low, and the effective heat exchange temperature difference is only about 10°C. The overall heat transfer coefficient is only 200–250 W/(m² K), and the heat transfer area required is large.

Considering the process condition of air cooler and after cooler, the use of surface evaporation air cooler is a better choice. The water cooler can be cancelled and the process medium can be directly cooled to 40°C. It can not only reduce the investment in equipment but also save the floor space.

The vapor process condition table at the fractionator OH of a coking unit is shown in Table 3.39. In the conventional design, the scheme of ordinary dry air cooler and water cooler is adopted. For economic considerations, 55°C is taken as the cut-off point between air cooler and water cooler.

Table 3.39 Process parameters of vapor at fractionator OH—Duty: 9.0 MW.

| Item | Hot fluid | Cold fluid | |
		Air cooler	Water cooler
Fluid name	Fractionator OH vapor	Air	Cooling water
Flow rate (kg/h)	84,460		
Allowable pressure drop (kPa)	20		100
Temperature (°C)	101 → 40	36 (inlet)	33 → 40
Gas density (kg/m³)	2.354 → 1.942		
Liquid density (kg/m³)	– → 754		
Gas viscosity (mN s/m²)	0.0112 → 0.0101		
Liquid viscosity (mN s/m²)	– → 0.573		
Gas thermal conductivity (W/(m K))	0.0279 → 0.025		
Liquid thermal conductivity (W/(m K))	– → 0.356		
Gas heat capacity (kJ/(kg K))	1.97 → 1.786		
Liquid heat capacity (kJ/(kg K))	– → 2.37		

If the evaporative air cooler is used instead of the OH air cooler and the water cooler of the fractionator, and 80°C is used as the cut-off point for the hot medium to enter the evaporation section, the condensation and cooling task can be accomplished in one equipment set.

The fractionator OH gas is cooled from 55°C to 40°C, the heat duty is 2.55 MW, and the cooling water temperature rise is calculated as 10°C. The cooling water at 32°C is about 220 t/h. As can be seen from Table 3.40, if the compound evaporative air cooler is used to directly cool the process medium to 40°C and cancel the water cooler, although the tempered water consumption is increased, a large amount of cooling water consumption can be saved and the purpose of water saving can be achieved. The equipment weight can be reduced by 55 tons, the metal consumption can be saved by 30%, and the basic construction cost and the later operation cost can be saved; also the investment in cooling tower, cooling water pump, cooling water network, and the steel frame platform of air cooler can be saved.

Table 3.40 Comparison of fractionator OH air cooler and water cooler scheme.

Item	Original option		Compound evaporative air cooler
	Dry air cooler	Water cooler	
Equipment size (mm)	9000×3000^a	1100×6000^b	9000×3000
Equipment (Set)	10	2	3
Consumption	Electricity: 300 kW	Cooling water: 220 t/h	Electricity: 177 kW Tempered water: 4 t/h
Heat transfer area (m^2)	1915	599	1900
Total weight of equipment (t)	143	22	110

[a] Air cooler size: tube bundle length × width.
[b] Water cooler size: shell diameter × length of straight pipe section of heat exchange tube.

2. Fractionator OH circulating oil system

The circulating oil extracted from the upper section of the fractionator is cooled by OH circulating oil/gas heat exchanger, OH circulating oil/tempered water heat exchanger, and air cooler and then returned to the fractionator.

The control thermal resistance in the OH circulating vapor/gas heat exchanger is the shell side gas. Due to the possibility of acoustic vibration, although the film heat transfer coefficient of the shell side is low, the film heat transfer coefficient of the shell side cannot be improved by increasing the flow rate. The appropriate heat transfer enhancement measure is to increase the secondary heat transfer surface, and the more effective heat transfer enhancement technology is to use the low-finned tube instead of the plain tube as the heat transfer tube. Taking a 2.9 Mt/a delayed coking unit as an example, the heat exchange tube can be shortened by 40% and the steel can be saved by 30% under the same heat exchanger diameter.

In the coking unit, the heat transfer performance of the OH circulating oil/tempered water heat exchanger is relatively good, and the control of thermal resistance is the shell side OH circulating oil. The helical baffle can be used to reduce the dead zone, and the low-finned tube can be used to expand the secondary heat transfer area to enhance the heat transfer, or the double-sided enhancing technology, such as spiral-twisted tube, can be used to enhance the heat transfer of the tube side and the shell side at the same time.

The circulating oil air cooler at the fractionator OH will cool the circulating oil to 80°C before entering the fractionator. If the plate air cooler is used, the floor space of the fractionator can be greatly reduced.

3. Diesel heat exchange equipment

The diesel extracted from the upper section of the fractionator is cooled by the feed heat exchange and divided into two series after heat exchange with the diesel/rich absorbed diesel heat exchanger, diesel/tempered water heat exchanger, and then through the air cooler. Diesel product is sent OSBL, and the absorbed diesel is cooled by the lean-absorbed diesel cooler and then sent to the diesel absorption tower.

The main thermal resistance of diesel/tempered water heat exchanger and lean absorbed diesel cooler is on the shell-side. The effective way to increase the film heat transfer coefficient of shell side is to increase the flow rate of shell side medium. Taking the diesel/tempered water heat exchanger of a delayed coking unit as an example, under the allowable pressure drop and maintaining the same area richness, it can simply reduce the shell diameter to increase the flow rate of the shell side medium. Three single-segmental baffle floating head heat exchangers with a diameter of 900 mm and a tube length of 6000 mm can be used to replace the original design of two heat exchangers with a diameter of 1200 mm and a tube length of 6000 mm, which can save 25% of steel usage. However, the increase in the number of heat exchangers will increase the capital construction cost and operation cost. A more suitable heat transfer enhancement technique is to use low-finned tube or transversally corrugated tube instead of plain tube bundles.

If the diesel is hot discharged, the outlet temperature of the diesel air cooler is about 100°C, and the ordinary dry air cooler can well complete the heat transfer. If the diesel is cold discharge, the diesel air cooler outlet temperature is about 55°C, and the heat transfer area required by the ordinary dry air cooler is large; therefore plate air cooling and other appropriate heat transfer technologies can be used.

The physical properties of the tube and shell side medium of diesel/rich absorbed diesel heat exchanger are close, and the film heat transfer coefficient is close. It is suitable to use spiral-twisted tube heat exchanger, corrugated tube heat exchanger, and other heat exchangers to enhance the shell and tube side liquid heat transfer technology. Taking the diesel/rich absorbed diesel heat exchanger of a coking unit as an example, the technical conditions are shown in Table 3.41.

Table 3.41 Process condition of diesel/rich absorbed diesel heat exchanger—Duty: 3.3 MW.

Item	Hot fluid	Cold fluid
Fluid name	Diesel	Rich absorbed diesel
Flow rate (kg/h)	169,931	44,025
Allowable pressure drop (kPa)	50	50
Temperature (°C)	$188.5 \rightarrow 160$	$58 \rightarrow 175$
Gas density (kg/m^3)	–	$10.6 \rightarrow 14.2$
Liquid density (kg/m^3)	$730 \rightarrow 753$	$779 \rightarrow 691$
Gas viscosity (mN s/m^2)	–	$0.012 \rightarrow 0.014$
Liquid viscosity (mN s/m^2)	$0.416 \rightarrow 0.526$	$0.784 \rightarrow 0.303$
Gas thermal conductivity (W/(m K))	–	$0.041 \rightarrow 0.039$
Liquid thermal conductivity (W/(m K))	$0.091 \rightarrow 0.096$	$0.111 \rightarrow 0.09$
Gas heat capacity (kJ/(kg K))	–	$2.12 \rightarrow 2.231$
Liquid heat capacity (kJ/(kg K))	$2.476 \rightarrow 2.381$	$2.032 \rightarrow 2.463$

Because the flow rate of rich absorbed diesel oil is only one-fourth of diesel, it is arranged to flow in the tube side of the heat exchanger, and the flow rate is increased by the method of multitube pass, so as to obtain relatively higher film heat transfer coefficient. Through thermodynamic simulation, it is found that even when six tube passes are used, the film heat transfer coefficient of the tube side is still lower than that of the shell side, accounting for 40% of the total thermal resistance, and the fouling resistance accounts for about 35% of the total thermal resistance. To enhance the heat transfer performance of the heat exchanger, the reducing of fouling resistance and improving the film heat transfer coefficient of the tube side should be considered. Better heat transfer enhancement results can be obtained by using a spiral-twisted tube with double-sided heat transfer enhancement capability instead of a plain tube. The medium in the tube side and shell side of the spiral-twisted tube is spiral flow, which can not only reduce the boundary layer and improve the film heat transfer coefficient but also effectively reduce the fouling thermal resistance. The comparison scheme is shown in Table 3.42.

Table 3.42 Comparison of diesel/rich absorbed diesel heat exchanger schemes.

Item	Original option	Enhanced option
Equipment size (shell diameter × straight tube length) (mm)	900×6000	700×6000
Set	2	2
Tube side pressure drop (kPa)	43	46
Shell side pressure drop (kPa)	38	9
Overall heat transfer coefficient (W/(m^2 K))	298	349
Heat transfer area (m^2)	388	332
Total weight of equipment (t)	18	10

As can be seen from Table 3.42, in order to not increase the number of equipment, the two-sided enhancing technique of spiral-twisted tube is used to effectively reduce the shell side pressure drop and improve the film heat transfer coefficient under unit pressure drop of shell side medium, but it fails to improve the absolute value of the shell side film heat transfer coefficient. However, in order to ensure the pressure drop of the tube side, the number of tube passes needs to be reduced, so the film heat transfer coefficient of the tube side is not improved. Although the overall heat transfer coefficient is not significantly improved and the heat transfer area is not hugely reduced, the baffle is not needed in the spiral-twisted tube bundle, and the tube bundle self-supporting structure improves the heat transfer area per unit volume. The weight of the equipment is reduced by 44% and the investment of the equipment is reduced.

The spiral flow, can prevent the settlement of particles, effectively destroy the fluid boundary layer, and weaken the trend of scaling. The cleaning cycle can be prolonged and the operation cost can be reduced.

4. Middle section oil heat exchange equipment

The middle oil extracted from the middle of the fractionator exchanges heat with the feed, then provides heat for the desorption tower bottom reboiler, and then partially flows to the steam generator. The partial middle oil from the steam generator mixes with the other middle oil without heat exchange and reaches the temperature of 225°C, which returns to the fractionator as reflux.

In the steam generator using middle oil as the heat source, the control thermal resistance is the tube-side middle oil. As the thermal conductivity of the middle oil is only $0.08 \, \mathrm{W/(m^2 \, K)}$, even if the flow velocity inside the tube reaches 1.5 m/s, the film heat transfer coefficient is only $1500 \, \mathrm{W/(m^2 \, K)}$, and the film heat transfer coefficient of the shell side steam/water is no less than $5000 \, \mathrm{W/(m^2 \, K)}$. In this case, in order to improve the performance of heat exchanger, the flow rate of tube side should be increased as much as possible under the allowable pressure drop. The suitable enhancement means is to use spiral-twisted tube, wave nodule tube or inner spiral tube, or another heat transfer enhancement tube instead of ordinary plain tube.

5. Gas oil heat exchange system

The gas oil is first used as the heat source of the stabilizer bottom reboiler, exchanges heat with feed, then generates steam in gas oil steam generator. The 220°C gas oil from the gas oil steam generator is divided into two streams, one of which returns to the fractionator as reflux, and the other is cooled from 143°C to 90°C via gas oil/deoxygenated water heat exchanger, gas oil/tempered water heat exchanger and air cooler successively before being sent OSBL as product gas oil.

The physical properties of the gas oil and the middle oil are close. The thermo-dynamic analysis of the gas oil steam generator is similar to that of the middle oil steam generator, which can adapt the same heat transfer enhancement technology. The gas oil is in the shell side of the gas oil/deoxygenated water heat exchanger and gas oil/tempered water heat exchanger, and the film heat transfer coefficient in the tube side is up to $5000\,\mathrm{W/(m^2\,K)}$, while the film heat transfer coefficient of the shell side is only $500\,\mathrm{W/(m^2\,K)}$. The appropriate heat transfer enhancement technology is to use spiral–twisted tube and corrugated tube to enhance the heat transfer coefficient of shell side and tube side at the same time and reduce the number of tube passes to reduce the pressure drop.

The effective heat exchange temperature difference of the gas oil air cooler can reach 90°C, and the heat transfer area is not much. Ordinary dry air cooler can achieve heat transfer well.

3.7.4.3 Desorption tower system heat exchange equipment

The reboiler in the middle section of the desorption tower is generally placed horizon-tally, and the reboiling liquid in the middle section flows in the shell side of the reboiler. The boiling range of the boiling medium is relatively long, and a temperature rise of 20°C is required for the state change from the bubble point to 8% vaporization. The film heat transfer coefficient on the boiling side is low, less than $1000\,\mathrm{W/(m^2\,K)}$. The heat source of the reboiler comes from the stabilized gasoline at the bottom of the stabilizer, which has good heat transfer performance. When the pressure drop is $15\,\mathrm{kPa}$, $8000\,\mathrm{W/(m^2\,K)}$ of film heat transfer coefficient can be obtained. The control heat resistance of the reboiler in the middle section of the desorption tower is on the shell side. In engineering, mechan-ical machined porous tubes on the surface are used to enhance boiling heat transfer.

Similar to the middle reboiler, the desorption tower bottom reboiler is mostly placed horizontally. The film heat transfer coefficient of the boiling side is not high, which is close to the film heat transfer coefficient of the medium without phase change. With the high temperature of the hot medium, the effective heat exchange temperature dif-ference is large, and since the heat flux is high, the heat transfer efficiency is within the acceptable range. As the fouling resistance of the bottom liquid of the tower is relatively high, generally $0.0006\,\mathrm{m^2\,K/W}$, resulting in fouling resistance, which becomes the con-trol thermal resistance of the reboiler. Heat transfer enhancement measures should be considered to reduce the scale formation of the medium in the tube.

3.7.4.4 Heat transfer of reflux oil in gasoline absorber

The first and second middle oils of the gasoline absorber are mainly gasoline, which is returned to the gasoline absorber after being cooled by the water cooler directly. The heat duty and flow rate of the two coolers are not large, but the effective temperature difference is relatively small; therefore, it is suitable for multiple series designs. Gasoline has good heat transfer performance, and water has excellent heat transfer performance. The control thermal resistance of water cooler is fouling resistance and shell side heat transfer. When further improving the heat transfer performance of the water cooler, low-finned tube and other heat transfer enhancement technologies to expand the secondary heat transfer surface can be used.

3.7.4.5 Stabilizer heat transfer system

① Stabilizer OH vapor heat exchange. The OH vapor is completely condensed in the stabilizer OH air cooler and then cooled to 48°C by stabilizer OH water cooler and flows into the reflux drum.

The heat duty of the stabilizer OH air cooler is large, there is no superheated or supercooled section in the air cooler, and the heat transfer coefficient is high. However, the temperature of the stabilizer OH vapor is only 68°C, and the effective heat exchange temperature difference of the air cooler is small. The use of compound efficient heat exchanger can save more than 40% of the floor area, about 20% of the comprehensive initial investment can be saved, 30%–60% energy saving and 40%–70% water saving are achieved [13].

The duty of the stabilizer OH cooler is not large, but because the stabilizer OH vapor needs to be cooled to 48°C, if the ordinary dry air cooler is used, the water cooler is also indispensable. Surface evaporation air cooling can be used to replace the stabilizer OH dry air cooler and water cooler.

② Stabilizer OH product cooling equipment. The extracted 50°C liquefied gas from the bottom of the reflux drum is sent to the liquefied gas cooler. After being cooled to 40°C with cooling water, it is sent to the desulfurization unit. Due to the low temperature, the effective heat exchange temperature difference is only about 6°C. The flow rate of liquefied gas is small. In order to increase the flowrate to obtain a higher film heat transfer coefficient, thin and long equipment should be used. Considering various factors, the liquefied gas cooler is suitable for two heat exchangers in series. If one set is used, it is recommended to use threaded pipe to expand the secondary heat transfer area.

③ Stabilizer bottom product stream. The stabilizer middle reboiler uses 3.5 MPa steam as heat source, and the heat transfer coefficient of steam condensation is very high. It is the heat transfer of boiling side that needs to be strengthened. The stabilizer middle reboiler is generally placed horizontally, and the steam is arranged on the tube side. Combined with the physical properties of boiling medium and the structure of reboiler, mechanical machined surface porous tubes can be used to enhance heat transfer at boiling side.

The stabilizer bottom reboiler generally boils on the shell side, and the designed vaporization rate is about 20%. The heat source comes from the high-temperature section of the gas oil in the fractionator. Similar to the desorption tower bottom reboiler, the film heat transfer coefficient on the boiling side is lower than tube side (without phase change), and the overall heat transfer coefficient of the reboiler is less than $400\,W/(m^2\,K)$. Considering the physical properties of the column bottom liquid, the inner porous surface tube can be machined and used to increase the vaporization core and enhance the boiling heat transfer coefficient of the shell side, so as to improve the overall heat transfer performance.

Stabilizer bottom gasoline, as supplement absorbent, is cooled by the stabilizer feed/stabilized gasoline heat exchanger, the desorption tower middle reboiler between, the desorption tower feed/stabilized gasoline heat exchanger/the stabilized gasoline/tempered water heat exchanger, the stabilized gasoline air cooler and water cooler. Then part of the stabilized gasoline is pumped into gasoline absorber as supplement absorbent and the rest is sent OSBL as product.

The main medium in the shell and tube side of stabilizer feed/stabilized gasoline heat exchanger is gasoline, with the same heat transfer performance, which means the two-sided enhanced heat transfer effect of the heat exchange tube, such as twisted pipe and corrugated pipe can be used instead of the plain tube. Thread tube can also be used to expand secondary heat transfer area or the helical baffle to reduce dead zone. Such measures can enhance heat exchanger heat transfer performance.

The specific analysis of the reboiler in the middle of desorption tower can be found in the desorption tower section. The heat transfer performance of the shell and tube side of the feed/stabilized gasoline heat exchanger in the desorption tower is similar. In order to ensure no excessive heat transfer, such heat exchanger is not suitable for the use of heat transfer enhancement technology.

Similar to the middle section cooler of the gasoline absorber, the control thermal resistance of the stabilized gasoline/tempered water heat exchanger and the stabilized gasoline water cooler is fouling resistance and shell side heat transfer thermal resistance. The heat transfer performance of gasoline is good, and the overall heat transfer performance of heat exchanger is good. When further improving the thermodynamic performance of heat exchanger, low-finned tube to expand the secondary heat transfer surface to enhance heat transfer technology can be used.

3.7.4.6 Heat transfer system around the vent tower

The $300–400°C$ vent of blowing steam from the coke tower is absorbed in the vent tower. The OH gas is cooled to $40°C$ by the vent tower OH/tempered water heat exchanger, the vent tower OH air cooler and water cooler, and free flows to the vent tower OH knockout drum for gas and liquid separation. The total allowable pressure drop is about $50\,kPa$.

Shell side condensation is the main control thermal resistance in the vent tower OH vapor/tempered water heat exchanger. Increasing the shell side flow rate cannot effectively improve the shell side film heat transfer coefficient. The technology that can quickly remove the condensate film, such as the cross grain tube, can effectively enhance the heat transfer performance.

Because the condensable part of the vent tower OH vapor is mostly condensed in the vent tower OH vapor/tempered water heat exchanger, the volume rate of the stream in the air cooler and water cooler is not large. Although the heat transfer duty of the air cooler and water cooler is not large, the water cooler must be set because the temperature of the knockout drum is 40°C required by the process.

Due to intermittent use, the water cooler with cooling water as the cooling medium is prone to scale formation, which leads to corrosion under scale and shortens the service life of the equipment.

Overall, the use of surface evaporation wet air cooler instead of vent tower OH air cooler and water cooler will achieve better results.

3.7.5 Summary

From the application effect of the above heat transfer enhancement technologies, the heat transfer enhancement technology in the process plays an important role in reducing the energy consumption of the unit and improving the long life cycle and stable operation of the unit. The heat transfer process of residue fractionator can be optimized to reduce fuel gas consumption. The application of heat transfer enhancement technology in the fractionator and the vent tower can improve the operation cycle of the unit and reduce the frequency of equipment maintenance. The thermal integration of the coking part and the absorption and stability part, and the reasonable utilization of the low-temperature heat of each side stream can not only produce low-pressure steam but also save the consumption of high-pressure steam, greatly reducing the energy consumption of the whole unit.

3.8 Adsorption desulfurization unit for catalytic cracking gasoline

3.8.1 Brief introduction of process

Vehicle gasoline contains a small amount of sulfide, the combustion of which forms SO_2, which is one of the causes of haze induction. Since 2010, China has implemented a gasoline quality upgrade process, and the sulfur content of gasoline has been reduced from not more than 150 to 10 mg/kg in 5 to 8 years in different regions and stages, reaching world-class level. The most important source of gasoline production in China is FCC gasoline, which accounts for about 70% of the gasoline pool, and its sulfur content accounts for more than 95%. Therefore, how to achieve ultradeep desulfurization of FCC gasoline with energy saving and high efficiency has become the key to upgrading the quality of gasoline.

Conventional selective desulfurization technology has many problems, such as high octane loss, high energy consumption, and high processing cost, when it is used for ultradeep desulfurization. To solve this problem, Sinopec acquired the intellectual property rights of ConocoPhillips' S Zorb technology in 2007. However, there are many serious problems in the industrial application of the original technology. For example, the continuous operation cycle of the unit is less than 6 months, which is far from meeting the requirements of industrial production. The technology is incomplete with high catalyst consumption, high energy consumption, and high octane loss, and ConocoPhillips does not have adsorbent preparation technology. To this end, Sinopec organized special scientific and technological breakthrough projects and, through unremitting efforts, successfully developed a new generation of energy-saving and efficient catalytic cracking gasoline adsorption desulfurization technology. This technology is based on the principle of adsorption to desulfurization of gasoline, through adsorbent selective adsorption of sulfur atoms in sulfur compounds to achieve the purpose of desulfurization. Compared with the selective hydrodesulfurization technology, this technology has the advantages of high desulfurization rate, small octane loss, low hydrogen consumption, high liquid yield, less olefin saturation, low operating pressure, and low operating cost. The sulfur content of catalytic gasoline can be reduced to less than 10 mg/kg, which fully meets the sulfur content requirements of national VI gasoline.

By October 2018, 35 units had been built in China using this technology, processing nearly 60% of the country's catalytic cracking gasoline annually, which is more than 46 million tons. The results of industrial application show that the technology has reached world-leading level. Compared with the original technology, the catalyst consumption is reduced by about 50%, the energy consumption is reduced by about 40%, the research octane number (RON) loss is lower, and the sulfur content of gasoline is controlled under 10 mg/kg for a long time. The unit has achieved a long life cycle and created a continuous operation record of 45 months.

The S Zorb unit consists of four parts: feed and desulfurization reaction section, adsorbent regeneration section, adsorbent circulation section, and product stabilization section. Taking a typical S Zorb process in China as an example, the schematic diagram of the process is shown in Fig. 3.44. The S Zorb technology adopts once-through desulfurization process for whole fractions of FCC gasoline without producing hydrogen sulfide. The sulfur in the feed is discharged as sulfur dioxide from the regenerated flue gas and then sent to the sulfur recovery unit or catalytic flue gas tail gas treatment facility for further treatment. Adsorption desulfurization reaction takes place in the reactor and oxidation reaction takes place in the regenerator. The reaction section is in a medium-pressure hydrogen environment, and the regeneration part is in a low-pressure oxygen environment, while the isolation of hydrogen and oxygen environment and the transport of adsorbent are realized by a lock–hopper step sequence control system. Fluidized bed reactor and regenerator are adopted.

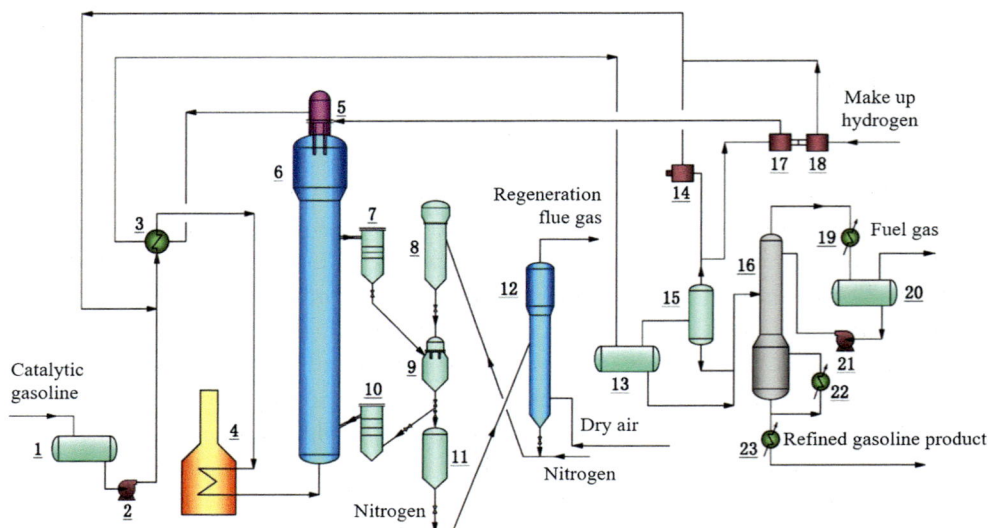

Fig. 3.44 Typical process flow schematic of S Zorb unit. (1) Feed surge drum; (2) adsorption reaction feed pump; (3) adsorption feed heat exchanger; (4) feed furnace; (5) reactor filter; (6) desulfurization reactor; (7) reactor receiver; (8) regenerator receiver; (9) lock hopper; (10) reducer; (11) regenerator feed drum; (12) regenerator; (13) hot product knockout drum; (14) recycle gas compressor; (15) cold product knockout drum; (16) stabilizer; (17) blowback hydrogen compressor; (18) make-up hydrogen compressor; (19) stabilizer OH water cooler; (20) stabilizer OH reflux drum; (21) stabilizer reflux pump; (22) stabilizer reboiler; (23) product cooler.

3.8.2 Analysis of heat energy characteristics

S Zorb unit has two aspects of heat energy utilization: desulfurization reaction heat release and regeneration reaction heat release.

The average comprehensive energy consumption of S Zorb units of Sinopec in 2022 is 5.56 kgoe/(t feed), and the lowest energy consumption is only 3.35 kgoe/(t feed).

3.8.3 Process enhancement and integration on heat transfer

3.8.3.1 Column bottom oil low-temperature heat utilization to enhance and optimize heat transfer

After heat exchange with the feed, the temperature of the stabilizer bottom oil is 140–150°C. In order to fully utilize the heat of the stream, the refined gasoline/hot water heat exchanger is set at the column bottom. The refined gasoline product temperature is reduced, so as to reduce the cooling duty of downstream air cooler and water cooler and achieve the purpose of energy saving. Taking a 1.5 Mt/a standardized S Zorb unit as an example, the heat exchanger can recover about 7.4 MW of heat.

The material stream can also be used for low-temperature power generation where low-temperature hot water is not available. Using ORC low-temperature thermal power generation technology, the thermoelectric conversion efficiency can reach 10%. In the case of a 1.5 Mt/a standardized S Zorb unit, this material can generate about 700 kW net power, accounting for more than 50% of the electrical duty of the unit.

3.8.3.2 The heat transfer enhancement and energy use optimization of the whole unit are of mainly five aspects

1. Heat recovery and utilization of reaction products

   In order to maintain desulfurization efficiency and reduce octane loss, the reaction temperature of S Zorb is generally in the range of 420–440°C. In order to achieve the high reaction temperature, apart from the feed furnace, the heat transfer between the reaction products and the reaction feed is essential for the furnace to raise the temperature of the reaction feed. This part of heat transfer accounts for the largest proportion of the energy consumption of the whole unit. Taking a 1.5 Mt/a S Zorb unit as an example, the total heat duty of such heat transfer system is 51.3 MW.

   Depending on the size of the unit and the nature of the product, the two heat exchange systems are usually designed in parallel, each of which has 3–5 sets of shell and tube heat exchangers in series. After heat exchange, the heated feed enters the furnace, so as to reduce the duty of the furnace. The cooled product enters the hot product knockout drum, which is conducive to the subsequent product separation.

2. Heat removal of regeneration air

   The optimized regeneration temperature of S Zorb is generally in the range of 520–530°C. The temperature of the reactor for catalyst to be regenerated is generally 400°C, and the feed temperature of the regeneration air is 260°C. The heat required for heating regeneration air from room temperature to 260°C comes from the exothermic heat of regeneration reaction. Due to the small amount of regeneration air (no more than 1500 m^3/h), the heat available for heat exchange is not high. Taking a 1.5 Mt/a S Zorb unit as an example, the heat duty of the regeneration air preheater is 96 kW.

3. Utilization of regeneration heat

   A hot coil is set in the regenerator to generate 0.45 MPa low–pressure steam. The low-pressure steam is superheated by the regeneration flue gas cooler, and the regeneration reaction heat is further recovered and utilized. Taking a 1.5 Mt/a S Zorb unit as an example, the total heat duty of the regeneration flue gas cooler is 150 kW.

4. Optimal utilization of heat in the stabilizer system

   The heat used in the stabilizer is derived from the heat of the regenerator. As shown in Fig. 3.45, steam and water closed-circuit circulation design is established within the unit; 1.0 MPa steam is used for stabilizer bottom reboiler as heat source, and the condensate continues heating the stabilizer feed. The heat exchanged low-temperature condensate is sent to regenerator to generate steam, which can be used as heat source for stabilizer bottom reboiler. It realizes high utilization and circulation of heat and water, reducing the heat consumption and water consumption.

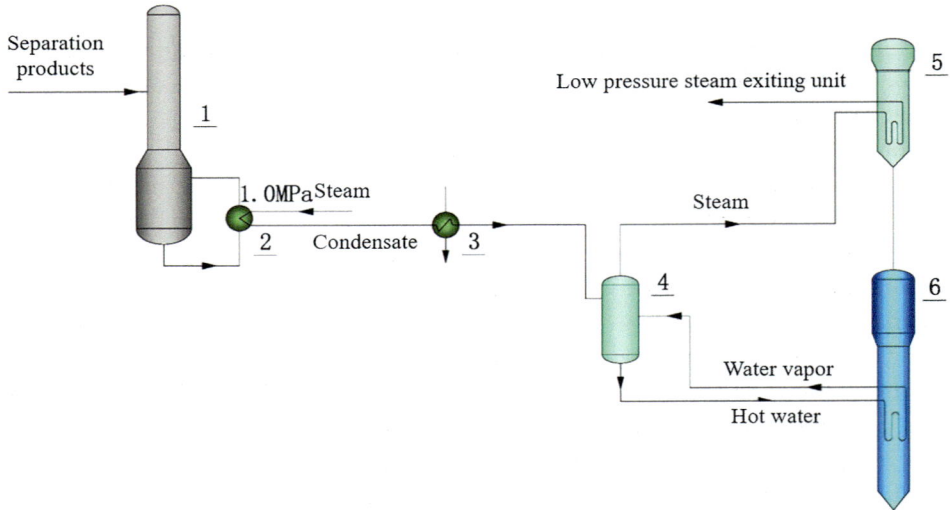

Fig. 3.45 Schematic diagram of heat optimization utilization process of stabilizer system. (1) Stabilizer; (2) stabilizer bottom reboiler; (3) stabilizer feed/condensate heat exchanger; (4) condensing water tank; (5) receiver for regenerator; (6) regenerator.

5. Heat utilization of the auxiliary system

The heat used in the auxiliary system includes heat of high–temperature hydrogen for blowback of the reaction filter, heat of high–temperature hydrogen for the adsorbent circulation system, heat of high–temperature nitrogen for the adsorbent circulation system, etc. The heat source of the high–temperature hydrogen used in the reaction filter blowback comes from the reaction product. After heating to the required 260°C, the reactor filter of the core equipment of the unit is back blown. The heat of high–temperature hydrogen in the adsorbent circulation system is provided by the reaction feed furnace and hydrogen electric heater, which increases the hydrogen temperature to 420°C. The heat of high–temperature nitrogen used in the adsorbent circulation system is provided by the nitrogen electric heater.

After the optimization of the above five aspects of heat utilization, S Zorb is in the world's leading level in terms of energy consumption.

3.8.4 Heat transfer enhancement technology using elements

3.8.4.1 Heat transfer enhancement of the reaction section

The feed heat exchange system in the S Zorb unit is used for heat exchange of reaction feed and reaction product to recover heat from the reaction product. The heat recovery of the heat exchange system is very large, and its heat duty is 6–7 times that of the reaction feed furnace, which has an important impact on the energy consumption and stable operation of the whole unit. The heat exchange system is characterized by high heat exchange duty, large temperature cross range, and small allowable pressure drop, and the cold medium is easy to scale, so the system design and optimization requirements are very strict. The system is usually two series in parallel, each with multiple heat exchangers in series.

Because the temperature cross range of the heat exchange system is large, if the conventional shell is used, the number of heat exchangers in series will be more, and the two shells with longitudinal separator can achieve pure countercurrent flow of shell side stream, significantly reducing the number of heat exchangers in series. In addition, in the reaction circulation loop of the heat exchange system, the pressure drop of the system should be controlled within a reasonable range, and the heat exchanger type should meet the heat transfer requirements with relatively small pressure drop. Since the heat transfer coefficient provided by double-segmental baffles per unit pressure drop is significantly higher than that of single-segmental baffles, it is more suitable for this heat exchange system.

The feed heat exchanger of a 2.4 Mt/a S Zorb unit is shown as an example, and the process conditions are shown in Table 3.43.

Table 3.43 Process condition of S Zorb feed heat exchanger—Duty: 82.0 MW.

Item	Shell side	Tube side
Fluid name	Adsorption product	Mix hydrogen materials
Flow rate (kg/h)	295,058	292,722
Allowable pressure drop (kPa)	100	75
Temperature (°C)	436 → 142	80 → 381
Gas density (kg/m^3)	33.39 → 22.46	9.565 → 47.22
Liquid density (kg/m^3)	– → 597.6	660.0 → –
Gas viscosity (mN s/m^2)	0.022 → 0.017	0.012 → 0.058
Liquid viscosity (mN s/m^2)	– → 0.173	0.269 → –
Gas thermal conductivity (W/(m K))	0.067 → 0.067	0.105 → 0.058
Liquid thermal conductivity (W/(m K))	– → 0.092	0.108 → –
Gas heat capacity (kJ/(kg K))	3.085 → 2.778	4.671 → 2.966
Liquid heat capacity (kJ/(kg K))	– → 2.618	2.286 → –

For the feed heat exchange system, the heat transfer enhancement heat exchanger is designed with remarkable effect. The comparison of schemes is shown in Table 3.44. The F-shell exchanger can reduce the total weight of the equipment by 474.5 t, save equipment investment and floor area, and reduce power consumption and subsequent operation costs.

Table 3.44 Comparison of S Zorb feed heat exchanger scheme.

Item	Original option	Enhanced option 1	Enhanced option 2
Equipment size (shell diameter × straight tube length) (mm)	BEU 1400 × 5000	BFU 1400 × 6500	BFU 1300 × 6500
Enhancement scheme	E-shell with plain tube	F-shell with plain tube	F-shell with spiral-twisted tube
Equipment (set)	16	10	8
Pressure drop (tube side + shell side) (kPa)	133.2	138.5	144.5
Overall heat transfer coefficient (plain tube) (W/(m² K))	281.0	325.4	371.2
Heat transfer area (m²)	8155.8	6167.2	5960.0
Total weight of equipment (t)	830.0	355.5	222.4

Both sides of the heat exchanger contain hydrogen and other noncondensing gas, which is easy to be enriched on the wall of the heat exchange tube. If the conventional plain tube heat exchanger is used, the film thermal resistance has a great influence on the heat transfer efficiency, resulting in a small film heat transfer coefficient. Therefore, the heat transfer enhancement technology of broken gas film should be adopted. The spiral-twisted tube can make the tube side and shell side stream into spiral flow, which has a good effect on the heat transfer of medium containing noncondensable gas. If the twisted heat exchange tube is used, the feed heat exchange system will only need eight sets of φ1300 mm×6500 mm heat exchangers, and the total weight of the equipment is only 222 t, which has obvious advantages over the investment and occupation of the plain tube heat exchanger.

The reaction products of S Zorb are separated by the "hot high-pressure separator" and "cold high-pressure separator" process. All of the reaction effluent first enters the hot separator, and the gas from hot separator OH is cooled by air cooler and water cooler before entering the cold separator. During the cooling process, the gas phase contains a large amount of noncondensable gas including hydrogen. The heat transfer coefficient of the process medium in the air cooler is relatively reasonable due to the high flow velocity at the tube side.

In the water cooler, the process medium is on the shell side. As the flow rate is low and contains a large amount of noncondensable gas, the heat transfer coefficient is low; the control thermal resistance is mainly on the shell side, which is necessary to enhance the heat transfer in the shell side of the water cooler. For this cooler, it is suitable to use low-finned tube heat exchanger for heat transfer enhancement.

The reaction product cooler of a 2.4 Mt/a S Zorb unit is shown as an example, and the process conditions are shown in Table 3.45.

Table 3.45 Process condition of S Zorb product cooler—Duty: 0.65 MW.

Item	Shell side	Tube side
Fluid name	Gasoline + hydrogen	Cooling water
Flow rate (kg/h)	45,733	55,597
Allowable pressure drop (kPa)	20	50
Temperature (°C)	$55 \rightarrow 43$	$33 \rightarrow 43$
Gas density (kg/m^3)	$8.606 \rightarrow 7.324$	
Liquid density (kg/m^3)	$617.9 \rightarrow 629.7$	$994.9 \rightarrow 991.2$
Gas viscosity (mN s/m^2)	$0.011 \rightarrow 0.010$	
Liquid viscosity (mN s/m^2)	$0.221 \rightarrow 0.223$	$0.749 \rightarrow 0.618$
Gas thermal conductivity (W/(m K))	$0.092 \rightarrow 0.098$	
Liquid thermal conductivity (W/(m K))	$0.107 \rightarrow 0.108$	$0.620 \rightarrow 0.632$
Gas heat capacity (kJ/(kg K))	$4.159 \rightarrow 4.702$	
Liquid heat capacity (kJ/(kg K))	$2.340 \rightarrow 2.278$	$4.178 \rightarrow 4.178$

For the feed heat exchange system, the heat exchanger designed with low-finned tube has obvious heat transfer enhancement effect. The comparison of the schemes is shown in Table 3.46.

Table 3.46 Comparison table of S Zorb product cooler scheme.

Item	Original option	Enhanced option
Equipment size (shell diameter × straight tube length) (mm)	BEU 1000 × 6000	BEU 700 × 6000
Type of heat exchange tube	Plain tube	Low-finned tube
Equipment (set)	1	1
Pressure drop (shell side) (kPa)	18.1	18.4
Overall heat transfer coefficient (plain tube) (W/(m^2 K))	352.7	655.3
Heat transfer area (m^2)	263.3	126.7
Total weight of equipment (t)	9.4	4.8

For such heat exchanger, if plain tube heat exchange is used in the design, a φ1000 mm×6000 mm heat exchanger will be needed. If the low-finned tube is used, a φ700 mm × 6000 mm can meet the requirements. When using low-finned tube, the shell diameter of heat exchanger can be reduced, which can also reduce the weight of the equipment and save floor area.

3.8.4.2 Heat transfer enhancement of stabilizer section

Due to the capacity and operating pressure of the stabilizer may fluctuate within a certain scope, sometimes the reflux will be adjusted accordingly. The stabilizer water cooler duty will also fluctuate within a certain scope; therefore, the application of enhanced heat transfer technology to deal with the potential flow rate and the heat duty fluctuation is very necessary. The low-finned tube and transverse groove heat exchange tube have the advantages of thin condensation liquid film formed in the condensation outside the tube and high condensation heat transfer coefficient. It is more suitable for enhancing the condensation effect on light oil and is suitable for enhancing the heat transfer of stabilizer OH vapor condensation.

Reboiler with 1.0 MPa (gauge pressure) steam as heat source is used for the stabilizer of S Zorb unit. The heat exchange temperature difference of 1.0 MPa (gauge pressure) steam is relatively small, so the required heat transfer area is correspondingly large; therefore the heat transfer enhancement should be carried out. Compared with the plain tube, the T-shaped grooved tube has a significantly lower boiling temperature difference, which can significantly improve the boiling heat transfer coefficient and heat flux and has a good enhancing effect on shell side boiling, so it is suitable for heat transfer enhancing of the stabilizer reboiler.

3.8.5 Summary

The energy consumption of S Zorb unit is only about 1/3 of that of similar gasoline desulphurization unit, and the overall process energy consumption is optimized by fully considering the conditions of reaction heat, regeneration heat release, and stabilizer heat usage.

3.9 Alkylation unit

3.9.1 Brief introduction of process

The alkylated oil has good antiknock and antiseismic properties, the octane number of research method (RON) can reach more than 96, and the octane number of motor method (MON) can reach more than 94. At the same time, alkylated oil has good sensitivity, low steam pressure, and wide boiling point range and does not contain aromatics or alkenes. On the one hand, with the improvement of the engine, the compression ratio of the engine continues to increase, and the requirements for the antiknock and antiseismic performance of gasoline continue to improve. On the other hand, from the point of view of preventing air pollution and protecting environment, the national standard requires that the content of aromatic hydrocarbons and olefins in gasoline are gradually reduced. Therefore, alkylation units play an increasingly important role in refineries [60,61].

The common alkylation processes are mainly hydrofluoric acid alkylation and sulfuric acid alkylation. However, due to the volatile, corrosive, and toxic properties of hydrofluoric acid, sulfuric acid alkylation has been basically adopted in newly built alkylation units in recent years. At present, the latest research progress of alkylation technology mainly focuses on the development of new catalysts, represented by ionic liquid and solid acid processes. The safety of these two technologies is relatively high, but the industrial application is still few. Therefore, this section mainly discusses the current mature alkylation process by sulfuric acid.

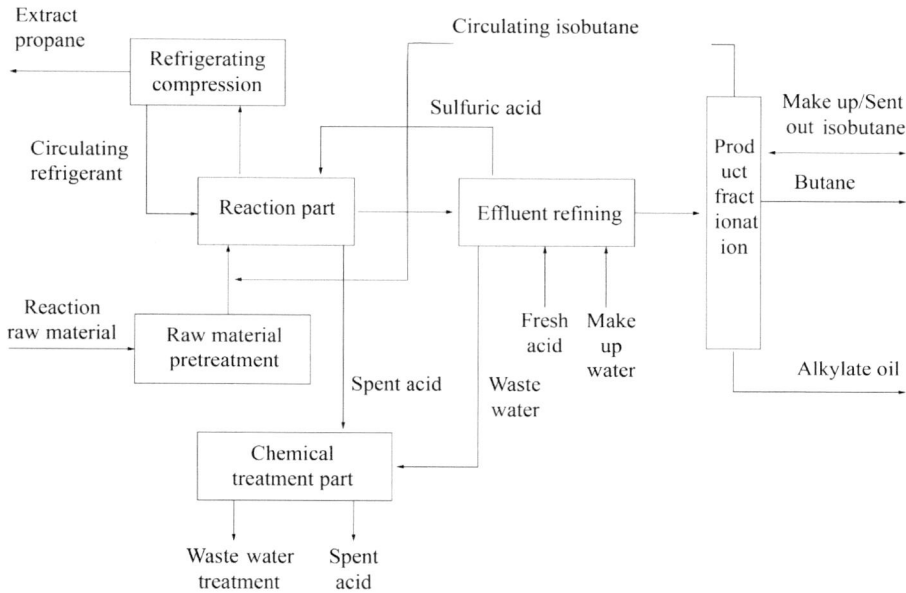

Fig. 3.46 Block diagram of typical process of sulfuric acid alkylation (effluent cooling).

The sulfuric acid alkylation process is mainly composed of C_4 feed hydrotreating, reaction section, refrigeration and compression, effluent refining, product fractionation, chemical treatment, etc., as shown in Fig. 3.46. In the feed hydrotreating section, butadiene is removed by selective hydrogenation in the C_4 fraction and sent to light hydrocarbon removal column to remove excess light components and most of the oxygen-containing compounds. Hydrotreated C_4 is well mixed with recycled isobutane and recycled coolant before sending to contact reactor for alkylation reaction. Reactor effluent flows into acid settling tank to separate acid and hydrocarbon. The separated acid is recycled back to reactor. Separated hydrocarbon is sent to flash drum for gas–liquid separation after being heated by reaction heat. The net reactor effluent is extracted by pump and cooled to 29°C before sending to effluent refining section. Hydrocarbon vapor from flash drum is compressed, condensed, and flashed by depressurization. The temperature of the hydrocarbon vapor is $-10°C$ and extracted by recycled coolant pump and sent to reactor feed section to mix with C_4 feed. On the one hand, the alkane and alkene ratio in the reactor can be guaranteed; on the other hand, the feed temperature of the reactor can be controlled in an appropriate range. The effluent is refined by deacidification, caustic washing, and water washing and sent to fractionation section to separate isobutane, n-butane, and alkylated oil. The chemical treatment section mainly involves the storage of new acid, waste acid, and caustic liquid and the treatment of acid or caustic containing wastewater, which do not involve heat transfer problems.

3.9.2 Analysis of heat energy characteristics

The alkylation unit is a unit with high energy consumption. The main heat energy characteristics can be summarized as follows: first, the reaction temperature is low; second, the fractionation section consumes a large amount of steam; and third, there is a lot of low-temperature heat, and the heat is not easy to be recycled. A brief analysis is as follows:

① Low reaction temperature. Although alkylation is an exothermic reaction, due to its low reaction temperature ($4–11°C$), the reaction heat not only is difficult to use but also needs coolant to remove the reaction heat to control the reaction temperature. Therefore it is necessary to configure a refrigeration system (effluent refrigerant and compressor system) and finally remove the reaction heat through air cooler or cooling water.

② Large steam consumption. Steam consumption accounts for more than 60% of the total energy consumption, among which the deisobutane column consumes the most energy, the bottom temperature is usually $120–150°C$, and 1.0 MPa steam is often used as the heat source.

③ Large amount of low-level temperature heat. The temperature of the fractionator OH of alkylation unit is low, which is only about $55°C$, so the heat is difficult to recover and utilize. Air cooler is usually used for condensation and cooling, so as to reduce the consumption of cooling water. The temperature of the products at the bottom of the column is not high ($100–150°C$). Apart from heating the column feed, the rest of the heat is cooled by air cooler or water cooler.

Table 3.47 shows the energy consumption design data of a 0.3 Mt/a alkylation unit (including the feed pretreatment section, and the refrigeration compressor is electrically driven). It can be seen that the main energy consumption of the unit is 1.0 MPa steam, accounting for 76.3% of the total energy consumption. Among them, 1.0 MPa steam is mainly used in the fractionator reboiler, including the light hydrocarbon removal column (9.1 t/h), the deisobutanizer (36.9 t/h), and the n-butane removal column (4.8 t/h). Therefore, reducing steam consumption is the key to optimizing the alkylation unit process and saving energy.

Table 3.47 Energy consumption index of 0.3 Mt/a alkylation unit in a refinery—Unit: kgoe/(t product).

Cooling water	Demineralized water	Electricity	1.0 MPa steam	Condensing water	Purified wind	Total energy consumption
3.2	0.8	26.6	110.2	−11.1	0.3	130

3.9.3 Process enhancement and integration on heat transfer

3.9.3.1 Streams heat transfer process optimization

The feed of alkylation unit at room temperature should be heated up and sent to the light hydrocarbon removal column and then sent into the low-temperature reaction section after cooling. The low-temperature reaction products are then heated up and sent to the deisobutanizer to meet the separation requirements. The process is shown in Fig. 3.47 [60]. The stream has gone through the process of heating up-cooling down-heating up. Such process will cause a significant loss of energy if the heat is not utilized properly.

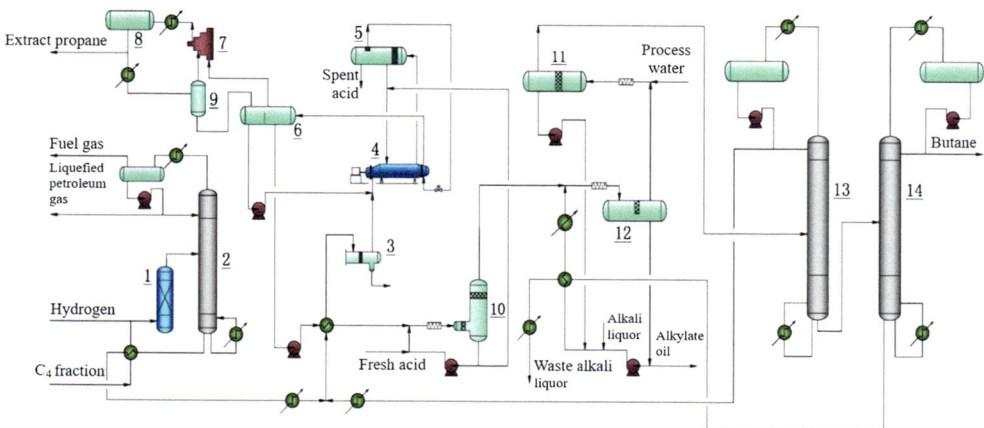

Fig. 3.47 The schematic of typical sulfuric acid alkylation process (effluent cooling). (1) Hydrotreating reactor; (2) light hydrocarbon removal column; (3) feed dehydrator; (4) reactor; (5) acid settler; (6) flash drum; (7) refrigeration compressor; (8) refrigerant tank; (9) energy saving tank; (10) acid wash tank; (11) water wash tank; (12) caustic wash tank; (13) deisobutanizer; (14) n-butane removal column.

In the first step of the heating process, feed can be heated to 70°C by the light hydrocarbon removal column bottom stream and flows into the light hydrocarbon column, so as to make full use of the heat of the light hydrocarbon removal column bottom stream.

The cooling process adopts the method of cascade heat exchange with the cold fluid to make full use of the cooling capacity. Firstly, heat transfer is carried out between the light hydrocarbon removal column bottom stream and the prehydrotreating reactor feed;

the temperature of the bottom stream is cooled to 80°C. It is further cooled by cooling water to 40°C and then heat is exchanged with the flashed liquid of reactor effluent. After the temperature drops to 10°C, the reaction feed is directly mixed with the coolant, thus cooling to the appropriate reaction temperature. The coolant is prepared by refrigeration compression circulation, and the main components are isobutane and propane. Through the above step cooling process, the light hydrocarbon removal column bottom C_4 exchanges heat with the corresponding coolant in the unit, so as to greatly reduce the duty of the refrigeration compressor.

The reaction effluent is throttled and becomes gas–liquid in two phases by the control valve depressurization. After the temperature is lowered, the reaction heat is removed by the reaction system. The temperature of the effluent after heat removal is about −3°C and raised to 30°C after heat exchange with reactor feed.

The heated reaction products are washed with caustic and water and sent to the fractionator. In order to make full use of heat, the high–temperature fractionator bottom stream is used to exchange heat with caustic. Through the above heat transfer process, the effluent cooling process can save the consumption of steam and cooling water as much as possible.

3.9.3.2 Heat transfer enhancement optimization of fractionator feed

The steam consumption in the fractionation section accounts for a large proportion of the total energy consumption of the unit. Therefore, the design optimization of the fractionator will have a great impact on the energy saving and consumption reduction of the unit. Especially for the deisobutanizer, the duty of overhead condenser and bottom reboiler can be significantly reduced by optimizing the feed inlet tray location. The sensitivity analysis method is mainly used to observe the duty changes of the reboiler and condenser by changing the location of the feed tray under the condition that the purity of circulating isobutane and the amount of circulating isobutane are unchanged, so as to find the optimal feed tray location. Taking a 0.3 Mt/a alkylation unit as an example, the relationship between the feed tray location and the duty of overhead condenser and bottom reboiler is shown in Fig. 3.48. It can be seen that when the feed location changes from the top of the tower to the bottom of the tower, the duty of overhead condenser and bottom reboiler decreases greatly and reaches the minimum at the 6th to 11th theoretical tray. As the feed tray continues to descend, the heat duty begins to increase again. Therefore, the feed location of the column can be selected from the 6th to the 11th theoretical tray. If other theoretical trays are selected, especially the trays below the 20th tray, the energy consumption of the unit will increase significantly.

In addition, under the premise of ensuring the circulation volume and purity of the column OH stream, the influence of changing the feed temperature on the duty of column can be investigated, and finally, the appropriate feed temperature can be found to meet the process requirements.

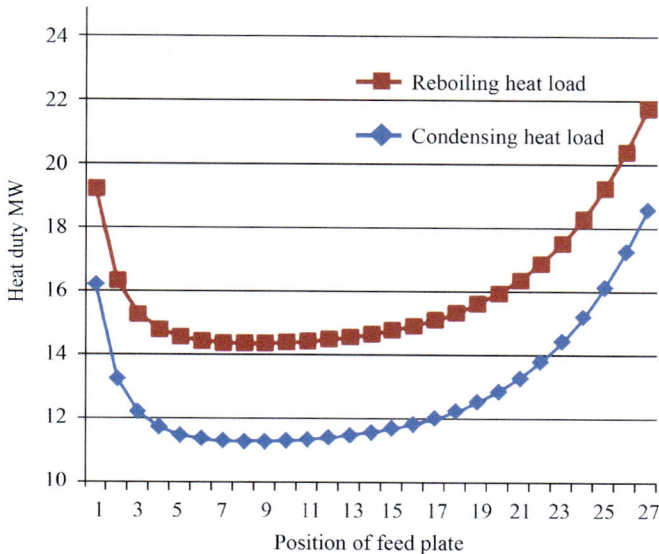

Fig. 3.48 Relationship between heat duty and feed tray location of a deisobutanizer.

3.9.3.3 Optimization of low-temperature heat utilization

The main energy consumption of the alkylation unit is on the reboilers, whose temperatures are not very high. The suitable low-temperature heat of the refinery can be used as the heat source of the reboilers to save steam consumption. For example, low-temperature heat sources in the refinery, such as FCC OH circulating oil, 150°C hot water, or condensate water from low-pressure steam, can be used as heat sources for the light hydrocarbon removal column bottom. According to the amount of heat supplied by 150°C hot water, the low-temperature heat energy consumption of the unit can be converted into standard oil. From the comparison of energy consumption before and after optimization of 0.3 Mt/a alkylation unit in Table 3.48, it can be seen that using 150°C hot water instead of 1.0 MPa steam as the heat source of the reboiler can save steam by 9.1 t/h, and the comprehensive energy consumption of the unit is reduced by about 10 kgoe/(t product), with remarkable energy saving effect. However, using low-temperature heat as reboiler heat source may lower the heat transfer temperature difference, which means more heat exchange area is needed, and therefore it will require more investment. The final judgment should focus on the maximum economic effect. In addition, when an intermediate reboiler is set in the isobutane removal column, since the cold side temperature of the reboiler is generally 60–70°C, almost all the low-temperature heat source in the refinery can be utilized.

Table 3.48 Comparison of energy consumption—Unit: kgoe/(t product).

	1.0 MPa steam	Condensing water	Low temperature heat	Others	Total energy consumption
Conventional energy consumption	110.2	−11.1	0	30.9	130
Low temperature heat utilization energy consumption	90.8	−9.2	7.2	30.9	119.7

3.9.4 Heat transfer enhancement technology using elements

The feed of alkylation unit is C_4 fraction, and the process medium in the unit has the characteristics of low viscosity and good heat transfer performance. Although the process flow of different licensors is different, and the requirements of feed and product are different, the enhancing measures of heat exchange equipment in alkylating unit can still be discussed through typical processes.

3.9.4.1 Heat exchange system of light hydrocarbon removal column system

The feed is heated by the C_4 fraction from the bottom of the light hydrocarbon removal column in the C_4/reactor feed heat exchanger and heated to 80°C by the heater, then enters the light hydrocarbon removal tower after hydrotreated in the reactor.

The temperature of the fluid from the column overhead is about 50–60°C; the low temperature makes it difficult to use it as a heat source. It is directly condensed by the overhead condenser, cooled to 40°C, and returned to the column by the reflux drum. The column OH vapor will need further cooling of 13°C after being condensed in the overhead condenser, and the duty of supercooled section accounts for about 20% of the total heat duty. In order to ensure the cooling effect, it is best to use two sets in series. The film heat transfer coefficient of the cooling water in the condenser is high, which can reach $6000 \, W/(m^2 \, K)$. The film heat transfer coefficient on the condensing side is relatively low, about $1400 \, W/(m^2 \, K)$. The column OH vapor mainly contains C_3 and C_4, and low-finned tubes can be used to increase the secondary heat transfer surface and enhance shell side condensation.

The C_4 fraction from column bottom is cooled to 40°C by the C_4/reactor feed heat exchanger and C_4 cooler and enters the alkylation section. Because of the clean medium, good heat transfer performance, and low fouling thermal resistance, the overall heat transfer coefficient of the heat exchanger can reach $600 \, W/(m^2 \, K)$ even if the ordinary segmental baffle plain tube and shell heat exchanger are used. When the flow velocity of cooling water in the tube is 1.5 m/s, the film heat transfer coefficient is more than

7000 W/(m^2 K). Compared with the tube side, the film heat transfer coefficient of shell side C$_4$ fraction is relatively low, about 1400 W/(m^2 K). In order to further reduce the specifications of heat exchangers and coolers, measures can be taken to enhance the heat transfer of the shell side by increasing the flow velocity of the shell side and increasing the Reynolds number of the shell side medium to enhance the heat transfer, or using heat exchange tubes with double-sided heat transfer enhancement effect such as corrugated tubes to enhance the heat transfer of the tube side and the shell side at the same time.

The reboiler is set at the bottom of the tower, which is a horizontal thermal siphon structure. Low-pressure steam is used as the hot stream in the tube, and the film heat transfer coefficient is more than 10,000 W/(m^2 K). Pure C$_4$ enters the column bottom reboiler, and the superheat of nuclear boiling on the outer wall of the heat exchange tube is relatively small, about 1.4°C. It is relatively difficult to generate bubbles on the smooth wall, and the boiling state of the shell is flow boiling. In this case, it is suitable to use inner porous surface tube instead of ordinary plain tube to promote the formation and growth of the vaporization core, to ensure the performance of heat exchanger.

3.9.4.2 Heat exchange equipment of alkylating reactor system

The alkylated C$_4$ fraction is cooled in the feed/effluent heat exchanger to a suitable temperature by the net reactor effluent before entering the reactor. The net effluent of the reactor contains acidic substances, and its flow velocity needs to be controlled to avoid corrosion and scaling.

Due to the temperature cross of cold and hot streams, it is necessary to adopt the design of a series multiple units, using ordinary segmental baffle plain tube heat exchanger, the overall heat transfer coefficient is 600 W/(m^2 K), and the heat transfer performance of the heat exchanger is good. If F-shell is used to enhance the shell side heat transfer and improve the effective heat exchange temperature difference, the heat transfer area can be reduced by 30%.

Heat transfer enhancement is the core technology of alkylating reactor, which is equipped with heat pipe bundle. However, patent products are mostly used at present, and the technical details have not been made public.

3.9.4.3 Deisobutanizer heat transfer equipment

The heat transfer equipment with the largest heat duty in the alkylation unit is the deisobutanizer system. The column OH temperature is 51–52°C. Although the heat duty is large, the temperature is low and it is not easy to utilize. The heat can only be removed by air cooler or water cooler.

The column bottom temperature is 120–150°C. The low-temperature heat in the refinery is also difficult to be used in the reboiler at column bottom, and steam is used as the heating medium.

Taking a 0.3 Mt/a alkylation unit as an example, the heat duty of column OH was 20.2 MW, and the medium was condensed and cooled to 48°C by air cooler. The technical conditions are shown in Table 3.49.

Table 3.49 Process condition of deisobutanizer OH—Duty: 20.2 MW.

Item	Hot fluid	Cold fluid Air cooler
Fluid name	Isobutane	Air
Flow rate (kg/h)	230,153	
Allowable pressure drop (kPa)	20	
Temperature (°C)	$52 \rightarrow 48$	31 (inlet)
Gas density (kg/m^3)	$17.49 \rightarrow -$	
Liquid density (kg/m^3)	$- \rightarrow 522$	
Gas viscosity (mPa s)	$0.0086 \rightarrow -$	
Liquid viscosity (mPa s)	$- \rightarrow 0.1433$	
Gas thermal conductivity (W/(m K))	$0.0194 \rightarrow -$	
Liquid thermal conductivity (W/(m K))	$- \rightarrow 0.0882$	
Gas heat capacity (kJ/(kg K))	$1.911 \rightarrow -$	
Liquid heat capacity (kJ/(kg K))	$- \rightarrow 2.632$	

According to the local meteorological conditions, the design temperature of the air in air cooler is 31°C. During the heat transfer design, it is found that the OH isobutane is completely condensed in the air cooler tube about 0.5 m from the outlet, and the cooling duty only accounts for 2% of the total duty. The tube film condensation heat transfer coefficient is less than 2400 W/(m^2 K). The effective way to enhance condensation in the tube is to increase the roughness of the tube wall, and the spiral inner rib tube can also be adopted. Although the internal plug inserted in the tube can also enhance the heat transfer of condensation, however, due to the pressure drop increasing rapidly, it is not suitable for this stage.

The water cooler adopts ordinary plain shell and tube heat exchanger, and the film heat transfer coefficient of the tube side cooling water exceeds 6000 W/(m^2 K), while the film heat transfer coefficient of vapor condensing on the shell side is less than 1400 W/(m^2 K). Moreover, when 52°C isobutane enters the water cooler, the effective temperature difference of the water cooler is only 9.6°C. Due to the limit of the allowable pressure drop of the OH isobutane, two heat exchangers with a shell diameter of 1800 and tube length of 8000 mm need to be connected in parallel to complete the heat exchange of 20.2 MW. For the condensation of isobutane outside the tube, the use of low-finned tube has a promoting effect. The external thread of the tube can give full play to the surface tension of the condensate, reduce the condensate film, and reduce the thermal resistance of the liquid film. The comparison scheme is shown in Table 3.50.

Table 3.50 Comparison of heat exchange equipment schemes of deisobutanizer OH.

Item	Dry air cooler	Water cooler	Water cooler
Equipment size (mm)	12,000 × 3000[a]	1800 × 8000[b]	1700 × 6000[b]
Tube type	High-finned tube	Plain tube	Low-finned tube
Equipment (set)	16	2	3
Consumption	Electricity: 592 kW	Cooling water: 1702 t/h	Cooling water: 1702 t/h
Heat transfer area (m²)	4162	2878	1928
Total weight of equipment (t)	288	85	60

[a]Air cooler size: tube bundle length × width.
[b]Water cooler size: shell diameter × length of heat exchange tube.

As can be seen from Table 3.50, 203 tons of steel can be saved if the ordinary segmental baffle plain tube water cooler is used instead of the air cooler. If the low-finned tube is used to replace plain tube in the water cooler, the equipment specification can be further reduced, 228 tons of steel can be saved, and the one-time equipment investment can be reduced by 80%.

The use of water cooler instead of air cooler can save electricity but increase the consumption of circulating water. The appropriate cooling scheme should be determined according to the site situation.

The heat duty of the deisobutanizer bottom reboiler is 22.4 MW. Limited by the harsh working conditions, low-temperature steam is used as heat source, and the effective temperature difference is 54°C. The process conditions are shown in Table 3.51.

Table 3.51 Process condition of the deisobutanizer bottom reboiler—Duty: 22.4 MW.

Project	Hot fluid	Cold fluid
Fluid name	Deisobutane tower bottom hydrocarbon	Diesel
Flow rate (kg/h)	39,665	871,524
Allowable pressure drop (kPa)	50	Thermal siphon
Temperature (°C)	The 175-175	112.4 → 128
Gas density (kg/m³)		− → 18.93
Liquid density (kg/m³)		569 → 566
Gas viscosity (mN s/m²)		− → 0.01
Liquid viscosity (mN s/m²)		0.134 → 0.131
Gas thermal conductivity (W/(m K))		− → 0.027
Liquid thermal conductivity (W/(m K))		0.080 → 0.077
Gas heat capacity (kJ/(kg K))		− → 2.25
Liquid heat capacity (kJ/(kg K))		2.676 → 2.732

According to the calculation of BJU type shell and tube heat exchanger with segmental-baffle and plain tube, the condensing film heat transfer coefficient of low-pressure steam on tube side exceeds 10,000 W/(m² K), and the boiling film heat transfer coefficient on shell side is less than 2000 W/(m² K). It is urgent to take measures to enhance the heat transfer of the shell side. The external low-finned tube and the T-shaped grooved tube have the effect of enhancing the boiling condition outside the tube, which is suitable for enhancing the boiling heat transfer of the deisobutanizer bottom medium.

From the calculation results of shell and tube heat exchanger, it can be seen that the core boiling superheat on the wall of the heat exchange tube is about 5°C. It is suitable to use high flux tube instead of ordinary plain tube to obtain a more vaporized core.

Common plain tube, low-finned tube, and high throughput tube are used as heat exchange tubes to calculate the reboiler. The simulation results are shown in Table 3.52.

Table 3.52 Comparison of reboiler schemes for the bottom of the deisobutanizer bottom.

Item	Plain tube	Low-finned tube	Porous surface tube
Equipment size (shell diameter × straight tube length) (mm)	1500 × 6000	1300 × 6000	1200 × 6000
Set	1	1	1
Tube side pressure drop (kPa)	5	13	10
Shell side pressure drop (kPa)	10	8.7	10
Overall heat transfer coefficient (W/(m² K))	812	1253	1315
Heat transfer area (m²)	670	425	400
Total weight of equipment (t)	20	14	13

It can be seen from Table 3.52 that using low-finned tube bundles instead of plain tube bundles can reduce the equipment weight by 30%. The metal consumption can be reduced by 35% when the inner porous surface tube is used instead of the plain tube bundle, which further reduces the equipment specification.

Low-finned tube can expand the heat transfer area. For comparison, the surface area of the base tube is used as the basis for calculating the overall heat transfer coefficient in Table 3.52. Because the thermal resistance of steam condensation is very small, the main thermal resistance is fouling resistance when using external low-finned tube and porous surface tube. The porous surface of the inner tube can maintain a stable and continuous bubble core, which helps to reduce the influence of fouling thermal resistance, thus obtaining a higher heat transfer coefficient.

3.9.4.4 n-Butane removal column heat transfer equipment

The overhead temperature of the *n*-butane removal column is 51–52°C, which is cooled by air cooler or cooling water. Its characteristics are the same as the deisobutanizer, and the appropriate heat transfer enhancement technology is the same as well.

The column bottom reboiler uses low-pressure steam. When simulating using the plain tube for shell and tube heat exchanger, the core boiling superheat of the heat exchange tube wall in the column bottom reboiler is only about 2°C. In order to ensure that sufficient vaporization core can be generated, it is appropriate to use high-flux tube instead of ordinary plain tube.

The alkylated oil extracted from the column bottom is about 160°C. It exchanges heat with the circulating caustic of the effluent caustic wash, recovers the heat, and then is cooled to 40°C by water cooler.

The heat duty of alkylated oil/circulating caustic heat exchanger is relatively small, the effective temperature difference is relatively large, the circulating caustic in the tube is dirty, and the fouling resistance is larger, which is up to $0.0017\,\mathrm{m^2\,K/W}$. During the design, the flow rate of the tube side should be increased as much as possible under the allowable pressure drop, in order to enhance the degree of caustic turbulence in the tube side and reduce the degree of scaling while enhancing heat transfer.

Taking a 0.3 Mt/a alkylation unit as an example, the duty of alkylated oil/circulating caustic heat exchanger is 2 MW with an effective temperature difference of 42°C. When simulating the heat exchanger with plain tube and single-segmental baffle, the alkylated oil can be cooled from 151°C to 71°C by two floating head heat exchangers with a diameter of 600 mm and a length of 7000 mm in series. When the flow velocity in the tube is 1.2 m/s, the film heat transfer coefficient of the tube side is close to $7000\,\mathrm{W/(m^2\,K)}$, and the film heat transfer coefficient of the shell side is less than $900\,\mathrm{W/(m^2\,K)}$. From the perspective of heat transfer enhancement, the heat transfer performance of shell side should be enhanced. Due to the high scaling property and easy corrosion of the circulating caustic in the tube, it is not suitable for further processing on the heat exchange tube.

When further reducing equipment size, F-shell can be used to reduce one heat exchanger, so as to reduce the equipment investment cost. However, due to the characteristics of caustic, the heat exchanger needs to be isolated very often for maintenance, cleaning, and even replacement of tube bundles. Therefore, a standby heat exchanger will be needed during design. Considering the cost of extracting core for cleaning or replacing the tube bundle, the economy of the F-shell heat exchanger may not necessarily be outstanding. Comprehensive analysis shows that alkylated oil/circulating caustic heat exchangers can improve the film heat transfer coefficient of shell side by using small diameter and multiple sets in series. However, it is necessary to consider the equipment investment, infrastructure cost, and subsequent operation cost to comprehensively evaluate the economy of different design schemes.

3.9.5 Summary

From the above analysis, it can be seen that the energy consumption of alkylation unit is high, mainly on steam consumption. Therefore, how to reduce the steam consumption of the unit has become the main goal of heat transfer enhancement technology in the process. By optimizing the location of the feed tray of the deisobutanizer, the duty of the overhead condenser and the bottom reboiler can be reduced, so as to reduce the consumption of cooling water and steam. The utilization of low-level temperature heat can also reduce steam consumption. Taking a 0.3 Mt/a alkylation unit as an example, by replacing 1.0 MPa steam with 150°C hot water, 9.1 t/h steam can be saved, and the energy consumption of the unit is reduced by about 10.0 kgoe/(t product). In addition, an intermediate reboiler of deisobutanizer can be adopted to reduce the steam consumption of the column by using hot water or other flows with appropriate temperature as the heat source.

The process medium temperature level in the alkylation unit is low. The heat exchanger and water cooler in the process have good heat transfer performance. However, it is still necessary to further reduce the equipment size and equipment numbers by enhancing heat transfer technology, so as to achieve the purpose of energy saving and consumption reduction and intensive floor area.

References

[1] Li ZHQ. Crude distillation process and engineering. Beijing: China Petrochemical Press; 2010. p. 378–80.
[2] Liu JM, Wang YC, Jiang RX. Refining plant process and engineering. Beijing: China Petrochemical Press; 2017. p. 93–7.
[3] Chi L, Yan CH. Atmospheric and vacuum distillation unit heat exchange network optimization design at tens thousands tons scale. Petrochem Ind Des 2008;25(4):1–3.
[4] Zhang JH, Chen YH. Energy saving technology and case analysis of oil refining unit. Beijing: China Petrochemical Press; 2011. p. 47–8.
[5] Wu JS. Application of high efficiency heat exchanger in the expansion and transformation of atmospheric and vacuum distillation unit. Petrochem Equip 2003;32(4):59–60.
[6] Yang QL, Yang M. Application of new technical modification for energy saving and consumption reduction of 100×10^{-4}t atmospheric decompression plant. Chem Ind Oil Gas 2007;36(4):306–9.
[7] Tian ZHF. Application of self-supporting spiral-twisted tube high efficiency heat exchanger in atmospheric and decompression plant. Guangdong Chem Ind 2014;41(8):131–2. 134.
[8] Gu JT, Ma GY, Wu Q. Research on plate air cooler and its application. Chem Equip Technol 2008;29(6):15–7.
[9] Luan HB, Tao WQ, Zhu GQ, et al. Review on the development of all-welded plate heat exchangers. Sci China Tech Sci 2013;43(9):1020–33.
[10] Xu CHY, Jin XR, Liu F. Application of all-welded plate heat exchanger in 10-million-ton grade oil refining plant. Petrochem Equip Technol 2016;37(5):52–7.
[11] Wang X. Overview of technical and economic performance and application of all-welded plate heat exchanger. Sci Technol Guide 2010;6:158–9.
[12] Huang ZY. Application of winding tube heat exchanger in atmospheric and vacuum distillation plant. Petrochem Corros Prot 2018;35(1):41–3.

[13] Tian GH. Application of compound efficient air cooler in petrochemical field. Ind Technol Innov 2014;1(5):600–3.

[14] Deng FY, Liu W, Guo HX, et al. Research and industrial application of corrugated tubes heat exchanger. Refin Technol Eng 2005;35(8):28–32.

[15] Liu PB, Zhang SHQ, Wu XQ. Application of helical baffle heat exchanger in atmospheric and vacuum distillation unit. Refin Technol Eng 2014;44(10):50–3.

[16] Zhou JX, Song BT, Chen SF, et al. Titanium plate and shell heat exchanger and its application at the top of atmospheric tower. Petrochem Equip 2006;35(4):46–7.

[17] Wang J, Yu YB. Application of four-sided detachable all-welded plate heat exchanger in atmospheric tower top condenser. Pet Petrochem Energy Conserv 2017;7(5):6–7.

[18] Liu YH, Chen YH, Liu XL, et al. Application of roof reduction plate air cooler in atmospheric decompression unit. Petrochem Equip 2005;34(6):59–60.

[19] Li ZHQ. Crude distillation process and engineering. Beijing: China Petrochemical Press; 2010. p. 920.

[20] Meng XP, Fan FF. Shell and tube surface condenser, vacuum condensers and vacuum system on top of vacuum column. 2013. CN 202822845U.2013-03-27.

[21] Han J, Wang W, Jiang HM, et al. A kind of surface condenser. 2017. CN 106643199A. 2017-05-10.

[22] Han J, Sun WH, Qu YL, et al. New condenser. 2015. CN 204085223U.2015-01-07.

[23] Liu JM, Wang YC, Jiang RX. Refining plant process and engineering. Beijing: China Petrochemical Press; 2017. p. 113–4.

[24] Wu L. Analysis of energy flow and energy saving in catalytic cracking. Petrochem Des 2018;35(1):55–9 [in Chinese].

[25] Li L. Industrial application of gas-controlled external circulation heat extraction technology. Refin Des 1999;29(9):11–7 [in Chinese].

[26] Sun FW, Zhang YM, Lu CX, et al. Large-scale cold model experiment on heat transfer and flow characteristics of external heat extractor in catalytic cracking. Acta Pet Sin (Pet Process) 2013;29(4):633–40.

[27] Lai ZP. Heat extractor. Petrochem Equip 1986;15(1):16–20.

[28] Yu LH. Calculation and analysis of actual energy consumption in catalytic cracking unit. Pet Refin Chem Ind 2004;35(3):53–6.

[29] Chen LP. Development and application of external heat extractor in residual catalytic cracking unit. Petrochem Equip 1992;21(6):1–5.

[30] Lai ZP. Heat removal technology in catalytic cracking process of heavy oil. Refin Des 1995;25(6):44–8.

[31] Zhang RK, Zhang RS, Gong H, et al. An external heat extractor. 2002. CN 2515637.2002-10-09.

[32] Johnson DR, Brandner KJ. FCC catalyst cooler. 1993. US 5209287.1993-05-11.

[33] Lai ZP, Liu HY, Yang QY, et al. External heat collector. 1992. CN 1016271B. 1992-04-15.

[34] Gu YZ, Hao XR, Li L, et al. Finned tube and finned tube external heat extractor. 2009. CN 201229140Y. 2009-04-29.

[35] Tang FY, Hao XR, Li L, et al. A finned tube and a finned tube external heat extractor. 2009. CN 201229141Y. 2009-04-29.

[36] Wang H, Zhang L, Zhao HJ, et al. Ribbed tube and ribbed tube external heat extractor. 2013. CN 202853448U.2013-04-03.

[37] Zhang RK, Zhang RSH. Development and application of external heat extractor with down-flow dense phase catalyst for heat transfer enhancement in FCC unit. Pet Refin Chem Ind 2006;37(4):50–4.

[38] Andersson E. Minimising refinery costs using spiral heat exchangers. Pet Technol Q 2008;13(2):75–84.

[39] Xu R. Industrial application of high heat flux heat exchangers in refineries. Petrochem Technol Appl 2001;19(4):248–9.

[40] Liang LH. Study on heat transfer performance and industrial application of T-shaped grooved tube reboiler. Refin Des 2001;31(4):20–3.

[41] Zhang Q, Wang XS, Chen QZ, et al. Process design and application of T-shaped finning tube reboiler for light diesel stripper bottom. Mod Chem Ind 2017;37(11):162–6.

[42] Ji XJ. Industrial application of corrugated tubes heat exchange tube bundle. Petrochem Equip Technol 2001;22(1):20–3.

[43] Liu JM, Wang YC, Jiang RX. Refining plant process and engineering. Beijing: China Petrochemical Press; 2017. p. 6.
[44] Han CHR. Hydrocracking process and engineering. Beijing: China Petrochemical Press; 2001. p. 1.
[45] He WF. First application of winding tube heat exchanger in hydrocracking unit. Petrochem Equip Technol 2008;29(3):14–7 [in Chinese].
[46] Chen YD, Chen XD, Jiang HF, et al. Design technology of large-scale coil-wound heat exchanger in refinery industry. Proc Eng 2015;130:286–97.
[47] Song D, Jian JH, Zhang YK. Research and performance analysis of twin-shell side heat exchanger with spiral-twisted tube. Petrochem Equip Technol 2012;33(5):1–3.
[48] Xue N. Design of the unit for producing European V diesel oil by ultra deep hydrodesulfurization (RTS) technology. In: Proceedings of 2011 national oil refining hydrogenation technology exchange conference. Beijing: Hydrodesulfurization Technology Information Station of Sinopec Group; 2011. p. 154.
[49] Lou WY. Application of winding tube heat exchanger in diesel hydrogenation plant. Ningbo Energy Conserv 2014;3(3):30–3 [in Chinese].
[50] Shi XL, Zhang HF. Development status and advantages of shell and plate heat exchangers. Chem Eng 2006;2:30–1.
[51] Barnes P. HDS benefits from plat heat exchangers. Pet Technol Q 2004;Q2:85–90.
[52] Shen WL, Li H. Application design comparison of shell and plate heat exchangers and tube heat exchangers in jet coal hydrogenation unit. In: Proceedings of 2012 SINOPEC hydrogenation technology exchange conference. Beijing: China Petrochemical Press; 2012. p. 658.
[53] Ding WB, Chen BD, Pang M. A new way to enhance heat transfer in finned tube of air cooler. Contemp Chem Ind 2004;33(2):112–5.
[54] Li DD, Nie H, Sun LL. Hydrotreating process and engineering. 2nd ed. Beijing: China Petrochemical Press; 2016. p. 792.
[55] Tan ZM, Luo YL, Zhang XY, et al. Application of new high efficiency heat exchanger in chemical production. Chem Equip Des 1993;1:30–3.
[56] Li H, Liu KX, Xue N. Clean diesel production technology to reduce carbon emission. Energy Conserv Emiss Reduction Pet Petrochem Ind 2012;2(1):8–13.
[57] Qian SW, Fang JM, Jiang N. Spiral-twisted tube and mixed bundle heat exchanger. Chem Equip Piping 2000;37(2):20–1.
[58] Qu GH. Delayed coking process and engineering. Beijing: China Petrochemical Press; 2007. p. 1.
[59] Hu YL. Technical manual of delayed coking unit. Beijing: China Petrochemical Press; 2013. p. 213.
[60] Liu JM, Wang YC, Jiang RX. Refining plant process and engineering. Beijing: China Petrochemical Press; 2017. p. 584–600.
[61] Xu CHM, Yang CHH. Petroleum refining engineering. 4th ed. Beijing: Petroleum Industry Press; 2009. p. 507–15.

CHAPTER 4

Heat transfer enhancement in a typical aromatic complex

Contents

The preparation of benzene, toluene, xylene, and other aromatic from crude oil is an important process in refining and petrochemical plants. Different technical routes and processes are used according to the raw materials and the requirements of products.

The aromatics complex with naphtha as the feedstock for the production of paraxylene (PX) is the most complete typical process, which usually includes naphtha hydrotreating, catalytic reforming, aromatics extraction, disproportionation, PX separation, xylene isomerization, xylene fractionation, etc., as shown in Fig. 4.1. Naphtha enters the catalytic reforming unit for cycloalkane dehydrogenation and alkane cyclization and dehydrogenation to produce aromatics with hydrogen as a by-product. Benzene, toluene, and nonaromatics are separated from the reformates through the aromatics extraction unit. Toluene and C_9/C_{10} aromatics undergo disproportionation and alkyl transfer reaction in the disproportionation unit to produce C_8 aromatics and benzene. o-Xylene and m-xylene in C_8 aromatics are differentiated in the isomerization unit to form PX, and ethyl benzene is converted to xylene or benzene. The heavy components in the mixed xylene produced by reforming, disproportionation, and isomerization are removed by xylene fractionation unit, and then the high-purity PX product was separated by PX separation unit. Other C_8 aromatics are returned to the isomerization unit for further conversion.

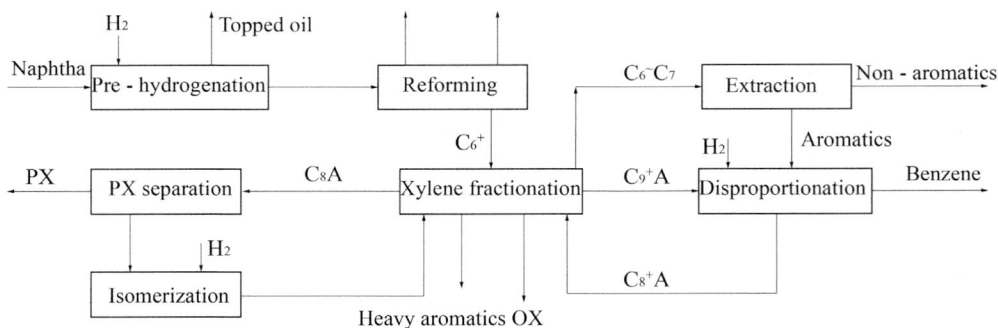

Fig. 4.1 Typical process flow diagram of the aromatics complex.

The process for the aromatics complex is long with equipment, and heat exchanges occur frequently. Therefore the heat transfer equipment plays an important role.

4.1 Naphtha hydrotreating and catalytic reforming unit

Catalytic reforming is the process of converting naphtha into aromatic-rich reformates and hydrogen gas with the action of catalyst, under a certain temperature, and pressure with hydrogen. The reformates can be used as the blending component of vehicle gasoline, as well as the feedstocks to produce benzene, toluene, and xylene. By-produced hydrogen is one of the important sources of hydrogen used in the hydrogenation unit of oil refineries. This unit is an important processing step in oil refining and petrochemical plants.

4.1.1 Brief description of processes [1]

The catalyst used in catalytic reforming is very sensitive to impurities such as sulfur, nitrogen, arsenic, lead, copper, etc., and the activity is also affected by olefin and water. These impurities are the poisons of the catalyst, which can make the catalyst temporarily or permanently deactivated. Therefore, the requirement for impurity content in the raw material is very strict. In addition, the distillation range of the feedstock usually needs to be limited according to the requirements of the reforming reaction. Therefore, the feed needs to be hydrotreated to remove harmful impurities and cut the appropriate fraction before it enters the reforming reaction system. The naphtha hydrotreating unit usually includes three parts: prefractionation, naphtha hydrotreating, and stripper. According to the positions of the prefractionation column in the pretreatment system, it can be divided into the distillation–hydrotreating process and the hydrotreating–distillation process. At present, the hydrotreating–distillation process is more widely used, and the typical process is shown in Fig. 4.2.

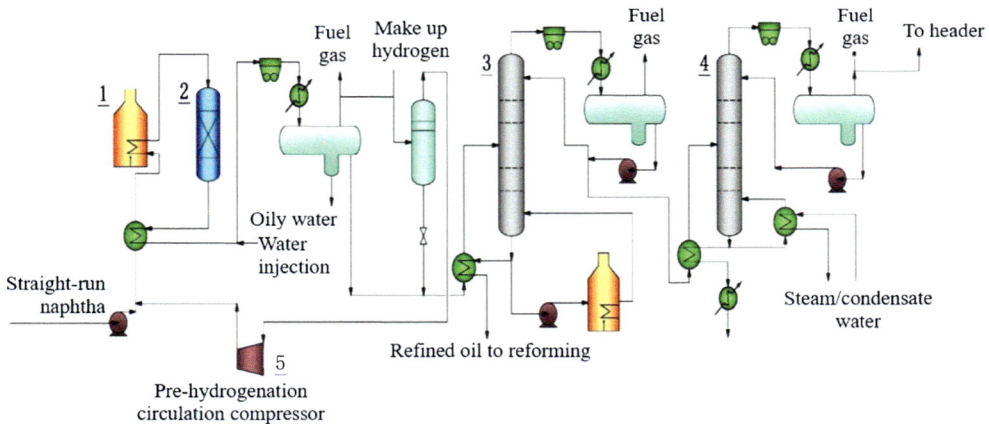

Fig. 4.2 Typical process flow diagram of naphtha hydrotreating unit. (1) Naphtha hydrotreating feed furnace; (2) naphtha hydrotreating reactor; (3) stripping column; (4) light naphtha stripper; (5) naphtha hydrotreating circulation compressor.

The catalytic reforming process can be divided into semiregeneration (fixed bed) and continuous regeneration (moving bed) reforming process according to the different regeneration methods of the catalyst. Semiregeneration reforming process has the advantages of simple process and low investment. However, in order to maintain a long operation cycle of the catalyst, the octane number of the product should not be too high, and the C_5^+ liquid yield and the hydrogen yield of the reforming product are also low. Continuous reforming process is more complicated; the corresponding investment is high, but the octane number of the product is high. The C_5^+ liquid yield, aromatics, and hydrogen yield of the reforming product are high, the startup cycle of the plant is long, and the operation flexibility is great. Usually, the larger the scale of the unit, the worse the raw material. The higher the reaction severity, the more prominent the advantages of continuous reforming. At present, the newly built large reforming plant mostly adopts continuous reforming technology.

Continuous catalytic reforming process is divided into the traditional continuous reforming process and countercurrent continuous reforming process. Universal Oil Products Company (UOP) and Institute Francis DuPetrole (IFP/Axens) hold traditional continuous reforming process. The reaction conditions of the two technologies are basically the same, and the level is roughly the same. There are only some differences in reactor layout, regeneration system control method, catalyst burning, and reduction technology. In order to solve the problem that the activity of the catalyst does not match the difficulty of the reaction in the traditional continuous reforming reaction, Sinopec has developed countercurrent continuous catalytic reforming process (SCCCR). The process has two main characteristics. ① The flow direction between reactors of the catalyst is opposite to the reaction material. The regenerated catalyst is first lifted to the last reforming reactor and then successively transported to the previous reactor. The spent catalyst is lifted from the first reforming reactor to the regenerator to complete the whole catalyst cycle. The difficulty of the reaction in the reactor matches with the activity of the catalyst, which optimizes the reaction kinetic conditions and is conducive to improving the yield of the product. ② The transportation of catalyst from low pressure to high pressure adopts the method of dispersing material sealing. The current pressure boosting method by lock hopper is abolished. The transportation system of catalyst is simplified, valve-free continuous operation is realized, the flow of catalyst is more stable, and the wear amount of catalyst is low. SCCCR has won the first prize in Sinopec Science and Technology Progress because of its advanced technology.

Catalytic reforming unit usually includes three parts: reforming reaction part, recontact purification and product separation part, and catalyst regeneration part. There is little difference in the process of reforming reaction among different technologies, and the process of recontact purification and product separation among different technologies is basically the same. The typical process of reforming reaction and product separation is shown in Fig. 4.3. For the continuous reforming process, most of the current average reaction pressure is 0.35 MPa (gauge pressure), and the pressure of the separator is only 0.24 MPa (gauge pressure). Under this pressure, a large number of light hydrocarbons enter the hydrogen–containing gas, the purity of hydrogen is low, and the loss of hydrocarbon is greater. A recontact system is usually set up to increase the purity of hydrogen, and the yield of reformed gasoline and liquefied gas can be improved. Low temperature is usually used for recontact, and the temperature is usually 4°C. A set of low-temperature freezing systems is required. The separation part of reformates products is generally set up with depentanizer and C_4/C_5 separation column. For units that are not for producing PX, the product separation part also includes dehexanizer, deheptanizer, and xylene column. The C_5^- on the top of the depentanizer is distributed to the debutanizer. The product form the debutanizer overhead is fuel gas and liquefied gas, and the bottom product is pentane. The C_6^+ fraction from the debutanizer bottom is fed into the xylene fractionation unit. A furnace is usually used to provide heat source at the bottom of the depentanizer. The debutanizer usually uses 1.0 MPa (gauge pressure) steam as the reboiler heat source.

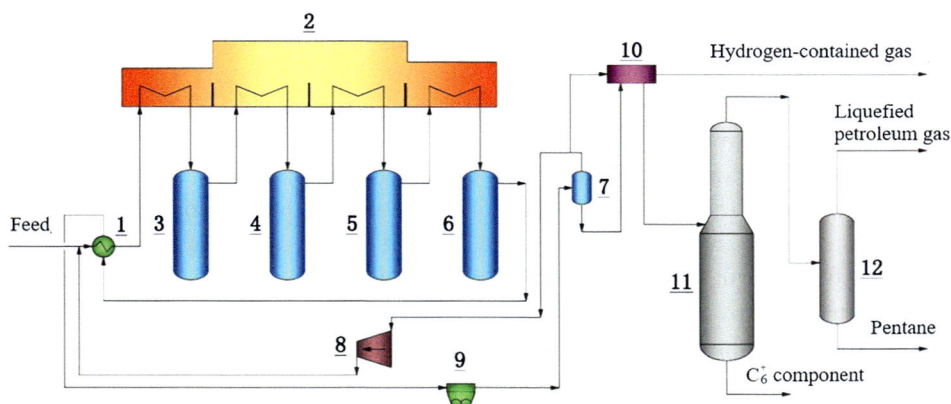

Fig. 4.3 Typical flow diagram of reforming reaction and product separation. (1) Feed/effluent exchanger; (2) reforming reaction furnace; (3) first reforming reactor; (4) second reforming reactor; (5) third reforming reactor; (6) fourth reforming reactor; (7) separator; (8) circulation hydrogen compressor; (9) effluent air cooler; (10) recontact system; (11) depentanizer; (12) C_4/C_5 separation column.

For the catalyst regeneration part, it generally includes coke burning, oxychlorination, drying and calcination, cooling, and reduction. In terms of regeneration loop, UOP uses hot cycle process, while other technologies all use cold cycle process. These technologies also vary on regeneration operation conditions, oxygen content control methods, heat exchange process, and dechlorination process of flue gas.

4.1.2 Analysis of thermal energy characteristics

The main energy consumption of naphtha hydrotreating unit is fuel consumption. How fuel consumption is reduced and the heat utilization efficiency of furnace is improved is the key to reducing the energy consumption of naphtha hydrotreating unit.

The duty of naphtha hydrotreating feed furnace depends on the heat exchange temperature of feed/effluent exchanger. The stripping fractionation column is generally heated by the reboiler heater. Increasing the number of trays in the column can reduce the reflux ratio and the duty of the reboiler heater, but the investment in the column should be increased. The air preheater is used to recover the heat from the flue gas discharged from the naphtha hydrotreating feed furnace and reboiler, and the thermal efficiency can reach more than 92%.

The catalytic reforming unit has more furnaces and much more fuel consumption, which is the largest thermal energy consumption unit in the refinery. Its energy consumption index is higher than that of the catalytic cracking, coking, and hydrogenation units, and about 10 times that of the atmospheric and vacuum distillation unit. Reforming reaction temperature is high (500°C), and the main reaction is naphthene dehydrogenation (reaction heat is more than 210kJ/mol), which needs to consume a lot of heat. Product separation also consumes a lot of heat. The main energy consumption of the catalytic reforming unit is the fuel consumption of the reforming reaction and the product separation section. The energy consumption of the catalyst regeneration part accounts for a very small proportion of the total energy consumption, generally not more than 5%. Energy saving is of great significance to the unit. The key to reducing the energy consumption of the catalytic reforming unit is to maximize the heat recovery by heat transfer enhancement.

The reforming reaction furnace consumes the most fuel, accounting for about 70% of the energy consumption of the whole unit. It includes two different scenarios:

① The first reforming furnace (feed furnace)—with a consumption of about 20% of the fuel, the size of the load mainly depends on the degree of heat exchange of the feed/effluent heat exchanger.

② The second, third, and fourth reforming furnace (intermediate furnace)—with a consumption of more than 70% of the fuel, the purpose is to compensate for the heat absorbed by reforming reaction. The higher the reaction severity, the greater the reaction heat and the greater the effective heat load required to be provided by the furnace. Therefore, the main means of energy saving for these furnaces is to strengthen flue gas waste heat recovery and improve the efficiency of furnaces.

4.1.3 Process enhancement and integration on heat transfer

4.1.3.1 Optimization of the heat transfer process through pinch point analysis method

The reforming feed/effluent exchanger is located in a special position, which is the key equipment of hydrogenation system. The equipment has a great influence on the pressure drop and the utilization of heat energy in hydrogenation system. The duty of the general reforming feed/effluent exchanger is as high as tens of megawatts, which is about five times the duty of the feed furnace. It plays a pivotal role in the utilization of heat energy. The amount of heat exchange of the exchanger is related to the structure and heat transfer area of the heat exchanger.

At early stage, reforming unit feed-effluent exchange adopted a number of U-tube heat exchangers, with low efficiency, larger area, and higher pressure drop. Then it was changed to countercurrent vertical heat exchanger, with hot end temperature difference further reduced, heat recovery increased, and pressure drop greatly reduced. With the further expansion of the scale of the unit, the fully welded plate and shell heat exchanger and coil-wound heat exchanger are widely used. In order to reduce the duty of the feed furnace as far as possible, the hot stream outlet temperature of the heat exchanger is determined by pinch point analysis method.

4.1.3.2 Process heat transfer optimization

In addition to optimizing the heat transfer process of the whole unit by pinch point analysis method, the influence of different process schemes on heat transfer enhancement was compared and analyzed by software of process simulation and heat transfer calculation, and the process scheme with both energy saving and significant economic benefits was selected.

The depentanizer is an important part of the reforming unit. Most of the gas from the depentanizer overhead is directly condensed and cooled by the air cooler and water cooler and then sent to the reflux drum for gas-liquid separation. With the increase of the scale of the unit, it is important to make full use of the low-temperature heat of the column overhead. The typical flow of heat transfer between the depentanizer feed and the column overhead is shown in Fig. 4.4.

Let's take the design of a unit as an example, to compare the difference between these two different process flows. Although one exchanger is added, after the heat exchange between the depentanizer feed and the gas from the column overhead, the temperature of the depentanizer feed is increased by 5°C, and load of the reboiler heater of the column bottom is reduced by 1.2 MW. Based on the fuel gas cost of 2500 yuan/t, the annual fuel cost can be saved by about 2.17 million yuan, and the investment of adding one heat exchange equipment is not more than 300,000 yuan.

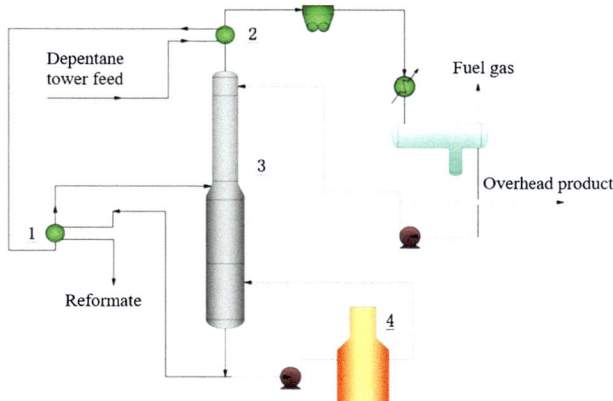

Fig. 4.4 Typical heat transfer flow diagram of depentanizer feed and overhead. (1) Depentanizer feed/bottoms heat exchanger; (2) depentanizer feed/overhead heat exchanger; (3) depentanizer; (4) reboiler heater.

As for reforming reaction furnace, not only its heat load is large, but the temperature of the material is also relatively high. The material does not enter the convection section but enters directly into the radiation section. The radiation section thermal efficiency is only about 60%, so the convection section heat must be considered for use, in order to improve the thermal efficiency of the furnace. At present, it is most commonly used for steam production. In order to further reduce the exhaust temperature and improve the efficiency of the furnace, it can be considered to set up two sections of plate air preheater. The air preheater system is divided into two separate equipment, the high-temperature section, and the low-temperature section, which can reduce the flue gas temperature to below 100, while the thermal efficiency of the furnace can be increased to more than 94%.

The bottom temperature of the depentanizer is 200–250°C. Generally, the reboiler heater is used for heating fluid, and sometimes 3.5 MPa steam is used. The reboiling heat load depends on the reflux flow of the column. Under the condition of meeting the fractionation requirements of the column, it is beneficial to reduce the reflux as far as possible for energy saving.

4.1.4 Heat transfer enhancement technology by improving elements

Heat transfer enhancement technology has been successfully applied in naphtha hydrotreating and catalytic reforming units. For example, corrugated tubes were successfully applied in the "de-bottleneck" revamp of units in 2000, reducing equipment investment by 53% [2]. Twisted tube exchangers, large fully welded plate and shell heat exchangers, and coil-wound heat exchangers have also been widely used in reforming units [3–10]. Plate evaporative air cooler and surface evaporation air cooler also have successful cases [11,12]. The composite high–efficiency air cooler has been applied in the cooling of the top medium of each column [13].

The specific process of different licensor is not completely the same, but the heat transfer enhancement technology of heat exchange element suitable for reforming unit can be studied by analyzing a typical unit. In this section, taking a continuous reforming unit as an example, the characteristics of cold and heat flows of each stage are analyzed according to the process sequence, and the appropriate enhanced option is selected.

4.1.4.1 Heat exchange equipment of naphtha hydrotreating unit

The most critical heat exchange equipment in the naphtha hydrotreating unit is the reaction feed/effluent exchanger. In the exchanger, the low-temperature raw material was heated to about 245°C, until its complete vaporization and superheated. The effluent was cooled from 306°C to 110°C, and 95% (mass fraction) of gas was condensed. The temperatures of cold and hot fluids highly coincide. If conventional shell and tube heat exchanger is used, 6–7 sets of heat exchangers need to work in series to avoid an internal temperature cross, but the total pressure drop of tube and shell side is as high as 600 kPa. Therefore, the F-shell exchanger combined with the double segmental is a kind of suitable heat transfer enhancement technology to avoid the internal temperature cross problem, reduce the number of equipment, and reduce the pressure drop. To further enhance heat transfer, the enhanced heat transfer tube can be used to strengthen the shell side heat transfer, such as corrugated tubes, instead of plain tubes.

The effluent is cooled from about 110°C to about 46°C in the air cooler, and the heat exchange load is relatively large. It is suitable to use surface evaporation air cooler to ensure the cooling effect, and it can reduce the heat transfer area by half. If the composite high-efficiency air cooler is used instead of the common high-finned tube air cooler, the occupied area can be saved by 40%, and the investment can be saved by 20% [13].

4.1.4.2 Heat exchange equipment of reforming system

1. Heat transfer equipment of feed and effluent system

    After the refined naphtha is mixed with hydrogen, it enters the reforming feed/effluent exchanger. It exchanges heat with the effluent from the fourth reactor and is heated by furnace, then enters the reactor. After heat exchange, the effluents are cooled to 40°C by the reforming effluent air cooler and enter the reforming product separator.

    The effluents contained about 81% (mole fraction) of H_2, C_1, C_2, and other noncondensed gas components, and the reactor feed contained 72% (mole fraction) of noncondensed gas components. The duty of reforming feed/effluent exchanger is large, and the process conditions are shown in Table 4.1. As can be seen from the table, the cold fluid temperature rises by 405°C and hot fluid temperature drops by 429°C in the heat exchanger, the temperature of the fluid changes greatly, and the temperature cross interval is large. If conventional multiple tube pass structure is selected, a number of exchangers need to be connected in series. The allowable pressure drop of cold and hot fluid is small, and the gas volume flow rate is large because of the small density, so it is not suitable for multiple tube pass structure. If shell and tube heat exchanger is selected, the only solution is a single tube pass with countercurrent flow heat transfer. The design results of conventional shell and tube heat exchangers and fully welded plate and shell heat exchangers are shown in Table 4.2.

Table 4.1 Process condition of reforming feed/effluent exchanger—Duty: 115.5 MW.

Item	Hot fluid	Cold fluid
Fluid name	Effluents	Feed
Flow (kg/h)	291,807	291,807
Allowable pressure drop (kPa)	The total pressure drop for hot and cold fluid does not exceed 100	
Temperature (°C)	$519 \to 90$	$79 \to 484$
Gas phase density (kg/m^3)	$1.39 \to 2.31$	$3.47 \to 3.17$
Liquid phase density (kg/m^3)	$- \to 796$	$676 \to -$
Gas viscosity (mN s/m^2)	$0.018 \to 0.0097$	$0.0094 \to 0.016$

Continued

Table 4.1 Process condition of reforming feed/effluent exchanger—Duty: 115.5 MW—cont'd

Item	Hot fluid	Cold fluid
Liquid viscosity (mN s/m^2)	$- \rightarrow 0.31$	$0.28 \rightarrow -$
Gas phase thermal conductivity (W/(m K))	$0.205 \rightarrow 0.111$	$0.115 \rightarrow 0.165$
Liquid phase thermal conductivity (W/(m K))	$- \rightarrow 0.114$	$0.107 \rightarrow -$
Gas relative heat (kJ/(kg K))	$2.78 \rightarrow 2.78$	$3.1 \rightarrow 3.63$
Liquid phase specific heat (kJ/(kg K))	$- \rightarrow 1.94$	$2.24 \rightarrow -$

Note: $\rightarrow$, Means from inlet to outlet. Left side of the arrow is the inlet data, and the right side is the outlet data.

Table 4.2 Scheme comparison of reforming feed/effluent exchanger.

Scheme	Tube and shell exchanger	Fully welded plate and shell exchanger
Equipment size (shell diameter × straight tube length) (mm)	$2200 \times 26{,}000$	$3523 \times 11{,}500$
Set	2	1
Tube side pressure drop (kPa)	20	
Shell side pressure drop (kPa)	83	
Heat transfer area (m^2)	9880×2	
Total weight of equipment (t)	471	199

As can be seen from Table 4.2, compared with the conventional shell and tube heat exchanger, the fully welded plate and shell heat exchanger can reduce the height of heat exchanger by 14.5 m, the metal weight of equipment is only 199 tons, and 60% of the metal consumption of material is saved. Due to the reduction of one piece of equipment, it can save the occupied area, civil construction and pipeline installation cost, and the operation cost and avoid the possible deflection problem. The gap between plates is smaller than the gap between the tubes. In actual operation, attention should be paid to controlling the feed impurities to avoid blocking. Cases of increased pressure drop in reforming feed heat exchanger caused by high metal impurities in hydrocracking heavy naphtha are described in relevant literature [14].

The application of coil-wound heat exchanger can also achieve better heat transfer enhancement effect. It is used to replace the shell and tube exchanger in the reforming unit, which can reduce the temperature difference of the hot end by 23.7°C, reduce the load of the reforming feed furnace by 23%, and reduce the energy consumption by 1.61 kgoe/(t PX). At the same time, the energy consumption of the air cooler of the reforming product is reduced by 20%, which brings considerable economic benefits [5]. The application of fully welded plate and shell exchanger and coil-wound heat exchanger can reduce the temperature difference at the hot end, so as to reduce the feed furnace load.

The outlet temperature of reforming effluent air cooler, compressor suction air cooler, and interstage air cooler is usually 40°C. If the conventional air cooler is used, the cooling water cooler should be matched. In order to avoid the leakage of cooling water in the water cooler into the process fluid, the surface evaporation air cooler is used to replace the conventional air cooler and water cooler. In this unit, the duty of the reforming product air cooler is 26.8 MW, and the product is condensed and cooled from 91°C to 40°C. Sixteen sets of 6000 mm × 3000 mm (tube length × bundle width) surface evaporative air coolers can replace 18 sets of tube bundles of 9000 mm × 3000 mm (tube length × bundle width) and 6 sets of water coolers of 1200mm × 6000mm (shell diameter × tube length). The annual electric consumption is saved is about 288,000 kWh, and the cooling water consumption is reduced by 2.568×10^6 t/h. Plate evaporative air cooler is also suitable for heat transfer enhancement, and the equipment size is further reduced. The plate evaporative air cooler is used to replace the original air cooler and water cooler in the reforming unit. Compared with the water cooler, the investment is saved by about 60.7% [11].

2. Recontacter feed and product heat exchange equipment

The bottom oil of the reforming product separator enters the air cooler of the recontact feed for cooling, then enters the recontact tank feed/tank top hydrogen heat exchanger, which is cooled to 38°C by 0°C hydrogen on the tank top, and then enters the recontact tank feed-tank bottom oil heat exchanger for heat exchange with recontact tank bottom oil.

In the recontact tank feed/tank top hydrogen heat exchanger, because hydrogen is a small molecule gas, which is easy to leak, the hydrogen is generally arranged in the tube side of the heat exchanger, and the U-shaped tube heat exchanger is used. The physical properties of cold and hot fluids in the recontact tank feed/tank bottom oil heat exchanger are close, and the recontact tank bottom feed is generally arranged on the shell side. Because the medium of the shell side contains noncondensing gas, which is not conducive to gas condensation, the control thermal resistance of the two heat exchangers is in the shell side, and the film heat transfer coefficient of the shell side is only half that of the tube side film. Measures such as corrugated tubes and twisted tubes to increase the disturbance of the shell side can be used to destroy the noncondensing gas film layer formed on the wall of the heat exchange tube.

The recontact precooler in a 2.0 Mt/a continuous reforming unit adopts twisted tube. The heat transfer coefficient increases by 53%, the weight of the exchanger decreases by 27%, and the pressure drop of the shell side decreases by 58% according to the calibrated data [3].

4.1.4.3 Depentanizer heat transfer system

1. Depentanizer feed heat exchange network

The bottom liquid of the recontact tank and the bottom liquid of the depentanizer undergo heat exchange several times and are heated to 138°C before entering the depentanizer.

In this heat transfer process, the physical properties and flow of cold and hot fluids are close to each other, and the heat transfer effect is better. Taking the first heat exchanger at the bottom of the depentanizer as an example, the process conditions are shown in Table 4.3. Although the heat transfer load is large, the temperature match is reasonable, and the effective temperature difference is 50°C. The comparison results of conventional tubular heat exchanger and enhanced heat transfer scheme are shown in Table 4.4.

Table 4.3 Process condition of depentanizer feed/bottom first heat exchanger—Duty: 8.36 MW.

Item	Hot fluid	Cold fluid
Fluid name	Bottom liquid	Feed
Flow (kg/h)	217,295	266,797
Allowable pressure drop (kPa)	70	70
Temperature (°C)	$210 \rightarrow 146$	$97 \rightarrow 138$
Gas phase density (kg/m^3)	–	$19.45 \rightarrow 21.96$
Liquid phase density (kg/m^3)	$639 \rightarrow 722$	$683 \rightarrow 671$
Gas viscosity (mN s/m^2)	–	$0.01 \rightarrow 0.01$
Liquid viscosity (mN s/m^2)	$0.12 \rightarrow 0.18$	$0.18 \rightarrow 0.15$
Gas phase thermal conductivity (W/(m K))	–	$0.029 \rightarrow 0.03$
Liquid phase thermal conductivity (W/(m K))	$0.098 \rightarrow 0.102$	$0.101 \rightarrow 0.097$
Gas relative heat (kJ/(kg K))	–	$2.09 \rightarrow 2.15$
Liquid phase specific heat (kJ/(kg K))	$2.59 \rightarrow 2.22$	$2.15 \rightarrow 2.30$

Table 4.4 Scheme comparison of depentanizer feed/bottom first heat exchanger.

Scheme	Plain tube	Low-finned tube	Spiral twisted tube
Equipment size (shell diameter × straight tube length) (mm)	1000×6000	1000×4400	900×6000
Set	1	1	1
Tube side pressure drop (kPa)	44	66	21
Shell side pressure drop (kPa)	32	58	8
Overall heat transfer coefficient (W/(m^2 K))	556	759	779
Basal tube heat transfer area (m^2)	421	308	300
Total weight of equipment (t)	9.89	7.89	7.2

The calculation results show that the film heat transfer coefficient is about $2800 \, W/(m^2 \, K)$ when the oil at the bottom of the depentanizer is cooled on the tube side using the ordinary shell and tube heat exchanger structure. The shell side medium is gradually vaporized, and its film heat transfer coefficient is only $1900 \, W/(m^2 \, K)$. The physical properties of cold and hot fluids are close, and the double-sided heat transfer enhancement technology can be used to improve the film heat transfer coefficient of the tube side and the shell side at the same time. The equipment size can also be reduced by secondary expansion such as low-finned tube. Because the feed is vaporized in the shell side, the low-finned tube can not only expand the secondary heat transfer area but also improve the heat transfer coefficient of the shell side film.

It can be seen from Table 4.4 that the length of heat exchanger can be shortened by 1.6 m, 20% metal consumption can be saved, and the primary investment cost can be reduced by using low-finned tube instead of plain tube. If the spiral twisted tube is used instead of the plain tube bundle, the heat transfer coefficient of the tube and shell side film can be improved, the diameter of the heat exchanger shell can be reduced, and the weight of the equipment can be reduced by 27%. Due to the self-supporting structure of the twisted tube bundle, the process of transverse scour of the shell side medium is eliminated, and the pressure drop of the shell side is reduced by 75%. It can be seen that the selection of twisted tube bundle can not only enhance the heat transfer effect but also reduce the power consumption.

In a 2.0 Mt/a continuous reformer unit, the feed/bottom heat exchanger of the depentanizer uses a twisted tube. The field calibration results show that the heat transfer efficiency is increased by 23%, the weight of the equipment is reduced by 12 tons, and the shell side pressure drop is reduced by 29% [3].

2. Heat exchange equipment around the top of depentanizer

The vapor from the depentanizer top is cooled to 40°C through the air cooler and the water cooler and then enters the reflux drum of the depentanizer. In addition to the part of reflux for the depentanizer, the other liquid from the depentanizer reflux drum bottom is used as a C_4/C_5 separation column feed, which is sent into the feed/bottom heat exchanger to be heated.

The process conditions of gas from the depentanizer top are shown in Table 4.5. Conventional air cooler combined with water cooler is usually adopted. The most effective method of heat transfer enhancement is to use surface evaporation air cooler or wet air cooler instead of the combination. It can not only save the amount of cooling water but also save the power consumption.

Table 4.5 Process condition of depentanizer overhead air cooler and water cooler—Duty: 9.0 MW.

Item	Air cooler process fluid	Water cooler process fluid
Fluid name	Depentanizer overhead gas	Depentanizer overhead gas
Flow (kg/h)	102,387	102,387
Allowable pressure drop (kPa)	70	70
Temperature (°C)	$70 \rightarrow 50$	$50 \rightarrow 40$
Gas phase density (kg/m^3)	$21.78 \rightarrow 19.6$	$19.6 \rightarrow 18.1$
Liquid phase density (kg/m^3)	$- \rightarrow 519$	$519 \rightarrow 523$
Gas viscosity (mN s/m^2)	$0.009 \rightarrow 0.0088$	$0.0088 \rightarrow 0.0087$
Liquid viscosity (mN s/m^2)	$- \rightarrow 0.11$	$0.11 \rightarrow 0.12$
Gas phase thermal conductivity (W/(m K))	$0.022 \rightarrow 0.021$	$0.021 \rightarrow 0.021$
Liquid phase thermal conductivity (W/(m K))	$- \rightarrow 0.09$	$0.09 \rightarrow 0.094$
Gas relative heat (kJ/(kg K))	$2.05 \rightarrow 1.96$	$1.96 \rightarrow 1.89$
Liquid phase specific heat (kJ/(kg K))	$- \rightarrow 2.67$	$2.67 \rightarrow 2.62$

The vapor at the top of the column is mainly C_5 fractions, in which H_2, C_1, C_2, C_3, H_2S, and other noncondensable gas components accounted for about 46% (mole fraction). When it is condensed on the shell side of the cooler, the interference of noncondensing gas is very obvious. When the film heat transfer coefficient of the cooling water on the tube side reaches 5000 W/(m^2 K), the coefficient of vapor on the shell side is only 1000 W/(m^2 K). And when the air cooler is set, the effective temperature difference of the water cooler is only 9.7°C. Because the inlet temperature of the process fluid is 70°C, the air cooler can be canceled and directly cooled with cooling water. The surface evaporation air cooler can also be used instead of the combination of air cooler and water cooler to reduce the occupied area. The comparison scheme is shown in Table 4.6.

Table 4.6 Scheme comparison of depentanizer overhead air cooler and water cooler.

Item	Regular option 1		Regular option 2	Enhanced option
	Conventional air cooler	Water cooler	Water cooler	Surface evaporation air cooler
Equipment size (mm)	9000×3000^a	1200×6000^b	1300×6000	6000×3000^a
Set	6	2	3	3
Consumption	Electricity: 180 kW	Cooling water: 232 t/h	776 t/h	Electricity: 150 kW Softened water: 4.5 t/h

Table 4.6 Scheme comparison of depentanizer overhead air cooler and water cooler—cont'd

| Item | Regular option 1 | | Regular option 2 | Enhanced option |
	Conventional air cooler	Water cooler	Water cooler	Surface evaporation air cooler
Energy consumption (baseline)	100%		87%	63.3%
Heat transfer area (m^2)	1152	735	1397	
Total weight of equipment (t)	85	27	48	105

[a]Air cooler size: tube bundle length × width.
[b]Water cooler size: shell diameter × straight tube length.

As can be seen from Table 4.6, compared with the conventional air cooler and water cooler scheme, the effect of surface evaporation air cooler on saving metal weight is not significant, but it can reduce power consumption and cooling water consumption and reduce energy consumption. At the same time, it can reduce the occupied area and infrastructure investment and reduce the operating cost. If the air cooler is canceled, all the heat load will be borne by cooling water, and the consumption of cooling water will increase by 234%. This scheme is not suitable for water shortage areas. If in the area where cooling water is not tight, the weight of the equipment is reduced by 64 tons after the cancelation of the air cooler, which can significantly reduce metal consumption, reduce equipment investment, and reduce energy consumption.

3. Depentanizer bottom heat exchange equipment

The fluid from the depentanizer bottom is divided into two routes: the first route is to exchange heat with depentanizer feed. The detailed analysis of process medium and heat exchange equipment is shown in the content of the feed part. The other route enters the depentanizer furnace or reboiler of the depentanizer bottom and returns to the depentanizer above the liquid level after being heated.

Taking a continuous reforming unit as an example, the specific process conditions of the reboiler at the bottom of the depentanizer are shown in Table 4.7, and steam is used as the heat source.

Table 4.7 Process condition of depentanizer bottom reboiler—Duty: 16.6 MW.

Item	Hot fluid	Cold fluid
Fluid name	3.3 MPa steam	Bottom liquid
Flow (kg/h)	33,033	66,466
Allowable pressure drop (kPa)		Thermal siphon
Temperature (°C)	240 → 240	225 → 234.6

Continued

Table 4.7 Process condition of depentanizer bottom reboiler—Duty: 16.6 MW—cont'd

Item	Hot fluid	Cold fluid
Gas phase density (kg/m^3)		$- \to 29.2$
Liquid phase density (kg/m^3)		$644.6 \to 640$
Gas viscosity (mN s/m^2)		$- \to 0.011$
Liquid viscosity (mN s/m^2)		$0.11 \to 0.11$
Gas phase thermal conductivity (W/(m K))		$- \to 0.031$
Liquid phase thermal conductivity (W/(m K))		$0.098 \to 0.098$
Gas relative heat (kJ/(kg K))		$- \to 2.1$
Liquid phase specific heat (kJ/(kg K))		$2.6 \to 2.645$

As can be seen from Table 4.7, the temperature difference between cold and hot fluids is relatively small. The effective temperature difference is only 8.9°C when calculation based on ordinary shell and tube heat exchanger, and the lowest temperature difference of heat transfer in boiling tube is less than 2°C. For such a low superheat, it is difficult to generate bubbles on the wall of the heat transfer tube. The porous surface tube can be used instead of the plain tube to obtain more gasification core.

Plain tube and porous surface tube were used as heat exchange tubes to simulate the reboiler. For plain tube, two shell types of J-shell and X-shell were used for simulation. The calculation results are shown in Table 4.8.

Table 4.8 Scheme comparison of depentanizer bottom reboiler.

Scheme	Plain tube		Porous surface tube
	BJU type	BXU type	BXU type
Equipment size (shell diameter × straight tube length) (mm)	1700×7500	1800×8000	1500×6000
Set	2	2	2
Overall heat transfer coefficient (W/(m^2 K))	722	591	1625
Heat transfer area (m^2)	1333×2	1579×2	824.5×2
Total weight of equipment (t)	92	103	56.6

According to the result, the shell side pressure drop of the X-shell is lower, and the corresponding column skirt height is also lower, which can reduce the installation cost of the column, but the boiling heat transfer coefficient is lower. If plain tube is used as the heat exchange tube, the J-shell structure is more reasonable.

When the porous surface tube is used instead of the plain tube, the vaporization core is increased and the boiling heat transfer coefficient on the shell side is greatly improved. The heat transfer coefficient of the steam condensing film in the heat exchange tube side is higher than $10,000\,W/(m^2\,K)$; the increase of boiling heat transfer coefficient directly increases the overall heat transfer coefficient of the reboiler. The application of porous surface tube can not only ensure the heat transfer performance of the reboiler more reliably but also save 35.4 tons of steel and further reduce the investment in civil construction.

4.1.4.4 Heat exchange equipment of C_4/C_5 separation column

The liquid from the depentanizer top is mixed with light naphtha from the naphtha hydrotreating unit, which is heated by the C_4/C_5 separation column feed/bottoms exchanger and enters the C_4/C_5 separation column.

The gas from the separation column overhead is cooled to $40°C$ by the air cooler and water cooler and then enters the reflux drum. The ideal method of heat transfer enhancement is to use surface evaporation air cooler instead of conventional air cooler and water cooler. Literature [6] describes the effect of using this scheme, saving 15%–30% of investment and 30%–70% of operation cost.

The C_5 fraction at the bottom of the separation column is divided into two routes. One route is cooled to $40°C$ through the C_4/C_5 separation column feed/bottoms heat exchanger and the C_5 fraction cooler and sent to the storage tank. The other route provides heat for the separation column by the reboiler. The fluid from the separation column bottom is relatively pure, and the fluid has a narrow boiling range. The wall superheat of the reboiler is $2°C$, and it is suitable to use porous surface tube instead of plain tube to obtain more vaporized core.

The process conditions of feed/bottoms heat exchanger in C_4/C_5 separation column of a unit are shown in Table 4.9. The heat transfer load is relatively small, but there is an internal temperature cross between cold and hot fluids.

Table 4.9 Process conditions of feed/bottoms heat exchanger—Duty: 1.2 MW.

Item	Hot fluid	Cold fluid
Fluid name	C_5 component	Heavy naphtha
Flow (kg/h)	20,716	35,457
Allowable pressure drop (kPa)	80	100
Temperature (°C)	$136 \rightarrow 59$	$44 \rightarrow 89$
Gas phase density (kg/m^3)	–	$- \rightarrow 25.4$
Liquid phase density (kg/m^3)	$503 \rightarrow 603$	$575 \rightarrow 516$
Gas viscosity (mN s/m^2)	–	$- \rightarrow 0.009$
Liquid viscosity (mN s/m^2)	$0.12 \rightarrow 0.2$	$0.16 \rightarrow 0.111$
Gas phase thermal conductivity (W/(m K))	–	$- \rightarrow 2.15$
Liquid phase thermal conductivity (W/(m K))	$0.082 \rightarrow 0.102$	$2.4 \rightarrow 2.81$
Gas relative heat (kJ/(kg K))	–	$- \rightarrow 0.025$
Liquid phase specific heat (kJ/(kg K))	$3.08 \rightarrow 2.4$	$0.1 \rightarrow 0.086$

Because the cold and hot fluids have a 30°C temperature coincidence range, the general design scheme of several heat exchangers in series is adopted. After simulation, it is found that if the multiple tube passes structure is adopted, three heat exchangers should be used in series. In the layout, two units in series occupy only one station position, while for the scheme of three units in series, two positions are required. The most effective way to solve this problem is to use countercurrent heat transfer. Considering the process conditions of the fluid, the flow rate of cold and hot fluids is not large; it is not suitable to use large-diameter equipment. If the fixed tube sheet heat exchanger with one tube-pass floating-head heat exchanger is used, two sets are needed in series. If the double-pipe exchanger is used, the heat transfer area is large with bad economy effect. The most suitable heat transfer enhancement technology is to use F-shell exchanger. The simulation results are shown in Table 4.10.

Table 4.10 Scheme comparison of feed/bottoms heat exchangers in C_4/C_5 separation columns.

Scheme	E-shell with plain tube	F-Shell with plain tube	F-shell with corrugated tubes
Equipment size (shell diameter × straight tube length) (mm)	400 × 5000	600 × 7000	500 × 7000
Set	3	1	1
Tube side pressure drop (kPa)	28	3	7
Shell side pressure drop (kPa)	15	14	31
Overall heat transfer coefficient (W/(m² K))	419	333	528
Heat transfer area (m²)	142	180	121
Total weight of equipment (t)	4.3	4.2	3.2

According to calculation, the shell side film heat transfer coefficient is only half of that of tube side when the E-shell heat exchanger is used, which is the control thermal resistance of the heat exchanger. The heat transfer effect of the heat exchanger should be enhanced by increasing the shell side fluid velocity and turbulence. It can be seen from Table 4.10 that the weight of one F-shell heat exchanger with plain tube is close to that of three ordinary E-shell heat exchangers, without saving the equipment cost. However, the F-shell exchanger saves the occupied area and the cost of pipeline installation and civil infrastructure construction. If the composite heat transfer enhancement technology of F-shell with corrugated tubes is adopted, one heat exchanger can be used instead of three conventional tubular heat exchangers, saving 25% of metal consumption and saving more than 50% of civil infrastructure construction costs and installation costs.

4.1.4.5 Heat exchange equipment of catalyst regeneration system

1. Regenerator feed heat exchange equipment

The regeneration circulating gas is divided into three routes into the regenerator: one route is heated to 484°C by the burning feed heat exchanger and electric heater, and it enters the top of the regenerator as the regeneration gas. The second route is heated to 520°C by drying and calcination feed heat exchanger and electric heater and enters the lower part of the regenerator as oxychlorination feed. The third route leads directly to the second-stage regeneration bed entrance.

The temperature of cold and hot fluids in the burning feed heat exchanger and the drying and calcination feed heat exchanger have a high degree of overlap. Only countercurrent heat transfer can be applied to achieve effective heat transfer. The flow rate and duty of the burning feed heat exchanger are 10 times larger than that of the drying and calcination feed heat exchanger. Floating head exchanger with one tube pass and fixed tube sheet exchanger with expansion joint are suitable equipment structures. When countercurrent heat transfer is applied in the drying and calcination feed heat exchanger, the effective temperature difference is 75.7°C, and the heat transfer area is small. The weight of the E-shell or double-pipe exchanger is basically the same. If the structure of E-shell with fixed tube sheet is used, the expansion joint must be set. Using E-shell with floating head structure, equipment manufacturing is more complicated. It is a reasonable choice to adopt hair-pin exchanger.

The possibility of fouling in cold and hot fluid of drying and calcination feed heat exchanger and burning feed heat exchanger is low; hence techniques such as low-finned tube can be used to enlarge the secondary heat transfer surface.

2. Regenerator discharge heat exchange equipment

The gas cooled by the burning feed heat exchanger and the drying and calcination feed heat exchanger is mixed and alkaline solution injected before entering the regeneration gas cycle subcooler.

In order to prevent salt crystallization deposition in the heat exchanger, the exchanger is generally installed vertically. The regeneration gas is arranged in the tube side of the exchanger, and the one tube pass structure is adopted. After the regeneration gas is cooled, it is directly leave the equipment from the bottom nozzle. Because the fluid temperature is not high, the general use of fixed tube sheet heat exchanger, tube bundle cannot be extracted, and the shell side is not easy to descaling. The shell side fluid is cooling water; the "accordion" effect of the low-finned tube is suitable for the removal of water fouling so that the shell side of the heat exchanger has a self-cleaning function.

4.1.4.6 Other heat exchange equipment

Fresh hydrogen gas from PSA is heated by a reduction gas heat exchanger and electric heater and goes into the reduction drum. The gas from the reduction drum top is cooled by the reduction heat exchanger and the lifting hydrogen heat exchanger and then sent to the fuel gas pipe network.

The characteristics of lifting hydrogen heat exchanger and reduction heat exchanger are the same as those of drying and calcination feed heat exchanger, and their applicable heat exchanger structure is the same.

4.1.5 Summary

The process fluid of reforming plant is relatively clean, and there are many heat exchangers bearing heavy load. The feed/effluent heat exchangers have their own unique characteristics. A variety of heat transfer enhancement technologies have been successfully applied, for the enterprise to save equipment investment and later operation costs.

4.2 Aromatics extraction unit

In aromatic raw materials, besides aromatics, there are also nonaromatic components such as chain alkanes, cycloalkanes, and trace alkenes. These nonaromatic components not only have very close boiling points with benzene, toluene, and xylene but also can form azeotropes. High-purity aromatic products cannot be obtained by ordinary distillation. Aromatics extraction is an important method to separate aromatics from nonaromatics in the production of aromatics.

4.2.1 Brief description of the process of aromatics extraction unit

According to the process principle, aromatics extraction technology is divided into two categories, one is the liquid-liquid extraction technology. Its principle is to use the difference in solubility of aromatic and nonaromatics in the solvent. The solubility of aromatics in solvents is higher than that of nonaromatics. By adding solvent to dissolve aromatics, solvent rich in aromatics is formed, and then the solvent rich in aromatics is distilled to obtain mixed aromatics and lean solvent. Lean solvent is recycled so that the separation of aromatics and nonaromatics can be realized. The other category is the extractive distillation process, which is often referred to as aromatics extraction distillation process and its principle is to use the different effects of selective solvents on the relative volatility of hydrocarbon components. By adding selective solvent, the relative volatility between aromatics and nonaromatics of the target product is increased, and the solvent rich in aromatics is separated by distillation. The separation of solvent and aromatics is the same as the liquid-liquid extraction process.

Compared with other solvent aromatics extraction technologies, sulfolane extraction technology has the characteristics of low investment, good quality of aromatics products, high yield of aromatics, and low energy consumption and is the mainstream technology in the construction of aromatics extraction in China.

4.2.1.1 Brief description of sulfolane aromatics liquid-liquid extraction process (SAE)

Fig. 4.5 shows the sulfolane liquid-liquid extraction process of aromatics using sulfolane as solvent (SAE).

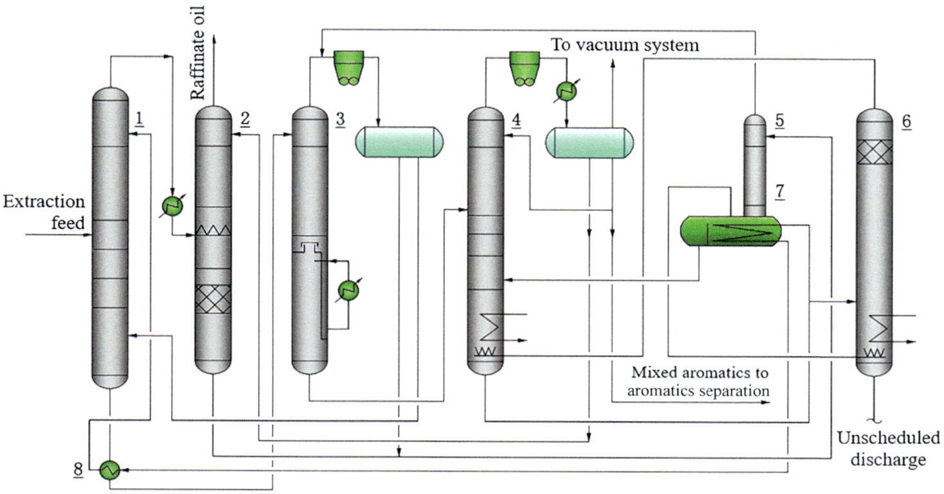

Fig. 4.5 Typical process flow diagram of sulfolane aromatics liquid-liquid extraction (SAE). (1) Extract column; (2) raffinate water scrubber; (3) stripping column; (4) recovery column; (5) water stripper; (6) solvent regeneration column; (7) water stripper reboiler; (8) lean/rich solvent heat exchanger.

The feedstock enters into the lower parts of extraction column; lean solvent and reflux aromatics enter into the extraction column at the first layer of the top tray and the first layer of the bottom tray, respectively. The nonaromatics (raffinate oil) containing a small amount of aromatics and solvents leave the top of the extraction column and enter the raffinate water scrubber to recover solvent after cooling to 40°C. The extract column and raffinate water scrubber are extraction operations, with no cooling at the top and no heat input at the bottom.

The rich solvent from the extraction column bottom is cooled by lean solvent, then enters the top tray of stripping column to extract the dissolved light nonaromatics. The column only has the stripping section. The evaporates overhead are generally condensed and cooled by air cooler and water cooler combined, and then oil and water are separated before the oil is sent to the extract column as reflux aromatics. The stripper adopts a one-pass reboiler, and the heat source is steam after temperature and pressure reduction.

The rich solvent from the bottom of the stripper enters the middle of the recovery column, and the aromatics and solvents are separated under negative pressure and stripping water. The fluid from column overhead is generally condensed and cooled by a combination of air cooler and water cooler and then sent to the reflux drum for oil–water separation. Part of the separated oil phase is used for the reflux of the column, and the rest is sent downstream for aromatics separation. The separated water is sent to the raffinate water scrubber as washing water. The recovery column adopts a stabbed–in reboiler, and the hot fluid is steam. The lean solvent from the column bottom is used as the heat source for the reboiler of the stripping column and then is heated by the rich solvent from extract column bottom to the appropriate temperature and returned to the top of the extract column for recycling.

The water containing solvent and hydrocarbon from the bottom of the raffinate water scrubber and the reflux drum of the stripper overhead is sent to the stripping column to remove the entrained and dissolved nonaromatics. The vapor containing a small amount of hydrocarbons from the stripper is sent to the air cooler to be condensated and cooled. A kettle reboiler is selected for the stripper. Most of the water is vaporized in the reboiler and then enters the solvent regeneration column to be used as stripping steam, and then enters the bottom of the recovery column to be used as stripped medium. A small amount of water containing solvent is extracted from the reboiler bottom and sent to the bottom of the recovery column to be used as the stripping water.

In order to remove the accumulated mechanical impurities and polymers in the solvent, a solvent regeneration column is set up. A small amount of solvent from the recovery column bottom is sent to the solvent regeneration column for regeneration. The stabbed–in reboiler is applied, and the steam after temperature and pressure reduction is used as the heat source.

4.2.1.2 Brief description of sulfolane aromatics extractive distillation (SED) process

A typical process flow diagram of aromatic extractive distillation using sulfolane as solvent (SED) is shown in Fig. 4.6. Two reboilers are usually set for the extractive distillation column bottom, one of which uses the lean solvent from the recovery column bottom as the heat source to maximize the heat recovery. The other one uses steam after temperature and pressure reduction as the heat source. The recycle lean solvent from the recovery column bottom is cooled to the appropriate temperature (95–115°C) by cooling water after heat is exchanged several times with process fluid, then divided into two routes into the extractive distillation column. The heated raw material is fed into the column from the middle of the extractive distillation column. Nonaromatics vapor leaves from the extractive distillation column overhead. Generally, it is condensed and cooled by air and cooling water and then sent to the reflux drum for oil–water separation. Part of the separated oil phase is used as the reflux of the column, and the rest is sent out as by-product. The extractive distillation column bottom is sent to the recovery column for the separation of aromatics and solvents.

The recovery columns of SED and SAE are exactly the same, but SED process has more lean solvent heat recovery processes. The lean solvent from the column bottom is used as the heat source for the reboiler of the middle section of the extractive distillation column and the reboiler of the nonaromatic distillation column. Then, it exchanges heat with the stripping water and feed of extractive distillation column in turns and returns to the extractive distillation column for recycling after being cooled in the lean solvent cooler.

The solvent regeneration of SED process is the same as that of SAE process.

The nonaromatics distillation column can be either a single column ("two–column process") or combined with the extractive distillation column ("single–column process"). The design is carried out according to the different requirements of raw material composition, aromatics, and solvent content in the raffinate. Typical "single-column process" is shown in Fig. 4.6, and "two-column process" is shown in Fig. 4.7.

The main differences between "single-column process" and "two-column process" are as follows: ① A column section is added above the solvent upper feed port of the extractive distillation column to remove the solvent in the nonaromatics. ② The independent reboiler and pump of the nonaromatic distillation column bottom are reduced.

In the process of waste heat recovery of lean solvent, compared with "single–column process," the "two–column process" has a nonaromatic distillation column reboiler between the reboiler of the middle section of the extractive distillation column and the lean solvent/stripping water heat exchanger.

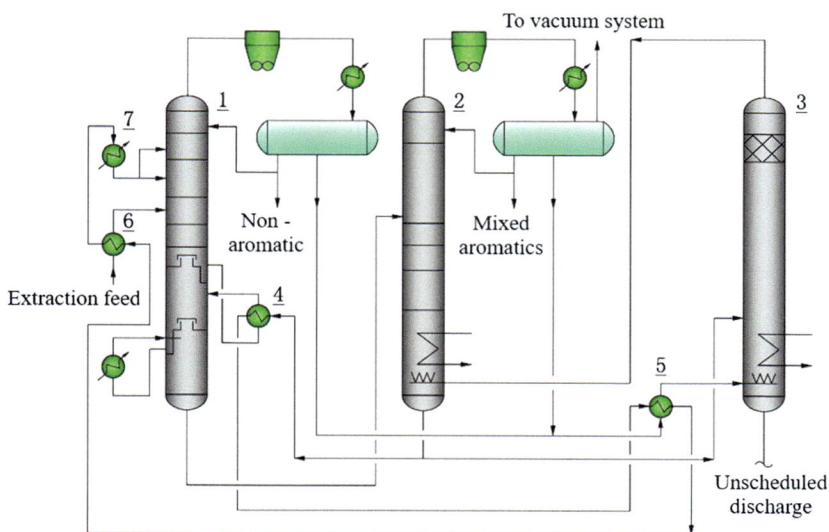

Fig. 4.6 Typical process flow diagram of sulfolane aromatics extractive distillation process (SED). (1) Extractive distillation column; (2) recovery column; (3) solvent regeneration column; (4) extractive distillation column middle section reboiler; (5) lean solvent/stripping water heat exchanger; (6) raw material/lean solvent heat exchanger; (7) lean solvent cooler.

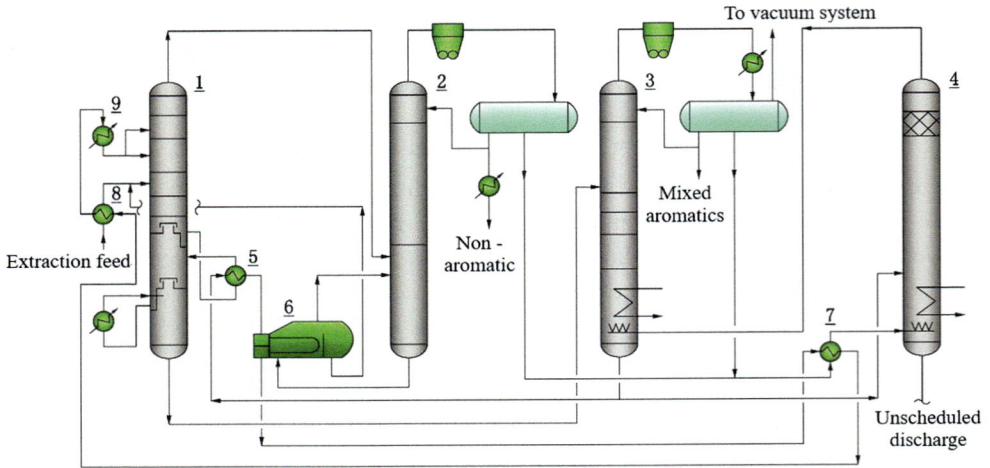

Fig. 4.7 Process flow diagram of SED with nonaromatic distillation column. (1) Extractive distillation column; (2) nonaromatic distillation column; (3) recovery column; (4) solvent regeneration column; (5) extractive distillation column middle section reboiler; (6) nonaromatic distillation column reboiler; (7) lean solvent/distilled water heat exchanger; (8) raw material/lean solvent heat exchanger; (9) lean solvent water cooler.

4.2.2 Analysis of thermal energy characteristics

The aromatics extraction process is a physical process, and the main energy consumption is the steam consumption of the reboiler in the distillation column, which accounts for about 90% of the energy consumption of the extraction unit. The energy consumption of the aromatics extraction unit is different due to the large differences in the aromatics content in the raw materials. Taking the BTX extraction unit of a reformed gasoline as the raw material as an example, the energy consumption distribution is shown in Table 4.11.

Table 4.11 Energy consumption index of aromatics extraction unit—kgoe/(t mixed aromatics).

Cooling water	Electricity	3.5 MPa steam	5.4 MPa deoxidizing water	Total energy consumption
0.49	3.15	58.48	0.87	62.99

The temperature of stripping column overhead is generally below 125°C, and the temperature of recovery column overhead is generally below 75°C. A large amount of low-temperature heat cannot be used and can only be cooled by air or cooling water.

The lean solvent temperature at the bottom of the recovery column is about 175°C, the operating temperature of the lean solvent going into extraction column in SAE process is 65–95°C, and the operating temperature of the lean solvent going into extraction distillation column in SED process is 100–115°C. The recovery and utilization of lean solvent heat is the main link to reducing the steam consumption of the unit. The sulfolane viscosity is shown in Table 4.12. Due to its high viscosity, heat transfer enhancement on the lean solvent side is the key to heat recovery. As for extraction units constructed at early stage, the heat transfer effect of the lean/rich solvent heat exchanger was poor, and the temperature of the lean solvent going into column could not reach the design control index, which affected the extraction operation effect. The units constructed after 2000 enhanced the heat transfer effect by efficient heat exchanger and overcame the defects of the early units.

Table 4.12 Sulfolane viscosity.

Temperature (°C)	50	100	150	200
Viscosity $(mN\,s/m^2)$	6.28	2.56	1.43	0.97

Sulfolane decomposes slowly at 220°C to produce SO_2 and unsaturated polymer. With increasing temperature, the degradation rate increases [15]. In order to reduce the degradation rate of sulfolane solvent, steam used in the reboilers of the stripper, recovery column, and solvent regeneration column bottom in SAE and SED process was typically let down from 3.5 MPa (gauge pressure) to 2.2–2.6 MPa (gauge pressure) for use. In practice, as long as the heat exchange area is large enough to maintain the required reboiling load, the lowest possible steam pressure should be considered to reduce the possibility of solvent degradation. Solvent degradation will form polymers, which will increase the fouling resistance of heat exchanger and reduce the heat transfer effect.

The air cooler of the recovery column overhead is operated under vacuum to allow a small pressure drop. It usually adopts a single tube-pass. The design of the air cooler needs to balance the heat transfer effect and the pressure difference.

4.2.3 Process enhancement and integration on heat transfer

4.2.3.1 SAE process enhancement and integration on heat transfer

The heat utilization of lean solvent in SAE process is relatively simple, and all of it is used for heating stripper feed besides meeting the duty requirement of the water stripper. The heat cascade recovery and utilization of lean solvent in a 0.45 Mt/a aromatics extraction unit is shown in Fig. 4.8, and all the lean solvent heat is recovered. The load distribution is shown in Table 4.13. The heat taken by the lean/rich solvent heat exchanger is large, and the temperature difference between the heat outlet and the cold inlet is about 10°C.

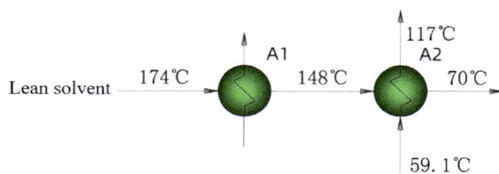

Fig. 4.8 Lean solvent heat utilization flow of SAE process. A1, Water stripper reboiler; A2, lean/rich solvent heat exchanger.

Table 4.13 Heat utilization distribution of lean solvent in SAE process.

Position	A1	A2
Heat exchange (kW)	2888	7766
Proportion (%)	27.1	72.9

4.2.3.2 SED process enhancement and integration on heat transfer

The lean solvent heat utilization integration of SED process can reach 4–5 positions. The heat cascade recovery and utilization of lean solvent in a 0.35 Mt/a aromatics extraction unit is shown in Fig. 4.9, and the load distribution is shown in Table 4.14. The lean solvent is successively used as heat source of the middle section reboiler of the extractive distillation column, used as heating stripping water and feed of the extractive distillation column in turns. The middle section reboiler of the extractive distillation column and the solvent/stripping water exchanger take a relatively large amount of heat. The temperature difference between the hot outlet and cold inlet of the reboiler is $-15°C$.

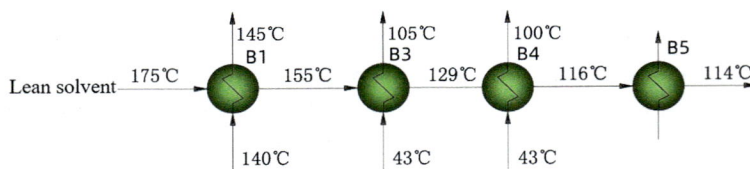

Fig. 4.9 Flow diagram of lean solvent heat utilization in SED with "single-column process." B1, Extractive distillation column middle section reboiler; B3, lean solvent/stripping water heat exchanger; B4, raw material/lean solvent heat exchanger; B5, lean solvent water cooler.

Table 4.14 Heat utilization distribution of lean solvent in SED with "single-column process."

Item	B1	B3	B4	B5
Duty (kW)	2033	2505	1267	184
Proportion (%)	33.9	41.8	21.2	3.1

The heat cascade recovery and utilization of lean solvent in a 0.75 Mt/a aromatics extraction unit with a separate nonaromatics distillation column is shown in Fig. 4.10. The lean solvent is used successively as heat source for the reboiler of the middle section of the extractive distillation column, the nonaromatic distillation column, and the stripping water heat exchanger. The load distribution is shown in Table 4.15. The heat taken by the middle section reboiler of the extractive distillation column and the solvent/stripping water heat exchanger is relatively large. The temperature difference between the hot outlet and cold inlet of the reboiler is about $15°C$.

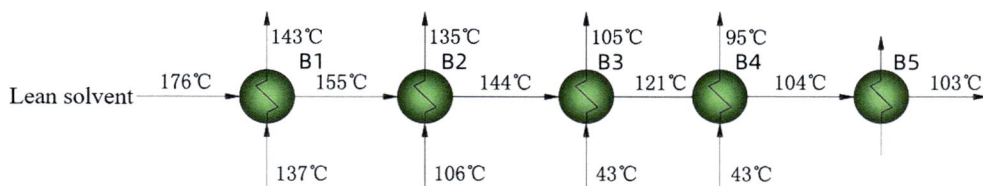

Fig. 4.10 Flow diagram of lean solvent heat utilization in SED with "two-column process." B1, Middle section reboiler of extractive distillation column; B2, nonaromatic distillation column reboiler; B3, lean solvent/stripping water heat exchanger; B4, raw material/lean solvent heat exchanger; B5, lean solvent water cooler.

Table 4.15 Heat utilization distribution of lean solvent in SED with "two-column process."

Position	B1	B2	B3	B4	B5
Heat exchange (kW)	3894	1456	4593	2767	239
Proportion (%)	30.1	11.2	35.5	21.4	1.8

There are many positions for lean solvent heat utilization in SED process. In engineering design, temperature control valve of hot fluid is usually set to adjust the matching temperature of cold and hot fluids, so as to make use of heat as much as possible while ensuring stable operation. Compared with 100% utilization of lean solvent heat in SAE process, in order to ensure stable operation, the heat utilization of lean solvent is slightly lower in SED process, and the temperature of lean solvent into column is adjusted by water cooler.

4.2.3.3 Heat transfer process enhancement of recovery column reboiler and solvent regeneration column reboiler

The recovery column and solvent regeneration column are operated under negative pressure and steam stripping conditions to decrease the temperature of the column bottom reboiler. Due to the high viscosity of sulfolane solvent, the stabbed-in reboiler is often applied. The stripping steam distribution pipe is arranged below the tube bundle of the reboiler, and the steam agitates the liquid, promotes the renewal of the outer wall surface of the tube, and enhances the heat transfer effect of the reboiler.

4.2.4 Heat transfer enhancement technology by improving elements

The process fluid in the extraction unit is relatively clean, and the heat transfer performance is better. In recent years, equipment with enhanced heat transfer characteristics, such as twisted tubes, corrugated tubes, and surface evaporative air cooler, has been applied in the extraction unit [16]. The plate heat exchanger has also been successfully used in the overhead cooler of the extraction unit.

Although the process flow of different licensors is different, it can still be analyzed and studied by typical process flow. In this section, a 1 Mt/a aromatics extraction unit designed by Sinopec with extractive distillation technology is taken as an example to analyze the characteristics of the heat transfer equipment in the unit and the applicable heat transfer enhancement technology.

4.2.4.1 Heat transfer equipment of extractive distillation column system

1. Heat transfer equipment around extractive distillation column

As for feed/lean solvent heat exchanger, the heat transfer performance of both tube side and shell side medium is good. The heat exchange tube with double-sided enhanced heat transfer function can be used instead of the plain tube to further reduce the heat exchanger size and reduce the metal consumption of the equipment.
2. Extractive distillation column reboiler

Generally, a once-pass reboiler is set in the middle of the extraction distillation column, and the heat source is the lean solvent from the recovery column bottom. The temperature levels of cold and hot fluids are close, and the effective temperature difference of reboiler is about 24°C. When horizontal heat exchangers are installed, the film heat transfer coefficients of tube side without phase change cooling and shell side boiling coefficients are close to each other, both exceeding $2000\,\text{W}/(\text{m}^2\,\text{K})$. The main thermal resistance of the reboiler is fouling resistance. The heat transfer enhancement should be focused on fouling prevention. Better heat transfer enhancement can be achieved by using twisted tube.

Steam is used as the heat source in the thermosiphon reboiler at the bottom of the extraction distillation column. Generally, the vertical installation is adopted. The heat transfer coefficient of the steam condensing on the shell side is close to $10,000\,\text{W}/(\text{m}^2\,\text{K})$, and the solvent-rich boiling coefficient on the tube side is about $1000\,\text{W}/(\text{m}^2\,\text{K})$. Increasing the surface bubble core can improve the boiling heat transfer coefficient. If the horizontal thermosiphon reboiler is selected, T-groove tube can be used to enhance the shell side boiling heat transfer.

4.2.4.2 Heat transfer equipment around nonaromatics distillation column

The gas of the nonaromatic distillation column overhead is condensed and cooled to 40°C through the air cooler and water cooler and then enters the reflux drum. The duty of the air cooler is not large, and the ordinary high-finned tube structure can meet the process requirements. Because the fluid overhead still needs to be subcooled by 20°C. It is recommended to use at least two tube-pass designs and only one row of tubes for the last tube-pass to control the fluid velocity to ensure the discharge supercooling degree.

If the vapor fraction at the reboiler outlet of the nonaromatic distillation column bottom is high, the kettle structure heat exchanger is generally selected. If the vapor fraction is low, the ordinary J-type heat exchanger can be used to complete the heat exchange. In the reboiler, the lean solvent is generally arranged on the tube side of the reboiler. Under the premise of making full use of the pressure drop, the film heat transfer coefficient of the tube side can reach $2500\,\text{W}/(\text{m}^2\,\text{K})$, the T-shaped finned tube can be used to enhance the boiling heat transfer on the shell side, and the twisted tube and corrugated tubes can be used to enhance the heat transfer both tube side and shell side at the same time.

4.2.4.3 Heat transfer system around recovery column

1. Heat exchange equipment of the recovery column overhead

Gas from the column overhead is condensed and cooled through the air cooler and water cooler and then enters the reflux drum. See Table 4.16 for the specific process conditions.

Table 4.16 Process conditions of fluid from the recovery column overhead—Duty: 24 MW.

Item	Hot fluid
Fluid name	Recovered column overhead gas
Flow (kg/h)	152,021
Allowable pressure drop (kPa)	20
Temperature (°C)	$71 \rightarrow 40$
Gas phase density (kg/m³)	$0.951 \rightarrow -$
Liquid phase density (kg/m³)	$- \rightarrow 862$
Gas viscosity (mN s/m²)	$0.009 \rightarrow -$
Liquid viscosity (mN s/m²)	$- \rightarrow 0.506$
Gas phase thermal conductivity (W/(m K))	$0.0157 \rightarrow -$
Liquid phase thermal conductivity (W/(m K))	$- \rightarrow 0.166$
Gas relative heat (kJ/(kg K))	$1.325 \rightarrow -$
Liquid phase specific heat (kJ/(kg K))	$- \rightarrow 0.195$

Since the process fluid is cooled from 50°C to 40°C, a separate water cooler must be set up to ensure the cooling effect. If the surface evaporation air cooler is used instead of conventional air cooler, the water cooler can be reduced and the area of air cooler is greatly reduced. The comparison results are shown in Table 4.17.

Table 4.17 Scheme comparison of recovery column overhead exchanger.

Item	Original option		Surface evaporation air cooler
	Air cooler	Water cooler	
Equipment size (mm)	9000×3000^a	900×4500^b	6000×3000^a
Set	18	1	6
Consumption	Electricity: 396 kW	Cooling water: 1150 t/h	Electricity: 300 kW Softened water: 27 t/h
Heat transfer area (m²)	1152	195	
Total weight of equipment (t)	216	7	210

[a]Air cooler size: tube bundle length × width.
[b]Water cooler size: shell diameter × straight tube length.

As can be seen from Table 4.17, surface evaporation air cooler can replace the conventional combination of dry air cooler and water cooler, which can reduce the cost of structure and pipeline. The softened water doesn't need to be discharged unless water quality is not satisfied. Different unit locations and external environments will affect the consumption of softened water. Calculated with annual working time of 8400 h, the annual power consumption can be reduced by 806,000 kWh, the consumption of cooling water can be reduced by 966×10^4 t, and energy consumption can be reduced by 20%.

The inlet temperature of process fluid is only 71°C, which can be directly cooled with cooling water. The cancellation of air cooler can reduce the investment cost of equipment, but due to the large heat load, the consumption of cooling water increases significantly, and the total energy consumption will increase instead.

2. Heat transfer system around the recovery column bottom

The lean solvent from the recovery column bottom is sent to the middle section reboiler of the extractive distillation column and the bottom reboiler of the nonaromatics distillation column as a heat source and then enters the lean solvent/stripping water heat exchanger, feed/lean solvent heat exchanger, and lean solvent water cooler to be cooled in turn. Finally, the lean solvent enters the top of the extractive distillation column.

The middle section reboiler of the extractive distillation column and the reboiler of the nonaromatics distillation column bottom have been analyzed in the previous section. Fluids are all liquid phase in the other three exchangers, with close film heat transfer coefficient of tube side and shell side. If further heat transfer enhancement is required, heat exchange tube with double heat transfer enhancement capacity can be used, such as twisted tubes and corrugated tubes.

The duty of the reboiler of the recovery column bottom varies greatly with different processes. If the heat load is large and the required vapor fraction is high, the kettle structure is adopted. When the heat load is small, the stabbed-in reboiler can be used. Whether it is kettle-type structure or stabbed-in reboiler, twisted tube and corrugated tubes can be used to enhance heat transfer, and threaded tube can be used to increase the secondary heat transfer area. When the kettle structure is selected, T-shaped finned tube and porous surface tube can also be applied to enhance the boiling process outside the tube.

4.2.4.4 Heat exchange equipment in solvent regeneration column system

The stripping water from the reflux drum of the nonaromatics distillation column and the reflux drum of the recovery column is mixed and then entered into the lean solvent/stripping water heat exchanger and then enters the solvent regeneration column bottom after heat exchange with lean solvent. The duty of the heat exchanger is about 6 MW. Due to the large difference in cold and hot fluid rates, when using the ordinary shell and tube heat exchanger structure, the lean solvent is generally arranged on the tube side to make use of the characteristics of flexible configuration of tube side number. When the structure design is reasonable, the film heat transfer coefficient of the tube and shell side is more than $2000\,W/(m^2\,K)$. If further heat transfer enhancement is required, then heat exchange tube with double heat transfer enhancement capacity can be used.

Column bottom reboiler load is small, with steam as heat source, which can be used in the column insert structure. The film heat transfer coefficient exceeds $10,000\,W/(m^2\,K)$ due to steam condensation on the tube side. It is the shell side boiling heat transfer which needs to be enhanced. The use of heat exchange tube with enhanced heat transfer effect can reduce the diameter of tube bundle. Twisted tube bundle and low-finned tube bundle have been successfully applied.

4.2.5 Summary

Heat recovery and utilization of lean solvent are the main means to reduce steam consumption in aromatics extraction unit. However, the high viscosity of sulfolane solvent and the polymer formed by degradation affect the heat transfer effect of related heat transfer equipment. Suitable heat transfer process enhancement is the key to making full use of solvent waste heat.

When designing heat exchangers with sulfolane in the process fluid, the metal wall temperature on the heat exchange surface should be avoided from being too high, so as to prevent the degradation effluent from blocking the heat exchanger, leading to the increase of pressure drop and even heat transfer failure.

4.3 Disproportionation unit

Disproportionation unit is an important unit of aromatics complex, with toluene and C_9/C_{10} as raw materials. C_8 aromatics are produced by intermolecular alkyl transfer reaction, so as to achieve the purpose of maximizing the production of PX.

4.3.1 Brief description of disproportionation unit process

The disproportionation process is mainly divided into two types: the first is toluene disproportionation and alkyl transfer, and the second is toluene shape-selective disproportionation. For process of toluene disproportionation and alkyl transfer, the main reaction is to increase the yield of C_8, which includes toluene disproportionation, alkyl transfer reaction between toluene and C_9/C_{10} aromatics, and C_9/C_{10} aromatics side-chain dealkylation reaction. Disproportionation reaction is a special form of alkyl transfer reaction.

Toluene shape-selective disproportionation process uses toluene as feedstock to produce benzene and xylene. The selectivity of PX in this reaction is high, and the content of PX in the effluent is as high as 82%–90%, but the yield of C_8 aromatics products is low (generally only about 40%). The single pass conversion rate is low (generally only about 30%), and only toluene can be used instead of C_9/C_{10} aromatics as feedstock, which cannot achieve the maximum production of PX.

Disproportionation and alkyl transfer process are mostly used in domestic dispropor-tionation units. This section takes disproportionation and alkyl transfer process as an example to analyze the utilization of thermal energy. The typical flow of this process is shown in Fig. 4.11.

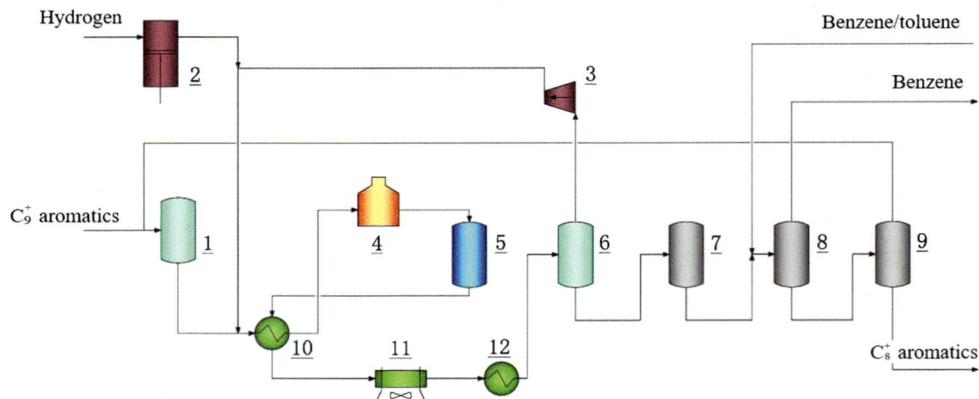

Fig. 4.11 Schematic diagram of disproportionation and alkyl transfer process. (1) Reaction feed buffer tank; (2) make-up hydrogen compressor; (3) recycle gas compressor circulating hydrogen compressor; (4) reaction feed furnace; (5) reactor; (6) separator for effluents; (7) disproportionation stripper; (8) benzene column; (9) toluene column; (10) reaction feed/effluent heat exchanger; (11) effluent air cooler; (12) effluent water cooler.

The feedstock of the disproportionation unit includes toluene and C_9/C_{10} aromatics. Toluene comes from the top of the toluene column, and C_9/C_{10} aromatics come from the top of the heavy aromatics column of the xylene separation unit. The two raw mate-rials are mixed in the feed buffer tank before entering the reaction part.

The disproportionation reaction raw material is pumped to the disproportionation reaction feed/effluent heat exchanger, to be mixed with gas containing hydrogen from the disproportionation recycle gas compressor, followed by heat exchange with the dis-proportionation effluent. Then it enters the feed furnace to heat up to the disproportion-ation reaction temperature and is sent to the disproportionation reactor. Due to the presence of cracking reaction, the reactor discharge will have a temperature rise of about 15°C compared with the reactor feed. The effluents exchange heat with the reaction feed and then enter the effluent separator for gas–liquid separation after cooled by air and cool-ing water. Most of the gas from the drum enters the disproportionation recycle gas com-pressor for pressurization and recycling, and the rest is discharged to the hydrogen recovery unit. In order to maintain the purity of circulating hydrogen in the reaction system, hydrogen should be added to the circulating hydrogen. The make-up hydrogen is usually boosted by the make-up hydrogen compressor and mixed with the circulating hydrogen at the outlet of the recycle gas compressor to enter the disproportionated reaction part together. The liquid from disproportionation effluent separator bottom enters into disproportionation stripper to strip light hydrocarbons. Mixed aromatics containing benzene and toluene from the stripper bottom and other units are fed into the benzene column. Benzene product is stripped from the side line of the column, cooled, and pumped to the product tank. Aromatics from benzene column bottom are sent to toluene column for treatment, and toluene from the toluene column overhead is returned to feed buffer tank as disproportionation reaction feed.

4.3.2 Analysis of thermal energy characteristics of the unit

The design energy consumption of the disproportionation unit in plant A is shown in Table 4.18:

Table 4.18 Design energy consumption unit of disproportionation unit in plant A—Unit: kgoe/t.

Item	Energy consumption (kgoe/t)	Proportion (%)	Item	Energy consumption (kgoe/t)	Proportion (%)
Electricity	5.44	9.5	Cooling water	0.08	0.1
3.5 MPa (gauge pressure) steam	6.66	11.6	Demin water	0.07	0.1
1.0 MPa (gauge pressure) steam	0.06	0.1	Condensate	0.26	0.49
Fuel gas	45.42	79	Total	57.47	100

The design energy consumption of the disproportionation unit in plant B is shown in Table 4.19:

Table 4.19 Design energy consumption of disproportionation unit in plant B—Unit: kgoe/t.

Item	Energy consumption (kgoe/t)	Proportion (%)	Item	Energy consumption (kgoe/t)	Proportion (%)
Electricity	5.7	15.6	Demin water	0.16	0.4
1.0 MPa (gauge pressure) steam	2.96	8.1	Deaerated water	0.66	1.9
0.5 MPa (gauge pressure) steam	1.78	4.9	Condensate	0.08	0.2
Fuel gas	28.73	78.5	Total	36.59	100
Cooling water	0.24	0.7			

As can be seen from the energy consumption table of the disproportionation units of the two different processes, the energy consumption focuses on fuel gas, accounting for about 80% of the unit energy consumption, followed by electricity and steam. Electricity is mainly used for the driving of rotating equipment, such as make-up hydrogen compressor, pump and air cooler, etc. Steam is mainly used for the driving of recycle gas compressor turbine. Fuel users of disproportionation unit include reaction furnace and distillation column reboiler. By enhancing heat transfer technology to reduce fuel consumption of furnace, energy consumption of disproportionation unit can be effectively reduced.

4.3.2.1 Effluent heat load

The catalytic activity of toluene disproportionation and alkyl transfer increased with increasing temperature. With the prolongation of the operation time, the activity of the catalyst decreases, and the conversion rate of the raw material can be maintained only by increasing the reaction temperature. The initial temperature of the disproportionation reaction is about 350°C, and the final temperature is about 460°C. Heating the material to the desired reaction temperature requires a lot of fuel. About 0.35 MW heat load is required to heat 1.0 t/h reaction feed to the desired reaction temperature at the end of the reaction, and 42 MW load for a disproportionation unit of 1.0 Mt/a. Therefore, it is necessary to set up the reaction feed/effluent heat exchanger, with high-temperature effluents to heat the reaction feed, so as to reduce the fuel consumption of the feed furnace. Enhancing the heat transfer process of the feed/effluent heat exchanger to increase the exchanged duty can further reduce the fuel consumption of the furnace, so as to reduce the energy consumption of the unit.

4.3.2.2 Heat at the top of fractionation column

The fractionation part of the disproportionation unit usually includes three column systems: stripper, benzene column, and toluene column. Stripper operating pressure is usually between 0.3 and 0.6 MPa (gauge pressure), and the benzene column and toluene column are usually operated at atmospheric pressure. Among the three columns, the heat load of the gas from the toluene column overhead is the largest, but the temperature level is not high under atmospheric pressure. The usual practice is to boost the pressure of the toluene column to increase the temperature level of the toluene column overhead, which is used as the heat source of the benzene column reboiler to reduce the energy consumption of the unit.

4.3.3 Process enhancement and integration on heat transfer
4.3.3.1 Effluent heat recovery

The temperature of the effluent from the disproportionation reactor is usually 360–480°C. The initial temperature is low, while the final temperature is high. Heat exchange with the reactor feed can effectively recover the heat of effluent and reduce the duty of the disproportionation reaction feed furnace. Vertical shell and tube heat exchanger is applied for disproportionation feed/effluent heat exchanger uses at an early time, but the heat transfer efficiency is low and the temperature difference between the hot ends is above 50°C. With the increase of the capacity of the unit, the disadvantage of low heat exchange efficiency is more obvious. Limited by the size of the structure, usually two parallel heat exchangers are needed to meet the heat exchange requirements. At present, efficient heat exchangers such as plate heat exchangers or coil-wound heat exchangers are usually used to replace the conventional shell and tube heat exchangers, which can further reduce the temperature difference at the hot end of the heat exchanger and recover more heat from the effluents. According to the typical temperature difference at the hot end of the two kinds of heat exchangers, the shell and tube heat exchangers and plate heat exchangers are compared. After applying plate heat exchangers, the duty of the reaction feed furnace is reduced by 55%, the pressure drop of the reaction system is reduced, and the steam consumption of the recycle gas compressor is reduced by 17% [17].

4.3.3.2 Benzene/toluene column heat integration and recovery

The toluene column at atmospheric pressure and higher pressure scheme were compared. In the two schemes, the benzene column is operated under atmospheric pressure and the number of trays is the same. In the pressurization scheme, the toluene column is designed with lifting pressure. The distillate gas from the column overhead is used as the heat source of the benzene column, and the excess gas is cooled by air. Compared with the atmospheric pressure scheme, the total required duty for boiling of the benzene column and toluene column can be reduced by 18% [17]. Usually, the condensation load of toluene column overhead gas is greater than the load required for benzene column reboiler. In order to reduce energy consumption, the excess heat load can produce 0.5 MPa (gauge pressure) low-pressure steam.

4.3.3.3 Low-temperature heat recovery

The low-temperature heat of disproportionation unit is mainly concentrated in the effluents and disproportionation stripper top, and this part of heat can be used to heat hot demineralized water for recovery. For the disproportionation unit of 0.9 Mt/a scale, it can heat the hot demineralized water from 70°C to 120°C, and the heat recovery energy is 4.8 MW.

Because benzene column operates under atmospheric pressure and the temperature of the gas from the column overhead is low, the gas cannot be used to heat hot demineralized water. It can be used to heat low-temperature media.

4.3.4 Heat transfer enhancement technology by improving elements

With the development of large-scale unit, it is inevitable to use high–performance components to reduce equipment size. In recent years, elements with enhanced heat transfer characteristics, such as porous surface tubes, fully welded plate and shell heat exchangers, and coil–wound heat exchangers, have been applied in disproportionation units [5,18–21]. The process fluid of the disproportionation unit is relatively clean, and the fully welded plate heat exchanger has been successfully applied in many positions in the disproportionation unit.

The process flow of different licensors is different, and the requirements of raw materials and products are different, which will also lead to the change in process flow. However, it may be useful to discuss the enhancement measures of heat transfer equipment in the disproportionation unit through a set of typical process. Taking an in-service disproportionation unit as an example, the characteristics of the heat exchange equipment and the appropriate heat transfer enhancement technology are analyzed.

4.3.4.1 Heat transfer equipment around disproportionation reaction part

1. Disproportionation feed heat exchanger

   Toluene from the top of toluene column is cooled from about 165°C to about 110°C by exchanging with liquid effluent from separator bottom, mixed with circulating hydrogen, and heated by disproportionation feed/effluent heat exchanger and furnace, then entering the disproportionation reactor. The disproportionation feed contains gas phase and liquid phase, in which the circulating hydrogen contains 80% hydrogen (mole fraction), and the rest is C_1, C_2, and C_3 components. The superheat hot fluid is 282°C at the heat exchanger inlet, the dew point of which is 215°C. Two-thirds of the enthalpy difference is in the pure gas cooling without phase transition region. When the cold flow is heated to 234°C, it has been completely vaporized, and it is superheated at 203°C in the heat exchanger. One-half of the enthalpy difference of the cold fluid occurs in the region without phase transition, where the pure gas phase heats up. The heat transfer coefficient of gas phase is relatively low, while the allowable pressure drop of cold and hot fluids in disproportionation feed/effluent heat exchanger is only 50 kPa per pass. Considering the low pressure drop allowed and vibration problem, it is determined that high gas velocity cannot be chosen, which leads to the further reduction of film heat transfer coefficient of cold and hot fluids. The specific process conditions are shown in Table 4.20. From the table of process conditions, it can be seen that the duty of the heat exchanger is very large, the temperature change of cold and hot fluids is large, and the internal temperature overlap is serious. If the heat transfer is completed in one exchanger, attention shall be paid to avoiding the problem of internal temperature cross. Because the allowable pressure drop of cold and hot flows is not large, and the gas density is small, the volume flow rate is large and the F-shell exchanger cannot be used to solve the temperature overlap problem. Plate heat exchanger is an ideal solution. If shell and tube heat exchanger is used, the best solution is single tube pass, pure countercurrent flow heat transfer. It is necessary to use a vertical installation of a diameter of 1800 mm and a length of 27,000 mm single tube-pass heat exchanger. The simulation results are shown in Table 4.21.

Table 4.20 Process condition of disproportionated feed/effluent heat exchanger—Duty: 68.53 MW.

Item	Hot fluid	Cold fluid
Fluid name	Effluent	Feed
Flow (kg/h)	211,844	211,844
Allowable pressure drop (kPa)	60	50
Temperature (°C)	$497 \rightarrow 140$	$92 \rightarrow 437$
Gas phase density (kg/m^3)	$13.72 \rightarrow 10.56$	$6.91 \rightarrow 16.03$
Liquid phase density (kg/m^3)	$- \rightarrow 747$	$797 \rightarrow -$
Gas viscosity (mN s/m^2)	$0.023 \rightarrow 0.015$	$0.013 \rightarrow 0.022$
Liquid viscosity (mN s/m^2)	$- \rightarrow 0.22$	$0.299 \rightarrow -$
Gas phase thermal conductivity (W/(m K))	$0.154 \rightarrow 0.072$	$0.085 \rightarrow 0.142$
Liquid phase thermal conductivity (W/(m K))	$- \rightarrow 0.103$	$0.116 \rightarrow -$
Gas relative heat (kJ/(kg K))	$3.07 \rightarrow 3.12$	$4.64 \rightarrow 2.981$
Liquid phase specific heat (kJ/(kg K))	$- \rightarrow 2.159$	$1.97 \rightarrow -$

Table 4.21 Thermodynamic simulation results of disproportionated feed/effluent heat exchanger.

Item	Original option	Enhanced option
Equipment size (shell diameter × straight tube length) (mm)	$1800 \times 27,000$	2162×9025
Set	1	1
Tube side pressure drop (kPa)	8	
Shell side pressure drop (kPa)	57	
Overall heat transfer coefficient (W/(m^2 K))	288	
Heat transfer area (m^2)	6410	
Total weight of equipment (t)	361	65

In Table 4.21, the manufacturer of fully welded plate and shell heat exchangers does not provide pressure drop and heat transfer coefficient data based on equipment performance due to proprietary confidentiality. The equipment has been running smoothly for many years and meets the requirements of the unit. Compared with the conventional shell and tube heat exchanger, the weight of the welded plate and shell heat exchanger is only 65 tons, reducing the weight of the equipment by 82%, and the saving of metal is very considerable. The height of the heat exchanger can be reduced by 18 m, so as to reduce the cost of piping and construction.

The application of plate heat exchanger can reduce the temperature difference of the hot end, so as to reduce the furnace load. Wang et al. [18] considered the disproportionation feed/effluent heat exchanger and furnace, the technical performance is compared from the aspects of equipment investment and fuel consumption, and the economic index is compared from the aspects of equipment investment, operation cost, depreciation cost, etc. The result of the comparison is that the fully welded plate and shell heat exchanger is superior to the shell and tube heat exchanger with plain tube. Cui et al. [19] described a successful case of replacing three φ2600 mm $\times$ 15,000 mm shell and tube heat exchangers with one 2300 mm $\times$ 12,000 mm fully welded plate and shell structure.

Some companies have used coil–wound heat exchangers in this position, which can also enhance heat transfer and reduce the load of the furnace, so as to achieve the purpose of reducing energy consumption. During the revamping for capacity expansion of the aromatics unit of Sinopec Tianjin Branch, the volume of disproportionation unit was increased by 30%. By replacing the original shell and tube heat exchanger with coil–wound heat exchanger, the temperature difference at the hot end is reduced from 53°C to 29°C, the load of the furnace is reduced, 288 tons of fuel gas is saved per month, and the energy consumption of the plant is reduced by 10.3 kgoe/tPX [5]. Xu et al. [20] described that one coil–wound heat exchanger was used to replace the original four shell and tube heat exchangers, and the feed temperature was raised from 379°C to 412°C. The actual operation results show that the use of coil–wound heat exchanger reduces the temperature difference at the hot end of heat exchanger, saves 1322 tons fuel per year, and reduces the load of air cooler and water cooler.

2. Disproportionation effluent heat exchanger

The effluent of disproportionation reactor is condensed and cooled to 40°C by disproportionation feed/effluent heat exchanger, effluent air cooler, and water cooler and then enters the disproportionation effluent separation tank.

The heat load of the disproportionation effluent air cooler is large, and the conventional air cooler needs a large heat transfer area. The process conditions are shown in Table 4.22. The suitable heat transfer enhancement technology is to use evaporative air cooler instead of conventional air cooler and water cooler, and the comparison of schemes is shown in Table 4.23.

Table 4.22 Process condition table of disproportionation reactor effluent—Duty: 17 MW.

Item	Hot fluid
Fluid name	Disproportionation effluents
Flow (kg/h)	236,319
Allowable pressure drop (kPa)	40
Temperature (°C)	$140 \rightarrow 40$
Gas phase density (kg/m^3)	$10.5 \rightarrow 7.17$

Table 4.22 Process condition table of disproportionation reactor effluent—Duty: 17 MW—cont'd

Item	Hot fluid
Liquid phase density (kg/m^3)	$747 \rightarrow 822$
Gas viscosity (mN s/m^2)	$0.015 \rightarrow 0.012$
Liquid viscosity (mN s/m^2)	$0.22 \rightarrow 0.37$
Gas phase thermal conductivity (W/(m K))	$0.072 \rightarrow 0.067$
Liquid phase thermal conductivity (W/(m K))	$0.103 \rightarrow 0.129$
Gas relative heat (kJ/(kg K))	$3.122 \rightarrow 4.31$
Liquid phase specific heat (kJ/(kg K))	$2.16 \rightarrow 1.77$

Table 4.23 Comparison of disproportionation reactor products air cooler and water cooler.

Item	Original option		Compound evaporative air cooler
	Conventional air cooler	Water cooler	
Equipment size (mm)	9000×3000^a	800×6000^b	9000×3000^a
Set	6	2	3
Consumption	Electricity: 180 kW	Cooling water: 136 t/h	Electricity: 118 kW Softened water: 8 t/h
Heat transfer area (m^2)	1152	635	1400
Total weight of equipment (t)	92	27	75

[a] Air cooler size: tube bundle length × width.
[b] Water cooler size: shell diameter × straight tube length.

The effluent contains noncondensing gas. When condensed in the shell side of the water cooler, the noncondensing gas will form a gas film on the wall of the heat transfer tube, hindering the condensation process and reducing the heat transfer effect of condensation. The film heat transfer coefficient of the cooling water on the tube side is about 5000 W/(m^2 K), and the heat transfer coefficient of oil–gas condensing on the shell side is only 1000 W/(m^2 K). The effective temperature difference of the water cooler is only 7.9°C. The superposition of the two factors leads to a large heat transfer area required.

According to the calculation results, when the disproportionation effluent gas is cooled from 50°C to 40°C, the heat load is 1.58 MW. The temperature rise of cooling water is calculated at 10°C, which means the 32°C cooling water required is about 136 t/h. If the compound evaporative air cooler is used to cool the process fluid directly to 40°C, the water cooler can be canceled. If the annual operation time is based at 8400 h, although the softening water consumption is increased by 6.7×10^4 t/a, the cooling water consumption still can be saved by 114×10^4 t/a. The water-saving advantages are even more pronounced in water-scarce areas. It can reduce the weight of the equipment by 44 tons and save the corresponding investment in the cooling tower, cooling water pump, cooling water pipe network, steel frame platform of air cooler, etc. It can save power consumption by 520,000 kWh and reduce comprehensive energy consumption by 40%.

4.3.4.2 Heat exchange equipment around disproportionation stripper

1. Disproportionation stripper feed heat exchanger

    Disproportionation stripper feed is heated by feed-effluent separator bottom heat exchanger and feed-bottom exchanger and then enters the disproportionation stripper.

    The common characteristics of the two heat exchangers are that the mass flow rates of cold and hot fluids are similar, and the effective temperature difference is higher by more than 50°C. The thermal conductivity of the hot fluid is lower than that of the cold fluid. When the hot fluid is placed on the tube side, higher film heat transfer coefficient can be obtained by using multiple tube passes. The overall heat transfer coefficient of the heat exchanger can be more than $600\,W/(m^2\,K)$. If heat transfer enhancement technology is to be considered, the secondary expanding heat transfer surface is more suitable for this position.

2. Disproportionation stripper top heat exchange equipment

    The vapor at about 130°C from the disproportionation stripper overhead is condensed and cooled to 40°C by the air cooler and water cooler and then enters the reflux drum of the disproportionation stripper.

    The air cooler bears most of the heat load. Due to the large effective temperature difference for the air cooler, the heat exchange area required is not large.

    Due to the existence of noncondensing gas, the film heat transfer coefficient of the shell side of the water cooler is low, generally not more than $1000\,W/(m^2\,K)$, so the overall heat transfer coefficient is only $300\,W/(m^2\,K)$ or so. The heat load borne by the water cooler is very small, although the overall heat transfer coefficient is not high and the equipment size is not large. To further reduce the size of the equipment, the effective method of heat transfer enhancement is, on the premise of avoiding vibration, to increase the turbulence level of the shell side as much as possible to destroy the gas film layer on the heat exchange surface that hinders the condensation process.

3. Disproportionation stripper bottom heat exchange equipment

    The reboiler at the bottom of the disproportionation stripper sometimes takes medium-pressure steam as the heat source, and the heat transfer performance of the tube and shell side fluid is good. Even if the ordinary shell and tube heat exchanger is used, the overall heat transfer coefficient of the reboiler can reach $1000\,W/(m^2\,K)$ and above. The fouling resistance becomes the control thermal resistance of the reboiler. If heat transfer needs to be further enhanced, heat exchange tube such as low-finned tube can be used to enhance boiling heat transfer and protect from fouling instead of plain tube.

4.3.4.3 Heat exchange equipment around benzene column

1. Feed heat exchanger

Before entering the benzene column, feedstock should be heated by the dispro-portionation clay processor feed/effluent heat exchanger. In the heat exchanger, the flow rate of feed and discharge is the same, the physical property is also very similar, and the fouling resistance is not high, generally at $0.00026\,m^2\,K/W$. If the structure design is reasonable, the film heat transfer coefficient of cold and hot fluids can be close to each other. Although the cold and hot fluids in the heat exchanger are liquid phase heat transfer, the overall heat transfer coefficient of the heat exchanger is also easily more than $600\,W/(m^2\,K)$. Combined with the high effective temperature dif-ference, although the duty of the heat exchanger is 6.4 MW, the heat transfer area required is only $200\,m^2$.

The effective way to enhance the heat transfer of the feed/effluent heat exchanger of the disproportionation clay processor is to use the threaded tube to increase the secondary heat transfer surface, which can lead to the cancellation of one heat exchanger, reducing the weight of the equipment by 15%.

2. Benzene column top heat exchange equipment

Gas from the benzene column overhead is condensed and cooled to about 80°C by the air cooler and enters the reflux drum. The duty of the air cooler is large, and 25 MW duty needs to be cooled. The film heat transfer coefficient of the tube side can reach $3000\,W/(m^2\,K)$. The outlet temperature of the process fluid is high, and it is not suitable for surface evaporation air cooler. Plate air cooler or elliptical tube with high fin can be used instead of round tube with high fin to enhance the heat transfer per-formance of the equipment.

3. Benzene product cooler

Benzene product is cooled to 40°C by a cooler and sent to the storage tank. The flow rate of cooling water in the cooler is twice that of benzene products. It is found that the film heat transfer coefficient of the cooling water in the tube exceeds $5000\,W/(m^2\,K)$, while the coefficient of the benzene product on shell side is less than $900\,W/(m^2\,K)$.

Because the product should be cooled to 40°C, which is the same as the outlet temperature of the cooling water, it is better to adopt the method of multiple sets in series, which can effectively improve the heat transfer temperature difference of the equipment. Multiple sets in series can reduce the diameter of the cooler and improve the flow velocity of the shell side, thereby improving the overall heat transfer coefficient and enhancing the performance of the equipment. The F-shell exchanger is also a reasonable means to improve the shell side flow velocity. However, due to the characteristics of cooling water, the equipment investment, the operation cost, and the tube bundle renewal cost caused by corrosion, the F-shell exchanger is not suitable.

4. Reboiler at the bottom of the benzene column

A reboiler is set at the bottom of the benzene column, and its process conditions are shown in Table 4.24. The fluid from the column bottom is relatively pure, and the boiling range is 12°C. The reasonable heat transfer enhancement technique is to use porous surface tube instead of plain tube to obtain more vaporized core. This enhanced option has been successfully applied [21]. The results of scheme comparison are shown in Table 4.25.

Table 4.24 Process condition of toluene column overhead/benzene column bottom reboiler—Duty: 29.2 MW.

Item	Hot fluid	Cold fluid
Fluid name	Toluene	Benzene column bottom
Flow (kg/h)	302,466	1,174,824
Allowable pressure drop (kPa)	50	Thermal siphon
Temperature (°C)	198 → 178	149 → 151
Gas phase density (kg/m³)	20 → −	− → 6.15
Liquid phase density (kg/m³)	− → 706	748 → 747
Gas viscosity (mN s/m²)	0.011 → −	− → 0.01
Liquid viscosity (mN s/m²)	− → 0.172	0.231 → 0.218
Gas phase thermal conductivity (W/(m K))	0.03 → −	− → 0.024
Liquid phase thermal conductivity (W/(m K))	− → 0.093	0.101 → 0.101
Gas relative heat (kJ/(kg K))	1.83 → −	− → 1.643
Liquid phase specific heat (kJ/(kg K))	− → 2.31	2.169 → 2.184

Table 4.25 Thermodynamic simulation results of toluene column overhead/benzene column reboiler.

Item	Original option	Enhanced option
Equipment size (shell diameter × straight tube length) (mm)	1800 × 5600	1750 × 4000
Set	2	1
Tube side pressure drop (kPa)	21	20
Shell side pressure drop (kPa)	6	14
Overall heat transfer coefficient (W/(m² K))	294	1189
Heat transfer area (m²)	2385	555
Total weight of equipment (t)	55	18

Calculation results show that for the horizontal thermosiphon reboiler option with plain tube, the heat transfer effect of the hot fluid in the tube can be better, and the film heat transfer coefficient is close to $3200\,W/(m^2\,K)$. The cold fluid on the shell side boils in the flow boiling region, and the heat transfer coefficient is less than $1600\,W/(m^2\,K)$; it cannot reach the nuclear-boiling region where with higher heat transfer coefficient. If the vertical thermosiphon reboiler is adopted, the heat transfer coefficient of hot fluid in the shell side is lower than $1500\,W/(m^2\,K)$. The cold fluid boils respectively in the convection boiling region, the transition boiling region, and the mist boiling region from the inlet to the outlet of the reboiler. In the convective boiling region, the heat transfer coefficient is about $1400\,W/(m^2\,K)$. The transition boiling zone is very unstable and the film heat transfer coefficient is maintained at $1200\,W/(m^2\,K)$. Once entering the mist boiling region, the wall of the heat exchange tube is dry wall, vaporization occurs only in the center of the gas phase, and the boiling heat transfer coefficient decreases rapidly to about 220–$270\,W/(m^2\,K)$.

It can be seen that under such conditions, neither the tube side nor the shell side of the plain tube can make the fluid reach the nuclear-boiling region. Considering that the boiling medium is relatively clean and not easy to scale, better results can be obtained by using porous surface to increase the bubbling core and enhance the nuclear boiling.

Considering the condensation process of the hot fluid in the reboiler, a high flux tube with a sintered porous surface inside the tube and a V-shaped longitudinal groove outside the tube can be used to enhance the heat transfer of both tube side boiling and shell side condensation. As can be seen in Table 4.25, the replacement of plain tube with high flux tube can reduce the weight of equipment by 67%. The option of one exchanger can avoid the possible "bias current" phenomenon, reduce the difficulty of piping design, and help to maintain operational stability.

4.3.4.4 Heat exchange equipment around toluene column

Part of the vapor from the toluene column overhead enters the toluene column overhead/benzene column bottom reboiler to be cooled and then enters the toluene column reflux drum. The remaining heat is generally cooled by air cooler in the early-stage unit, and it is mostly used to generate steam at present.

The fluid inlet temperature of the air cooler of toluene column top is high, the effective temperature difference is large, the overall heat transfer coefficient is relatively high, and the heat transfer area required by the ordinary high–fin air cooler is not large. If the equipment size is to be further reduced, heat transfer enhancement technologies such as plate air cooler and elliptical tube with high fin air cooler can be used.

For steam generator, control heat transfer resistance lies in the tube side, but the overall equipment heat transfer performance is better. To further reduce the size of the steam generator, the traverse corrugated tube can be used to enhance the heat transfer effect of the tube side and the shell side at the same time, and the thread tube can also be used to expand the secondary heat transfer surface.

4.3.5 Summary

Because of the high temperature of disproportionation reaction, it is necessary to recover and utilize the heat load of the effluents effectively. The condensation load of toluene column overhead is very considerable, and the energy consumption of the unit can be effectively reduced by thermal integration.

The process fluid in the disproportionation unit is clean and the heat transfer performance is excellent. A variety of equipment with heat transfer enhancement has been successfully applied. Proper selection of heat transfer enhancement technology can effectively reduce metal usage and fuel gas consumption. The goal of energy saving, consumption reduction, and green development can be achieved.

4.4 Xylene isomerization unit

C_8 aromatics from catalytic reforming, disproportionation, and alkyl transfer are mixtures, including the four isomers of ethylbenzene, PX, m-xylene, and o-xylene, in which the content of PX is generally not more than 24%. Through the separation of PX technology, the required PX can be separated from the C_8 aromatic mixture. The rest of the material needs to be restored to the close thermodynamic equilibrium of C_8 aromatic mixture by xylene isomerization technology, so as to increase the yield of PX.

4.4.1 Brief introduction of process

The basic function of xylene isomerization catalyst is to isomerize m-xylene and o-xylene and to convert ethylbenzene. According to the different conversion methods of ethylbenzene, the catalyst is divided into ethylbenzene conversion type and ethylbenzene dealkylation type. The prominent feature of the conversion process of ethylbenzene is that the ethylbenzene in C_8 aromatics is isomerized into xylene, and the resources of C_8 aromatics are fully utilized to maximize the production of p-xylene. This process is the best choice when raw material sources are tight. The disadvantage of this process is that the single path conversion rate of ethylbenzene is low, at only about 25%, and the reaction requires a "bypass" between naphthenes. As a result, the circulation of ethylbenzene and C_8 nonaromatics was large in adsorption separation, isomerization, and xylene fractionation unit. The ethylbenzene dealkylation process is characterized by the dealkylation of ethylbenzene in the raw material to benzene, with single path conversion rate of up to 70%, and the circulation of cycloalkanes is not required for the reaction, greatly reducing the load of adsorption separation, xylene fractionation, and isomerization unit. This process is suitable for the unit with abundant raw materials, which needs to increase the production of benzene and paraxylene, or the revamp project of expanding the processing capacity of the existing unit. For the newly designed unit, compared with the conversion process of ethylbenzene, the ethylbenzene dealkylation process can reduce the engineering investment and utilities consumption, but the production rate of PX per raw material is decreased.

The ethylbenzene dealkylation process is similar to the process of conversion of ethylbenzene, only using different catalysts and operating conditions. The typical process is shown in Fig. 4.12.

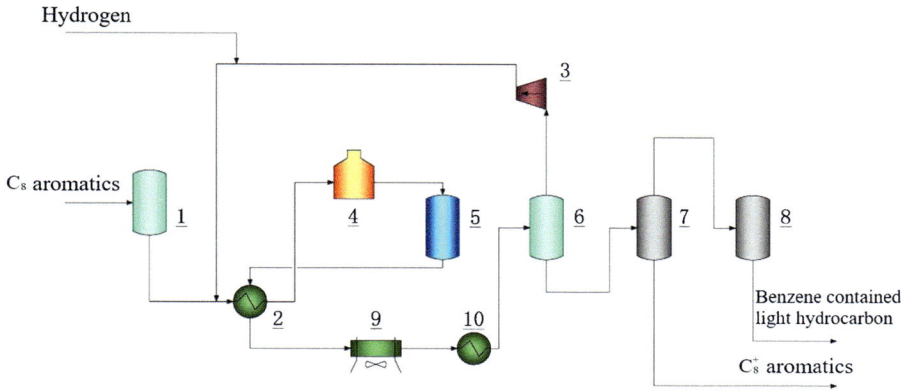

Fig. 4.12 Isomerization process flow diagram. (1) Reaction feed buffer tank; (2) reaction feed/effluent heat exchanger; (3) recycle gas compressor; (4) reaction feed furnace; (5) reactor; (6) separator; (7) deheptanizer column; (8) stripping column; (9) effluent air cooler; (10) effluent cooler.

The raffinate from the adsorption separation unit is pumped from the feed buffer tank pump and mixed with the circulating hydrogen from the compressor in the isomerization feed heat exchanger to exchange heat with the effluent. After heat transfer, the material enters the isomerization feed furnace to heat up to the reaction temperature and then enters the isomerization reactor. The isomerization reaction of C_8 aromatics in the raw material occurs under the action of catalyst. After being cooled by the feed, the effluent enters the isomerization effluent separation tank for gas–liquid separation. The gas containing hydrogen from the tank is sent to the recycle gas compressor as the isomerization circulating hydrogen and then recycled for use. In order to maintain the purity of the circulating hydrogen in the reaction system, fresh hydrogen should be made up to the circulating hydrogen system. The liquid at the bottom of the isomerization effluent separator enters the deheptanizer column, and C_7-component is removed from the top of the deheptanizer column. The bottom oil of the deheptanizer column is sent to the xylene column of the xylene separation unit, the liquid group from the deheptanizer column top is sent to the isomerization stripper, and the material from the isomerization stripper bottom is sent to the aromatics extraction unit.

4.4.2 Analysis of thermal energy characteristics of the unit

The design energy consumption of isomerization unit in plant A is shown in Table 4.26.

Table 4.26 Design energy consumption of isomerization unit in plant A—kgoe/t.

Item	Energy consumption (kgoe/t)	Share (%)	Item	Energy consumption (kgoe/t)	Share (%)
Electricity	0.87	6.1	Cooling water	0.01	0.1
3.5 MPa (gauge pressure) steam	1.35	9.6	Condensate	0.04	0.3
1.0 MPa (gauge pressure) steam	0.35	2.5	Total	14.13	100
Fuel gas	11.59	82			

The design energy consumption of isomerization unit in plant B is shown in Table 4.27.

Table 4.27 Design energy consumption of isomerization unit in plant B—kgoe/t.

Item	Energy consumption (kgoe/t)	Share (%)	Item	Energy consumption (kgoe/t)	Share (%)
Electricity	0.98	5.9	Cooling water	0.02	0.2
3.5 MPa (gauge pressure) steam	7.6	45.6	Condensate	0.1	0.6
1.0 MPa (gauge pressure) steam	0.41	2.4	Total	16.67	100
Fuel gas	7.76	46.5			

From the above energy consumption table, it can be seen that the energy consumption of the isomerization unit is mainly concentrated in fuel, followed by steam. The fuel users of the isomerization unit are the furnace of the isomerization reaction and the reboiler furnace of the distillation column of the fractionation system. The steam is mainly used to drive the compressor turbine. The fuel consumption of the furnace can be reduced by enhancing heat transfer, so the energy consumption of the isomerization unit can be effectively reduced.

4.4.2.1 Effluent heat

Similar to the disproportionation unit, the catalyst activity decreases with the prolongation of the catalyst operation time, and the reaction temperature needs to be increased to maintain the raw material conversion rate. Because the end reaction temperature of the isomerization unit is above 400°C, it is necessary to set the reaction feed/effluent heat exchanger to heat the reaction feed with high-temperature effluents, so as to reduce the fuel consumption of the reaction feed furnace. By enhancing the heat transfer process of the reaction feed/effluent heat exchanger, the heat exchange load may be increased, so as to further reduce the fuel consumption of the reaction feed furnace, thus reducing the energy consumption of the unit.

4.4.2.2 Fractionation column top heat

The fractionation part of the isomerization unit usually consists of two column systems, the deheptanizer column, and the stripping column. The condensation load at the top of the deheptanizer column is relatively large, especially for the project of ethylbenzene dealkylation-type technology. It can exchange heat with the feed of the deheptanizer column or set low-temperature hot water to recover the condensation load. Due to the low temperature at the top of the stripping column and the low heat load, the heat recovery is generally not considered.

4.4.3 Process enhancement and integration on heat transfer

4.4.3.1 Effluent heat recovery

The temperature of the effluent from the isomerization reactor is usually between 370°C and 430°C (start stage-end stage), and the reaction feed/effluent heat exchange can effectively recover the heat. Usually, fully welded plate and shell heat exchanger or coil-wound heat exchanger is used to reduce the temperature difference at the hot end of the heat exchanger and reduce the duty of the isomerization reaction feed furnace, so as to further reduce the energy consumption of the unit. According to the typical temperature difference at the hot end of the plate heat exchanger and the actual operating conditions of the shell and tube heat exchanger, the simulation calculation is carried out. The duty of the reaction feed furnace can be reduced by 56% after the plate heat exchanger is used [22].

4.4.3.2 Low-temperature heat recovery

The low-temperature heat of the isomerization unit is mainly concentrated on the top of the fractionation column, which can be used to heat the hot water for recovery. For 2.7 Mt/a isomerization unit using ethylbenzene conversion process, the heat recovery can be 6.5–7.9 MW.

The effluents from the isomerization feed/effluent heat exchanger can be further utilized from the point of view of temperature level, but from the point of view of protecting the isomerization noble metal catalyst, it is not recommended to use hot water for heat recovery.

4.4.4 Heat transfer enhancement technology by improving elements

The process of isomerization unit is relatively short, and the heat exchange equipment is relatively less. Because the fluid in the isomerization unit is relatively clean, the fully welded plate and shell heat exchanger has been successfully used in many positions in the isomerization unit. In recent years, coil-wound heat exchangers and porous surface tubes with enhanced heat transfer characteristics have been applied in isomerization units [4,10].

The processes of different licensors are not the same, but the physical properties of the process fluid in the isomerization unit are similar. A typical process can be taken as an example to analyze the characteristics of heat transfer equipment in the isomerization unit and the applicable heat transfer enhancement measures.

4.4.4.1 Heat transfer system of isomerization reaction part heat transfer system

After the raw material is cooled by the isomerization feed/separator bottom heat exchanger, it exchanges heat with the mixed hydrogen in the isomerization feed/effluent heat exchanger and enters the isomerization reactor after further heated by the furnace. The effluents are cooled in the isomerization feed/effluent heat exchanger to be the gas-liquid two-phase fluid, then enter the product air cooler for further cooling.

The heat transfer performance of the tube side and shell side of the isomerization feed/separator bottom heat exchanger is relatively good. The overall heat transfer coefficient can reach $600\,W/(m^2\,K)$ even in the conventional segmental-baffle shell and tube heat exchanger. The control thermal resistance lies in the shell side. If heat transfer enhancement measures are further required, elements which can improve the heat transfer performance of the shell side can be used instead of the plain tube, or low-finned tube and traverse corrugated tube can be used to expand the secondary heat transfer area.

The duty of the effluent air cooler is large, and the effective temperature difference of the air cooler is relatively large due to the high temperature of the process fluid inlet. In spite of the existence of noncondensing gas, the overall heat transfer coefficient of the air cooler is still up to $460\,W/(m^2\,K)$ when the conventional high-fin air cooler is used, which shows good heat transfer performance. Considering the requirements of allowable pressure drop and heat transfer, and considering the low fouling resistance of isomerization effluent, surface evaporation air cooler and plate air cooler can be used to enhance the heat transfer effect and further reduce the heat transfer area, metal usage, and the number of motors.

The isomerization feed/effluent heat exchanger bears a high heat load, which directly affects the fuel consumption of the furnace of the isomerization unit, so it is a key research position. The hot fluid is overheated 253°C in the heat exchanger, and two-thirds of the heat load is in the region without phase transition where the pure gas phase is cooled. When the cold fluid is superheated at 233°C, half of the heat load occurs in the region without phase transition where the gas phase is heated up. The process conditions are shown in Table 4.28.

Table 4.28 Process condition of isomerized feed/effluent heat exchanger—Duty: 92.5 MW.

Item	Hot fluid	Cold fluid
Fluid name	Product	Feed
Flow (kg/h)	320,938	320,938
Allowable pressure drop (kPa)	50	70
Temperature (°C)	$470 \rightarrow 146$	$102 \rightarrow 420$
Gas phase density (kg/m^3)	$13 \rightarrow 7.48$	$5.55 \rightarrow 16.87$
Liquid phase density (kg/m^3)	$- \rightarrow 748$	$790 \rightarrow -$
Gas viscosity (mN s/m^2)	$0.016 \rightarrow 0.011$	$0.011 \rightarrow 0.015$
Liquid viscosity (mN s/m^2)	$- \rightarrow 0.2$	$0.27 \rightarrow -$
Gas phase thermal conductivity (W/(m K))	$0.118 \rightarrow 0.12$	$0.138 \rightarrow 0.11$
Liquid phase thermal conductivity (W/(m K))	$- \rightarrow 0.103$	$0.112 \rightarrow -$
Gas relative heat (kJ/(kg K))	$2.81 \rightarrow 2.95$	$4.21 \rightarrow 2.72$
Liquid phase specific heat (kJ/(kg K))	$- \rightarrow 2.15$	$2.0 \rightarrow -$

The heat transfer coefficient of gas phase heat transfer is relatively low. Considering the allowable pressure drop limitation and vibration problems, the gas flow velocity should not be too high during the design, which further reduces the film heat transfer coefficient of gas. The temperature of the cold and hot fluids in the heat exchanger varies greatly, and the internal temperature cross is serious. Only pure countercurrent can solve this problem. At the same time, the volume flow rate is large, and it is impossible to select an F-shell exchanger constrained by the allowable pressure drop. If the shell and tube heat exchanger is used, the best solution is that fluids countercurrent flow in a one-pass exchanger. The simulation results are shown in Table 4.29.

Table 4.29 Thermodynamic simulation results of isomerized feed/effluent heat exchanger.

Item	Original option	Enhanced option
Equipment size (shell diameter × straight tube length) (mm)	$2500 \times 24,000$	$2354 \times 13,125$
Set	2	1
Tube side pressure drop (kPa)	10	
Shell side pressure drop (kPa)	66	
Overall heat transfer coefficient (W/(m^2 K))	325	
Heat transfer area (m^2)	12,027	
Total weight of equipment (t)	323	88

It can be seen from Table 4.29 that the equipment weight of fully welded plate and shell heat exchanger is only 27% of that of conventional shell and tube heat exchanger, which reduces the equipment weight by 235 tons and the primary investment of equipment. The plate and shell structure can shorten the length of heat exchanger by 11 m, reduce the basic investment of civil construction, and reduce the cost of pipeline configuration and installation.

The use of fully welded plate and shell heat exchanger can also reduce the temperature difference of the hot end and reduce the load of the furnace and fuel consumption. Literature [22] introduced a successful case of replacing the conventional shell and tube heat exchanger with fully welded plate and shell heat exchanger. After adopting the welded plate and shell structure, the temperature of the reaction feed at the furnace inlet is increased by 32.9°C, and the duty of the furnace is reduced by 3.2 MW. According to the average yield of "BTX" in the aromatics complex of 42 t/h, the comprehensive energy consumption of the unit is reduced by 6.6 kgoe/t, and the investment payback period is only 2 years. Literature [23] introduced the successful application of fully welded plate and shell heat exchangers in 3.3 Mt/a isomerization unit. The comprehensive performance of plate and shell heat exchangers produced at home and abroad is compared. The comparison results show that the heat transfer effect and pressure drop are basically the same.

The purpose of heat transfer enhancement can also be achieved by using coil-wound heat exchangers. Literature [10] analyzed the characteristics and influencing factors of the coil-wound heat exchanger according to the field operation, and the excellent heat transfer performance is illustrated by the calibration data. Literature [24] described the application effect of coil-wound heat exchanger in isomerization feed/effluent exchanger. By replacing the conventional vertical shell and tube heat exchanger, the coil-wound heat exchanger reduces the fuel gas consumption of the furnace by 160 kg/h. At the same time, three air coolers can be shut down, and power consumption is reduced. On account of only this two items, the annual cost savings reaches 2,662,800 yuan.

4.4.4.2 Heat transfer equipment around deheptanizer column

1. Feed heat exchanger

   After the heat is exchanged in turn in isomerization feed/separator bottom heat exchanger, the deheptanizer column feed/overhead heat exchanger and the deheptanizer column feed/bottom heat exchanger, the isomerization effluent enters the deheptanizer column as raw material.

   Isomerization feed/separator bottom heat exchanger and deheptanizer column feed/column bottom heat exchanger have the same characteristics: cold and hot fluids are relatively clean, heat transfer effect is very good, and the film heat transfer coefficient of tube side and shell side is close. When the structure design is reasonable, the overall heat transfer coefficient can reach 650 W/(m² K), even if the ordinary shell and tube heat exchanger is used. Considering the large allowable pressure drop of the hot fluid in the tube path, low-finned tube, corrugated tube, or transversally corrugated tube with double-sided heat transfer enhancement effect can be used to further reduce the equipment size.

The cold and hot fluids in the feed/overhead heat exchanger of the deheptanizer column are relatively clean, and the heat transfer effect is acceptable. Due to the influence of the operating pressure in the column, the allowed pressure drop of vapor condensation at the column top is small, which limits the heat transfer of condensation. The low-finned tube can be used instead of the plain tube to improve the heat transfer effect of condensation and reduce the equipment size.

2. Deheptanizer column top heat exchange equipment

The temperature of overhead gas drops to about 145°C after cooling in the deheptanizer column feed/overhead heat exchanger, and the fluid still contains about 44% (mass fraction) of the gas phase. The gas–liquid two-phase medium enters the air cooler and water cooler to be condensed and cooled to 40°C.

Similar to the isomerization effluent air cooler, the effective temperature difference is relatively large due to the high temperature of the process fluid. The heat transfer performance of the fluid is good, and the overall heat transfer coefficient of the air cooler is still up to 400 W/(m² K) even when the conventional high–fin air cooler is selected. The required heat transfer area is not large. Considering the cooling requirements of the process fluid and the low fouling resistance of the fluid, plate air cooler can be applied to enhance the heat transfer effect, further reducing the heat transfer area, metal usage, and the number of motors.

A company in the Republic of Korea replaced the air cooler with a fully welded plate and shell heat exchanger to recover heat load and reduce the fuel consumption of the furnace. A domestic unit replaced the air cooler with all-welded plate heat exchanger, which has been in good running condition for many years

3. Deheptanizer column bottom heat exchange equipment

The liquid from the deheptanizer column bottom is sent into the deheptanizer column bottom/xylene column feed heat exchanger and deheptanizer column feed/bottom heat exchanger. These two heat exchangers have common characteristics, i.e., good heat transfer effect of cold and hot fluids, and an overall heat transfer coefficient of 700 W/(m² K) can be achieved even with conventional shell and tube heat exchangers. The size of the exchanger can be further reduced by replacing plain tube with low-finned tube, corrugated tube, or corrugated low-finned tube with double-side heat transfer enhancement.

Deheptanizer column is heated by furnace or reboiler. The composition of deheptanizer column bottom medium is relatively simple; the boiling range is narrow. So the porous surface tube is suitable to replace the plain tube. The reboiler of the deheptanizer column bottom in a 1.0 Mt/a aromatics complex adopts a high flux tube, and the enhancement effect is remarkable [21]. Considering the characteristics of cold and hot fluids, when the reboiler is installed vertically, technical measures for enhancing heat transfer on both tube side and shell side can be adopted. Taking a 2.8 Mt/a isomerization unit as an example, the process conditions of the reboiler at the column bottom are shown in Table 4.30, and the scheme comparison results are shown in Table 4.31.

Table 4.30 Process conditions of deheptanizer column bottom reboiler—Duty: 7.7 MW.

Item	Shell side	Tube side
Fluid name	C_8 aromatics	Hydrocarbon
Flow (kg/h)	99,211	288,373
Allowable pressure drop (kPa)	70	Thermal siphon
Temperature (°C)	$248 \rightarrow 239$	$217 \rightarrow 218$
Gas phase density (kg/m^3)	$28.3 \rightarrow -$	$- \rightarrow 17.3$
Liquid phase density (kg/m^3)	$- \rightarrow 642$	$668 \rightarrow 667$
Gas viscosity (mN s/m^2)	$0 \rightarrow -$	$- \rightarrow 0.01$
Liquid viscosity (mN s/m^2)	$- \rightarrow 0.168$	$0.184 \rightarrow 0.183$
Gas phase thermal conductivity (W/(m K))	$0 \rightarrow -$	$- \rightarrow 0.026$
Liquid phase thermal conductivity (W/(m K))	$- \rightarrow 0.0973$	$0.1 \rightarrow 0.1$
Gas relative heat (kJ/(kg K))	$2.15 \rightarrow -$	$- \rightarrow 2.02$
Liquid phase specific heat (kJ/(kg K))	$- \rightarrow 2.628$	$2.532 \rightarrow 2.531$

Table 4.31 Scheme comparison of deheptanizer column bottom reboiler.

Item	Original option	Enhanced option
Equipment size (shell diameter × straight tube length) (mm)	1500×6000	1300×4000
Set	1	1
Shell side pressure drop (kPa)	5	5
Static head (m)	4.1	3
Overall heat transfer coefficient (W/(m^2 K))	475	1009
Heat transfer area (m^2)	728	345
Total weight of equipment (t)	24	11

As can be seen in Table 4.31, after replacing the plain tube with an enhanced heat transfer tube with a sintered porous surface inside the tube and a V-shaped longitudinal groove outside the tube, both boiling inside the tube and condensation outside the tube are enhanced, the overall heat transfer coefficient of the reboiler is doubled, the heat transfer area is reduced by 50%, and the weight of the equipment is reduced by 54%. The static pressure head required by the reboiler is reduced from 4.1 to 3 m, and the skirt height of the deheptanizer column can also be reduced by 1 m, which can reduce the cost of civil engineering and installation.

4.4.5 Summary

Although the reaction heat of the isomerization unit is small, the reaction temperature is high. And the feeding scale of the unit is much larger than that of other units. The recovery and utilization of the heat as much as possible can significantly reduce energy consumption. The fractionation column top is difficult to effectively utilize due to its light component and low temperature, so it can be considered for recovery by hot water.

The process flow of the isomerization unit is relatively short with less heat exchange equipment. Because the fluid in isomerization unit is relatively clean, a variety of efficient heat exchange equipment has been successfully applied, bringing obvious economic benefits to enterprises.

4.5 Paraxylene separation and xylene fractionation unit

Mixed xylene contains four kinds of C_8 aromatic isomers, namely ethylbenzene, p-xylene (PX), m-xylene, and o-xylene. The density between them is close and the boiling point difference is small, which makes it difficult to separate using conventional distillation technology. At present, the industrial technologies for the separation of PX include crystallization method and adsorption separation method. Before the industrialization of the adsorption separation method, the cryogenic crystallization method was the only method for the separation of PX. The freezing point of PX is much higher than that of other C_8 aromatics. The cryogenic crystallization method achieves the separation of PX based on this property. But under low temperatures, the C_8 aromatics isomers will form eutectoid, coupled with the influence of crystal grain size and mother liquid state on separation efficiency, the single pass yield of PX is low, at only about 65%, far less than the yield of simulated moving bed adsorption separation technology. Hence at present, the adsorption separation method is the mainstream PX separation method.

4.5.1 Brief description of process
4.5.1.1 Brief description of the process of adsorption separation unit

Adsorption separation method is based on the fact that the surface affinity of adsorbent molecular sieve for PX is much higher than that of other components to separate PX. The industrialization of simulated moving bed technology makes it possible to obtain high-purity paraxylene by adsorption separation technology. In the adsorption column, the adsorbent and C_8 aromatics mixture simulated continuous reverse flow, forming a repeated mass transfer process so that PX was gradually concentrated on the adsorbent, and the high-purity product was produced. Due to the simple process and high single pass recovery rate, the adsorption separation method has developed rapidly after industrialization. At present, most industrial paraxylene production units adopt this method. Up to now, the purity of PX products is generally above 99.7%, and the highest purity can reach 99.9%. The single pass yield of PX is usually 95%–98%, and the highest can reach more than 99%.

At present, the industrial adsorption separation technologies for paraxylene include Parex process developed by UOP, Eluxyl process developed by Axens (IFP) [25], and SorPX process developed by Sinopec. These processes are different in the implementation of simulated moving bed, mainly in the bed pipeline switching scheme, bed pipeline flushing scheme, and so on. The function of the simulated moving bed is realized by periodically switching the material in and out of position. Parex process controls the drive of the rotary valve by adsorption separation control system (ACCS) and then realizes the periodic switching of material in and out of the adsorption column. Eluxyl process realizes periodic switching of incoming and outgoing materials by controlling 144 program-controlled valves through the adsorption column sequence control system (SCS). SorPX process controls the switch of 192 program-controlled valves to realize the periodic switching of incoming and outgoing materials through moving bed control system (MCS). As the recycling of material in and out of the adsorption column moves from top to bottom, different materials pass through the same bed pipeline and the internal components of the adsorption column connected with it at different times. In order to get high-purity products and improve the single pass recovery rate of products, it is necessary to wash the bed pipeline and internal components of the adsorption column. Different processes use different flushing schemes. Parex process adopts three flushing schemes. Eluxyl process adopts bypass flushing scheme. SorPX process adopts four flushing schemes, which can realize accurate control of the composition and flow of washing flows, improve the adsorption separation efficiency, and reduce the circulating flow of desorbent by more than 10%.

The adsorption separation unit usually includes adsorption column, raffinate column, extract column, PX column, and desorbent rerun column. The typical process flow using heavy desorbent is shown in Fig. 4.13.

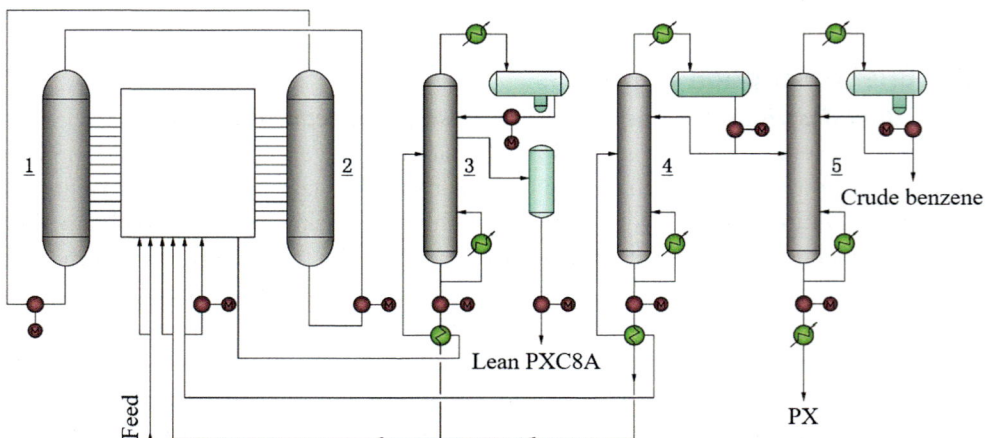

Fig. 4.13 Schematic diagram of typical process flow of adsorption separation unit. (1) Adsorption column; (2) adsorption column; (3) raffinate column; (4) extract column; (5) PX column.

A typical simulated moving bed adsorption separation technology has 2 adsorption columns and 24 adsorption beds. Two circulating pumps are used to connect the two adsorption columns head to tail so that the 24 beds form a closed loop. Under the action of the circulating pump, the liquid flow circulates around the 24 beds periodically. The rotating valve or program control valve periodically changes the inlet and outlet position of each material, which is equivalent to changing the position of each area so that the area circulates periodically along the bed to achieve the purpose of continuous adsorption and separation.

The mixed xylene feed passes through the filter to remove trace solid particles and enters the adsorption zone of the adsorption column. In the adsorption area, PX is adsorbed on the adsorbent, and the raffinate (the mixture of unadsorbed C_8 aromatics and desorbent) flows out from the lower part of the adsorption area under pressure control and enters the raffinate column after heating up. The C_8 aromatics were separated from the desorbent by distillation in the raffinate column. C_8 aromatics were drawn out from the sideline of the column and were sent to the isomerization unit as raw materials. The products at the bottom of the column, that is, the desorbent, are boosted by pump and mixed with the desorbent from the extract column bottom after heat exchange with feed, which generally can be used as the heat source of the PX column reboiler. Most of them are recycled into the desorption area of the adsorption column after removing trace solid particles through the desorbent filter, while the rest are sent to the desorbent rerun column for treatment.

The extract (adsorbed PX and desorbent) in the lower part of the purification area of the adsorption column leaves from the column under the flow control and enters the extract column after being heated up to separate the PX from desorbent. The effluent from the top of the column is PX containing part of crude toluene, which is sent to the PX column for treatment. The desorbent from the bottom of the extract column is boosted by pump, mixed with the desorbent from the raffinate column bottom after heat exchange with the column feed, and returned to the adsorption column for recycling.

The crude PX from the top of the extract column enters the PX column, and the crude toluene from the top of the column is pumped to the disproportionation unit as raw material. The material at the bottom of the column, which is a PX product cooled to 40°C by the paraxylene cooler and paraxylene trim subcooler and then pumped to the product tank after passing the online chromatography analysis.

A small part of the desorbent from the raffinate column and the extraction column bottom is sent to the desorption rerun column for treatment. The material of the column bottom is a metamorphic desorbent, which is pumped and cooled and then sent out of the unit. The fluid from the column top is the regenerated desorbent, which is sent into the raffinate column for recycling.

4.5.1.2 Brief description of the xylene fractionation unit process

The xylene fractionation unit is composed of a olefin removal unit and a number of fractionating columns, usually including raw material refining, reformate splitter, xylene column, and heavy aromatic column. If *o*–xylene (OX) products need to be produced, *o*–xylene columns need to be set up. The feedstock for the xylene fractionation unit usually includes C_6^+ reformates supplied by the reforming unit, fluids from deheptanizer bottom of isomerization unit, and toluene column bottom of disproportionation unit.

The function of the reformates splitter is to separate C_6^+ reformates fraction into C_6/C_7 fraction and C_8^+ aromatics. The C_6-C_7 fraction from the top of the column is used as raw materials for the aromatics extraction unit. The bottom is mixed with C_8^+ aromatics produced by the disproportionation unit and isomerization unit and then was sent to xylene column, o-xylene column, and heavy aromatic column to obtain C_8 aromatics, o-xylene products, C_9/C_{10} aromatics, and heavy aromatics. C_8 aromatics are sent to the adsorption separation unit as raw materials. C_9/C_{10} aromatics are sent to the disproportionation reaction part as raw material, and a small amount of heavy components containing C_{11}^+ can be sent to hydrocracking or catalytic cracking units.

In some aromatics complex, the reformate splitter is arranged in the continuous reforming unit, and the heavy aromatics column is arranged in the disproportionation unit.

The typical process flow of the xylene fractionation unit is shown in Fig. 4.14.

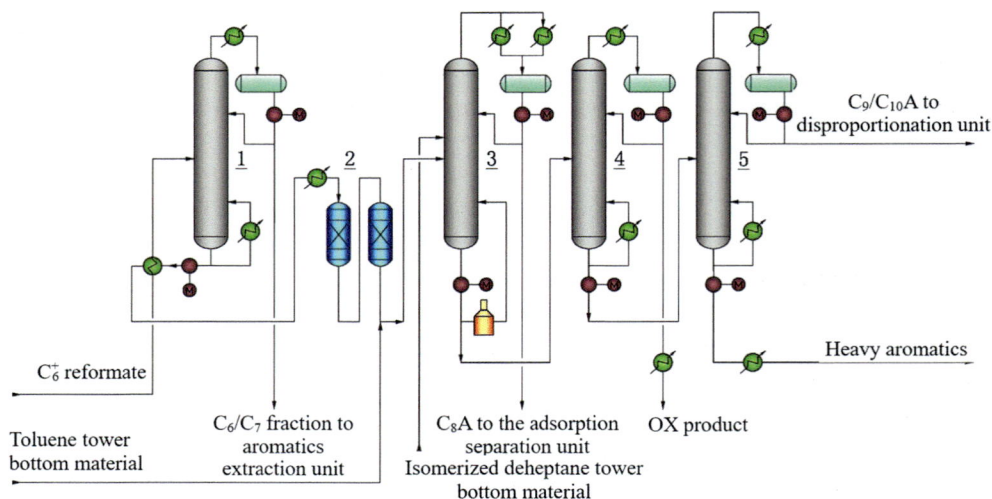

Fig. 4.14 Schematic diagram of typical process flow of xylene fractionation unit. (1) Reformate splitter; (2) olefin removal processor; (3) xylene column; (4) o-xylene column; (5) heavy aromatic column.

The C_6^+ reformates fraction enters the reformate splitter after heat exchange with the fluid from the column bottom. The fluid from the column top is C_6-C_7 fraction, which is cooled and sent to the aromatics extraction unit for the separation of aromatics from non-aromatics. C_8^+ material from the reformate splitter bottom is pumped and heated by the column feed and the olefin removal processor heater, then enters into the olefin removal processor to remove the olefin. If olefins have been removed from the reformate upstream, the olefin removal processor here may be omitted.

After olefin removal, the C_8^+ material from the reformate splitter bottom is mixed with the C_8^+ material from the disproportionated toluene column bottom. The mixture exchanges heat with fluid from the xylene column bottom and then enters the xylene column. The fluid from the isomerized deheptanizer bottom is also sent to different positions in the xylene column. The product from the top of the xylene column enters the adsorption separation feed buffer tank. It is sent to the adsorption separation unit as raw material after pump boost.

For unit producing *o*-xylene, part of *o*-xylene is separated from the xylene column bottom along with the C_9^+ aromatics and is sent to the *o*-xylene column for separation; the *o*-xylene product from column overhead is sent out of the unit. The C_9^+ aromatics from the column bottom are sent to the heavy aromatics column.

For units that do not produce *o*-xylene, *o*-xylene goes into the top of the xylene column along with other C_8 aromatics. The C_8^+ material from the xylene column bottom is directly sent to the heavy aromatic column after being cooled by the feed. The C_9/C_{10} aromatics material from the heavy aromatics column overhead is sent to the disproportionation unit as the raw material for disproportionation reaction, and the C_{11}^+ from the column bottom is sent out of the unit as heavy aromatics product after being boosted by pump and cooled.

4.5.2 Analysis of thermal energy characteristics

4.5.2.1 Multiple distillation columns with many heating and cooling processes and high energy consumption

In order to feed disproportionation and aromatics extraction unit and produce products such as PX and *o*-xylene, the adsorption separation unit and xylene fractionation unit are usually set up with eight distillation columns. Due to the strict requirements of product purity and unit feed for impurity content, the separation accuracy of the distillation column is very high, and a large amount of heat load needs to be put in. On the other hand, there are a large number of materials at the column top that need to be condensed and cooled, which is one of the main reasons for the high energy consumption of the aromatics unit. Taking an aromatics unit with a production of 650 kt/a paraxylene as an example, the scale of the upstream supporting reforming unit is 2000 kt/a, with by-product capacity of 50 kt/a *o*-xylene. The heat load input to the distillation column of the adsorption separation unit and xylene fractionation unit is 107.66 MW and 180.12 MW, respectively. The duty of the distillation column of the two units is shown in Table 4.32.

Table 4.32 Duty distillation reboiler of the two units.

Adsorption and separation unit (MW)		Xylene fractionation unit (MW)	
Raffinate column	70.18	Reformate splitter	22.32
Extract column	24.46	Xylene column	130.58
PX column	10.97	o-Xylene column	15.52
Desorbent rerun column	2.05	Heavy aromatics column	11.7
Total	107.66	Total	180.12

4.5.2.2 Low-temperature heat

There is a large amount of condensation heat at the top of the distillation column in both the adsorption separation unit and the xylene fractionation unit. And the temperature level is generally low, and the temperature of reflux is usually between 66°C and 131°C. Because of the low temperature, the condensation heat is difficult to be used in the unit, usually cooled by air cooler. Taking the aforementioned unit as an example, the condensation heat of this part is as high as 143.77 MW. The condensation heat at the top of the rectifying column of the two units is shown in Table 4.33.

Table 4.33 Condensation heat at the top of the column.

	Reformate splitter	o-Xylene column	Raffinate column	Extract column	PX column	Total
Temperature (°C)	$109 \rightarrow 71$	$158 \rightarrow 131$	$145 \rightarrow 121$	$145 \rightarrow 121$	$116 \rightarrow 66$	
Heat load (MW)	20.43	15.77	72.45	25.03	10.09	143.77

4.5.3 Process enhancement and integration on heat transfer

4.5.3.1 Heat recovery technology at the top of the column

1. Moderate pressurization operation of xylene column for heat integration with raffinate column and extract column

   The xylene column is the integration center of material and heat in the aromatics complex, with large feed scale, high reflux ratio, and large condensing load at the top of the column. In order to recover the condensation load of the fluid from the xylene column top, the xylene column is usually designed to integrate with other distillation columns, and the condensation load is used as the heat source of other distillation column reboilers. The specific heat integration scheme needs to be matched according to the actual heat balance.

   The traditional heat integration process is: the xylene column operates under moderate pressure, while the raffinate column and the extract column operate under normal pressure. The condensation latent heat of the gas phase flow from the xylene column overhead is used as the heat source of the reboilers at the bottom of the raffinate column and the extract column. The fluids from raffinate column and extract column overhead flow back to the column as hot reflux after being cooled by air cooler. The selection of the operating pressure of the xylene column depends on the temperature requirements of the reboiler at the bottom of the raffinate column and the extract column.

Taking the aforementioned aromatics complex as an example, when the xylene column is properly pressurized, the column operating pressure is 848 kPa, the condensing heat load on the top is 87.39 MW, and the fluid is cooled to 239°C. The load at the bottom of the raffinate column and the extract column is 70.38 MW and 26.50 MW, respectively, and the temperature at the bottom of the columns is 216°C and 209°C, respectively. The condensate fluid from the xylene column overhead can be used as the reboiling heat source of the raffinate column and the extract column. The operating conditions of the xylene column, the raffinate column, and the extract column in the traditional process are shown in Table 4.34.

Table 4.34 Operating conditions of columns.

Operating conditions	Xylene column	Raffinate column	Extract column
Top pressure (kPa)	848	29	22
Top temperature (°C)	247	145	145
Bottom temperature (°C)	288	216	209
Top load (MW)	87.39	71.17	26.11
Bottom load (MW)	91.68	70.38	26.50

If heat integration technology is not applied to recover the condensation duty at the top of the xylene column, the xylene column can be operated under normal pressure, and the condensation duty at the column top is 76.8 MW. Therefore, 76.8 MW heat load can be recycled by applying heat integration technology in the unit.

2. Xylene column is further pressurized, and steam is generated by moderately pressurizing the raffinate column and the extract column

As can be seen from Table 4.34, although the heat integration technology is adopted for the xylene column and the raffinate column as well as for the xylene column and the extract column, and the condensation duty is utilized by raising the fluid temperature level at the top of the column, a large amount of condensation duty is still not recovered because of the low-temperature level at the top of both the raffinate column and the extract column. In order to further recover the condensate load, the SorPX technology of Sinopec takes appropriate pressure lifting on the raffinate column and the extract column to improve the temperature level of the fluid from the column overhead and produce 0.5 MPa saturated steam. After being superheated the steam can be directly utilized with high efficiency, such as driving compressor turbine or as heat source, or can also be used to generate electricity. In order to meet the temperature requirements at the bottom of the raffinate column and the extract column operated under the pressurized condition, the xylene column needs to be further pressurized. The tray number of the raffinate column and the extract column should be increased appropriately to decrease the reboiling load. See Table 4.35 and Fig. 4.15 for the operating conditions and flows of heat recovery for the three columns.

Table 4.35 Operating conditions of column operated under further pressurization.

Operating conditions	Xylene column	Raffinate column	Extract column
Top pressure (kPa)	1300	250	280
Top temperature (°C)	273	190	195
Bottom temperature (°C)	313	252	253
Top load (MW)	105.54	71.51	26.31
Bottom load (MW)	109.94	79.95	28.84

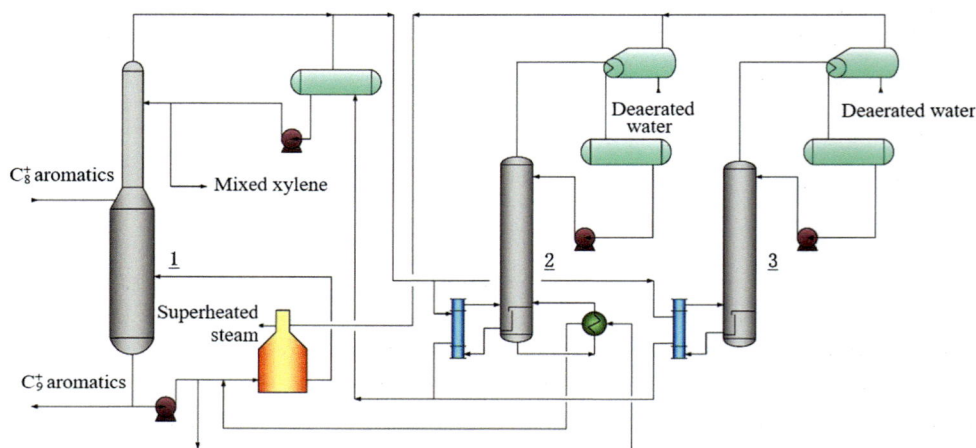

Fig. 4.15 Schematic diagram of heat integration process of xylene tower, raffinate column, and extract column. (1) Xylene column; (2) raffinate column; (3) extract column.

By applying deep heat integration technology to recover the condensation load from the top fluid of the three columns, the unit further saves 85.8 MW of heat load.

3. Heavy aromatic column operated under moderate pressurized and heat integration with reformate splitter

In aromatics complex, the function of the heavy aromatics column is to separate C_9/C_{10} aromatics at the column top as raw materials for disproportionation and alkyl transfer and remove the recombination components at the column bottom. In the traditional design, the heavy aromatics column is operated under atmospheric pressure, and the top fluid temperature is about 184°C, to be directly cooled to about 150°C

When the pressure of the heavy aromatics column is raised to 200 kPa, the temperature of the top of the heavy aromatics column is correspondingly increased to 218°C. The gas phase fluid from the heavy aromatics column overhead is sent to the reboiler of the reformate splitter as heat source, and the heat recovery is realized. Since the condensate heat recovered from the heavy aromatics column overhead cannot meet the demand of the reformate splitter bottom, a steam reboiler is usually set up to supplement the heat of the reformate splitter and to adjust the operation of the column in normal operation. The heat integration operation parameters of the heavy aromatic column and the reformate splitter are shown in Table 4.36.

Table 4.36 Operation parameters of heat integration of heavy aromatics column and reformate splitter.

Operating conditions	Heavy aromatics column	Reformate splitter
Top pressure (kPa)	200	40
Top temperature (before/after cooling) (°C)	218/193	109/71
Bottom temperature (°C)	269	178
Top load (MW)	12.13	20.43
Bottom load (MW)	11.70	22.32[a]

[a]The heavy aromatics column can provide 12.13 MW heat for the reformate splitter, and the insufficient part is supplemented by the steam reboiler.

For the aforementioned aromatics complex, by adopting the heat integration technology between the top of the heavy aromatics column and the reformate splitter, the steam consumption can be reduced by 21 t/h, and the air cooler power consumption can be saved by 49 kW, which is converted into energy consumption of 24 kgoe/t PX.

4.5.3.2 Stage-by-stage heat transfer for energy cascade utilization

The mixed xylene from the top of the xylene column is the raw material for adsorption separation unit, usually at 240–265°C. And the feed of the adsorption separation unit is usually between 160°C and 180°C. The feed carries more heat load. In order to make full use of the heat load, heat transfer can be done step by step according to the different temperature requirements of the materials to be heated in the unit, so as to achieve the best economy in heat recovery and heat transfer equipment settings. In the complete set of aromatics technology of Sinopec, this fluid is used to heat the feed of xylene column and the feed of deheptanizer and as the heat source of the PX column in turn. Cooled to the required temperature, it is returned to the adsorption column.

The circulating amount of desorbent in the adsorption separation unit is also relatively large, usually 0.95–1.3 times that of the feed, which also carries a large amount of heat load. In the conventional process, the desorbent is separated at the raffinate bottom and the extract column bottom. The desorbent exchange heat with the feed and then as the heat source of the reboiler of the PX column; finally it is returned to the adsorption column.

4.5.3.3 Low-temperature heat recovery by hot medium water

Using hot medium water to recover low-temperature heat is one of the widely used low-temperature heat recovery technologies in recent years. The temperature of hot medium water entering unit is about 60°C, and the temperature of leaving unit is about 90°C, so the low-temperature heat of about 100°C in the unit can be all recovered. For the adsorption separation unit and xylene fractionation unit, such level of low-temperature heat includes the condensation heat of PX column top, the cooling heat of PX product, and the condensation heat of reformate splitter top. For the aforementioned aromatics complex, the above heat is 25.8 MW. After the heat is recovered by the hot medium water, it can be used for centralized refrigeration, heating, and power generation and can also be supplied to the chemical unit with the matching temperature level.

4.5.3.4 Low-pressure steam power generation and hot medium water power generation technology

1. Low-pressure steam power generation technology

In the adsorption separation unit, the raffinate column and the extract column are normally designed as atmospheric pressure columns. The temperature of the fluid from the column overhead is 145–150°C and to be cooled to 121–130°C. The temperature at the top of the column is slightly different with the column operating pressure. The condensation heat from the top of the raffinate column and the extract column is major low-temperature heat resources in the aromatics complex. Taking a 600 kt/a aromatics complex as an example (same as in below), the duty of the two positions is 97.4 MW, which translates into energy consumption greater than 100 kgoe/t PX. If this part of low-temperature heat can be partially recovered, it will play a great role in energy saving and consumption reduction of the aromatics complex.

When the raffinate column and extract column are operated at atmospheric pressure, 0.2 MPa steam can be generated at the top of the column, and the pressure is generally about 0.15 MPa after superheated. If the steam of this quality is used to generate power, the conversion efficiency of steam and electricity is low. In addition, it is not very reasonable from technical economy as it has large-scale generator, large occupied area, and long payback period. Therefore, appropriately increasing the operating pressure of the two columns can improve the economy of low-temperature heat recovery at the top of the column.

When the pressure of the raffinate column and the extract column is pressurized to 0.25 and 0.28 MPa, the saturated steam of 0.5 MPa can be generated, the net generating power is 18 MW, and the net thermoelectric conversion efficiency can reach 15%. By low-pressure steam power generation technology, the energy consumption per ton of paraxylene products can be reduced by about 55 kgoe.

2. Hot medium water power generation technology

The low-temperature heat source lower than 130°C in the aromatics complex is difficult to be used to generate steam. Desalted water can be used as heat medium, and the process materials can exchange heat with the low-temperature desalted water respectively. The recovered low-temperature heat is concentrated in the hot water generator set for power generation.

In recent years, Carina cycle technology (ammonia water mixture) and Rankine cycle technology (Freon, pentane and other hydrocarbons) with low boiling point organic medium have been promoted, which can achieve low-temperature thermal power generation at lower temperatures.

One of the main factors affecting the efficiency of hot water power generation is the temperature of the hot medium water. The low-temperature heat source in the unit is mainly the condensation heat from the top of the column and the reactor effluents being cooled. Comprehensive analysis of the temperature level of the heat source in the unit, reasonable heat transfer process, and efficient heat transfer equipment can improve the final heat transfer temperature of the hot medium water and then improve the utilization efficiency of low-temperature heat of the unit. Taking a 1.0 Mt/a aromatics complex as an example, the low-temperature heat in four positions of the unit is used to heat hot medium water. Different heat transfer processes are adopted to obtain hot medium water with different final temperatures, and the net power generation is different. The higher the hot water temperature, the higher the thermoelectric conversion efficiency. The comparison of thermoelectric conversion efficiency of different heat transfer processes is shown in Table 4.37, and the optimized heat transfer process is shown in Fig. 4.16.

Table 4.37 Comparison of thermoelectric conversion efficiency with different heat transfer processes.

Plan	Low-temperature heat load (MW)	Number of heat exchangers, set	Hot medium water flow rate (t/h)	Final temperature of hot medium water (°C)	Net power generation (kW)
Plan 1 parallel heat transfer	31.06	5	634	112	2448
Plan 2 optimize heat transfer	31.06	5	530	120	2671

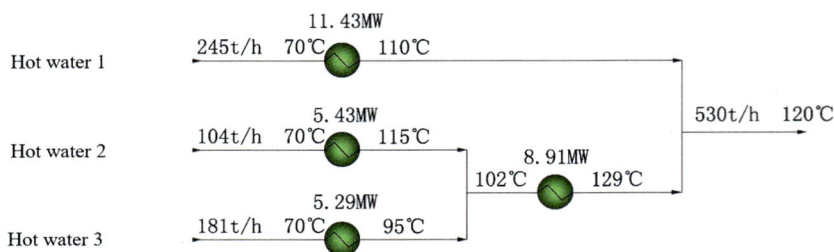

Fig. 4.16 Optimized hot water heat transfer process.

4.5.3.5 The application of efficient heat transfer equipment
1. High flux tube

For the column that provides heat for other columns using the condensation heat of its overhead material, such as the aforementioned xylene column and heavy aromatics column, the operating pressure can be reduced and the bottom heat load can be reduced by the application of high flux tube. Because the high flux tube has the characteristics of efficient heat transfer at a small temperature difference, which leads to the relatively low temperature requirement for hot fluid. When the process fluid is used to generate steam, the temperature of hot fluid required for generating steam of the same grade is lower if high flux tube is applied. It can also lead to a decrease in the column operating pressure and the column bottom heat load. Taking the aforementioned 650 kt/a aromatics complex as an example, when steam of the same grade is generated in conventional tubular heat exchanger and high flux tube heat exchanger with the hot medium from the top of the raffinate column and the extract column, the comparison of operation conditions of the columns with different types of heat transfer tube is shown in Table 4.38.

Table 4.38 Comparison of the operating conditions of columns.

Item	Plain tube	High flux tube
Raffinate column operating pressure (kPa)	250	200
Raffinate column top operating temperature (°C)	190/182	182/173
Raffinate column top heat load (MW)	71.51	71.04
Raffinate column bottom heat load (MW)	79.95	77.87
Extract column operating pressure (kPa)	280	215
Extract column top operating temperature (°C)	195	186
Extract column top heat load (MW)	26.31	25.53
Extract column bottom heat load (MW)	28.84	27.52
0.5 MPa vapor saturation temperature (°C)	158	158

As can be seen from Table 4.38, when steam is generated in high flux tube heat exchanger, the operating conditions are different. The operating pressure of the raffinate column is reduced by 50 kPa, and the heat load at the top and bottom of the column is reduced by 0.47 MW and 2.08 MW, respectively. The operating pressure of the extract column is reduced by 65 kPa, and the heat load at the top and bottom of the column is reduced by 0.78 MW and 1.32 MW, respectively. The energy consumption is reduced by about 3.4 kgoe/t PX.

2. Plate heat exchanger and coil-wound heat exchanger

Plate heat exchanger and coil-wound heat exchanger are efficient heat transfer equipment. Because the heat exchanger hot end temperature difference is smaller than tubular heat exchanger, when used for the feed-effluent heat exchange, it can effectively recover heat from the effluents, reduce the load of the furnace, and, at the same time, reduce load of the air cooler and water cooler. Hence the energy consumption of the reaction system is decreased. It can increase the final temperature of hot water and improve the efficiency of hot water power generation when used for heat exchange between column top fluid and low-temperature hot water.

The complete set of aromatics technology of Sinopec won the National Science and Technology Progress Prize, including all the above process enhancement and integration on heat transfer technology, which greatly reduces the energy consumption of PX unit and improves the technical competitiveness.

4.5.4 Heat transfer enhancement technology by improving elements

The process fluid of adsorption separation unit and xylene fractionation unit is relatively clean, and the heat transfer performance is superior. The conventional plain shell and tube heat exchanger is used, and the overall heat transfer coefficient of the equipment is also relatively high. The heat transfer enhancement technology, such as porous surface tube, is most used in the unit [21].

Different licensors have different process flows. However, it is still instructive to discuss the enhancement measures of heat exchange equipment in the unit through a set of typical process flows. Taking an in-service adsorption separation unit and xylene fractionation unit as an example, the characteristics of the heat transfer equipment and the appropriate heat transfer enhancement technology were analyzed.

4.5.4.1 Heat transfer equipment in extract column system

1. Extract column feed heat exchanger

The extraction liquid from the adsorption column enters the extract column after being heated in the extract column feed/bottoms heat exchanger. The cold and hot fluids in the heat exchanger are all liquid phase, and heat transfer performance is good. The overall heat transfer coefficient of the conventional shell and tube structure will exceed 600 W/(m^2 K) if the structure design is reasonable. Because the heat transfer property of medium is good, the general technology of enhancing liquid phase heat transfer can be used in this heat exchanger.

2. Extract column top heat exchanger

The gas volume from the extract column overhead is large, carrying large heat load. The vapor in most early units is condensed and cooled to 130°C by air cooler and enters the top reflux drum of the extract column. The components from the extract column overhead are relatively pure, and the vapor-liquid equilibrium is greatly affected by pressure. In order to maintain the pressure drop, it is not suitable to adopt multitube-pass structure, so the plate air cooler can be used to enhance the heat transfer. With the development of new technology and adaptation to the requirements of energy utilization, the air cooler has been canceled now, to make full use of the condensation heat.

The bottom liquid of the reflux drum is divided into two routes: one route is used as reflux to return to the extract column, and the other route way is sent into the PX column after being heated in PX column feed/bottoms heat exchanger. The duty of the PX column feed/bottoms heat exchanger is very small; the heat transfer effect is good, so the size of the exchanger is small.

3. Reboiler of extract column bottom

Two reboilers are set at the bottom of the extract column. The heat source of the reboilers comes from the top of the xylene rerun column and the top of the xylene column separately. The process fluid and heat transfer characteristics of the two reboilers are basically the same. The medium at the bottom of the extract column is aromatics which is close to pure substances, and the boiling process is very sensitive to pressure and temperature. The process conditions are shown in Table 4.39. Hot fluid from the column overhead is subcooled by 20°C after condensation in the reboiler, and the effective temperature difference between cold fluid and hot fluid is 36°C. Through heat transfer calculation, it was found that the condensation effect of the hot fluid on the shell side was not good, and the film heat transfer coefficient was 1376 W/(m² K). The cold fluid boils on the tube side and the tube side heat transfer coefficient is only about 2240 W/(m² K) because of boiling in the convection region. The overall heat transfer coefficient of the reboiler is not high. The reasonable heat transfer enhancement technology is to use the high flux tube instead of the plain tube, so as to enhance the heat transfer of boiling inside the tube and condensation outside the tube simultaneously. The results of scheme comparison are shown in Table 4.40.

Table 4.39 Process conditions of xylene rerun column overhead/extract column bottoms reboiler—Duty: 17.3 MW.

Item	Hot fluid	Cold fluid
Fluid name	Aromatics	Aromatics
Flow (kg/h)	197,339	39,665
Allowable pressure drop (kPa)	50	Thermal siphon
Temperature (°C)	258 → 234	214 → 214
Gas phase density (kg/m³)	33.45 → −	− → 7.24
Liquid phase density (kg/m³)	− → 653	698 → 697
Gas viscosity (mN s/m²)	0.012 → −	− → 0.009

Table 4.39 Process conditions of xylene rerun column overhead/extract column bottoms reboiler—Duty: 17.3 MW—cont'd

Item	Hot fluid	Cold fluid
Liquid viscosity (mN s/m²)	-- → 0.13	0.19 → 0.189
Gas phase thermal conductivity (W/(m K))	0.036 → --	-- → 0.026
Liquid phase thermal conductivity (W/(m K))	-- → 0.08	0.086 → 0.096
Gas relative heat (kJ/(kg K))	2.15 → --	-- → 2.03
Liquid phase specific heat (kJ/(kg K))	-- → 2.54	2.46 → 2.46

Table 4.40 Thermodynamic simulation results of xylene rerun column overhead/extract column bottoms reboiler.

Item	Original option	Enhanced option
Equipment size (shell diameter × straight tube length) (mm)	1800 × 4500	1500 × 4000
Set	1	1
Tube side pressure drop (kPa)	Thermosiphon	Thermosiphon
Shell side pressure drop (kPa)	7	7
Overall heat transfer coefficient (W/(m² K))	510	1135
Heat transfer area (m²)	1080	387
Total weight of equipment (t)	32	20.5

By comparison, it can be found that the heat transfer of tube side boiling and shell side condensation was enhanced when high flux tube instead of plain tube, the weight of equipment can be reduced by 36%, and the equipment investment cost can be reduced.

In the 1.0 Mt/a aromatics complex, high flux tube was used to replace the plain tube in the reboiler of the raffinate column and the reboiler of the extract column, and better economic benefits were obtained [21].

4.5.4.2 Heat exchange equipment in the PX column system

1. PX column feed heat exchanger

   The heat load borne by PX column feed/bottoms heat exchanger is not large, generally about 1.0 MW, and the heat transfer performance is very good. Using conventional shell and tube heat exchanger, the overall heat transfer coefficient of the equipment can exceed 600 W/(m² K) with small heat exchange area required.

2. PX column top air cooler

   The 125°C oil vapor from the PX column overhead is condensed and cooled to 70°C in the air cooler and enters the reflux drum. Part of the liquid from the bottom of the reflux drum is used as the reflux of the PX column, and the rest is used as feed of disproportionation reaction.

If the two tube pass structure with the same flow area is selected in the design of the PX column top air cooler, the flow velocity of the process fluid into the heat transfer tube is 40 m/s, and only 0.2 m/s after complete condensated. The heat transfer effect in the cooling section is not good. When the flow area of the first tube-pass is five times that of the second tube-pass, the velocity of the process fluid into the first tube-pass is 30 m/s, and 0.8 m/s after complete condensation, which improves the heat transfer efficiency of the cooling section. In this way, the number of air coolers, fans, and motors is reduced, and the power consumption of air cooler is reduced. Heat transfer enhancement technology can be used for PX column top air cooler, such as plate air cooler and other structures, to enhance heat transfer inside and outside the tube at the same time.

3. PX column bottom heat exchange

PX product from column bottom is fed through the PX column feed/bottoms heat exchanger, air cooler, and water cooler to the storage tank after cooled to 40°C.

Two reboilers are set at the bottom of the PX column. The heat source of one of the reboilers is aromatics from the top of xylene rerun column and xylene tower. The heat source of the other reboiler is the desorbent, and the hot fluids of the two reboilers are all in the liquid phase.

The reboiler is usually arranged with horizontal thermosiphon, and the paraxylene is boiled on the shell side. When conventional shell and tube heat exchangers are used, the heat transfer coefficient of shell side is about 2500 W/(m^2 K), which is comparable to the film heat transfer coefficient of the tube side without phase change. Considering the properties of process fluid, porous surface is suitable for enhancing the boiling process of PX. The heat transfer coefficient of boiling side can be increased by about 10 times by using sintered porous surface tube instead of plain tube, so as to improve the overall heat transfer performance of the reboiler.

The heat load in the PX column feed/bottoms heat exchanger is relatively small, and the overall heat transfer coefficient of the conventional shell and tube heat exchanger structure is more than 600 W/(m^2 K). The heat transfer coefficient of the tube side and shell side is close. The heat exchanger specification can be effectively reduced by the way of applying double-side enhanced heat transfer tube, for example, the corrugated tube and twisted tube.

The effective temperature difference of the PX product in the air cooler is about 45°C, and the heat transfer area required is not large. If further reduction of equipment size is needed, ellipse base tube with high fin can be used instead of round tube with high fin to improve heat transfer.

4.5.4.3 Heat exchange equipment around the raffinate column

1. Raffinate column feed heat exchanger

The raffinate is heated in the raffinate column feed/bottoms heat exchanger and then enters the middle section of the raffinate column. The cold fluid and hot fluid in the exchanger are liquid phase, the film heat transfer coefficient on both the tube side and shell side is more than 2000 W/(m² K), the overall heat transfer coefficient of the heat exchanger is more than 600 W/(m² K), with good heat transfer effect. But the heat transfer area required is large because of the great heat load; double-sided heat transfer enhancement heat tubes, such as corrugated tubes, twisted tubes, and transverse tubes, can be used instead of plain tube to reduce equipment size. And plate heat exchanger can be applied, too.

2. Raffinate column overhead air cooler

   The 160°C vapor from the raffinate column overhead is cooled to 130°C in the air cooler and then enters the reflux drum.

   Due to the high flow rate and high temperature of process fluid in the overhead air cooler of raffinate column, the heat load of air cooler is high. Taking this unit as an example, the flow rate of the process fluid is 153,420 m³/h, and 88 MW heat load shall be condensed and cooled. Considering that the outlet temperature of the fluid is 130°C, conventional dry air cooler can be used in order to avoid air side fouling. Due to the large gas rate, it is suitable to adopt a single tube pass design to ensure the pressure drop of the fluid. If heat transfer enhancement technology is used, the plate air cooler is the appropriate choice.

3. Heat transfer equipment around raffinate column bottom

   The raffinate containing mainly desorbent from the raffinate column bottom is pumped into the raffinate column feed/bottoms heat exchanger to be cooled, then it is cooled to about 178°C in the desorbent/PX column reboiler, and then cooled to 160–178°C by the air before proceeding to the desorption column. The two heat exchangers have been analyzed in detail before. The process fluid leaves the air cooler at a high temperature, the effective temperature difference of the air cooler reaches 120°C, and the heat transfer area required is very small.

   Two reboilers are set at the bottom of the raffinate column. The heat source is from xylene rerun column overhead and xylene column overhead separately.

   The flow rate and duty of the raffinate column reboiler are relatively large; the boiling medium is a pure component, and the hot fluid is aromatics condensation. The hot fluids in the two reboilers of the raffinate column come from different columns, but their temperature and physical properties are the same, only with different heat loads. Table 4.41 shows the process conditions of the xylene rerun column overhead/raffinate column reboiler. The duty of the reboiler is 35.6 MW, and the hot fluid flow rate is 411.668 t/h. The rest of the physical property data are consistent with the data in Table 4.42.

   Using the vertical thermosiphon design of plain tube, the condensation effect of hot fluid in the shell side is not good, and the film heat transfer coefficient is 1405 W/(m² K). The cold fluid in the tube side boils in the convection region, and the heat transfer coefficient is only 2224 W/(m^2 K). The appropriate means of heat transfer enhancement are: tube side porous surface to increase the bubbling core and strengthen the nuclear boiling, and the longitudinal groove surface outside the tube to strengthen the vertical condensation process of the shell side, which can obtain the effect of double-sided enhancement. The results of the scheme comparison are shown in Table 4.42.

Table 4.41 Process condition of xylene rerun column overhead/raffinate column reboiler.

Item	Hot fluid	Cold fluid
Fluid name	Aromatics	Aromatics
Flow (kg/h)	569,625	2,522,021
Allowable pressure drop (kPa)	50	Thermal siphon
Temperature (°C)	258 → 236	222 → 222.4
Gas phase density (kg/m^3)	33.45 → –	– → 8.5
Liquid phase density (kg/m^3)	– → 650	689 → 688
Gas viscosity (mN s/m^2)	0.012 → –	– → 0.01
Liquid viscosity (mN s/m^2)	– → 0.13	0.18 → 0.18
Gas phase thermal conductivity (W/(m K))	0.036 → –	– → 0.027
Liquid phase thermal conductivity (W/(m K))	– → 0.079	0.084 → 0.084
Gas relative heat (kJ/(kg K))	2.15 → –	– → 2.06
Liquid phase specific heat (kJ/(kg K))	– → 2.559	2.5 → 2.5

Table 4.42 Scheme comparison of raffinate column reboiler.

Item	Xylene rerun column overhead/raffinate column reboiler		Xylene column overhead/raffinate column reboiler	
	Original option	Enhanced option	Original option	Enhanced option
Equipment size (shell diameter × straight tube length) (mm)	2200 × 4500	1950 × 4000	2500 × 4700	2350 × 4000
Set	2	2	1	1
Tube side pressure drop (kPa)	Thermosiphon	Thermosiphon	Thermosiphon	Thermosiphon
Shell side pressure drop (kPa)	9	7	9	7
Overall heat transfer coefficient (W/(m^2 K))	519	1132	531	1140
Heat transfer area (m^2)	3360	1370	2203	998
Total weight of equipment (t)	100	60	68	44

According to the comparison in Table 4.42, it can be found that the heat transfer of boiling inside the tube and condensation outside the tube is enhanced at the same time when the porous surface tube is used instead of the plain tube. The two reboilers can reduce metal consumption by 36%, the primary investment of equipment, and the cost of civil construction and installation.

4.5.4.4 Heat exchange equipment in desorbent rerun column system

The feed of the desorbent rerun column comes directly from the bottom of the raffinate column. The bottom of the desorbent rerun column is equipped with a thermosiphon reboiler, and the heat source is aromatics from xylene rerun column bottom. The duty of the reboiler is not large, the effective temperature difference is about 35°C, and the heat transfer area required is not large. If the heat exchange effect wishes to be further enhanced, T-shaped finned tube is a suitable choice.

4.5.4.5 Heat exchange equipment around reformate splitter

1. Reformate splitter feed heat exchanger

The C_6^+ reformate from the depentanizer is heated in the feed/bottoms heat exchanger and sent to the reformate splitter.

The duty of the reformate splitter feed/bottoms heat exchanger is not large, and the cold fluid and hot fluid are heat-exchanged without phase change. The heat transfer performance of both the tube side and shell side is good, and the overall heat transfer coefficient is more than $600\,W/(m^2\,K)$ even in the conventional shell and tube heat exchanger. If considering to further improve the heat transfer performance of the equipment, spiral groove tube, corrugated tube, and other enhanced heat transfer tube can be used to enhance the heat transfer on both side.

2. Reformate splitter bottom

Three reboilers are set at the bottom of the reformate splitter. Medium-pressure steam is used as heat source in the reformate splitter steam reboiler. Xylene gas from the xylene column overhead is the heat source in the xylene column overhead/reformate splitter reboiler. Xylene rerun column bottom/reformate splitter reboiler uses fluid from the xylene column bottom as heat source.

The column bottom liquid is heated in the reformate splitter feed/bottoms heat exchanger and then into the xylene clay treatment heater, followed by entering into the xylene fractional clay treater. The two heat exchangers were respectively analyzed in the reformate splitter feed and the xylene column bottom.

The total duty of the three reboilers is 33 MW, and the steam reboiler accounts for 60% of the total duty. The horizontal thermosiphon option is generally adopted, and the heat transfer performance of the reboiler is very good. The film heat transfer coefficient of the steam condensing in the tube side exceeds $10,000\,W/(m^2\,K)$, and the film heat transfer coefficient of boiling on the shell side is about $3000\,W/(m^2\,K)$. The overall heat transfer coefficient can exceed $1000\,W/(m^2\,K)$, even considering the influence of fouling resistance. The sintered porous surface tube can be used to enhance the boiling on the shell side and further reduce the size of the reboiler.

The heat transfer coefficients of cold and hot fluid of the other two reboilers are basically the same; the effective temperature difference is large, and the heat transfer area required is small. If considering to further improve the heat transfer performance, enhanced heat transfer structures such as spiral groove tube and twisted tube can be used to simultaneously improve the heat transfer performance of tube side and shell side together.

4.5.4.6 *Heat transfer equipment around the xylene rerun column*

1. Xylene rerun column feed heat exchanger

    The feed from the isomerization effluent clay treater is sent to the upper part of the xylene rerun column after being heated in the deheptanizer bottom/xylene rerun column feed exchanger. The fluid from the xylene fractionated clay treater is heated in the xylene rerun column feed/bottoms heat exchanger and then enters the middle section of the xylene rerun column.

    Cold fluid and hot fluid in the two heat exchangers are liquid phase, the film heat transfer coefficient of the tube side is slightly lower than that of the shell side, and the overall heat transfer coefficient is more than $600 \, W/(m^2 \, K)$. Heat transfer enhancement technology, such as low-finned tube, can be used to expand the secondary heat transfer area or spiral twisted tubes and corrugated tubes can be used simultaneously to improve the heat transfer performance of tube and shell side together. Taking the deheptanizer bottoms/xylene rerun column feed heat exchanger in an unit as an example, the diameter of the heat exchanger can be reduced from 1000 to 900 mm with the same tube length by using low-finned tube or twisted tube bundle instead of plain tube.

2. Xylene rerun column overhead air cooler

    The 255°C vapor from the xylene column overhead is condensed and cooled to about 240°C in the air cooler and then enters the reflux drum. The effective temperature difference of the air cooler is as high as 160°C, and the heat exchange area required is not large. Because of the high temperature of the process fluid, the key to the design of the xylene rerun column overhead air cooler is to avoid the air leaving the air cooler bundle with high temperature. It is suggested that fewer tube rows should be used instead of six tube rows to ensure equipment performance.

3. Heat exchangers around xylene rerun column bottom

    Part of the xylene rerun column bottom liquid is heated through the xylene rerun column feed/bottoms heat exchanger. The rest materials at the column bottom are sent in sequence to the desorbent rerun column reboiler, the heavy aromatics column reboiler, the xylene rerun column bottom/reformate splitter reboiler, and the xylene clay treatment heater to exchange heat, then heated in the reboiler heater, finally return to the bottom of the xylene rerun column.

The effective temperature difference of the xylene clay treatment heater is as high as 116°C, and that of the xylene rerun column feed/bottoms heat exchanger is close to 70°C. The heat transfer performance of the two heat exchangers is good. The film heat transfer coefficient of the shell side is more than $2500\,W/(m^2\,K)$, and it is $2000\,W/(m^2\,K)$ on tube side. If further enhancement of the heat exchanger performance is required, double-side heat transfer enhancement technology such as twisted tube and corrugated tubes can be used. The rest of the heat exchanger and reboiler have been analyzed in detail in the corresponding position in the front, and the reboiler of the heavy aromatics column will be analyzed in the part of the heavy aromatics column bottom.

4.5.4.7 Xylene column heat exchanger network
1. Xylene column feed heat exchanger

The aromatics from the deheptanizer bottoms is fed into the xylene column after being heated by the o-xylene column feed/deheptanizer bottoms exchanger.

The cold fluid and hot fluid in the o-xylene column feed/deheptanizer bottoms exchanger are all liquid, and the film heat transfer coefficient is higher than $2500\,W/(m^2\,K)$ on tube side and slightly lower than $2000\,W/(m^2\,K)$ on shell side even if the conventional shell and tube heat exchanger is used. Heat exchange tubes such as twisted tube and corrugated tube can be used to enhance the heat transfer of tube and shell side at the same time, so as to reduce the heat exchanger size.
2. Xylene column overhead heat exchange system

Xylene column overhead is divided into four routes, namely, xylene column overhead/reformate splitter reboiler, xylene column overhead/extract column reboiler, xylene column overhead/raffinate column reboiler, and o-xylene column reboiler for heat exchange, and then enters the reflux drum of xylene column. The corresponding parts in the front of this chapter have been analyzed in detail.

In some cases a little part of xylene column overhead gas is cooled to 240°C in the air cooler and sent to the xylene column reflux drum. The load of the xylene column overhead air cooler is not large, and the effective temperature difference is relatively large, which can be as high as 180°C. The heat exchange area required is not large. Due to the high temperature of the process fluid, it is recommended to use fewer heat exchange tube rows to reduce the air outlet temperature.

4.5.4.8 Heat exchange equipment around o-xylene column
The gas from the column overhead is cooled to about 147°C in the air cooler and enters the o-xylene column reflux drum. Except for the reflux, the rest of the liquid from the drum bottom is cooled to 40°C by the o-xylene column overhead product air cooler and

water cooler. Although the duty of the *o*-xylene column overhead air cooler is large, the inlet and outlet temperatures of the process fluid are high, and the effective temperature difference is relatively large. The duty of the *o*-xylene column overhead product air cooler is very small. The heat transfer area required by these two air coolers is not much, and the cost performance of ordinary dry high fin tube air coolers is relatively high.

The column bottom reboiler can be designed as horizontal thermosiphon type. The heat source is the gas from the xylene column overhead, and the *o*-xylene boils in the shell side. The film heat transfer coefficients are 2000–3000 W/(m² K) in both the tube side and shell side, and the overall heat transfer coefficient of the reboiler is relatively high. To further improve the performance of the reboiler, the appropriate heat transfer enhancement technology is a T-shaped finned tube to enhance boiling heat transfer. If a vertically mounted thermosiphon reboiler is applied, the tube with a longitudinal groove outside the tube and porous surface inside the tube can enhance heat transfer effect on both sides.

4.5.4.9 Heat exchange equipment around the heavy aromatics column

The overhead vapor temperature is about 190°C. Unit designed in the early stage is generally cooled to 166°C by the heavy aromatics column overhead air cooler and then enters the reflux drum. With the application of high-efficiency heat exchangers such as porous surface tubes, the vapor is mostly used as heat source for the reformate splitter reboiler or for steam generator.

Although the air cooler needs to complete the heat exchange of 19 MW, the heat transfer area required is not much because of the high temperature of the process fluid and the large effective temperature difference. Revamping for capacity expansion of existing unit, dry air coolers such as plate air coolers and elliptical base tubes with high fins can be used to enhance heat transfer technology.

The thermosiphon reboiler is set at the bottom of the column, which takes a large amount of heat and has a sufficient driving force of temperature difference. It is suitable for using low-finned tube structure to enhance heat transfer.

4.5.5 Summary

The adsorption separation unit and xylene fractionation unit require more heating and cooling processes due to the number of rectification columns. The low-temperature heat in the unit is difficult to recover, so it consumes higher energy in the aromatics complex. In view of the above characteristic, through the column heat recovery and heat integration technology as well as process fluid cascade heat transfer and energy cascade utilization technology, coupled with low-temperature hot water power generation and low-pressure steam power generation and the appropriate application of efficient heat exchange equipment, heat transfer enhancement can be effectively achieved with comprehensive energy consumption of the adsorption separation unit and xylene fractionation unit reduced.

The process fluid of adsorption separation unit and xylene fractionation unit is relatively clean, which is suitable for the application of various heat transfer enhancement technologies. The use of efficient heat transfer equipment will save energy and money for the enterprise.

References

[1] Liu JM, Wang YC, Jiang RX. Refining plant process and engineering. Beijing: China Petrochemical Press; 2017.
[2] Yu HM, Cai YB. Application of corrugated tubes in heat exchangers for reforming products. China Pet Mach 2002;30(11):48–51.
[3] Zhang FF, Gao LP. Application of a new type of high efficiency heat exchanger in continuous reforming unit. Petrochem Equip Technol 2017;38(4):1–5.
[4] Yu LJ, Zhang YF, Zhou JX. Application of domestic super large fully welded plate and shell heat exchanger in petrochemical plant. Petrochem Equip 2010;39(5):69–73.
[5] Lu LY, Zhou LJ. Application of coil-wound tube heat exchanger in aromatics plant. Energy Conserv Emiss Reduct Pet Petrochem Ind 2013;3(6):29–32.
[6] Tian HF. Application of coil-wound heat exchanger in reforming plant. Petrochem Technol Econ 2017;33(6):29–33.
[7] Liu Y. Application of coil-wound heat exchanger in continuous reforming plant. Petrochem Technol Econ 2016;32(6):32–5.
[8] Ruan FJ. Application of coil-wound heat exchanger in continuous reforming plant. Chin Foreign Energy 2017;22(11):84–9.
[9] Chen CHG. Development and industrial application of coil-wound heat exchanger for continuous reforming. Press Vessel 2011;28(5):41–7.
[10] Chen YD, Chen XD. Technology development of large heat exchangers in China. J Mech Eng 2013;49(10):134–43.
[11] Zhang F, Chen SF, Chen M, et al. Application of compact plate evaporative air cooler in reforming unit. Petrochem Equip 2016;45(4):72–5.
[12] Wei LD, Ma J. Application of surface evaporation air cooler in reforming unit. Petrochem Equip 2003;32(6):52–3.
[13] Tian GH. Application of compound efficient air cooler in petrochemical field. Ind Technol Innov 2014;1(5):600–3.
[14] Chen XC. Cause analysis of pressure drop increase of plate heat exchanger in reforming plant. Qilu Petrochem Ind 2010;38(4):306–7.
[15] Li QM, Zhao M, Ma HJ, et al. Analysis and countermeasure of corrosion structure of heat exchanger in aromatic extraction Unit. Corros Corros Prot 2008;29(7):418–20.
[16] Gu JT, Ma GY. Research on plate air cooler and its application. Chem Equip Technol 2008;29(6):15–7.
[17] Yang WSH, He LB. Development of energy saving process for toluene disproportionation unit. Refin Technol Eng 2009;39(5):12–4.
[18] Wang GSH, Zhou JX. Selection of fully welded plate and shell heat exchanger in disproportionation unit. Petrochem Equip 2008;37(5):86–7.
[19] Cui SC, Yang WS, Liu WJ, et al. Application of plate and shell heat exchanger in heat recovery of toluene disproportionation unit. Chem Eng 1999;27(3):56–7.
[20] Xu WD. Application of a new type of coil-wound heat exchanger in aromatics plant. Sci Technol Wind 2014;9:101.
[21] Zhao L, Zhang YF, Chen SHF. Application and prospect of high throughput tube heat exchanger in aromatics plant. Petrochem Equip 2010;39(6):68–70.

[22] Liu T. Discussion on energy saving of fully welded plate and shell heat exchanger used in aromatics isomerization reaction system. Oil Refin Technol Eng 2013;43(4):41–3.

[23] Qin YQ. Application of domestic large fully welded plate and shell heat exchanger in aromatics complex. Petrochem Equip Technol 2017;38(3):1–5.

[24] Ye SH, Sui YG, Gao F. Energy saving analysis of a new type of coil-wound heat exchanger applied in aromatics isomerization unit. Henan Sci Technol 2017;9:60–2.

[25] Dai HL. Aromatics technology. Beijing: China Petrochemical Press; 2014. p. 18.

CHAPTER 5

Heat transfer enhancement in typical ethylene plants and downstream units

Contents

5.1 Steam cracking

Steam cracking is a key technology in the petrochemical industry. Its products such as ethylene, propylene, butadiene, and BTX (benzene, toluene, xylene) are the most basic raw materials for petrochemical industry. Ethylene gross production, plant capacity, and process technology usually indicate the development level of a country's petrochemical industry. Generally, ethylene plant is composed of steam cracking unit and ethylene separation unit. This section introduces steam cracking unit, and ethylene separation unit will be introduced in Section 5.2.

5.1.1 Brief description of process

The main function of steam cracking is to produce cracking gas including olefin and aromatics by steam cracking from gas feedstocks such as ethane and refinery dry gas and/or liquid feedstocks such as naphtha and HVGO. After separation and refining, on–spec low-carbon olefin and aromatics can be obtained. Cracking furnace is the core of ethylene plant. At present, including Sinopec, there are six major steam cracking technology licensors in the world [1].

Steam cracking is usually achieved by cracking furnace, which is generally composed of radiation section, convection section, and quench system. The high–temperature thermal cracking reaction is carried out in the radiation section, and the heat required for the reaction is provided by fuel combustion. The heat from the high-temperature flue gas is recovered in the convection section, and part of it is used for preheating and vaporizing feedstocks, which are overheated to the initial reaction temperature and then sent to the radiation section for thermal cracking reaction. The remaining heat is used for preheating boiler feed water and superheating superhigh-pressure steam. The quench system recovers the heat from high-temperature cracking gas, most of which is used for generating saturated superhigh-pressure steam, and the remaining heat is recovered for utilization through the ethylene separation unit.

Cracking furnace is the biggest energy consumer in ethylene plant, and its energy consumption accounts for more than 50% of the total energy consumption of the whole ethylene plant. How to reduce the energy consumption of steam cracking through heat transfer enhancement and energy–saving approach has been one of the focuses of the research and development of cracking furnace technology [2].

The typical process flow of steam cracking is shown in Fig. 5.1.

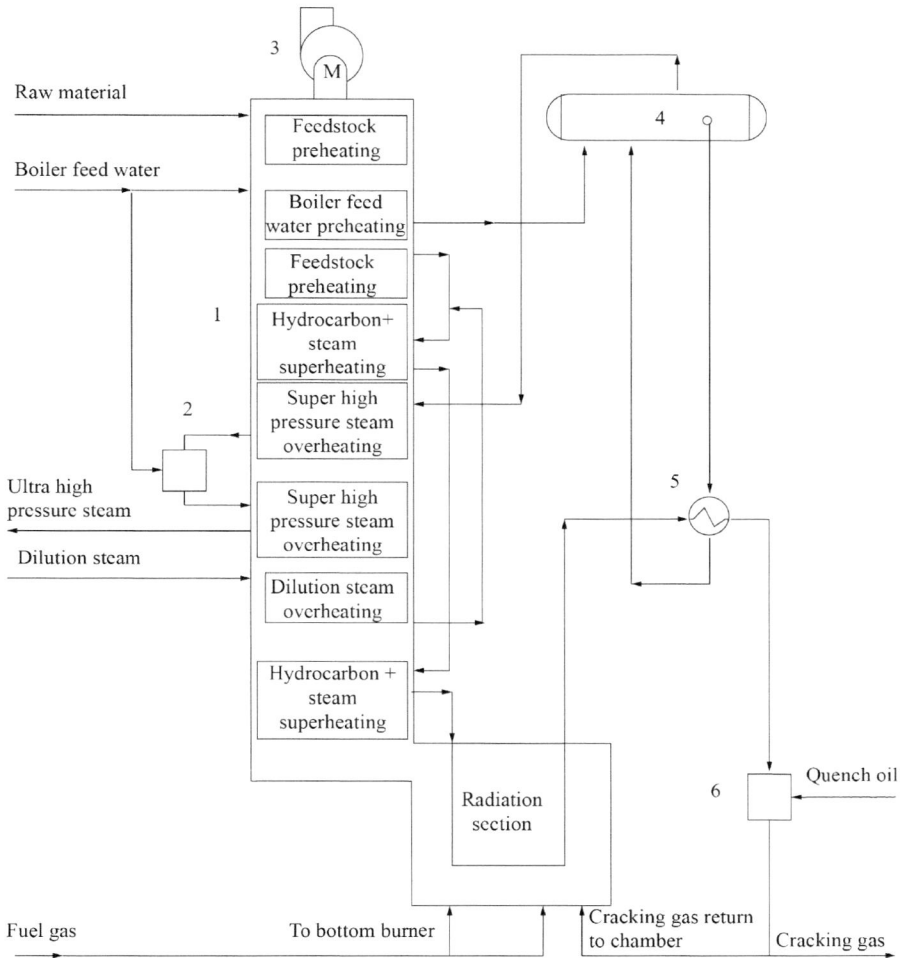

Fig. 5.1 Typical process flow of steam cracking. (1) Cracking furnace; (2) desuperheater; (3) induced draft fan; (4) high-pressure steam drum; (5) transfer line exchanger; (6) quench fitting.

5.1.2 Analysis of characteristics of thermal energy

1. Radiation section

Thermal cracking reaction is a combination of free radical reaction and molecular reaction, which determines that only under the conditions of high temperature, short residence time, and low hydrocarbon partial pressure can higher yields of target olefin products be obtained and energy utilization efficiency be improved, thus reducing the energy consumption per unit product [3].

Heat transfer in the radiation section of furnace includes radiation heat transfer, convection heat, and heat conduction, of which radiation heat transfer dominates, and its heat transfer proportion accounts for more than 80% of the total heat transfer in the radiation section. The heat transfer in radiation section is affected by a variety of factors, such as furnace structure and size, fuel type and heating mode, burner type, and furnace tube arrangement.

There are many ways to improve the heat utilization efficiency of radiation section. One way is to optimize the configuration of radiation tube to better meet the needs of cracking reaction, increase the production of target olefin products, and reduce the energy consumption per unit product. Another way is to optimize and increase the temperature of the feedstock entering the radiation section and increase the proportion of reaction heat absorption in radiation section in the total heating so that the heat is more used in the cracking reaction rather than passive recovery. The third way is to enhance heat transfer by swirling element of radiant tube and other measures inside and outside the tube to improve the heat utilization efficiency. The fourth way is to use energy-saving refractory materials to reduce the heat loss of the outer wall of radiation section.

2. Convection section

After 40%–45% of the total heat release of fuel is absorbed by the radiation section, the high-temperature flue gas at about 1100°C carries the rest of the heat into the convection section and recovers most of the heat before being discharged into the atmosphere. The main function of the convection section is to maintain the heat balance of the cracking furnace on the premise of meeting the requirements of the cracking process, recover the waste heat in flue gas as much as possible by optimizing the arrangement of heat exchange tubes in the convection section, improve the thermal efficiency of the furnace, and reduce the investment.

According to the characteristics of flue gas temperature distribution in the convection section, different heat exchange sections are set to recover the heat according to the grade. With the highest heat grade, the flue gas in the lower part of the convection section is generally used for superheating the mixture of hydrocarbon and dilution steam so that the mixture can reach or approach the initial temperature of cracking reaction and then enter the radiation section for cracking reaction. Heat at high energy level is generally used for superheating the saturated superhigh-pressure steam generated by the transfer line exchanger. Heat at intermediate energy level is generally used for superheating the raw material hydrocarbon and dilution steam. The flue gas heat with a lower grade is used for preheating the boiler feed water. Heat at the lowest energy level is generally used for preheating cracking feedstocks [4].

In the convection section, radiation heat transfer is the main heat transfer mode in the bottom part, and convection heat transfer is the main heat transfer mode in the rest parts.

3. Transfer line exchanger

Due to the different feedstocks, the temperature of cracking gas from the radiation section is usually between 800°C and 870°C. At this temperature, secondary reaction will occur and the yield of the target product will be reduced. Therefore, the cracking gas should be quenched to below the secondary reaction temperature as soon as possible. There are usually two ways of cracking gas quenching. In the early days, water or quench oil was directly injected into the furnace. In modern cracking furnaces, cracking gas is cooled by transfer line exchanger. The two methods recover the same amount of heat, but the energy grade is very different. Using the heat recovered from transfer line exchanger to generate superhigh-pressure steam can recover the highest grade energy, and the superhigh-pressure steam generated can be used to drive the downstream compressor turbines. It is the mainstream technology at present.

The transfer line exchanger is the key equipment in cracking furnace. On the one hand, it rapidly quenches the high-temperature cracking gas to below the secondary reaction temperature to reduce the loss of target olefins, and at the same time, it recovers heat from the high-temperature cracking gas to generate high-grade thermal energy. In transfer line exchanger, generally high-temperature cracking gas passes through the tube side, and the cooling medium saturated water passes through the shell side, generating superhigh-pressure steam by boiling heat transfer.

5.1.3 Process energy optimization and heat transfer network optimization

After the material of heat transfer element in each unit of cracking furnace is determined, the thermal resistance becomes a fixed value, and the heat transfer is basically enhanced by increasing the heat transfer area and increasing the temperature gradient.

5.1.3.1 Enhancement and integration of heat transfer process in convection section

Convective heat transfer is the main heat transfer mode in convection section, and radiation heat transfer also exists in the lower high-temperature section. Materials in convection section are subject to multistage heat transfer, such as preheating, vaporization, and superheating, which involves the complex heat transfer problems of single-phase flow and two-phase flow. The inner film heat transfer coefficient and influencing factors of flows with different flow patterns are different.

The heat transfer intensity in the convection section can be improved by increasing the temperature gradient of hot and cold flows, that is, the heat transfer temperature difference, selecting and controlling the flow velocity and flow pattern to improve the heat transfer coefficient, and increasing the heat transfer area. The following enhancement and integration methods can be adopted.

1. Increasing temperature gradient and optimizing heat exchange network

In order to recover and utilize flue gas heat reasonably, different convection section arrangements are designed for different cracking raw materials.

For gas feed or light oils such as naphtha, raffinate oil, and topped oil, vaporization is not necessary or the vaporization process is short, and the convection section is relatively simple. It is usually divided into heat exchange tube sections of feed preheating, boiler feed water preheating, mixed superheating of hydrocarbon and dilution steam (DS), and superheating of superhigh–pressure steam. The typical convection section arrangement is shown in Fig. 5.2. The superheating of superhigh–pressure steam is divided into two sections mainly to adjust the superheating temperature through water injection, so as to meet the needs of steam turbine users of ethylene separation compressor.

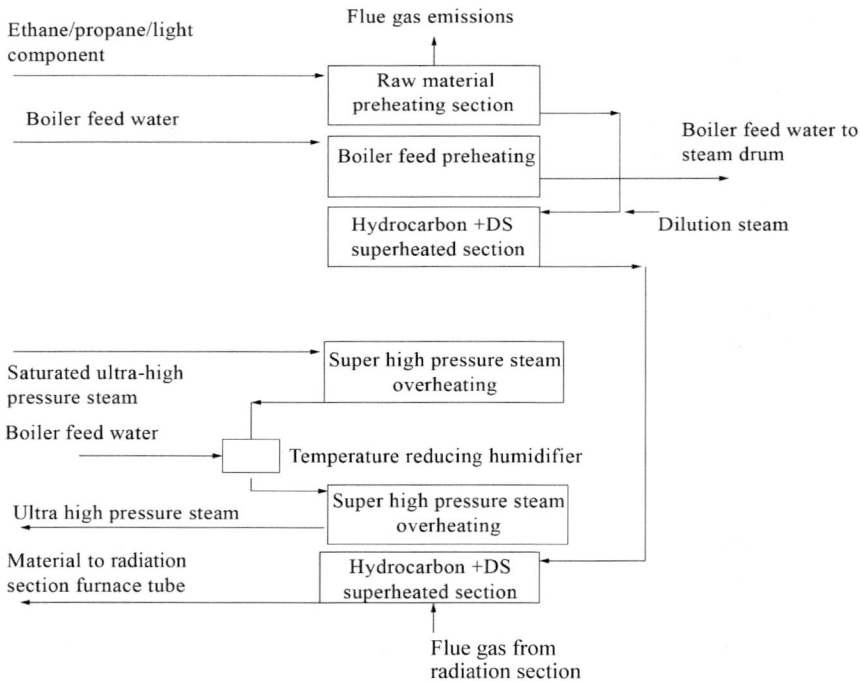

Fig. 5.2 Convection section arrangement for gas and light component feed.

For heavy oils such as diesel oil and HVGO, the vaporization point is high, the distillation range is long, and the vaporization time is long when the oils are preheated in the convection section. In order to avoid coking, the design of convection section is complex. It is usually divided into heat exchange tube sections of feed preheating, boiler feed water preheating, mixed superheating of hydrocarbon and dilution steam, superheating of dilution steam, and superheating of superhigh-pressure steam. The typical convection section arrangement is shown in Fig. 5.3. Usually, the dilution steam is injected in two stages. By controlling the proportion of the primary and secondary steam injection and the superheating temperature of steam, the heavy feed can be completely vaporized and have a certain superheat before being mixed with dilution steam for superheating, so as to avoid coking in the convection section. During heavy oil cracking, the heat load in convection section changes obviously at the start of run and the end of run. The output of superhigh-pressure steam varies greatly. In some cases, it is difficult to balance the heat load in convection section and the superheating temperature of superhigh-pressure steam with a single injection of water, so superheating water is usually injected twice. In this case, the superheating of superhigh-pressure steam is divided into three sections [4].

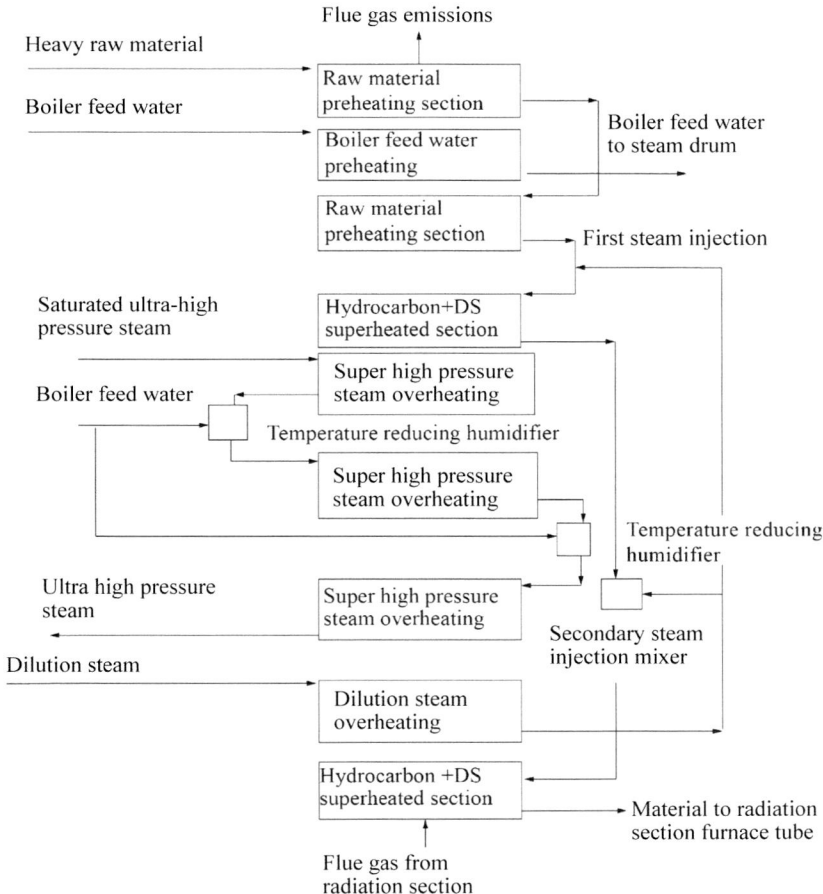

Fig. 5.3 Convection section arrangement for heavy raw materials.

As the cracking feedstocks become heavier, the design of convection section becomes more difficult, the division of heat exchange tube section is more precise, the number of tube rows is increased, and the investment is increased. Generally, the heat exchange tube section is reasonably arranged according to the temperature at flue gas side and the requirements of material heating to make full use of the flue gas heat. Under the premise of meeting the process requirements, the number of tube rows is reduced, the heat exchange area is reduced, especially the area of high–grade material heat exchange tube row is reduced, so as to minimize the investment cost while guaranteeing a certain amount of heat exchange.

In recent years, with the improved heat utilization efficiency and stricter requirements for energy conservation and emissions reduction in radiation section, the temperature of flue gas entering the convection section has dropped, the flue gas temperature in convection section has been reduced, and the heat transfer demand for process materials in convection section has been increased, which increased the difficulty of heat recovery in convection section, and pinch point is easy to appear especially in the feed preheating and boiler feed water preheating sections at the end of the convection section. In engineering design, especially the revamped design of units, using "pinch point" technology to analyze the temperature level of flue gas and the temperature range of the heated medium and to optimize the segment division in the convection section can achieve a balance in heat transfer efficiency and tube arrangement and reduce the investment in equipment while ensuring the heat transfer efficiency in convection section.

By increasing the average temperature difference of heat transfer, the heat transfer area can be reduced at a given amount of heat exchange, thus saving the investment cost. The flue gas and the heated medium usually flow countercurrently; that is, the flue gas flows upward and the heated medium enters at the horizontal upper tube row and leaves from the bottom. In the countercurrent arrangement, high-temperature flue gas contacts with high-temperature material first for heat exchange, the wall temperature of heat exchange tubes is correspondingly high, and there is a potential coking risk in some segments or will increase the design temperature of heat exchange tube row, leading to material upgrade and resulting in unreasonable implementation of the project. At this time, the cocurrent flow of flue gas and feed should be used to control the tube wall temperature, so as to effectively control the total cost of heat exchange tube while increasing the heat exchange area a little.

The lowest rows of heat exchange tubes in the convection section are located at the junction of the radiation section and the convection section. Flue gas temperature is usually about 1100°C. Radiation heat transfer is the main heat transfer mode. These tubes are generally known as shield tube. In order to reduce the wall temperature of shield tube, conventional stainless steel tube and cocurrent heat transfer method can be used. Considering the characteristics of radiation heat transfer, in order to make full use of radiation heat transfer energy, plain tube is usually selected for these heat exchange tubes.

2. Improving heat transfer coefficient

For all kinds of heat transfer tubes in cracking furnace, improving the heat transfer efficiency focuses on improving the inner and outer film heat transfer coefficient of the heat transfer tube.

① Measures on the flue gas side. The outer film heat transfer coefficient of plain tube is related to parameters such as tube bundle arrangement, tube pitch, tube outer diameter, flue gas flow rate, flue gas thermal conductivity, and viscosity. The outer film heat transfer coefficient of finned tube is not only related to the above parameters but also related to fin height, fin clearance, and other parameters. Under the condition that the flue gas composition is determined, the parameters such as flue gas flow rate, tube bundle arrangement, and fin specification are optimized to improve the outer film heat transfer coefficient, thus improving the heat transfer efficiency.

The induced draft fan is used at the top of the convection section to provide enough power to overcome the flue gas flow resistance and improve the flow rate. High flow rate can improve the heat transfer coefficient and reduce the heat transfer area, but it requires high power–induced draft fan and increases the equipment investment and operating costs. Therefore, the flue gas flow rate must be considered from both aspects.

In order to make full use of the radiation heat transfer of the shield tubes in the convection section, the lower heat exchange tubes are generally equivalent to the length of the radiation section, and the upper part is reduced. On the one hand, the flue gas flow rate is appropriately increased, the heat transfer coefficient on the flue gas side is increased, and the heat transfer area is reduced. On the other hand, it is beneficial to uniformly distribute flue gas temperature in the horizontal direction and reduce the influence of uneven temperature distribution on heat transfer.

The pyrolysis furnace is generally fueled by the methane off-gas produced by the ethylene plant. Ash is not easily accumulated on the tube to affect heat transfer. With the operation of the plant, part of the dust in the combust air and the pulverized refractory fiber material peeled off from the surface in the radiation section will accumulate outside the tube wall of the convection section to improve the fouling coefficient outside the tube and reduce the heat transfer efficiency. Appropriate arrangement and spacing of heat exchange tubes should be considered in the design, and soot blower or steam/chemical cleaning can be used to clean the heat exchange tubes [5].

② Measures on the material side. In addition to the vaporization section of liquid feed, most of the medium in the convection section of furnace, including the preheating of some feeds, the preheating of boiler feed water, the mixed overheating of part of hydrocarbons and dilution steam, and the overheating of superhigh-pressure steam, is single-phase flow.

The inner film heat transfer coefficient of single-phase flow is related to the tube diameter, flow rate, and the specific heat capacity, viscosity, and thermal conductivity of the flow. In the case of a certain medium, the choice of tube diameter and flow rate has a great impact on the heat transfer effect and should be considered according to the pressure drop of the convection section system. The diameter of the heat exchange tube in the convection section should be optimized according to the number of flow paths, and the flow rate inside the tube should be increased as far as possible under the premise of satisfying the system pressure drop, so as to improve the heat transfer efficiency. For the tube row of the boiler feed water preheating and superhigh-pressure steam superheating in convection section, flows are distributed through the manifold pipe. The pressure drop requirement of the superhigh-pressure system is relatively loose, and there is a large room for optimization of this part of heat exchange tube section.

The tube rows in the convection section of the furnace are all horizontally arranged. With the heat transfer, the vaporization rate of the liquid feed increases, and the flow pattern of the two-phase flow gradually transits from laminar flow to bubble flow and finally completely vaporizes. In the design of the convection section of furnace, it is necessary to avoid and reduce the duration of specific flow patterns in the tube as much as possible. On the one hand, the influence of system pressure drop is taken into account, and on the other hand, the coking of hydrocarbons in the convection section is to be avoided. In the vaporization section of heavy liquid hydrocarbons, it is necessary to control the vaporization rate at the outlet of the heat exchange tube and the tube wall temperature of the heat exchange tube so that the vaporization process of some heavy hydrocarbons can be moved to the outside of the tube in convection section, so as to avoid the coking caused by the high wall temperature in the convection section. A variety of dilution steam injection methods have been developed and studied by different cracking technology licensors to optimize the heat transfer in the convection section to solve this problem.

3. Application of finned tube in convection section

The application of swirl flow tube in convection section of furnace is very common. With finned tube, the heat transfer area is increased, and the convective heat transfer coefficient is increased to enhance heat transfer, so as to effectively reduce the equipment cost. The selection of fin size should take into account its influence on the outer film heat transfer coefficient, the pressure drop on flue gas side and the design of supporting tube plate. At the same time, the influence of ash deposit on the heat transfer and the pressure drop at the flue gas side during long-term operation should also be taken into account.

4. Application of other heat transfer enhancement measures

The improvement in energy saving makes the flue gas temperature of cracking furnace lower and lower. In order to make use of the low-level heat of ethylene plant, the feed is usually preheated by quenching water first, which leads to the reduction of heat transfer temperature difference and heat transfer power in the preheating tube section of feed at the tail of convection section, and the metal temperature of heat exchange tube wall decreases with it, gradually entering the temperature range of low-temperature acid dew point corrosion.

In order to reduce the effect of acid dew point corrosion, in addition to the selection of metal materials resistant to acid dew point corrosion, there are also attempts to introduce phase change heat transfer into the convection section. Phase change heat exchanger uses the principle of heat pipe and indirect heat transfer, and the phase change section is arranged in the convection section. The heat of the flue gas is absorbed through the medium vaporization in the tube, and the temperature of the heat exchange tube is basically constant. When the medium in the tube vaporizes, the inner film heat transfer coefficient is reduced, and good heat transfer effect can be achieved under the limited heat transfer temperature difference. The wall temperature can be adjusted to a certain extent according to the dew point of flue gas, which can effectively avoid the occurrence of dew point corrosion.

5.1.3.2 Enhancement of heat transfer process in radiation section of furnace

Heat transfer in radiation section is mainly radiation heat transfer, accounting for 80%–85% of the total heat transfer, and the rest is basically convection heat transfer. Radiation heat transfer is more complex, including both the direct radiation heat transfer from combustion flame and high-temperature flue gas and the indirect radiation heat transfer from high-temperature wall to furnace tube. In general, radiation heat transfer is affected by the type and layout of burner, the fuel composition, the size and structure of radiation furnace, the type and layout of furnace tube, and other factors.

With the furnace wall material and furnace tube material selected, the temperature distribution of radiation section, the heat transfer temperature difference between the flue gas and the medium, the heat transfer area of furnace tube, and the tube spacing and angle factor of furnace tube are the enhancement direction of radiation heat transfer.

1. Reasonable temperature distribution

    The uniformity of temperature distribution in the furnace is beneficial to the improvement of radiation heat transfer rate and affects the evolution of heating mode of cracking furnace. The initial furnace generally used sidewall burners for heating, and the uniformity of temperature distribution in the furnace was good, but the number of burners was large, the cost was high, and maintenance was difficult. Modern cracking furnace gradually turns to the joint heating of hearth burners and sidewall burners, and the proportion of bottom heating is increasing. A variety of cracking technologies adopt the option of all hearth burners. The optimization and adjustment of the heating option reduces the investment and the operation and maintenance, along with meeting the heating requirements of cracking reaction.

In the initial calculation of radiation section of cracking furnace, isothermal model was used to calculate radiation heat transfer. With the in–depth understanding of the principle of cracking reaction and the simulation calculation of combustion process, the idea of reasonable heat supply according to the requirements of cracking reaction is gradually put forward. After entering the radiation section, the feedstock needs to be heated up rapidly. The initial reaction that produces free radicals requires the most heat. With the progress of the cracking reaction, the heat demand gradually decreases. After a period of time, the concentration of the target olefin reaches its maximum, and the side reactions begin to increase. The heat absorption of this part should be reduced to reduce the intensity of the side reactions and reduce the loss of the target product. According to the characteristics of heat absorption in radiation section reaction, all steam cracking licensors are optimizing their burner layout options and putting forward the requirements of heat flux distribution in furnace according to the requirements of cracking reaction, which is the key point of cracking furnace design.

In recent years, computational fluid dynamics (CFD) technology has been widely used in the simulation of the radiation section of cracking furnace to realize the coupling calculation of combustion, radiation heat transfer, and cracking reaction in the radiation section. By using this technology, the actual operation of furnace can be analyzed to solve the production problems, and the design can be guided to optimize the technical scheme of furnace from the source [6].

2. Increasing the heat transfer temperature difference and increasing the temperature gradient

The temperature difference between cold and hot flows can be increased and the heat transfer rate can be increased by increasing the furnace temperature. Over the years, researchers have developed a series of heat-resistant furnace tubes. The nickel content of radiant tube material has been continuously increased, and the maximum wall temperature of radiant tube has been increased from 1040°C to 1100°C, and up to 1125°C of the commonly used 35Cr45Ni alloy material. Due to the limitation of metallurgical technology, the wall temperature of conventional high–chromium nickel alloy furnace tube has been rising very slowly.

On the one hand, licensors and manufacturers add aluminum, silicon, rare earth elements, and other materials to improve the temperature class of radiation furnace tube from the perspective of metallurgical technology. On the other hand, nonmetallic materials are used; for example, the use of nonmetallic ceramic materials can greatly improve the temperature resistance of furnace tube. There is no catalytic coking of metal elements on the surface, which helps to extend the operation cycle of cracking furnace. Due to engineering factors, it is only in the experimental stage; there is no large-scale industrial application yet.

3. Increasing the heat transfer area

According to the characteristics of radiation heat transfer, it is not suitable to use ordinary finned tube to increase the heat transfer area, but other ways can be used.

Centrifugal casting radiation furnace tube is commonly used. According to the requirements of the casting process, external surface processing is not required in the application. Retaining the surface roughness has a certain effect on improving the heat transfer efficiency.

Technip has previously published a patent for its swirl flow tube. Radiant furnace tube adopts swirl flow tube, which can increase heat transfer area and increase heat transfer rate at the same furnace height. And the change in flow pattern of material in the tube also has a certain thinning effect on the boundary layer, which is conducive to increasing the heat transfer coefficient of the inner surface. However, due to the limitations of engineering such as thermal stress and furnace size, this option has only been tested in industry and has not been applied at industrial scale.

4. Optimizing tube arrangement

According to the principle of cracking, high temperature, short residence time, and low hydrocarbon partial pressure are beneficial to the cracking reactions for producing ethylene, propylene, and other target products. To increase the cracking temperature at the same cracking severity, the resident time must be reduced to increase the thermal strength of the furnace tube. Increasing heat transfer area per unit reaction volume by reducing tube diameter is an effective means to increase heat transfer.

Small diameter furnace tube has a large specific surface area, feedstock is heated up rapidly in the tube, and the cracking process is in line with the requirements of high temperature and short residence time, which is beneficial to improving the selectivity and yields of target products. However, if the tube diameter is too small, the pressure drop increases, the power consumption increases and the coking intensifies, which will shorten the operation cycle of the cracker.

In order to overcome the disadvantage of short operation cycle of small-diameter furnace tubes, branch coil technology has been applied. Branch coil is used at the inlet pipe where the feedstock conversion rate is low and a large amount of heat absorption is needed. Furnace tube with smaller diameter can increase the specific surface area and increase the heat transfer. At this stage, due to the low conversion rate, the secondary reaction is less and coking is not obvious. In areas with high conversion rate, using large-diameter tube can reduce the sensitivity of coking, effectively extend the operation cycle, and improve the yield of target products.

The radiative heat transfer rate is related to the outer diameter of the furnace tube, the distance between the tubes, and the angle factor of radiation tube relative to furnace wall. The radiation furnace tube is arranged in the center of the furnace to receive double-sided radiation. If the furnace tube is arranged in double row or too dense, it will cause a reduction in radiation angle factor and the shelter between heat exchange tubes, thereby reducing the radiation heat transfer rate. Therefore, it is necessary to consider the utilization efficiency of radiant furnace space and arrange radiant furnace tubes reasonably when arranging tubes.

5. Increasing the heat transfer coefficient

With the progress of radiation reaction, a layer of coke has gradually deposited on the inner surface of furnace tube, and the heat transfer coefficient of the coke layer is much lower than that of metal. With the increase of the thickness of coke layer, more heat needs to be provided to maintain the same reaction depth, which increases the tube metal wall temperature and shortens the operation cycle of the cracker. Coke inhibition technology is an important means to prevent the reduction of heat transfer coefficient in the tube.

There are mainly three coking mechanisms of hydrocarbon cracking. The first one is the catalytic coking effect of the metal composition of furnace tube, which are mainly nickel and iron atoms. The second one is the coke generated by free radical reactions during hydrocarbon pyrolysis. The third one is the coke formed by the polycondensation of unsaturated hydrocarbons and polycyclic aromatics generated during hydrocarbon pyrolysis. In addition to the properties of feedstock and the operating conditions such as cracking temperature, hydrocarbon partial pressure, and residence time, the factors affecting the coking of hydrocarbons include the surface material of furnace tube.

In addition to reducing the composition of aromatics in the feedstock and optimizing the operating conditions, the following technologies are mainly used to inhibit and delay coking:

① Adding coking inhibitors to feedstock, surface modification of furnace tubes, or improving furnace tube materials can inhibit homogeneous and heterogeneous coking reactions or change the physical form of coke, making it loose and easy to clean, which can reduce the carburization of furnace tubes, thus prolonging the operation cycle of cracker. There are many kinds of coking inhibitors. Organic sulfide, organosilane, alkali metals, alkaline earth metal compounds, organic phosphorus, and organic sulfur phosphorus compounds are commonly used.

② Coating on the inner surface of the radiation furnace tube or in situ generation of oxide film can reduce the catalytic activity on the surface of furnace tube and completely inhibit or reduce the formation of catalytic coke. An example is the surface modification of thermal-oxidized spinel, which uses technologies such as the decomposition of organosilane under hot conditions to form SiO_2 coating. A layer of coating can be formed on the inner surface of the furnace tube to insulate the metal elements such as Fe and Ni, thus greatly reducing the catalytic coke production.

③ The use of nonmetallic ceramic materials for the furnace tube or the addition of specific elements such as Al, Ca, Ba, Be, Li, and other metal elements during casting can effectively reduce the catalytic coke on the surface of furnace tube. Foreign companies have tried it and the test results were good, but it has not been commercialized on large scale due to the cost-effectiveness ratio.

6. Improving the blackness of heat exchange tube and furnace wall materials

Improving the blackness of heat exchange tube and furnace wall materials can effectively increase the radiation heat transfer rate and heat transfer intensity. At present, there are technologies to apply metal or nonmetallic nanomaterials to coat the surface of metal and refractory materials, which can improve the material blackness, improve the radiation heat transfer rate, and achieve the purpose of energy saving. However, limited by the material properties and high-temperature operating conditions, it has a certain energy-saving effect but short service life.

5.1.3.3 Enhancement of heat transfer process in quench heat exchanger

Transfer line exchanger rapidly quenches the cracking gas to below the termination temperature of the secondary reaction to reduce the loss of the target olefin products, avoid coking, and recover the high-temperature potential heat of the cracking gas to generate superhigh-pressure steam.

During operation, the cracking gas in transfer line exchanger will coke. There are two reasons for the coking: high-temperature gas phase coking and low-temperature condensation coking. High-temperature gas phase coking mainly occurs in the high-temperature section, generally in the parts above 600°C, and is mainly affected by the secondary reaction. For low-temperature condensation coking, condensation occurs when the heavy components in cracking gas meet the wall of low-temperature heat exchange tube whose temperature is lower than its dew point temperature, and then coke fouling is gradually formed. In terms of process design, the loose coke fouling is difficult to accumulate in the tube and the coking trend can be slowed down to a certain extent by rationally choosing the diameter of heat exchange tube and the proper single-tube processing capacity and maintaining the high mass flow rate in the tube. Early traditional transfer line exchanger mostly uses small-diameter heat exchange tubes; the heat exchange effect is good, but the coking sensitivity is high and the coke is not easy to clean online, so it is gradually replaced by larger-diameter heat exchange tubes. The heat exchange tube of linear quench heat exchanger is directly connected with the radiation furnace tube, the inlet head is canceled, the volume of adiabatic section is effectively reduced, and there is no cracking gas flow distribution problem. It has certain advantages over the traditional transfer line exchanger in terms of process performance.

For gas feed cracking, the content of heavy components in the cracking gas is less and the dew point temperature is low. Two-stage or three-stage transfer line exchangers can be used to reduce the temperature of cracking gas before sending it to the ethylene separation quench area, so as to effectively recover the high energy heat and reduce the energy consumption of ethylene plant.

For the boiling heat transfer of high-pressure water outside the heat exchange tube of transfer line exchanger, the main heat transfer optimization is to prevent the boiler water outside the tube from vaporizing and scaling. On the one hand, the pH value of boiler water is adjusted by dosing. On the other hand, the structure form of the downcomer entering the water side of heat exchanger is improved, the bottom water circulation dead zone is reduced, and both the scale accumulation and the temperature difference between the water inlet side and the back side of heat exchange tube are reduced by hydrocyclone, so as to improve the heat transfer effect.

5.1.4 Heat transfer enhancement technology using elements

In addition to process measures to enhance heat transfer, a variety of heat transfer elements play a key role in the heat transfer enhancement in steam cracking, as described below:

① Mitsubishi Corporation of Japan has used elliptic tubes in its cracker. For cross-sections of the same size, the specific surface area of elliptic tube is larger than that of circular furnace tube, and because the short axis of the elliptic tube is perpendicular to the heat source and the long axis faces the heat source, the radiation heat transfer is larger than that of ordinary circular tube, thus enhancing heat transfer.

Because the cross-section of the tube is elliptical, the requirements for tube material are high and the manufacturing is complex, it is no longer used for industrial purpose.

② Some companies have used the method of plum blossom tube and inner spiral blossom plum tube to enhance the heat transfer of the radiant section furnace tube.

The plum blossom tube can effectively increase the heat transfer area and improve the convective heat transfer of materials in furnace tube. Cracking furnaces with this kind of furnace tube structure have been used in some units abroad.

In order to improve the short operation cycle of millisecond furnace, KBR Company adopts inner spiral plum blossom tube for its furnace, which effectively increases the heat transfer area and the radiation heat transfer rate and prolongs the operation cycle of the cracker from 7–14 days to about 25 days.

Compared with the plain tube, the MERT tube technology of Kubota Company in Japan has higher heat transfer coefficient and higher material pressure drop. The improvement of the second-generation technology mainly reduces the pressure drop while maintaining the heat transfer area and heat transfer rate.

Plum blossom tube and inner spiral plum blossom tube have thick walls, heavy weight, and high processing and manufacturing difficulty. Most of them are patented products with high cost of furnace tube, so they are not widely used at present.

Lummus Company published the patent for enhanced heat transfer furnace tube by adding ribs inside the tube and nail head outside the tube. Due to the easy damage of the ribs and nail head added, this technology has not been applied in industry.

③ Heat transfer enhancement technology of swirling element radiant tube (SERT) was developed by Sinopec Beijing Research Institute of Chemical Industry.

According to the heat transfer mechanism, the maximum resistance of furnace tube heat transfer is the boundary layer of the inner wall of radiant coil. If the resistance of the boundary layer can be reduced, the heat transfer of the radiant coil will be greatly enhanced. The SERT technology is based on this principle.

After adding SERT, the flow is forced to rotate from the piston flow, and the circumferential flow rate is greatly increased, scouring the tube wall. As a result, the boundary layer with high thermal resistance is greatly thinned and the overall heat transfer coefficient of furnace tube is increased. With the decrease of the wall temperature, the coking slows down, and the overall heat transfer coefficient of radiant coil is further increased.

The pressure drop of the flow passing through radiant coil will be increased when SERT is added to the tube. Experimental results show that the negative effect of increased pressure drop on cracking is far less than the positive effect of enhanced heat transfer.

Heat transfer enhancement technology of SERT has been widely used in Sinopec, and it can effectively prolong the operation cycle of cracking furnace. According to the actual operation of the radiant coil in the furnace of an ethylene plant, the wall temperature of the radiant coil decreases by 20°C after adding swirling element. Under the same load conditions, the operation cycle of furnace can be extended by 30%–50% with the SERT. Under the same product yield conditions, the feeding load of furnace can be increased by 5%–7%, and the fuel consumption can be decreased by 1% [7].

At present, aiming at the problems in the first-generation swirling element such as high-pressure drop, poor passability, and easy local thermal stress concentration, a second-generation swirling element has been developed. The comparison of the passability of the two generations of swirling elements is shown in Fig. 5.4.

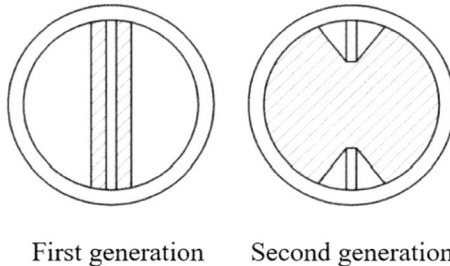

First generation   Second generation

Fig. 5.4 Comparison of passability between two generations of swirling element.

For radiant coil using the first-generation and second-generation swirling element, there is almost no big difference between the CFD simulation results of temperatures in the length direction and cross-sectional direction, and the heat transfer enhancement effect of the two generations of swirling element is similar.

The industrial test of the second-generation swirling element began in 2014. The experimental results show that there is basically no difference between the wall temperature of radiant coil using the second-generation swirling element and that using the first-generation swirling element. The pyrolysis products obtained using the two generations of swirling element were sampled and analyzed respectively, and the deviation of the results was within the deviation range of sampling and analysis. Its effect on yield was basically the same as that of the first-generation swirling element, but the pressure drop of radiant coil was greatly reduced. The second-generation swirling element has good passability and small local thermal stress, which significantly reduced its manufacturing difficulty. At present, the second-generation swirling element has been gradually put into large-scale industrial applications.

This heat transfer enhancement element has the advantages of simple structure and light weight. In addition to being widely used in radiant coils, it has a wide application in places such as the shelter tube row in convection section and the heat exchange tubes in transfer line exchanger.

5.1.5 Summary

Steam cracking is a major energy consumer in ethylene plant. In recent years, a variety of methods including various heat transfer enhancement technologies have been adopted to optimize energy utilization. On the one hand, the process performance of steam cracking (cracking product yield and selectivity) is improved, and the operation cycle is prolonged. On the other hand, the energy consumption is reduced, more reasonable and sufficient heat recovery is achieved, the flue gas temperature of cracking furnace is reduced, the thermal efficiency is improved, the steam production is increased, and the energy saving level reached a new height.

5.2 Ethylene separation
5.2.1 Brief description of process

The cracking gas produced by steam cracking is a mixture of hydrogen, methane, ethane, ethylene, propane, propylene, mixed C_4, mixed C_5, pyrolysis gasoline, and hundreds of other components. In addition, it contains a small amount of carbon dioxide, hydrogen sulfide, carbon monoxide, alkyne, and other impurities, as well as a large amount of dilution steam/water. In order to meet the needs of downstream processing units, the cracking gas must be separated and refined to produce on-spec ethylene and propylene products. At present, there are five main ethylene separation licensors in the world, including Sinopec Engineering Incorporation.

Ethylene separation process includes three process routes, which are sequential separation (separate from light to heavy in sequence according to carbon number), front–end depropanization, and front–end hydrogenation (first separate C_3 and C_4 and then separate the two flows from light to heavy in sequence according to carbon number) and front–end deethanation and front–end hydrogenation (first separate C_2 and C_3 and then separate the two flows from light to heavy in sequence according to carbon number). Sequential separation is the most traditional and widely used process, which can be subdivided into sequential low-pressure demethanation, sequential medium-pressure demethanation, and sequential high-pressure demethanation. Sequential low-pressure demethanation is more widely used in industry, and its process mainly includes gasoline fractionation and cracking gas quenching, cracking gas compression and caustic scrubbing, cracking gas drying and chilling, demethanation, deethanation and acetylene hydrogenation, ethylene distillation, depropanization and methyl acetylene/propadiene (MA/PD) hydroconversion, propylene distillation, debutanization, propylene refrigeration, and binary refrigeration units, which are divided into four areas, that is, quench area, compression area, cold separation area, and thermal separation area. The process flow of typical sequential separation (low-pressure demethanation) process is shown in Fig. 5.5.

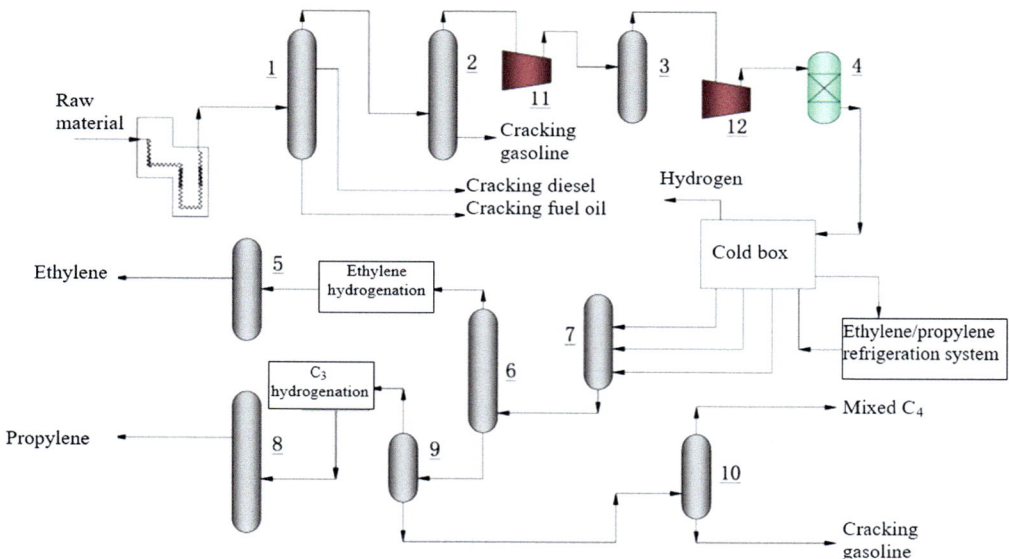

Fig. 5.5 Schematic diagram of ethylene sequential separation (low-pressure demethanation) process. (1) Quench oil tower; (2) quench water tower; (3) caustic scrubber; (4) cracking gas dryer; (5) ethylene distillation tower; (6) deethanation tower; (7) demethanation tower; (8) propylene distillation tower; (9) depropanization tower; (10) debutanization tower; (11, 12) cracking gas compressor.

5.2.2 Analysis of characteristics of thermal energy

Ethylene separation has long process flow and large temperature variation amplitude. The operating temperature of process material ranges from above 210°C to below −170°C. The utility materials include 520°C superhigh-pressure steam and −140°C binary refrigerant. There are many heat exchangers of different types. The heat exchange network is complex, including the waste heat recovery and utilization in quench area, the feed/effluent heat exchange of dryer and reactor, and the cold energy integration of cold separation system.

5.2.2.1 Quench area

1. Heat recovery and utilization of oil scrubber system

    Oil scrubber is set up in the quench area, mainly with liquid cracking raw materials to cool the cracking gas to about 105°C. The temperature of cracking gas from steam cracking is between 370°C and 400°C. It can be seen that the cracking gas has a temperature drop of about 300°C in the oil scrubbing process. Due to the large temperature drop span of the process, the heat load is high. For a 1.0 Mt/a ethylene plant with naphtha as the main cracking raw material, the heat load is up to 157 MW. The design of the oil scrubber is generally divided into three sections of heat removal. The first section uses circulating quench oil, which is used to generate 170°C dilution steam, equivalent to replacing the 280°C medium pressure steam. In the second section, pan oil is used instead of the 210°C low-pressure steam as the heat source for downstream process users. In the third section, the pyrolysis gasoline reflux from scrubber is used. After the gasoline is vaporized, it is returned to the scrubber for condensation. Finally, the heat is transferred to the low-level circulating quench water and provides heat for downstream process users. The heat distribution ratio of these three parts is generally about 70% of quench oil, 20% of pan oil, and 10% of gasoline reflux. Because of different temperature levels, the high-temperature heat of cracking gas can be used reasonably and efficiently.

2. Heat recovery and utilization of scrubber system

    The function of scrubber is to cool the cracking gas from about 105°C to about 40°C, condense the dilution steam and heavy pyrolysis gasoline at the same time, and separate them out. Since most of the heavy pyrolysis gasoline returns to the quench oil tower as reflux for repeated vaporization and condensation, the initial temperature of cracking gas actually cooled by quench water tower should be about 135°C.

    According to different cracking raw materials, the bottom temperature of quench water tower varies from 80°C to 88°C, and the heavier the raw material is, the higher the bottom temperature can be. Through the circulation of quench water, this part of the heat can be used for a variety of low-temperature heating users. The quench water returned has a temperature of 60°C to 65°C and is further cooled to a 55°C stream and a 37°C stream finally by circulating cooling water or air, which return to the scrubber respectively [8].

5.2.2.2 Dryer and reactor system

1. Dryer regeneration system

    There are 3–4 groups of dryers for ethylene separation, which share the same regeneration system. The regeneration gas is self-produced methane, which comes from the outlet of cold box and the temperature is 30°C. It needs to be heated to more than 210°C for the regeneration of the dryer. After returning, it needs to be cooled to room temperature to separate the entrained moisture.

2. Methanator

    Hydrogen methanation reaction is divided into high-temperature method and low-temperature method. The initial reaction temperature is 160°C for low-temperature method and 288°C for high-temperature method. The temperature rise after reaction varies from 20°C to 50°C according to the content of CO. The feed temperature of crude hydrogen is normal temperature, and the reaction effluent should be cooled to about 15°C before drying.

3. Acetylene hydrogenation reactor

    In the sequential separation process and some front-end deethanation process, the C_2 feed of acetylene hydrogenation reaction comes from the deethanation tower overhead, and the effluent is sent to the cold separation area or ethylene distillation tower with a temperature of dozens of degrees below zero, but the reaction temperature of acetylene hydrogenation is about 60–90°C.

    It can be seen that for the feed/effluent of the above systems, heat transfer enhancement is very important to save energy.

5.2.2.3 Cold separation zone

In ethylene separation, 95% (mole fraction) of crude hydrogen is separated from methane at −163°C. The separated methane is throttled and depressurized to low–pressure methane with a temperature of below −170°C, and the separated methane from high-pressure demethanizer is throttled and depressurized to high-pressure methane with a temperature of below −140°C. These materials are at normal temperature when they are extracted from the cold separation system.

The ethane separated from the ethylene column is liquid phase from the bottom of column, and the temperature is about −10°C. It needs to be vaporized and superheated to 60°C before returning to the cracker for cyclic cracking. The operating temperature of demethanation tower and ethylene distillation tower ranges from dozens of degrees below zero to −130°C. The overhead reflux requires to be condensed by low-temperature cold energy, and the reboiling at the bottom of tower can also recover cold energy to reduce the energy consumption.

5.2.2.4 Heating and refrigeration

For ethylene separation, heating is needed for the evaporation and overheating of cracking raw materials, the generation of dilution steam, and the heating of caustic scrubbing feed. The required heat is mainly provided by the heat recovery of internal flows and supplemented by external steam.

As for the cold energy required for ethylene separation, room temperature grade is provided by circulating cooling water, 40°C to −37°C grade is provided by propylene (or methane/ethylene/propylene ternary) refrigeration system, −37°C to −98°C grade is provided by ethylene (or methane/ethylene binary, or methane/ethylene/propylene ternary) refrigeration system, and below −98°C grade is provided by methane, binary, ternary, or methane tail gas expander refrigeration system.

5.2.3 Process enhancement and integration on heat transfer

Cracking gas has a large temperature span in the hydrogen/methane separation process. From the perspective of heat transfer, it can be divided into high-temperature and low-temperature heat transfer and recovery, refrigeration, and cold recovery. With the development of technology, the hot energy and cold energy of ethylene separation can be efficiently recovered. Pinch point analysis and heat transfer network analysis technology have been widely used. Firstly, the cold energy and hot energy should be recovered through process flows, and the intermediate heat transfer medium should be avoided as much as possible to minimize the exergy loss in the process. Secondly, appropriate utilities should be selected as the cooling or heating media to minimize the heat transfer temperature difference within a reasonable range. Thirdly, according to the cooling requirements of the process, multitemperature refrigerant is used to reduce the power of refrigerator and the energy consumption of unit. The overall heat transfer process of ethylene separation system is shown in Fig. 5.6.

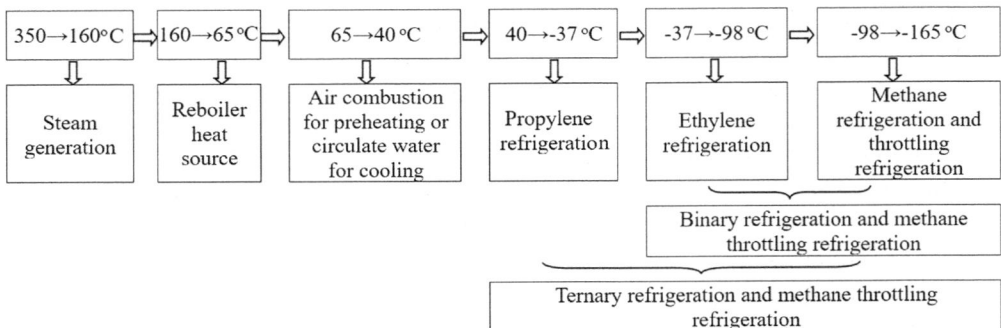

Fig. 5.6 Schematic diagram of heat transfer in ethylene separation system.

5.2.3.1 High-temperature heat transfer and heat utilization

Quench oil heat removal system: the cracking gas from steam cracking is directly injected into the circulating quench oil via quench fitting for cooling and then enters the bottom of quench oil tower and is further cooled with the circulating quench oil after dilution steam generation. After heating up, the quench oil is extracted from the bottom of quench oil tower. After decoking and boosting by quench oil circulating pump, the quench oil is recycled after being used for generating dilution steam.

5.2.3.2 Low-temperature heat transfer and heat utilization

1. Pan oil heating system

    In the pan oil circulation section of the quench oil tower, the low-temperature pan oil cools the cracked gas and condenses the light fuel oil, while the high-temperature pan oil provides heat sources for the reboiler of process water stripper and the reboiler of depropanizer in the downstream process. After cooling itself to about 120°C, the pan oil circulates back to the middle of the quench oil tower.

2. Quench water system

    The cracking gas comes out from the quench oil tower overhead and enters the quench water tower, where it directly contacts the circulating quench water and is further cooled to close to room temperature. In the quench water tower, the dilution steam in cracking gas and the heavy cracked gasoline distillate are condensed, the condensed hydrocarbon and water are separated in the tower or separation tank, and the separated process water is used as the feed to dilution steam generation system after stripping. The separated quench water is sent to the heater or reboiler of each process through the quench water circulation pump to provide heat, and then it is further cooled to different temperatures through the cooling water and returned to different positions of the quench water tower in two cycles. In order to make effective use of the heat of quench water, it is necessary to set a reasonable tower kettle temperature, 83–85°C being the appropriate temperature range. In the early stage, the main users of quench water were the reboiler of propylene tower. The heat utilization rate of quench water was low, which not only lost heat but also required cooling water to cool the quench water later, thus increasing the consumption of utilities. At present, in addition to the reboiler of propylene tower, there are naphtha raw material heater, LPG vaporizer, and combustion air preheater of cracking furnace, which make full use of the low-temperature potential heat of quench water.

3. The use of feed/effluent heat exchanger

    For the regenerative system of dryer, the feed needs to be heated and the effluent needs to be cooled, which consumes energy in both directions. By setting up the feed/effluent heat exchanger, the feed and effluent can exchange heat to the maximum extent, and the insufficient part can be supplemented by heating or cooling, realizing two-way energy saving.

It is very important for energy saving to enhance heat transfer of feed/effluent heat exchangers. For example, two feed/effluent heat exchangers of methanation reactor are set in series so that the feed and the effluent can exchange heat as far as possible, greatly saving the consumption of medium-pressure steam. The setup of feed/effluent heat exchanger in the regeneration system of the dryer and the heat transfer enhancement reduces the consumption of high-pressure steam and cooling water. The setup of feed/effluent heat exchanger in acetylene hydrogenation reactor and the heat transfer enhancement reduce the cold energy consumption.

5.2.3.3 Recovery and utilization of cold energy

Ethylene separation systems involving cold energy include cracking gas drying and component precutting. The process flow is shown in Fig. 5.7. The energy consumption of cold utilities in ethylene plant accounts for about 25% of the total energy consumption of the unit. The heat transfer enhancement during the use of cold energy is very important for the energy saving of the unit.

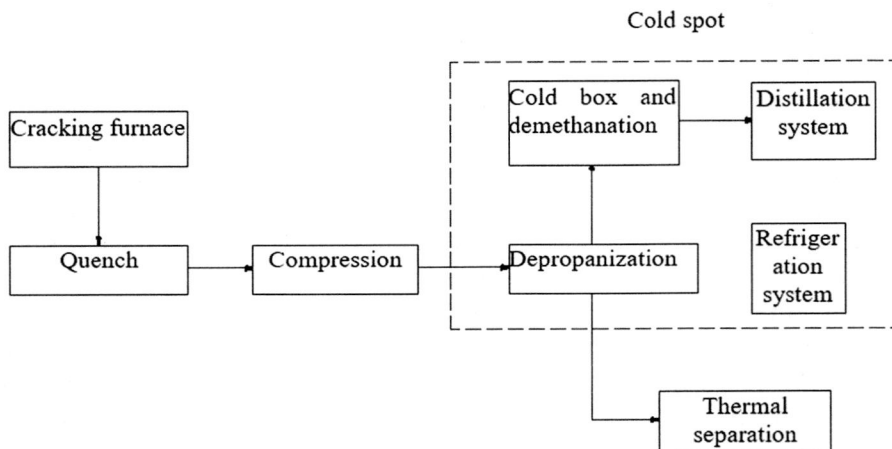

Fig. 5.7 Schematic diagram of the range of cold separation in ethylene unit.

1. Cold energy recovery of tail gas such as crude hydrogen

   Flows such as crude hydrogen are still at low temperature when they are separated in cryogenic system, and their cold energy is fully recycled, which is very important to reduce energy consumption and reduce the material quality of the product system. The cold energy of tail gas such as crude hydrogen is recovered in a series of cold box heat exchangers through condensing and cooling the cracking gas.

   Cold box (see Fig. 5.8) contains a number of plate-fin heat exchangers and separation tanks, filled with perlite for cold insulation. It is a kind of efficient heat transfer enhancement equipment with high heat transfer coefficient, large heat transfer area per unit volume, and a heat transfer temperature difference as low as 1.0°C. At the same time, due to centralized cold insulation, its cold loss is small.

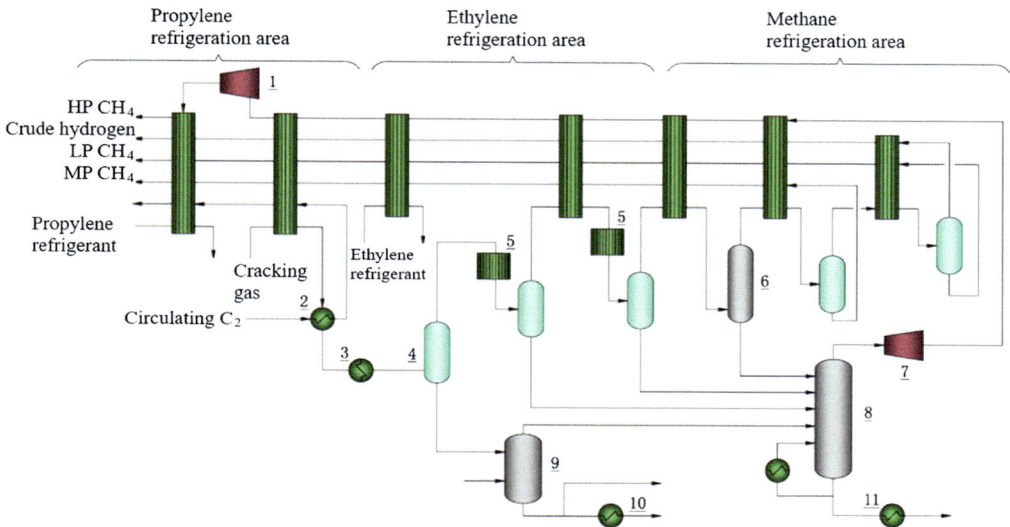

Fig. 5.8 Flow diagram of cold box and demethanation system. (1) Recompressor; (2) circulating ethane vaporizer; (3) feed cooler; (4) C_3 scrubber; (5) feed quencher; (6) C_2 scrubber; (7) tail gas expander; (8) demethanation tower; (9) premethanation tower; (10) 1# depropanization tower cooler; (11) bottom liquid vaporizer.

2. Cold energy recovery of circulating ethane

The low-temperature liquid ethane from the ethylene column kettle is sent to the vaporizer to cool the cracking gas and vaporize itself into low-temperature gas. After cooling the cracking gas and propylene refrigerant in the cold box, it is heated to room temperature and finally sent to the furnace area for circulation cracking after overheating with quench water. In the process, the flow changes from liquid to gas state and is heated from close to −40°C to 60°C. It passes through different types of heat exchangers, such as kettle type, plate-fin, and tubular heat exchangers, to adapt to different heat exchange needs.

3. Cold energy recovery of demethanizer bottom liquid

The demethanation tower is the tower with the lowest operating temperature in ethylene separation. In the sequence separation process of low-pressure demethanation, the overhead temperature of the tower is about −130°C and the bottom temperature of the tower is −50°C. In a front-end depropanization process, the −9°C bottom liquid from the demethanation tower is divided into three strands, which provide cold energy for different users in three heat exchangers and then merge into the ethylene tower after vaporization. The process setting is an effective measure to enhance heat transfer, because the temperature of demethanation tower bottom liquid after throttling expansion is about $-57°C$, and the reflux of ethylene tower is condensed by open heat pump or cold flow of about $-35°C$. The former is ethylene refrigeration level, and the latter is propylene refrigeration level. From the perspective of cold level, it is beneficial to energy saving.

4. Recovery of reboiling cold energy

    The operating temperature of reboilers of the rectification towers in the cold separation zone is below zero, so it is necessary to recycle the cold energy.

    ① Demethanizer. In a front-end depropanization process, the reboiler of the demethanizer uses cracking gas as the heat source, the $-9°C$ column bottom liquid is heated and vaporized by cracking gas, and the cracking gas is cooled from $4°C$ to $-4°C$ to recover the cold energy more effectively.

    ② Ethylene tower. In a front-end depropanization process, a stream of ethylene gas is extracted from the third stage of ethylene heat pump/refrigeration compressor as the heat source of the boiler in the low-pressure ethylene rectification tower. After the outlet gas of the fourth stage of compressor is successively desuperheated by multistage propylene refrigerant, part of the gas can be used as the heating medium for the reboiler of ethylene rectifying tower (Fig. 5.9). In the two groups of heat exchangers, the gas phase ethylene releases heat and is condensed, so as to recover the cold energy. The setting of middle reboiler can effectively reduce the power consumption of the compressor. The two groups of heat exchangers adopt plate-fin type to reduce heat exchange temperature difference, improve heat exchange efficiency, reduce the equipment size, and reduce the footprint.

5. Optimization of front-end depropanization (or front-end deethanization) process

    During the first component separation of the cracking gas, the combination of the front-end depropanization tower (or front-end deethanization tower) with the last stage of the cracking gas compressor can save equipment, such as the reflux pump, and increase the condensation temperature of the cracking gas from the tower upper, which is conducive to reducing the refrigerator power.

5.2.3.4 Refrigeration system

1. Propylene refrigeration

    Propylene refrigeration uses the heat absorption effect of propylene liquid vaporization to achieve cooling. In ethylene separation, propylene refrigeration system provides cold energy at various temperature levels above $-40°C$ for ethylene separation. Its users include cracking gas precooling, process flow condensation, and so on. The cold consumption of each part varies with the separation process.

As shown in Fig. 5.9, a four-stage throttling refrigeration cycle of propylene is used for ethylene separation. A four-stage centrifugal compressor driven by steam turbine is used for circulating refrigeration in the loop, and liquid propylene is throttled and vaporized under different pressures to provide the process users with four grades of propylene refrigerant including $-40°C$, $-21°C$, $-1°C$, and $25°C$. The purpose of multistage refrigeration is to use different refrigeration temperatures for users with different temperatures to ensure the effective heat exchange temperature difference and maximize the power saving of propylene refrigerant cyclic compression.

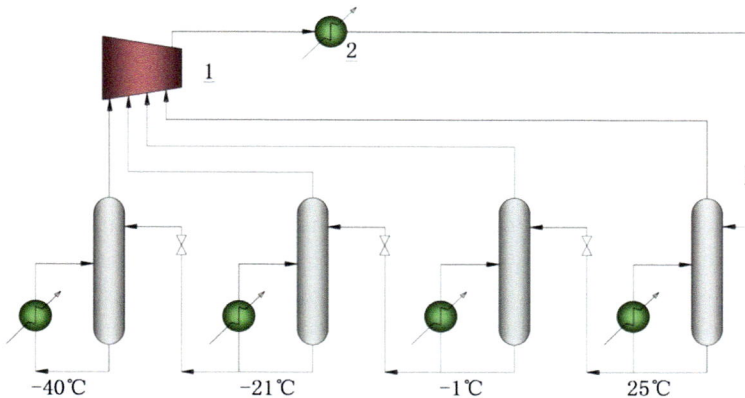

Fig. 5.9 Propylene four-stage throttling refrigeration system. (1) Propylene compressor; (2) condenser.

2. Ethylene refrigeration

In ethylene separation, ethylene refrigeration system provides cold for various temperature levels from $-40°C$ to $-101°C$ required by ethylene separation. For example, the open heat pump system combined with ethylene rectifying tower and ethylene refrigeration compressor of an ethylene plant can provide refrigerant at three temperature levels of $-62°C$, $-80°C$, and $-101°C$. After desuperheating, part of the gas phase discharged from the fourth stage of the ethylene heat pump/refrigeration compressor is used as the heat source of the ethylene tower reboiler to recover the cold energy and be liquefied, and the remaining is condensed by $-40°C$ propylene refrigerant. At the same time, a stream of hot ethylene gas is directly extracted from the outlet of the third stage to be used as the heating medium of the middle boiler in the ethylene rectifying column to recover the cold energy at lower temperature level in the middle section of the ethylene rectifying column. The ethylene refrigerant obtained by the two can be used as the reflux of the ethylene column after supercooling or directly. By doing so, the equipment can be reduced, the reboiling heat source of the tower can be saved, and the consumption of propylene refrigerant needed for ethylene condensation can be reduced, thus saving investment and reducing energy consumption.

The purpose of the graded use of ethylene refrigerant is to save refrigerator power by enhancing heat transfer, just like propylene refrigerant.

3. Mixed refrigeration

In the sequential low-pressure demethanation process, there are three kinds of traditional single-component refrigerants: methane, ethylene, and propylene. Two components or three components are mixed together in a certain proportion to provide two or all temperature levels of cold flow for ethylene plant [9,10]. The currently used binary refrigerant mainly refers to methane-ethylene binary refrigerant. Methane refrigeration compressor and ethylene refrigeration compressor are relatively small compressor units. Combining them into one larger compressor unit is conducive to energy saving, investment reduction, and footprint reduction.

Ternary refrigeration is to mix methane, ethylene, and propylene in a certain proportion and compress them in a refrigeration compressor. Conventional single-component refrigeration generally requires three refrigeration units, i.e., propylene refrigerator, ethylene refrigerator, and methane refrigerator, while ternary refrigeration only requires one refrigerator. Therefore, ternary refrigeration can reduce the amount of equipment, save investment, and reduce land occupation.

The cooling curve of material flow during cracking gas cooling is continuous and smooth, while the cooling curve of refrigerant in traditional single-component cascade refrigeration is discontinuous and done step by step, as shown in Fig. 5.10.

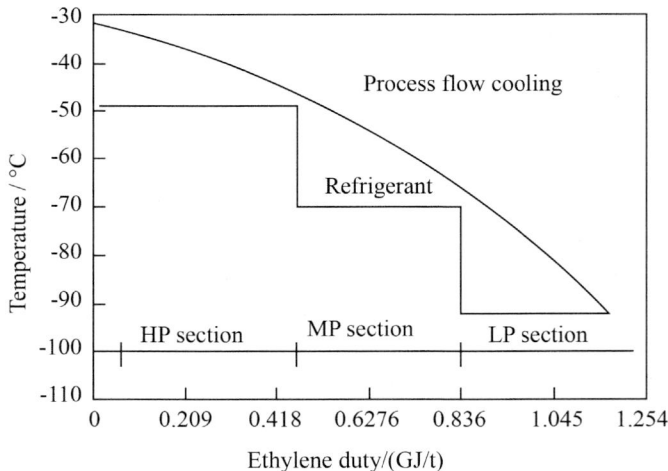

Fig. 5.10 Relationship between temperature level of refrigerant in ethylene refrigeration cycle and cooling curve of process flow.

In the heat exchanger, the average heat transfer temperature difference between the material and the refrigerant is large, the heat transfer process is irreversible, and the energy utilization efficiency is relatively low. Using mixed refrigerant and making the evaporation curve of refrigerant close to the cooling curve of process flow (as shown in Fig. 5.11) can reduce the average heat transfer temperature difference, improve the energy utilization efficiency, and reduce the energy consumption [11]. Binary refrigeration and ternary refrigeration are based on this principle and designed according to the level of ethylene separation refrigerant.

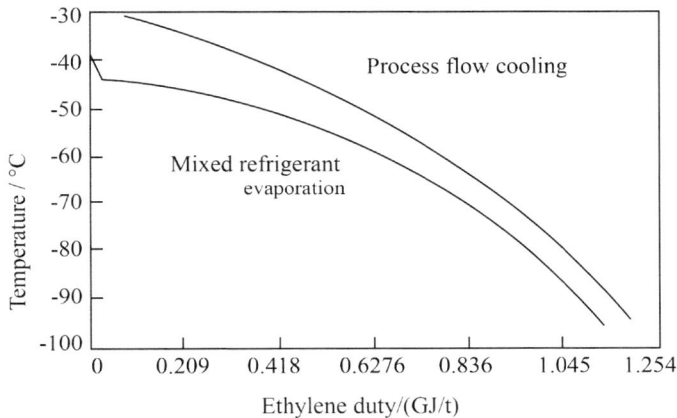

Fig. 5.11 Relationship between evaporation curve of mixed refrigerant and cooling curve of process flow.

According to preliminary investigation, under stable operation, the total power of ethylene separation refrigerator can be reduced by about 2% using binary refrigeration and by nearly 5% using ternary refrigeration.

5.2.4 Heat transfer enhancement technology using elements

According to the requirements of ethylene separation heat exchange unit, different heat transfer enhancement technologies are adopted.

5.2.4.1 Key heat exchanger of quench system

Quench system heat exchangers mainly serve for dilution steam generation system, quench water system, and quench oil system and are characterized by large equipment size and large number of equipment. If the water and oil separation in the system is not complete, the heat transfer efficiency of the heat exchanger will be seriously reduced and corrosion leakage will be caused. Advanced separation process, optimized operating conditions, and heat exchanger design can effectively reduce the ethylene separation energy consumption and material consumption and minimize the possibility of corrosion leakage.

The quench water at the bottom of quench water tower is sent to the downstream process users and the raw material preheater as the heat source and then returned to different positions of the quench water tower as reflux after being cooled by the first quench water cooler and the second quench water cooler. Spiral baffle heat exchanger has been successfully applied in quench water cooler. The high-speed rotating medium flow in the spiral channel can wash away particles and sediments, eliminate the shell side flow dead zone, reduce the possibility of shell side quench water fouling deposition, increase the effective heat transfer area, reduce the harm of leakage caused by corrosion, so as to reduce the operation cost, avoid heat exchanger vibration, and improve the heat transfer capacity of a single heat exchanger. However, it should be noted that when large ethylene plants use large-diameter spiral baffle heat exchangers, if the internal structure design is not reasonable, the heat transfer efficiency will not be as good as that of the ordinary bow baffle.

As an important heat carrier for ethylene separation unit, quench oil is used to cool the cracking gas and generate the dilution steam needed for cracking. The viscosity of quench oil is an important technical index of quench system, which determines the operation cycle and heat transfer efficiency of quench oil/dilution steam generator. If the quench oil is well controlled and no mechanical cleaning is required, it is recommended to adopt vertical tube thermalsiphon quench oil/dilution steam generator. Compared with the horizontal thermalsiphon, the vertical tube has no dead zone and the retention time of process water is short. In this way, even if the pH value of process water is not qualified, it is not easy for the process water to cause corrosion leakage of equipment at high-temperature parts, which ensures that the quench oil/dilution steam generator maintains high heat transfer efficiency.

5.2.4.2 Key heat exchanger of compression system

The typical cracking gas compression for ethylene separation uses multistage compression, with aftercooler set between sections. The first stage aftercooler of cracking gas compressor and the second stage aftercooler of cracking gas compressor have low allowable pressure drop on the cracking gas side, so they are prone to vibration. The composite enhancement measures of rod baffle and low-finned tube can be used to solve the pressure drop and vibration problems and enhance the heat transfer effect of the equipment.

5.2.4.3 Key heat exchanger of cold separation system

The core heat exchangers in cryogenic zone are cold box and plate-fin heat exchanger. In-vessel heat exchanger is used for ethylene separation demethanation tower and ethylene distillation tower. The inner core of the heat exchanger is plate-fin heat exchanger, and the inner core is immersed in the liquid phase of the vessel during operation. In-vessel heat exchanger using plate-fin inner core has better heat transfer effect than that using tube bundle inner core. It can save half of the investment and reduce the volume by more than half, and its weight is only one-quarter of the ordinary kettle shell and tube heat exchanger.

The total pressure drop in ethylene cold box is demanding. Fins with high heat transfer performance and low resistance should be selected. The size of the guide plate should be specially designed. The design of plate-fin heat exchanger requires careful analysis of the state of each stream and comprehensive consideration of different stream to ensure the rational heat transfer design and structure design of cold box. The No. 1 cold box of an ethylene separation unit has eight streams for heat transfer, and the process parameters are shown in Tables 5.1 and 5.2. Comparative analysis between heat transfer enhancement by plate-fin heat exchanger and normal tube and shell heat exchanger technology is shown in Table 5.3.

From the comparison of schemes in Table 5.3, it can be clearly seen that only one plate and fin heat exchanger is needed, while 21 conventional shell and tube heat exchangers are needed. Plate-fin heat exchanger has obvious advantages, for their compact structure can save land, pipelines, valves, insulation, and foundation.

5.2.4.4 Key heat exchanger of thermal separation system

The important heat exchange equipment of thermal separation system is the reboiler and condenser in high- and low-pressure depropanizer. In ethylene separation process, high- and low-pressure towers are used for depropanization. The minimum end temperature difference of heat exchanger is only 3.0°C, the total heat transfer temperature difference of heat exchanger is small, and the boiling intensity of the propylene side is low. If the conventional shell and plain tube heat exchanger are used, boiling under small temperature difference cannot be achieved, and the heat source grade has to be improved. Using high flux tube instead of plain tube bundle improves heat transfer intensity and reduces heat transfer area, and equipment size is significantly reduced. Low-grade heat source can be applied to achieve the purpose of energy saving.

The cold flow of the overhead condenser of propylene distillation tower is provided by cooling water. The condensing load is high. With the use of shell and plain tube heat exchanger, the total heat transfer area required for a megaton ethylene plant reaches tens of thousands of square meters. The amount of equipment, occupation area, and investment are high. It is suitable to use efficient condensing heat exchanger to enhance condensation heat transfer and optimize equipment layout.

5.2.4.5 Key heat exchanger of refrigeration system

The propylene refrigerant condenser set at the outlet of propylene refrigeration compressor is used to condense all the gas phase propylene at the outlet of the propylene refrigeration compressor. The process conditions of the propylene refrigerant condenser in an ethylene plant are shown in Table 5.4. Propylene is very clean and condenses outside the tube. The use of low-finned tube or high-efficiency condensing heat exchange tube has the effect of enhancing condensation heat transfer outside the tube and fully eliminating the surface tension of condensate, so as to reduce the thickness of condensate film and reduce the thermal resistance of film. The comparison of schemes is shown in Table 5.5.

Table 5.1 Process conditions of No.1 heat exchanger in No.1 cold box—Duty 3.3 MW.

Item	Cold flow 1	Cold flow 3	Cold flow 4	Cold flow 5	Cold flow 6	Hot flow 7
Fluid name	High-pressure methane	Hydrogen	Tail gas	High-pressure methane	Recycling ethane	Propylene refrigerant
Flow rate (kg/h)	29,764.8	5224.8	5872.8	14,725.2	24,722.4	62,998.8
Allowable pressure drop (kPa)	4	13	10	10	18	14
Temperature (°C)	$18.41 \rightarrow 30$	$-26 \rightarrow 30$	$-26 \rightarrow 30$	$-26 \rightarrow 30$	$-26 \rightarrow 30$	$34.01 \rightarrow -21.33$
Gas phase density (kg/m^3)	$3.732 \rightarrow 3.559$	$4.367 \rightarrow 3.546$	$1.03 \rightarrow 0.782$	$4.616 \rightarrow 3.659$	$13.611 \rightarrow 10.176$	
Liquid density (kg/m^3)						$489.05 \rightarrow 572.621$
Gas phase viscosity (mN s/m^2)	$0.011 \rightarrow 0.0113$	$0.0078 \rightarrow 0.0089$	$0.0096 \rightarrow 0.0014$	$0.0095 \rightarrow 0.0113$	$0.0082 \rightarrow 0.0098$	$0.0773 \rightarrow 0.1416$
Liquid phase viscosity (mN s/m^2)						
Gas phase thermal conductivity (W/(m K))	$0.0393 \rightarrow 0.0411$	$0.1421 \rightarrow 0.1641$	$0.0381 \rightarrow 0.047$	$0.031 \rightarrow 0.039$	$0.0177 \rightarrow 0.0243$	$0.0984 \rightarrow 0.1273$
Liquid phase thermal conductivity (W/(m K))						
Gas phase specific heat (kJ/(kg K))	$2.4053 \rightarrow 2.4229$	$2.6493 \rightarrow 2.5965$	$2.4006 \rightarrow 2.531$	$2.27 \rightarrow 2.3714$	$1.7521 \rightarrow 1.875$	
Liquid phase specific heat (kJ/(kg K))						$2.9434 \rightarrow 2.2688$

Table 5.2 Process conditions of No.2 heat exchanger in No.1 cold box—Duty 0.8 MW.

Item	Cold flow 2	Cold flow 3A	Cold flow 4A	Cold flow 5A	Cold flow 6A	Hot flow 8
Fluid name	High-pressure methane	Hydrogen	Tail gas	High-pressure methane	Recycling ethane	Cracking gas
Flow rate (kg/h)	29,764.8	5224.8	5872.8	14,725.2	24,722.4	194,085.6
Allowable pressure drop (kPa)	7	13	7	10	11	14
Temperature (°C)	$-40 \to -26$	$-40 \to -26$	$-40 \to -26$	$-40 \to -26$	$-38.05 \to -26$	$-18 \to -19.96$
Gas phase density (kg/m^3)	$2.801 \to 2.583$	$4.645 \to 3.367$	$1.144 \to 1.03$	$4.973 \to 3.616$	$14.886 \to 13.611$	$41.977 \to 41.505$
Liquid density (kg/m^3)						$\to 467.471$
Gas phase viscosity (mNs/m^2)	$0.009 \to 0.0095$	$0.0075 \to 0.0078$	$0.0091 \to 0.0096$	$0.009 \to 0.0095$	$0.0078 \to 0.0082$	$0.0106 \to 0.0106$
Liquid phase viscosity (mNs/m^2)						$\to 0.0796$
Gas phase thermal conductivity (W/(mK))	$0.0308 \to 0.0328$	$0.1365 \to 0.1421$	$0.0359 \to 0.0381$	$0.0291 \to 0.031$	$0.0164 \to 0.0177$	$0.0357 \to 0.0359$
Liquid phase thermal conductivity (W/(mK))						$\to 0.1159$
Gas phase specific heat (kJ/(kgK))	$2.2914 \to 2.3019$	$10.575 \to 10.595$	$2.4 \to 2.4227$	$2.27 \to 2.2859$	$1.7432 \to 1.7521$	$2.4314 \to 2.4412$
Liquid phase specific heat (kJ/(kgK))						$\to 2.9314$

Table 5.3 Comparison of schemes for No.1 cold box.

Scheme		Diameter × length (mm)	Set piece	Overall heat transfer coefficient (W/(m² K))	Total heat transfer area (m²)	Total weight of equipment (t)
Shell and tube type (plain tube)	E-1 (heat exchange between flow 1 and 7)	1500 × 8000	21	50	10,223 (plain tube)	266.3
			4		4313 (plain tube)	120
	E-2 (heat exchange between flow 3 and 7)	800 × 6000	2	95	620 (plain tube)	20
	E-3 (heat exchange between flow 4 and 7)	900 × 6000	2	85	800 (plain tube)	25
	E-4 (heat exchange between flow 5 and 7)	900 × 7000	4	85	1840 (plain tube)	50
	E-5 (heat exchange between flow 6 and 7)	900 × 8500	4	90	2230 (plain tube)	30
	E-6 (heat exchange between flow 2 and 8)	800 × 3000	1	218	150 (plain tube)	7
	E-7 (heat exchange between flow 3A and 8)	600 × 3000	1	360	80 (plain tube)	4.6
	E-8 (heat exchange between flow 4A and 8)	400 × 3000	1	210	35 (plain tube)	1.9
	E-9 (heat exchange between flow 5A and 8)	600 × 3000	1	200	80 (plain tube)	4.6
	E-10 (heat exchange between flow 6A and 8)	500 × 4000	1	245	75 (plain tube)	3.2
Plate–fin type		10,800 (length) × 1100 (width) × 1297 (height)	1		11,411 (including the secondary expansion area of fin)	20

Table 5.4 Process conditions of propylene refrigerant condenser—Duty: 84 MW.

Item	Hot flow	Cold flow
Fluid name	Propylene refrigerant	Cooling water
Flow rate (kg/h)	837,363	12,084,725
Allowable pressure drop (kPa)	14	70
Temperature (°C)	70.8 → 44.0	33.0 → 39.0
Gas phase density (kg/m^3)	31.85 → −	
Liquid density (kg/m^3)	− → 477	
Gas phase viscosity (mN s/m^2)	0.01 → −	
Liquid phase viscosity (mN s/m^2)	− → 0.0723	
Gas phase thermal conductivity (W/(m K))	0.0219 → −	
Liquid phase thermal conductivity (W/(m K))	− → 0.0959	
Gas phase specific heat (kJ/(kg K))	1.95 → −	
Liquid phase specific heat (kJ/(kg K))	− → 3.156	

Table 5.5 Comparison of schemes for propylene refrigerant condenser.

Item	Plain tube	Low-finned tube	High-efficiency condensation tube
Equipment size (shell diameter × straight tube length) (mm)	2700 × 11,000	2600 × 12,000	2200 × 12,000
Number of equipment (piece)	6 in parallel	4 in parallel	4 in parallel
Consumption (t/h)	Cooling water: 12,085	Cooling water: 12,085	Cooling water: 12,085
Heat transfer area (m^2)	29,547	20,412	13,383
Total weight of equipment (t)	1005	600	465

According to the comparative analysis in Table 5.5, choosing the efficient condensing tube instead of plain tube can enhance the propylene condensation heat transfer by about 10 times, reduce the total heat transfer area by 50%, optimize the equipment layout, and reduce the investment in civil construction and pipelines. The space in ethylene plant is tight, the equipment used is large, and there is requirement for piping layout; hence the existing space is unable to meet the demand for increased number of equipment. From the perspective of reducing the number of equipment, optimizing the layout, saving land, and reducing capital construction costs, the use of efficient condensing heat exchange tube to enhance heat transfer has obvious advantages.

5.2.4.6 Application of high flux tube in ethylene separation

In ethylene separation, heat exchanger using high flux tube has obvious advantages compared with that using conventional plain tube. As an effective enhancement method in boiling heat transfer with small temperature difference, high flux tube can effectively solve the problem of insufficient boiling heat intensity using ordinary plain tube. The effective heat transfer temperature difference of some heat exchangers is less than 3°C. Because of the high flux heat exchange tube, high boiling film heat transfer coefficient is maintained, and the small temperature difference heat transfer is possible. The kettle reboiler is generally operated in the nucleated boiling area, and the flow needs a certain thermal intensity to boil. When the heat transfer temperature difference is small, there are few or even no nucleate occurrence points, and the heat transfer rate is significantly lower than the extrapolated value of the nucleate boiling curve. High flux tube can provide a large number of stable vaporization cores under small temperature difference, and its minimum heat transfer temperature difference can reach 2.0°C, or even close to 1.0°C. In this case, high flux tube is the best choice. High-flux tube has been successfully used in large ethylene plants in Fujian, Tianjin, and Zhenhai.

The use of high flux heat exchange tubes can expand the range of nucleate boiling in heat exchanger in ethylene separation. At present, the high flux heat exchange tubes used in ethylene separation are mainly externally sintered or externally coated high flux tubes. Under small temperature difference, the boiling film heat transfer coefficient of high flux heat exchange tube can be increased by 10–30 times compared with that of plain tube, the overall heat transfer coefficient can be increased by 3–5 times, and the critical heat flux is high. The processing technology of high flux tube is mature, and it has been running for more than 10 years in large ethylene plants. The porous layer does not fall off and the heat transfer performance does not decrease, which verifies that the sintered surface of high flux heat exchange tube is tough and can resist the corrosion and erosion of the fluids during operation.

5.2.4.7 Application of high-efficiency enhanced condensation heat exchange tube in ethylene separation

The condensation heat transfer efficiency is high if efficient enhanced condensation heat exchange tube is used in ethylene separation. For example, the total heat transfer load of propylene refrigerant condenser (see Table 5.5 for comparison analysis) and propylene distillation column condenser in the ethylene plant is about 80–110 MW, with a total heat transfer temperature difference of 10°C. If ordinary shell and tube heat exchanger is used, the total heat transfer area needs to be 20,000–30,000 m². The heat transfer area can be reduced by half and the investment can be reduced by choosing high-efficiency enhanced condensation heat exchange tubes for high-efficiency heat transfer.

The enhancement effect of the high-efficiency enhanced condensation heat exchange tube on propylene condensate film heat transfer coefficient is 7–10 times that of the common plain tube. Efficient enhanced condensation heat exchange tube has highly enhanced condensation heat transfer efficiency under small temperature difference and expands the condensation surface area. Therefore, the number of equipment is reduced, which is conducive to the optimization of equipment layout and the investment saving. The inner wall of the efficient enhanced condensation heat exchange tube is smoother and less prone to dirt deposition, so its antifouling performance is better than that of ordinary plain tube.

5.2.5 Summary

In large-scale ethylene plant, there are a large number of equipment with large dimensions. Conventional shell and tube heat exchanger cannot meet the process operation conditions and will bring difficulty in equipment layout, especially in cold separation system. The manufacturing technology of shell and tube heat exchanger is mature, but the heat transfer efficiency of conventional shell and tube heat exchanger is low, and the material and energy consumption are large. High-efficiency heat exchanger has higher heat transfer efficiency and less heat transfer area, which helps to fully improve the heat recovery in ethylene separation process and reduce the consumption of utilities. It is more advantageous in low temperature difference and large heat exchangers.

5.3 Ethylene oxide/ethylene glycol unit

5.3.1 Brief description of process

With ethylene and oxygen as basic raw materials, ethylene oxide/ethylene glycol unit produced important organic chemical products, such as ethylene oxide, monoethylene glycol, diethylene glycol, and triethylene glycol. Until the coal-to-ethylene glycol unit [12] was put into operation in 2009, the "ethylene–ethylene oxide–ethylene glycol" process route has been adopted for ethylene glycol production, and ethylene oxide can be co-produced in the same unit. This section focuses on the application of heat transfer enhancement technology in the production of ethylene oxide/ethylene glycol by direct oxidation process.

At present, the major licensors of direct oxidation process to produce ethylene oxide/ethylene glycol are SD (Scientific Design), SHELL, DOW, and BASF. For these licensors, the basic process routes are very similar, which all include ethylene oxidation reaction, ethylene oxide absorption and stripping, carbon dioxide absorption and desorption, MEG recovery and impurity removal, hydration reaction ratio, light component removal, ethylene oxide refining, ethylene glycol hydration reaction, multieffect evaporation, monoethylene glycol refining, and diethylene glycol and triethylene glycol refining, but unique characteristics remained in implementation methods between different licensors. The typical process block diagram is shown in Fig. 5.12 [13,14].

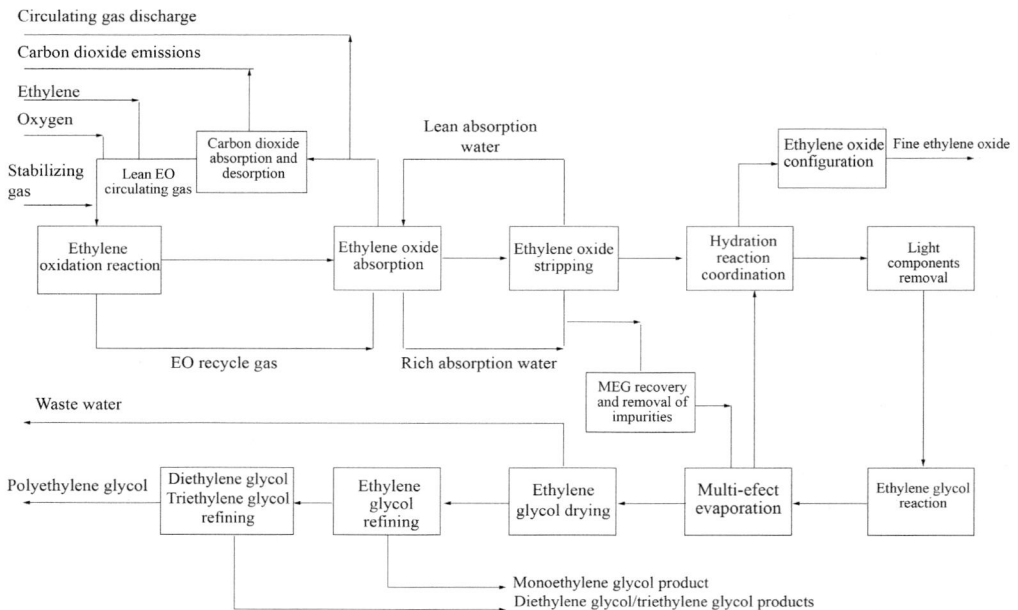

Fig. 5.12 Block diagram of direct oxidation process for ethylene oxide/ethylene glycol production.

5.3.2 Analysis of characteristics of thermal energy

The utility consumption of ethylene oxide/ethylene glycol unit is slightly different with different technical routes. Table 5.6 shows the energy consumption data of main utilities of typical ethylene oxide/ethylene glycol unit. It can be seen that the energy consumption of ethylene oxide/ethylene glycol unit is mainly concentrated on steam, cooling water, and electricity.

Table 5.6 Main utility consumption of ethylene oxide/ethylene glycol unit.

4.0 MPa steam t/(t EOE)	0.4 MPa steam t/(t EOE)	Cooling water t/(t EOE)	Demin water t/(t EOE)	Electricity kW h/(t EOE)
1.680	1.082	459	1.586	411

Note: EOE is the amount of ethylene oxide produced by ethylene oxidation reactions, and the energy consumption of ethylene plant is usually measured using the amount of utilities consumed per ton of ethylene oxide produced by the ethylene oxidation reactor.

The heat input of ethylene oxide/ethylene glycol production process is mainly used for reaction raw material preheating, component stripping, multieffect evaporation, and product refining. For example, the reaction raw materials for the oxidation of ethylene to ethylene oxide and the hydration of ethylene oxide to ethylene glycol need to be preheated to a certain temperature to trigger the reactions. The ethylene oxide and carbon dioxide contained in the ethylene oxide-rich recycle gas at the outlet of the ethylene oxide reactor are absorbed by lean absorption water and carbonate solution, respectively, and then a large amount of heat is required to extract the ethylene oxide and carbon dioxide from the rich absorption water and rich carbonate solution. The crude ethylene oxide stripped by the ethylene oxide stripper needs to be refined to separate ethylene oxide from water to get the refined ethylene oxide product. The refining process requires heat input. In order to improve the yield of the target product monoethylene glycol, ethylene glycol reaction is carried out with excess water, but after the completion of the reaction, a large amount of heat is needed to separate the excess water from the reaction product. At the same time, in order to get high-purity monoethylene glycol, diethylene glycol, and triethylene glycol products, these products need to be separated one by one, which also requires a lot of heat.

The cold consumption of ethylene oxide/ethylene glycol unit is mainly used for reaction heat removal, absorption, and condensation. For example, the oxidation of ethylene to ethylene oxide is a strong exothermic reaction. In order to ensure the performance of the catalyst and maintain appropriate reaction operating conditions, the reaction heat needs to be removed in time. The lean absorption water at the bottom of ethylene oxide stripper and the lean carbonate solution at the bottom of carbon dioxide desorption tower should be reused as absorbent and should be cooled to a suitable absorption temperature. The overhead condensers of ethylene oxide stripper, carbon dioxide desorption tower, and ethylene glycol refining tower need a large amount of cooling water to maintain the normal operation of the towers. For the process with reabsorption tower, a large amount of cooling water is needed to remove the absorption heat. For the process with condensate recovery and water distribution, the refrigerant is needed to directly cool the gas phase ethylene oxide to liquid. There are many materials in the unit to be cooled and utilized or stored at low temperatures, so the cold consumption is high.

The main electrical equipment of ethylene oxide/ethylene glycol unit is compressor, pump, and air cooler. The single pass conversion of ethylene oxidation reaction is low, there is a large amount of recycle gas, and the power consumption of recycle gas compressor is large. Due to the large amount of circulating fluid of the lean absorption water and lean carbonate solution and the high pressure difference, the power consumption of lean absorption water pump and lean carbonate solution pump is large.

In conclusion, there are many unit operation processes such as raw material preheating, high-temperature reaction, low-temperature absorption, stripping, evaporation, and refining in the ethylene oxide/ethylene glycol unit, the temperature of materials in the process changes greatly, and there is repeated cooling-heating-cooling of materials during absorption and desorption, which requires a large amount of cold and hot energy. Therefore, how to make full use of the heat released by the reaction and the energy of material flows in the unit, enhance the heat transfer between materials, enhance the energy utilization option, optimize the heat transfer network, and reduce the consumption of steam and cooling water is the focus of the process design and heat transfer process enhancement of the unit.

5.3.3 Process energy optimization and heat transfer network optimization

5.3.3.1 Optimizing process operating conditions, reasonably designing reactor type and enhancing heat transfer effect of ethylene oxidation reactor

Ethylene oxidation reaction is a strong exothermic reaction, and the activation energy of epoxidation reaction, the main reaction, is lower than that of combustion reaction, the side reaction. When the reaction temperature increases, the rates of the main and side reactions increase simultaneously, but the rate of the side reaction increases faster than that of the main reaction. At the same time, ethylene conversion increases, but selectivity decreases. Because the thermal effect of the side reaction is about 12 times that of the main reaction, with the increase in temperature, more heat is released. If the heat cannot be removed in time, it will be difficult to control the temperature, resulting in temperature runaway. It is necessary to optimize the process operation conditions, design the reactor type reasonably, and enhance the heat transfer of the reactor. Ethylene oxidation reaction usually adopts tubular fixed bed reactor. The following measures can be taken to enhance the heat transfer in the shell side of the reactor:

① Appropriate refrigerant is used to remove the reaction heat in time and reduce the axial and radial temperature difference of the reaction tube. Ethylene glycol units in the 1970s and 1980s used hot oil as refrigerant. When hot oil was the heat removal medium, the radial temperature difference of the reaction tube could reach about 30°C, and the larger the tube diameter, the greater the radial temperature difference, causing the catalyst particles in the center of the tube to overheat and deactivate [15]. Later studies show that if boiling water vaporization is used instead of hot oil heat removal, the shell side temperature is basically constant and the reaction heat is easier to be taken away, to ensure that the reaction can be carried out under the approximate constant temperature as far as possible. Boiling water vaporization heat removal is characterized by a uniform temperature distribution in the reactor tubes [16]. In the use of boiling water heat removal, radial temperature difference is only more than 10°C. It is difficult for the catalysts in the center of the tube to overheat, so larger diameter of the reactor tube can be selected compared with hot oil heat removal, so as to reduce the investment of reactor equipment under the same scale. Therefore, the current ethylene oxide/ethylene glycol units use boiling water vaporization heat removal. The reaction temperature can be adjusted by adjusting the pressure in the drum on the shell side of reactor.

② New catalyst carriers with good heat transfer performance are developed to enhance the heat transfer effect of catalysts and reduce the radial temperature difference of the reactor.

③ Reactor structure design is optimized to avoid "tail burning" in the reactor. "Tail burning" refers to the phenomenon that the gas at the outlet of the reactor continues to burn after leaving the catalyst bed, resulting in temperature runaway. In order to avoid the accumulation of catalyst powder at the lower end of the reactor, the lower head of the reactor can be conical. The reactor and the gas cooler at the reactor outlet can be made into one, and the material at the reactor outlet directly enters the gas cooler after leaving the reactor to reduce the residence time of high-temperature gas at the reaction outlet, reduce the outlet gas temperature, and reduce the possibility of "tail burning."

④ Regulator is added to improve the selectivity of catalysts, effectively inhibit side reactions, and reduce reaction heat release. For example, adding a small amount of monochloroethane or dichloroethane can improve the selectivity of catalysts so that more ethylene oxide is generated and the heat release of the reaction is reduced.

5.3.3.2 Enhancement and integration of heat transfer process in typical ethylene oxide/ethylene glycol process

1. Enhancement and integration of heat transfer process in ethylene oxidation reaction unit

The enhanced utilization and integration process of heat in a typical ethylene oxidation reaction unit is shown in Fig. 5.13.

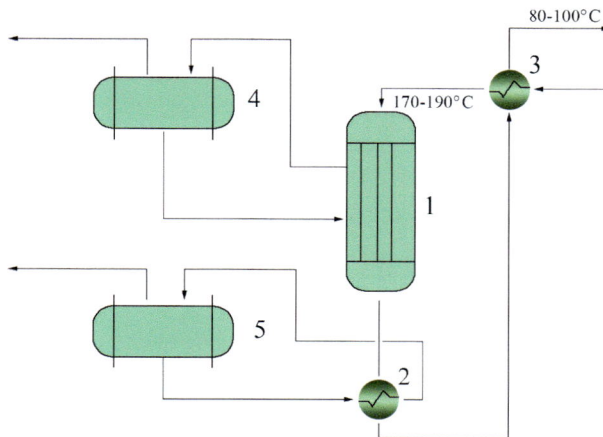

Fig. 5.13 Diagram of enhanced utilization and integration of heat in ethylene oxidation unit. (1) Ethylene oxidation reactor; (2) ethylene oxidation reactor outlet cooler; (3) reactor feed/effluent heat exchanger; (4) high-pressure drum; (5) medium-pressure drum.

The reactor feed gas exchanges heat with the reactor outlet gas in the reactor feed/effluent heat exchanger 3. The reactor feed gas is preheated in the reactor feed/effluent heat exchanger 3 and then flows downward through the ethylene oxidation reactor 1 and the reactor outlet cooler 2. Part of the ethylene is converted into ethylene oxide and by-products CO_2 and water. In the upper part of the reactor, the feed gas is heated to the reaction trigger temperature and then sent to the reaction section for reaction [17]. The heat of reaction is removed by boiling water on the reactor shell side. Water is recycled by thermalsiphon, and the vaporization rate of water is 1.5%–2.5% (mass fraction). The reactor temperature is adjusted according to the conditions of the catalyst, usually by adjusting the drum pressure.

The reactor outlet gas is cooled in the ethylene oxidation reactor outlet cooler 2 by steam generation. After that, it is cooled in the reaction feed/effluent heat exchanger 3. By setting the reactor feed/effluent heat exchanger, reactor feed preheating is integrated with reactor effluent cooling, saving energy. By using steam from reaction heat, the amount of steam introduced from OSBL is saved.

The amount of by-product steam in the reactor is closely related to the selectivity of silver catalyst for ethylene oxidation reaction. At the early stage of catalyst use, the catalyst selectivity was high, the reaction heat release was low, a small amount of by-product steam was generated, and a large amount of steam was introduced from OSBL. With the increase of the catalyst's service duration, the catalyst selectivity gradually decreased, the heat released gradually increased, the amount of byproduct steam gradually increased, and the steam introduced from OSBL decreased. As can be seen from Table 5.7, the catalyst selectivity decreased from 87.6% to 85% in different periods (1 year interval), the 4.0 MPa steam consumption decreased from 1.68 to 1.31 t/(t EOE) and the 0.4 MPa steam consumption decreased from 1.08 to 0.97 t/(t EOE) with the unit capacity remaining unchanged. The catalyst selectivity decreased, the steam consumption decreased, but the material consumption increased. The ethylene consumption increased from 0.73 to 0.75 t/(t EOE). When the cost of raw material ethylene is relatively high, material consumption has a greater impact on the economic benefits of the plant than energy consumption. Therefore, when assessing the ethylene oxide/ethylene glycol unit, energy consumption and material consumption should be considered comprehensively.

Table 5.7 Operation data of a typical ethylene oxide/ethylene glycol unit—t/(t EOE).

Year	4.0 MPa steam consumption	0.4 MPa steam consumption	Ethylene consumption	Ethylene oxidation catalyst selectivity (%)
First year	1.68	1.08	0.73	87.6
Second year	1.31	0.97	0.75	85

2. Enhancement and integration of ethylene oxide absorption and stripping heat transfer processes

There are many enhancement and integration options for ethylene oxide absorption and stripping heat transfer processes. There are different heat utilization options according to the different temperatures of lean absorbed water, the temperature difference between rich and lean absorbed water, and the differences in process settings, but all of them are based on the maximum recovery of the heat of the material itself. Figs. 5.14 and 5.15 list several typical heat integration options.

As shown in the heat utilization diagram I of ethylene oxide absorbed water (Fig. 5.14) [13], the circulating gas at the outlet of the reactor, after exchanging heat with rich absorbed water in the circulating gas/rich absorbed water heat exchanger 3, enters the ethylene oxide absorber 1 and makes reverse contact with the lean absorbed water to absorb ethylene oxide. Water by-product from the reactor and the ethylene oxide are absorbed and condensed in the absorber. Rich absorbed water is first heated in circulating gas/rich absorbed water heat exchanger 3, and then heat is exchanged with lean absorbed water in lean/rich absorbed water heat exchanger 4. It enters the ethylene oxide stripper 2 to recover the ethylene oxide. After two steps of heat exchange, the temperature of the rich absorbed water increases, reducing the heat input of the reboiler in ethylene oxide stripper 2. However, the temperature of the rich absorbed water rises, and the hydration reaction between ethylene oxide and water begins in the rich absorbed water. In order to prevent the concentration of generated ethylene glycol in the absorbed water from being too high, which will lead to system foaming, it is necessary to control the temperature of the rich absorbed water and the residence time in the equipment and pipelines. Usually industrial lean/rich absorbed water heat exchanger 4 uses high-efficiency plate heat exchanger to enhance heat transfer, and it is located as close as possible to ethylene oxide stripper 2 to shorten the residence time of high-temperature materials. The absorbed water absorbs part of the ethylene and methane in the circulating gas while absorbing ethylene oxide, and these gases will gradually be desorbed from the rich absorbed water as the temperature of the water increases. Therefore, the influence of noncondensing gas on heat transfer coefficient should be fully considered during the design of lean/rich absorbed water heat exchanger 4. At the same time, it is necessary to design reasonable flow direction and fully consider the problem of noncondensing gas emission so that the heat exchanger can achieve the best heat transfer effect.

After the rich absorbed water enters the ethylene oxide stripper 2, the ethylene oxide is stripped out from the tower upper. The lean absorbed water at tower bottom is used to preheat the rich absorbed water after pressurization and then enters the lean absorbed water cooler 5 to be cooled to 38°C before entering the ethylene oxide absorber to realize the recycling of lean absorbed water. In order to maintain the system water balance and eliminate the accumulated ethylene glycol in the cooling water loop, a blowdown stream is usually extracted out and sent to the ethylene glycol reaction/evaporation unit. The lower the temperature of the lean absorbed water, the more beneficial it is to reduce the ethylene oxide content in the circulating gas from the ethylene oxide absorber upper. Therefore, in industrial production, some units use lower absorbed water temperature. Based on the limit of the water supply temperature of the cooling water, it is necessary to replace the refrigerant with a lower temperature refrigerant, such as adding an ice machine, if the lower temperature of the lean absorbed water is used.

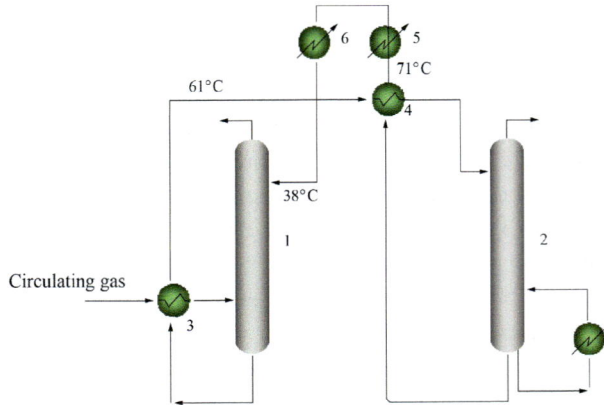

Fig. 5.14 Heat utilization diagram I of ethylene oxide absorbed water. (1) Ethylene oxide absorber; (2) ethylene oxide stripper; (3) circulating gas/rich absorbed water heat exchanger; (4) lean/rich absorbed water heat exchanger; (5) lean absorbed water cooler; (6) ice machine.

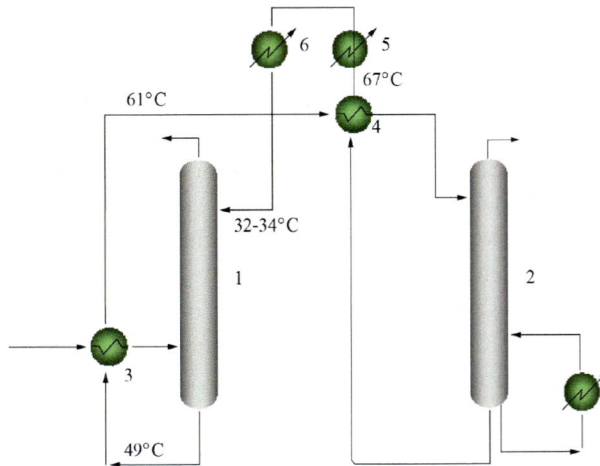

Fig. 5.15 Heat utilization diagram II of ethylene oxide absorbed water. (1) Ethylene oxide absorber; (2) ethylene oxide stripper; (3) circulating gas/rich absorbed water heat exchanger; (4) lean/rich absorbed water heat exchanger; (5) lean absorbed water cooler; (6) ice machine.

According to the design of the ice machine, hot water or low-pressure steam should be provided as the heat source of the ice machine. Combined with the situation of ethylene oxide/ethylene glycol unit, the heat of the overhead gas of the regeneration tower [13] can be used as the heat source of the ice machine.

For the quench process of the circulating gas at the reactor outlet to remove impurities [18], the circulating gas at the reactor outlet first exchanges heat with the rich absorbed water in the upper column of the ethylene oxide absorber, then enters the cryogenic section to remove impurities, and then enters the upper section of the ethylene oxide absorber for reverse contact with the lean absorbed water to achieve the separation of ethylene oxide from circulating gas. The ethylene-oxide-rich absorbed water in the upper tower of the ethylene oxide absorber is firstly heated by the circulating gas at the reactor outlet and then passes through the lean/rich absorbed water cooler I and the lean/rich absorbed water cooler II in turn. After being heated by lean absorbed water, it enters the ethylene oxide stripper, and the ethylene oxide is extracted by steam at the top of the tower. The lean absorbed water in the tower bottom is cooled step by step to recover heat and then sent back to the ethylene oxide absorber upper for recycling. The discharge temperature of lean absorbed water from the bottom of ethylene oxide stripper is 120°C, the temperature required for the final entry into the ethylene oxide absorber is only 25°C, and the flow rate of lean absorbed water is relatively high, so lean absorbed water can provide more heat. In addition to providing heat for rich ethylene oxide absorbed water, it is also used as the heating source of the reboiler in ethylene oxide refining tower and the reboiler in light component removal column. After fully recovering heat, it is cooled to the target temperature by the lean absorbed water cooler and the lithium bromide ice machine. Because the operating conditions of the ethylene oxide stripper and the material composition at the tower bottom are different, the temperature difference and heat that can be provided by lean absorbed water are also different. Different heat utilization options correspond to the specific processes.

3. Low-temperature heat utilization of condensate in feed preheater of ethylene glycol hydration reaction [13]

The tower bottoms of reabsorber directly enter the light component removal column. Too low feed temperature in the light component removal column will increase the steam consumption of the tower, while too high feed temperature will increase the content of ethylene oxide at the upper tower. Usually, the temperature of tower bottoms of the reabsorber is about 40°C, and the process condensate of the ethylene glycol reaction feed preheater can be used to preheat the stripper feed, which can raise the feed temperature to about 51°C and reduce the steam consumption of the light component stripper. Before the tower bottoms of the stripper enter the reactor, it is preheated through the feed preheater of the ethylene glycol hydration reactor, the heating medium is the process condensate, and the temperature is 113°C. After heat transfer through the feed preheater of the reactor, the temperature of the condensate drops to 67°C. Then, it raises the feed temperature of the light component stripper to about 51°C, and its own temperature drops to 57°C.

4. Multieffect evaporation

Multieffect evaporation is used to remove excess water in ethylene glycol reaction products, and the concurrent process flow is generally adopted. The advantages are as follows: pump is not required for the flow of solution between effects, because the pressure of the latter effect evaporation chamber is lower than that of the former effect, and the solution can automatically flow from the former effect into the latter effect through the pressure difference. The boiling point of the solution of the former effect is higher than that of the latter effect, so the solution of the former effect becomes overheated upon entering the latter effect and immediately evaporates by itself, which can vaporize more solvents. In order to facilitate the manufacturing and installation of the equipment, the heating area of each effect evaporator is basically the same. According to this principle, the operating pressure of each effect and the final concentration of each effect solution are adjusted and determined by the total temperature difference formed by the pressure of the first effect heating steam and the final effect vacuum. The determination of the number of effects for multieffect evaporation depends on the comprehensive consideration of energy consumption and investment. Triple-effect, four-effect, five-effect, and seven-effect evaporation techniques are commonly used.

For the evaporation system with more than five effects, the final effect evaporation tower is operated under vacuum, and all the steam at the top of the tower is condensed. This process is referred to as condensing multieffect evaporation for short. The more the evaporation tower, the less the steam consumption. However, with the increase in the number of evaporation towers, the effect of reducing steam consumption is weakened. With more evaporation towers, the temperature difference of the intermediate reboiler is reduced accordingly and the reboiler area is too large, so the number of evaporation towers cannot be increased without limit.

Using triple-effect evaporation system, the final effect evaporation tower is operated under positive pressure, and the steam from the upper tower can be used as the heat source for other process flows. The multieffect evaporation uses the energy level difference of the heating steam, and it is referred to as backpressure multieffect evaporation for short. However, due to the high final effect pressure, the total temperature difference of the multieffect evaporation system becomes smaller, so the comprehensive utilization of energy should be balanced in the design. In order to increase the total temperature difference of the multieffect evaporation unit, high temperature and high pressure can also be used in the ethylene glycol hydration reaction, which not only is conducive to the operation of the multieffect evaporation unit but also shortens the residence time of the reaction. Because the heat transfer temperature difference of the multieffect evaporation unit is very small, high flux heat transfer tubes are used to enhance the heat transfer.

5. Other energy-saving methods

For the process that only produces refined ethylene oxide, the tower bottoms of ethylene oxide stripper are used as the heat source for the reboilers of ethylene oxide refining tower and carbon dioxide stripper to save energy [19]. Air cooler can be used instead of cooling water cooler for the overhead condensers of carbon dioxide desorption tower and ethylene oxide stripper tower to save cooling water consumption. In plants with guaranteed steam supply, steam turbine can be used to drive compressor to save power consumption. Hydraulic turbine is used to recover high-pressure hydrostatic pressure energy.

5.3.4 Heat transfer enhancement technology using elements

The heat transfer characteristics of each unit of ethylene oxide/ethylene glycol plant are different, so different heat transfer elements should be used to enhance heat transfer. This section takes a 900 kt/a ethylene glycol unit as an example to analyze the characteristics of heat exchange equipment and the enhancement measures.

5.3.4.1 Heat exchange equipment of ethylene oxidation reaction unit

The reaction unit is equipped with feed/effluent heat exchanger to recover the heat energy of product gas. The process conditions of the reactor feed/effluent heat exchanger are shown in Table 5.8. The main characteristics of the reactor feed/effluent heat exchanger include high heat load, low allowable pressure drop, highly coincident temperature interval of cold and hot flows, and small heat transfer temperature difference. All the existing units use fixed tube sheet shell-and-tube heat exchangers, and the shell-side heat transfer can be enhanced by changing the baffle type. Single segmental baffles are not commonly used because they are prone to bundle vibration. Rod baffle and no-tube-in-window (NTIW) can eliminate the vibration of tube bundle and reduce the pressure drop. The pressure drop on the shell side of rod baffle is low, and the saved pressure drop can be used in the tube. Under the premise of ensuring that the total pressure drop on tube side and shell side does not exceed the total allowable pressure drop, the heat transfer in the tube is enhanced. The comparison between the single segmental baffle–NTIW option and the rod baffle option is shown in Table 5.9.

Table 5.8 Process conditions of reactor feed/effluent heat exchanger—Duty: 60.92 MW.

Item	Shell side	Tube side
Fluid name	Product gas	Feed gas
Flow rate (kg/h)	922,860	922,860
Allowable pressure drop (kPa)	40	35
Temperature (°C)	$185.6 \rightarrow 60.6$	$41.4 \rightarrow 168.3$
Gas phase density (kg/m^3)	$9.41 \rightarrow 13.24$	$16.82 \rightarrow 11.6$
Gas phase viscosity (mN s/m^2)	$0.017 \rightarrow 0.013$	$0.013 \rightarrow 0.017$
Gas phase thermal conductivity (W/(m K))	$0.049 \rightarrow 0.032$	$0.03 \rightarrow 0.047$
Gas phase specific heat (kJ/(kg K))	$2.06 \rightarrow 1.75$	$1.736 \rightarrow 2.02$

Table 5.9 Comparison of reactor feed/effluent heat exchanger options.

Item	Single segmental baffle—NTIW	Rod baffle
Equipment size (shell diameter × straight tube length) (mm)	4500 × 20,000	3000 × 20,000
Quantity of equipment (piece)	1	1
Pressure drop (kPa)	35 (total)	46 (total)
Total heat transfer temperature difference (°C)	15.9	18.3
Overall heat transfer coefficient (W/(m² K))	180	237
Heat transfer area (m²)	24,158	13,920
Total weight of equipment (t)	600	306

As can be seen from Table 5.9, due to the reduced shell diameter, the overall heat transfer coefficient of rod baffle heat exchanger is 36.1% higher than that of heat exchanger with NTIW, and its equipment weight is 49% lower than that of NTIW exchanger, saving the investment and floor space.

In addition, at the oxygen mixer, ethylene and oxygen in the feed gas may undergo complete oxidation reaction, resulting in an instantaneous rise in reaction temperature to 800°C. Similar combustion of the product gas may occur at the outlet of the ethylene oxidation reactor. These two extreme conditions will have a serious impact on the gas/gas heat exchanger and put forward more demanding requirements for the design of expansion joint of fixed tube-sheet heat exchanger. Coil-wound heat exchanger not only enhances heat transfer but also solves the problems in design of expansion joint.

5.3.4.2 Heat exchange equipment of carbon dioxide absorption and desorption unit

The absorption and desorption unit is equipped with lean/rich carbonate solution heat exchanger to recover heat energy. The rich carbonate solution is heated up in the heat exchanger. Affected by the vibration of tube bundle, gas phase CO_2 may be flashed out. CO_2 will reside in the dead zone of the heat exchanger, resulting in gas resistance and reducing the heat transfer capacity of the heat exchanger. The heat transfer characteristics of lean/rich carbonate solution heat exchanger and reactor feed/effluent heat exchanger are the same, and the heat transfer enhancement method is the same.

Taking the lean/rich carbonate solution heat exchanger in a 900 kt/a ethylene oxide/ethylene glycol unit as an example, the equipment weight can be reduced by 15% by using rod baffle and 36% by using coil-wound heat exchanger compared with the double segmental baffle option.

5.3.4.3 Heat exchange equipment of ethylene oxide absorption, stripping, and refining unit

The process conditions of lean/rich absorbed water heat exchanger with high heat load are shown in Table 5.10, and the temperature ranges of cold and hot flows highly coincide. The medium characteristics require that there should be no flow dead zone in the equipment. Foreign licensors recommend the use of rod baffle structure. The comparison of options is shown in Table 5.11.

Table 5.10 Process conditions of lean/rich absorbed water heat exchanger—Duty: 159.56 MW.

Item	Hot side	Cold side
Fluid name	Lean cooling water	Rich cooling water
Flow rate (kg/h)	2,963,970	3,081,810
Allowable pressure drop (kPa)	140	140
Temperature (°C)	$113 \rightarrow 67$	$56 \rightarrow 101$
Liquid phase density (kg/m^3)	$952 \rightarrow 980$	$974 \rightarrow 958$
Liquid phase viscosity (mPa s)	$0.256 \rightarrow 0.433$	$0.499 \rightarrow 0.279$
Liquid phase thermal conductivity (W/(m K))	$0.682 \rightarrow 0.657$	$0.629 \rightarrow 0.658$
Liquid phase specific heat (kJ/(kg K))	$4.21 \rightarrow 4.17$	$4.10 \rightarrow 4.14$

Table 5.11 Comparison of rich/lean absorbed water heat exchanger options.

Item	Rod baffle option	Fully welded plate option	Coil-wound option
Equipment size (shell diameter × straight tube length) (mm)	$1600 \times 12,000$	1600×3200	$2700 \times 12,000$
Number of equipment (piece)	4	4	1
Pressure drop (kPa)	225 (total)	271 (total)	275 (total)
Overall heat transfer coefficient (W/(m^2 K))	1450	4096	1560
Heat transfer area (m^2)	9535	3380	8850
Total weight of equipment (t)	190	162	115

It can be seen that even if small–diameter heat exchange tube with an outer diameter of 16 mm is used, the equipment weight and floor area of rod baffle heat exchanger are still large. At present, four-sided detachable all-welded plate heat exchangers are mostly used to enhance heat transfer. However, its advantage of compactness is not obvious due to the difficulty in equipment large-scaling, large number of equipment, large floor area, and complex piping. Coil-wound heat exchanger is suitable for equipment large-scaling because one piece of equipment can meet the heat transfer requirements and the equipment weight is 30% less than that of all-welded detachable plate heat exchanger.

The temperature of lean cooling water needs to be further cooled by cooling water from 67°C to below 38°C. Plate and frame heat exchangers can be used to enhance heat transfer and reduce the heat exchange area by about 85%.

5.3.4.4 Heat exchange equipment of multieffect evaporation unit

Increasing the number of energy efficiency stages of multieffect evaporation can save energy, but it also raises higher requirements for heat exchange equipment. The more the multieffect evaporation stages, the smaller the heat transfer temperature difference at each stage. The low-temperature difference decreases the boiling intensity and reduces the overall heat transfer coefficient, and the heat transfer area is very large. The process conditions of the fifth effect reboiler of a unit are shown in Table 5.12.

Table 5.12 Process conditions of the fifth effect reboiler—Duty: 75.39 MW.

Item	Shell side	Tube side
Fluid name	Steam, ethylene glycol steam	Ethylene glycol solution
Flow rate (kg/h)	130,730	942,710
Allowable pressure drop (kPa)	3	Thermosiphon
Temperature (°C)	$156.9 \rightarrow 156.9$	$146.8 \rightarrow 147.8$
Gas phase density (kg/m³)	$2.833 \rightarrow -$	$2.15 \rightarrow 2.15$
Liquid phase density (kg/m³)	$- \rightarrow 918$	$957 \rightarrow 957$
Gas phase viscosity (mN s/m²)	$0.014 \rightarrow -$	$0.014 \rightarrow 0.014$
Liquid phase viscosity (mN s/m²)	$- \rightarrow 0.173$	$0.24 \rightarrow 0.24$
Gas phase thermal conductivity (W/(m K))	$0.03 \rightarrow -$	$0.029 \rightarrow 0.029$
Liquid phase thermal conductivity (W/(m K))	$- \rightarrow 0.675$	$0.523 \rightarrow 0.523$
Gas phase specific heat (kJ/(kg K))	$1.92 \rightarrow -$	$1.92 \rightarrow 1.92$
Liquid phase specific heat (kJ/(kg K))	$- \rightarrow 4.29$	$3.8 \rightarrow 3.8$

All multieffect evaporations use vertical thermosiphon reboilers, with the medium boiling in the tube. Under the condition of low temperature difference, the boiling intensity of conventional shell and tube heat exchangers is greatly reduced, and stable boiling cannot be produced in the inlet section. The porous structure of the inner sintered porous and outer longitudinal groove high flux tube increases the bubble nucleation center, which can produce stable boiling at 1–3°C. The self-cleaning mechanism reduces the thermal resistance of fouling in the tube. The longitudinal groove outside the tube is helpful to rapidly discharge the condensate, reduce the thickness of liquid film on the outer wall of the tube, and improve the film heat transfer coefficient of condensation. The Cu–Ni alloy has strong heat conduction ability, which further reduces the metal thermal resistance of the heat exchange tube. Under the action of various factors, the high flux tube greatly reduces the heat transfer area of the reboiler. The comparison option between ordinary plain tube and high flux tube is shown in Table 5.13.

Table 5.13 Comparison of the fifth effect reboiler options.

Item	Plain tube option	High flux tube option
Equipment size (shell diameter × straight tube length) (mm)	2900 × 5000	2800 × 4000
Number of equipment (piece)	4	1
Overall heat transfer coefficient (W/(m² K))	1025	6050
Total heat transfer area (m²)	8950	1481
Total weight of equipment (t)	205	44

It can be seen from Table 5.13 that the overall heat transfer coefficient of the heat exchanger with high flux tube is about five times that of the plain tube, and the corresponding heat transfer area and equipment weight are also greatly reduced.

The bottom materials of the ethylene glycol refining tower are mainly monoethylene glycol, diethylene glycol, and triethylene glycol and they also contain a variety of impurities. It has high heat sensitivity and is operated under vacuum. Vertical thermosiphon reboiler is commonly used, but falling film evaporator is more suitable for this working condition because of its small liquid holdup, short residence time, no dead zone, low metal wall temperature, and enhanced evaporation of high viscosity materials. Table 5.14 shows the process conditions of ethylene glycol refining tower reboiler, and Table 5.15 shows the comparison of heat transfer options.

Table 5.14 Process conditions of ethylene glycol refining tower reboiler—Duty: 50.8 MW.

Item	Shell side	Tube side
Fluid name	Steam	Monoethylene glycol, diethylene glycol, triethylene glycol
Flow rate (kg/h)	97,100	1,044,200
Allowable pressure drop (kPa)	5	Thermosiphon
Temperature (°C)	212.7	$161 \rightarrow 164$
Gas phase density (kg/m³)	–	0.38
Liquid phase density (kg/m³)	–	1009.7
Gas phase viscosity (mN s/m²)	–	0.011
Liquid phase viscosity (mN s/m²)	–	1.113
Gas phase thermal conductivity (W/(m K))	–	0.021
Liquid phase thermal conductivity (W/(m K))	–	0.195
Gas phase specific heat (kJ/(kg K))	–	1.85
Liquid phase specific heat (kJ/(kg K))	–	2.92

Table 5.15 Comparison of ethylene glycol refining tower reboiler options.

Item	Vertical thermosiphon option	Falling film evaporator option
Equipment size (shell diameter × straight tube length) (mm)	3500 × 4000	3200 × 5000
Number of equipment (piece)	1	1
Overall heat transfer coefficient (W/(m² K))	515	688
Heat transfer area (m²)	2375	1829
Total weight of equipment (t)	97.5	80.8

It can be seen from Table 5.15 that the heat transfer area of the falling film evaporator is reduced by 23% and the equipment weight is reduced by 17%. In addition, falling film evaporator also has the advantage of long operation cycle. Shell–and–tube heat exchanger with corrugated tube can also be used in ethylene glycol distillation [20].

5.3.5 Summary

Taking the ethylene oxidation reaction part, the ethylene oxide absorption and stripping part, and the multieffect evaporation part as examples, the application of heat transfer enhancement technology in ethylene oxide/ethylene glycol plant is mainly introduced from the perspectives of enhanced integration of reaction heat and heat utilization between process materials. Through the examples of the process energy utilization options, it can be seen that the energy utilization options of ethylene oxide/ethylene glycol plant can be optimized by taking measures such as reasonably designing the process flow, enhancing the heat exchange between hot and cold flows, improving the heat transfer effect of heat transfer equipment, and using high-efficiency heat exchange equipment, so as to improve the efficiency of integrated utilization of heat and reduce the comprehensive energy consumption of the unit.

5.4 Propylene oxide unit

5.4.1 Brief description of process

Propylene oxide (PO) is produced by cooxidation process (Haakon process), direct oxidation process, or chlorohydrin process using propylene as raw material [21]. It is mainly used for the production of polyether polyol and propylene glycol and also as the major raw material for nonionic surfactant, oil demulsifier, pesticide emulsifier, and propylene carbonate. Cooxidation has become the main direction for the development of propylene oxide at present due to its advantages such as large unit capacity, high production stability, good product quality, and reliable environment protection measures. The cooxidation method is divided into ethylbenzene cooxidation method (PO/SM method), isobutane cooxidation method (PO/MTBE method), and isopropyl benzene method (CHP method) according to different raw materials and coproduced products. This section mainly takes PO/SM method as an example to introduce the application of heat transfer enhancement technology in propylene oxide unit.

PO/SM process takes propylene and ethylbenzene as the raw material to produce PO, and SM is the coproduct. It consists of four main reaction units, such as ethylbenzene peroxidation unit, propylene epoxidation unit, phenylethyl alcohol dehydration unit, and acetophenone hydrogenation unit, as well as related raw material and product distillation unit. For the flow diagram of typical propylene oxide production process by ethylbenzene cooxidation, see Fig. 5.16.

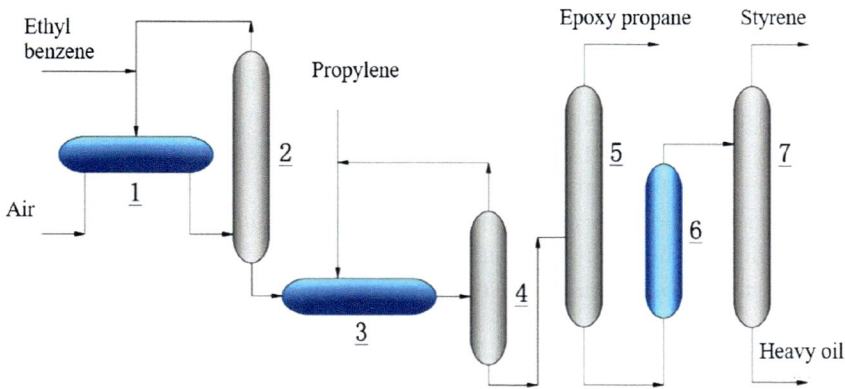

Fig. 5.16 Flow diagram of typical propylene oxide production process by ethylbenzene cooxidation. (1) Peroxidation reactor; (2) concentration tower; (3) epoxidation reactor; (4) propylene separation tower; (5) propylene oxide separation tower; (6) dehydration reactor; (7) styrene separation tower.

In ethylbenzene peroxidation unit, ethylbenzene is oxidized with the air to produce ethylbenzene hydrogen peroxide, which is further concentrated and then sent to epoxidation unit. In the epoxidation unit, propylene reacts with ethylbenzene hydrogen peroxide to produce crude propylene oxide, phenylethyl alcohol, and acetophenone. Unreacted propylene is recycled, and crude PO is sent to PO refining unit to obtain finished propylene oxide. Phenylethyl alcohol is dehydrated in the dehydration unit to produce crude styrene, which is refined to obtain styrene product. Acetophenone is hydrogenated in the hydrogenation unit to produce phenylethyl alcohol, which is recycled to the dehydration unit.

5.4.2 Analysis of characteristics of thermal energy

The cooxidation process has complicated process flow and high energy consumption due to multiple reactions and high separation requirements. According to the total product volume of propylene oxide and styrene, the energy consumption is about

250–350 kgoe/(t product), and the steam consumption can reach 15–20 t/(t PO product). The energy consumption distribution of typical propylene oxide unit by ethylbenzene cooxidation is shown in Table 5.16.

Table 5.16 Energy consumption distribution of typical propylene oxide unit by ethylbenzene cooxidation.

Item	Energy consumption (kgoe/t product)	Distribution (%)
Steam	200–280	70–75
Electricity	15–30	6–9
Cooling water	30–50	15–20
Others	10–20	3–6

Note: The "t product" in the table is the total amount of PO products and coproducts.

As can be seen from Table 5.16, the energy consumption of steam is the largest, followed by the energy consumption of cooling water. Therefore, the main direction of energy saving and consumption reduction and heat transfer enhancement of the unit is to reduce the consumption of steam and cooling water as much as possible.

The propylene oxide process by cooxidation has the following characteristics of thermal energy: firstly, the reaction heat effect is outstanding but is difficult to recycle; secondly, the steam consumption is large, but the proportion of low-grade steam is high; and thirdly, the utilities consumption of condensation and cooling is large.

The peroxidation reaction and epoxidation reaction are both strong exothermic reactions, but the reaction temperature is not higher than 150°C, so the heat is difficult to recycle. Dehydration reaction is a strong endothermic reaction, which needs to be heated by high-temperature heat source. The thermal effect of hydrogenation reaction is relatively small and can be ignored.

The steam consumption is large, because the temperature of the process side is mostly below 180°C; medium-pressure steam basically meets the requirements. For heat-sensitive media such as the easily decomposed ethylbenzene hydrogen peroxide and the easily polymerized styrene, the temperature of the process side is generally limited to below 120°C from the point of process safety, and low-pressure steam or even low-low-pressure steam must be selected for the heating.

Due to the temperature limit of ethylbenzene hydrogen peroxide and styrene, the separation is mostly vacuum operation, the condensation temperature is mostly in the range of 50–90°C, and cooling water or air cooling is generally adopted. The heat is difficult to use, and the heat load is high.

Based on the above thermal energy characteristics, the general principles for energy consumption and heat transfer enhancement of propylene oxide unit are as follows: the reaction heat is mainly solvent evaporation heat removal and direct contact heat transfer, so as to enhance the heat removal effect and heat transfer efficiency; the cascade utilization of steam is mainly used, and steam energy is recovered through backpressure turbine on the premise of meeting the demand of low-pressure steam; within the scope of process safety allowance, the heat integration degree of separation system is improved; the use range of air cooling is expanded and the consumption of cooling water is reduced; and the low-temperature heat of the steam condensate produced by the unit is fully utilized.

5.4.3 Process energy optimization and heat transfer network optimization

According to the energy characteristics of the aforementioned unit, the commonly used heat exchange optimization measures are as follows:

1. Heat recovery and heat transfer in the reaction system

   In view of the characteristics of large thermal effect and low temperature and high temperature control requirements of peroxidation and epoxidation reactions, heat removal by solvent evaporation or excessive solvent recycling is usually used to improve the temperature distribution in the reactor. The contact cooling between evaporation solvent and cold feed can eliminate heat transfer temperature difference, improve heat removal ability, and enhance heat transfer effect.

2. Cascade utilization of steam

   According to the process requirements, especially the safety requirements of ethylbenzene hydrogen peroxide and styrene, the steam grade is subdivided on the premise of ensuring the necessary heat transfer temperature difference. Secondly, on the basis of subdividing the steam grade, the cascade matching of all steam grades, especially the rational utilization of steam turbine, should be achieved to improve the overall energy efficiency of the unit.

3. Heat integration of the system

   Within the scope of process safety allowance, by adjusting the tower pressure, heat integration between different separation systems can be realized, or the use of high-grade energy can be reduced, thus reducing the comprehensive energy consumption of the unit. In HPPO process, since a large amount of circulating methane is to be recovered in the subsequent units, high-pressure and low-pressure dual-tower process flow is adopted and the overhead steam from high-pressure methanol tower is used as the heat source for the reboiler of low-pressure methanol tower, so the steam consumption per ton of product is reduced by 35.7%, achieving remarkable energy saving effect [22].

4. Recovery and utilization of low-temperature heat

   According to the characteristics of the large amount of steam condensate in the unit, lithium bromide refrigeration process can be used to replace the refrigeration propylene of some users with chilled water to reduce the load of propylene refrigeration unit. It is estimated that the comprehensive energy consumption of a 300 kt/a unit can be reduced by about 25–35 kgoe/(t PO) after the lithium bromide refrigeration process is adopted.

5. Expansion of the use range of air cooler

Because the cooling load of the unit is concentrated in the range of 50–90°C, if only water cooling is used, the consumption of cooling water is large. It is estimated that for a 300 kt/a unit, expanding the use range of air cooling can reduce the consumption of cooling water by 10,000–15,000 t/h.

5.4.4 Heat transfer enhancement technology using elements

According to the above analysis, the production process of propylene oxide determines the technical focus of heat transfer enhancement, which is to solve the problems in large-scale equipment caused by high vacuum and low temperature difference and to meet the safe operation requirements of heat-sensitive media. Heat transfer enhancement technology is of great significance to improve the level of process technology and engineering technology.

Spiral plate heat exchanger is suitable for occasions where temperatures are highly crossed because of its pure countercurrent, and the flow is spiral flow with violent turbulence, so it has a certain self-cleaning ability and is not easy to scale. The temperatures of the hot medium and cold medium in a feed/effluent heat exchanger are highly crossed, the media are heat sensitive, and there is a tendency of coking and scaling. Heat transfer analysis shows that the spiral plate heat exchanger is suitable for this scenario. The process conditions of this heat exchanger are shown in Table 5.17. The comparison between spiral plate heat exchanger and conventional shell and tube heat exchanger is shown in Table 5.18.

Table 5.17 Process conditions of feed/effluent heat exchanger—Duty: 9.455 MW.

Item	Hot side	Cold side
Fluid name	Hot feed	Cold feed
Heat load (MW)	10.0	10.0
Allowable pressure drop (kPa)	70	60
Temperature (°C)	$210 \rightarrow 100$	$50 \rightarrow 190$
Liquid phase density (kg/m^3)	$800 \rightarrow 910$	$920 \rightarrow 830$
Liquid phase viscosity (mN s/m^2)	$0.259 \rightarrow 0.90$	$1.79 \rightarrow 0.35$
Liquid phase thermal conductivity (W/(m K))	$0.098 \rightarrow 0.122$	$0.132 \rightarrow 0.105$
Liquid phase specific heat (kJ/(kg K))	$2.326 \rightarrow 1.82$	$1.585 \rightarrow 2.18$

Table 5.18 Comparison of feed/effluent heat exchanger options.

Item	Shell and tube exchanger option	Spiral plate exchanger option
Equipment size (shell diameter × straight tube length) (mm)	600×6000	2000×2320
Number of equipment (piece)	6	1
Pressure drop (kPa)	70	60
Overall heat transfer coefficient ($\text{W}/(\text{m}^2\,\text{K})$)	290	360
Heat transfer area (m^2)	650	480
Total weight of equipment (t)	30	48

As can be seen from Table 5.18, one spiral plate heat exchanger can replace six shell and tube heat exchangers. Although the weight of the equipment is slightly larger, it saves valuable floor area and prolongs the maintenance period.

There are more heat-sensitive materials in the propylene oxide process, and boiling and condensation are mostly carried out under vacuum. Corresponding heat transfer enhancement measures should be selected according to the characteristics of the process. For shell side condensation, limited by the allowable pressure drop of the vacuum system, the gas phase flow rate is low, the vapor–liquid interface shear force is small, and the condensate film is thick. Therefore, using low-finned tube for condensation on horizontal tubes and longitudinal groove tube for condensation on vertical tubes can help rapidly discharge the surface condensate, reduce the thickness of liquid film, and enhance heat transfer. Compared with baffle, rod baffle has low pressure drop, which is convenient for the discharge of noncondensing gas, and it is more suitable for vacuum condensation. For tube-side boiling, the static pressure difference affects the bubble point temperature under vacuum condition. The conventional thermosiphon reboiler not only has unfavorable heat transfer but also has high metal wall temperature and long liquid residence time, which is more likely to aggravate the decomposition and polymerization of heat-sensitive materials. Vertical falling-film reboiler [23] not only eliminates the influence of static pressure difference but also has high liquid film surface flow rate, small heat transfer thermal resistance, and short liquid residence time, which is especially suitable for the evaporation operation of heat-sensitive materials under vacuum. Table 5.19 shows the process conditions of reboiler in a tower, including shell–side condensation and tube-side evaporation.

Table 5.19 Process conditions of reboiler in a tower—Duty: 30.2 MW.

Item	Shell side	Tube side
Fluid name	Hot flow	Cold flow
Heat load (MW)	35	35
Allowable pressure drop (kPa)	3	2
Temperature (°C)	$100 \rightarrow 90$	$65 \rightarrow 68$
Gas phase density (kg/m^3)	1.41	0.387
Liquid phase density (kg/m^3)	$824.8 \rightarrow 802.2$	$894.4 \rightarrow 909.8$
Gas phase viscosity (mN s/m^2)	$0.008 \rightarrow 0.009$	0.008
Liquid phase viscosity (mN s/m^2)	$0.334 \rightarrow 0.32$	$0.563 \rightarrow 0.60$
Gas phase thermal conductivity (W/(m K))	0.018	0.0143
Liquid phase thermal conductivity (W/(m K))	0.112	0.123
Gas phase specific heat (kJ/(kg K))	1.55	1.41
Liquid phase specific heat (kJ/(kg K))	$2.086 \rightarrow 1.98$	$1.94 \rightarrow 1.97$

The comparison of thermalsiphon exchanger and rod baffle falling film heat exchanger is shown in Table 5.20. The comparison results show that the heat exchange area can be reduced by more than 20% and the polymerization loss of heat-sensitive materials can be improved at high temperatures by using the rod baffle falling film option.

Table 5.20 Comparison of reboiler options in a tower.

Item	Thermalsiphon option	Falling film option
Equipment size (shell diameter × straight tube length of heat exchange tube) (mm)	3600×5000	3000×7500
Specifications of heat exchange tube (outer diameter × wall thickness) (mm)	38×2.5	19×2
Number of equipment (piece)	2	2
Pressure drop (kPa)	20	2
Overall heat transfer coefficient (W/(m^2 K))	210	280
Heat transfer area (m^2)	5500	4000
Total weight of equipment (t)	200	150

5.4.5 Summary

The thermal energy of cooxidation propylene oxide process is characterized by large consumption of steam and cooling water, but the low-grade thermal energy and reaction heat in the unit are not easy to recover and utilize. Therefore, the reasonable arrangement of heat exchange network and the comprehensive application of energy-saving measures are important to improve the energy consumption level of the unit. The process characteristics determine the technical focus of heat transfer enhancement, which is to solve the problems in large-scale equipment caused by high vacuum and low temperature difference and to meet the safe operation requirements of heat-sensitive material. The heat transfer analysis shows that falling film reboiler is more suitable for the evaporation operation of heat-sensitive materials such as peroxide under vacuum condition, and rod baffle structure is beneficial to improving the shell-side condensation under vacuum condition, while spiral plate heat exchanger is suitable for heat transfer operation with highly crossed temperatures and coking tendency because of its pure countercurrent and self-cleaning characteristics.

5.5 Styrene unit

5.5.1 Brief description of process

Styrene (SM) is an important aromatic chemical raw material used in the production of polystyrene (PS), styrene rubber (SBR), styrene series resins, and other products. Styrene is mainly produced from benzene and ethylene or directly extracted from pyrolysis gasoline. There are three main processes for the production of styrene: ethylbenzene-styrene method, propylene oxide-styrene coproduction method, and styrene extraction method. This section mainly introduces the commonly used ethylbenzene-styrene process and styrene extraction process.

5.5.1.1 Ethylbenzene-styrene process

This method produces ethylbenzene from benzene and ethylene through alkylation reaction and then produces styrene from ethylbenzene through dehydrogenation reaction. It is the most important process route for styrene production at present. The typical process flow is shown in Fig. 5.17.

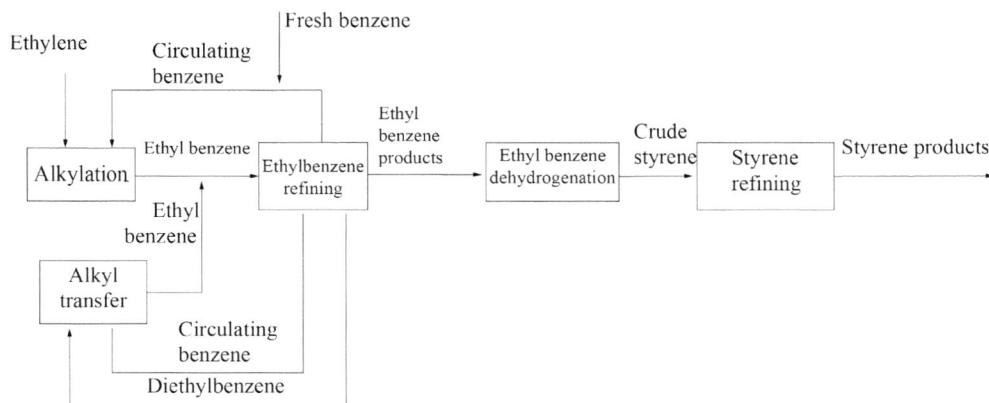

Fig. 5.17 Block diagram of ethylbenzene-styrene process.

5.5.1.2 Styrene extraction process

The by-product pyrolysis gasoline from steam cracking unit with naphtha and hydrogenation tail oil as the raw materials contains 3%–5% (mass fraction) styrene, which can be separated by extractive distillation. The typical process flow is shown in Fig. 5.18.

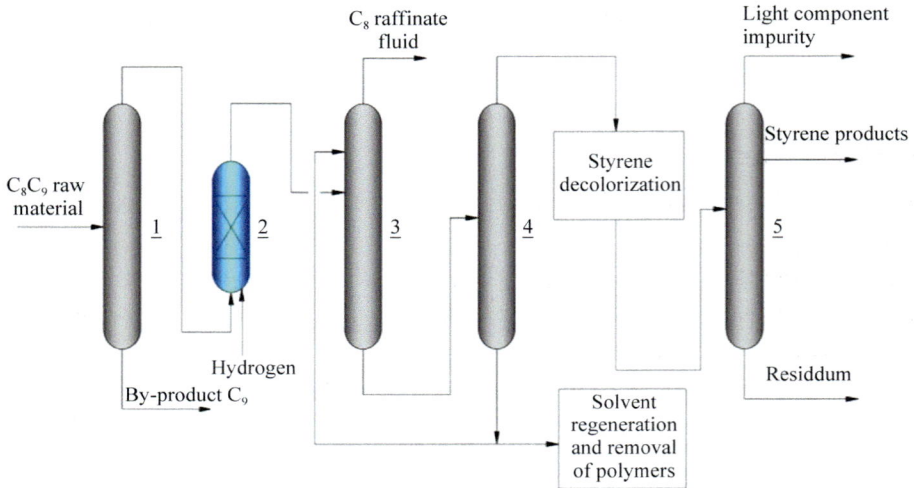

Fig. 5.18 Flow chart of styrene extraction process. (1) C_8 cutting tower; (2) phenylacetylene hydrogenation reactor; (3) extractive distillation column; (4) solvent recovery tower; (5) styrene finished product tower.

5.5.2 Analysis of characteristics of thermal energy

5.5.2.1 Ethylbenzene-styrene process [24–28]

1. Ethylbenzene

    The alkylation reaction of benzene and ethylene to ethylbenzene and polyethylbenzene is a strong exothermic reaction. The alkylation reaction mainly exports heat to the outside, and the reaction heat is recovered by generating steam and being used as the hot feed to benzene distillation tower. The heat source required for ethylbenzene distillation is provided by high-pressure steam, and the heat of overhead gas phase of each distillation tower is recovered by generating low-pressure steam.

2. Ethylbenzene dehydrogenation

    Ethylbenzene dehydrogenation to styrene is a strong endothermic reaction, and its energy consumption is the inherent energy consumption of ethylbenzene dehydrogenation to styrene process, which depends on the performance of dehydrogenation catalyst and is difficult to reduce and optimize through traditional heat coupling utilization.

3. Styrene distillation

The energy consumption of styrene distillation accounts for about 1/3 of that of the whole unit, and the heat recovery rate can reach 40%–50% by using multistage pressure swing heat integration technology.

5.5.2.2 Styrene extraction process

The self-polymerization of styrene material is a spontaneous reaction, which is aggravated with the increase of operating temperature and residence time of the material, so the operating temperature should be reduced as far as possible and the flow should be accelerated [29]. In order to reduce the polymerization of styrene material, the heat transfer temperature difference between the heat source and the material is as small as possible, and the material in the equipment needs to have a certain flow rate to avoid flow dead zone and local overheating. It is difficult to enhance heat transfer, optimize the type of heat transfer equipment and the design of internal structure, and maintain high heat transfer efficiency under low temperature difference [30].

The consumption of 0.6 MPa (gauge pressure) saturated low-pressure steam accounts for a large proportion of energy consumption (Table 5.21). It is mainly used as the heat source of reboiler for distillation and separation operation. The temperature of saturated steam under this pressure is about 170°C, which not only avoids overheating of materials but also ensures a certain heat transfer temperature difference. However, limited by the setting of steam grade in the whole plant, most of them are 1.0–1.6 MPa (gauge pressure) superheated steam, which must undergo temperature and pressure reduction saturation, resulting in energy waste. The energy consumption distribution of a typical styrene extraction unit is shown in Fig. 5.19. The energy consumption of the unit can be greatly reduced if it can rely on the advantages of the whole plant integration, replace high-grade steam with low-temperature heat source, and adopt efficient heat exchange equipment.

Table 5.21 Main energy consumption composition of styrene extraction unit—kgoe/(t raw material).

Cooling water	Electricity	0.6 MPa (gauge pressure) Steam	0.4 MPa (gauge pressure) low pressure steam	Total energy consumption
5.59	6.83	63.83	6.73	82.98

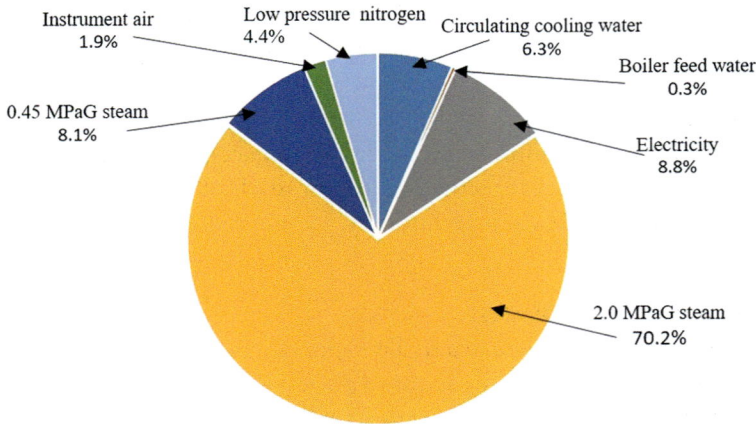

Fig. 5.19 Energy consumption distribution of a typical styrene extraction unit.

5.5.3 Process energy optimization and heat transfer network optimization

5.5.3.1 Ethylbenzene-styrene process

Through the thermal integration of heat exchange process, the reaction heat and heat sources at all levels of ethylbenzene-styrene unit can be rationally and fully utilized, and the energy consumption of the unit can be reduced. The high–pressure steam generator in the effluent exchanger series of styrene dehydrogenation reactor can provide part of the high–pressure steam required by the reboiler of each tower in ethylbenzene distillation system. At the same time, the steam generated in ethylbenzene reaction and distillation can be used for low pressure steam users in the unit. The rational utilization of different grades of steam avoids the loss of energy.

① In ethylbenzene part, heat can be recovered by enhancing heat transfer, heating the circulating benzene, and circulating the feed to benzene tower in addition to the generation of low–grade steam.

② The three-stage effluent exchanger of ethylbenzene dehydrogenation reactor is a shell and tube heat exchanger. The dehydrogenation reaction liquid goes through the tube side, the heat energy is fully recovered, and the heat loss is reduced [24].

③ Styrene distillation system adopts sequential separation, and the ethylene benzene is separated by two–stage pressure swing thermal integration technology. Two distillation towers with different operating pressures are used, and the material from the high–pressure tower overheads is condensed as the heating medium of the reboiler of the low-pressure tower, which improves the recovery rate of heat energy and reduces the steam consumption [24].

5.5.3.2 Styrene extraction process

1. Heat transfer process and heat transfer network pinch calculation

The heat of styrene extraction recycle solvent and steam condensate are rich and can be recovered by heat exchange. The recycle solvent of the solvent recovery column contains more heat, and the waste heat is recovered through a series of heat exchangers. The waste heat of steam condensate is used to generate stripping steam and then used to generate tracing hot water and tracing warm water, which can save about 0.8–1.2 MW heat, accounting for 4%–6% of the total energy consumption of the unit.

Pinch point analysis of the above process shows that the heat transfer pinch point is 83.8°C according to the combination curve of Fig. 5.20, and the operating temperature of all materials to be heated is not more than 140°C. The cooling of steam exhaust generated by steam condensate flashing violates the pinch point heat transfer principle. The improvement measure is to use the steam exhaust to heat the process material. However, due to the unstable steam exhaust and low temperature, the utilization effect is not good, and the energy-saving effect is not obvious.

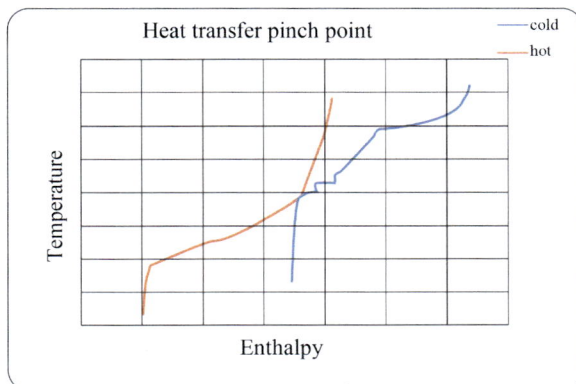

Fig. 5.20 Composite curve of heat for styrene extraction process.

2. Thermal coupling of extraction process

Thermal coupling extraction process is one of the energy-saving methods (Fig. 5.21). For the solvent extraction-desorption process, the existing experience shows that this technology can reduce the energy consumption of solvent desorption tower by about 25%, and the overall energy saving is about 4%.

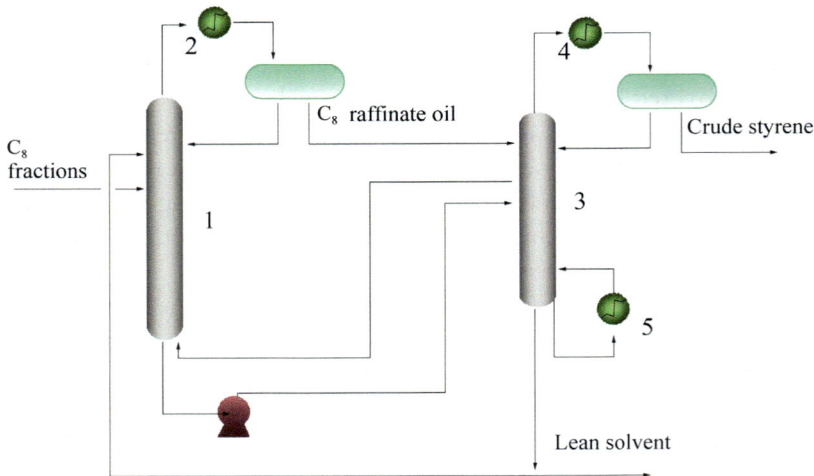

Fig. 5.21 Flow diagram of thermal coupling for extraction-solvent recovery tower system. (1) Extractive distillation column; (2) condenser; (3) solvent recovery tower; (4) condenser; (5) reboiler.

According to the calculation, the solvent extraction–desorption process of styrene extraction process adopts thermal coupling, and the solvent extraction column can save energy by 26.7%, accounting for about 7% of the total energy consumption of extraction-solvent recovery process. Its energy-saving effect is consistent with that of general extraction thermal coupling. However, this thermal coupling process increases the complexity of operation in engineering, and the difficulty of recovery and adjustment after operation fluctuation is higher than that of the normal process, producing adverse effects on stabilizing the purity of styrene products, and the overall energy-saving effect is not obvious, which needs further engineering research.

3. Heat transfer optimization design for hydrogenation of phenylacetylene

The phenylacetylene hydrogenation reactor involves the hydrogenation exothermic reaction and heat conduction, which requires the reaction to be as mild as possible. Phenylacetylene, which accounts for 0.5% (mass fraction) of the total material, is reacted via hydrogenation, and styrene, which accounts for 30%–50% (mass fraction) of the total material, does not undergo hydrogenation reaction, so the reaction temperature is low and the pressure is low with small heat release. It is necessary to reduce the temperature rise of the hydrogenation reaction and take away the heat released by the reaction.

From engineering point of view, with the selection of fixed bed hydrogenation reactor with lower inlet and upper outlet, the liquid feed and hydrogen enter the reactor from the bottom. The liquid phase in the reactor is continuous phase, and the hydrogen is dispersed phase. The hydrogen content on the catalyst surface is low, and it bubbles from bottom to top. A certain amount of nitrogen can be bubbled into the reactor to enhance the dispersion effect. On the one hand, the dispersion characteristics of the gas phase are taken into account. It differs from the liquid phase in flow rate, which is conducive to enhancing the mass transfer of material on the surface of catalysts to avoid excessive local reaction and to enhance the heat transfer on the surface of catalysts. On the other hand, the liquid phase has a large capacity and can absorb reaction heat, which is beneficial for reducing reaction temperature rise and inhibiting overhydrogenation reaction.

The heat transfer characteristics and equipment of different industrial processes of styrene vary greatly. It is necessary to pay attention to the easy polymerization of styrene and take measures to avoid the fouling of equipment and pipelines, which will affect their long-term operation. As for the ethylbenzene-styrene process route, through the effective use of alkylation reaction heat, the reasonable recovery of heat of the products from ethylbenzene dehydrogenation reaction and the optimization of heat transfer separation process of styrene distillation system, it can better achieve energy optimization and reduce operating costs. For styrene extraction process, the unit capacity is limited by the amount of pyrolysis gasoline, so the total energy consumption is low and the operating temperature is not high. Using its own heat to optimize heat transfer, the utilization effect is limited. If the OSBL low-temperature heat supply can be adopted, the efficient heat transfer technology with small heat transfer temperature difference can greatly save the energy consumption of the unit.

5.5.4 Heat transfer enhancement technology using elements

5.5.4.1 Heat transfer enhancement technology for ethylbenzene-styrene process

Dehydrogenation of ethylbenzene to styrene is a strong endothermic reaction under high temperature and low pressure. The reaction feed temperature is 620°C, and the reaction effluent temperature is 580°C. In order to recover the heat energy of the effluent, a series of heat exchangers are set up for heating the feed or generating by-product steam. The feed/effluent heat exchanger in high-temperature section is the most typical.

Taking a 720 kt/a ethylbenzene-styrene unit as an example, the process conditions of feed/effluent heat exchanger in high-temperature section are shown in Table 5.22. The cold and hot flows have the characteristics of high coincidence of temperature interval, large volume flow on both sides, and small allowable pressure drop. The rod baffle heat exchanger has small pressure drop and high heat transfer coefficient per unit pressure drop, which is suitable for low-pressure conditions and helps to enhance heat transfer under such conditions. The option comparison is shown in Table 5.23.

Table 5.22 Process conditions of feed/effluent heat exchanger in high-temperature section—Duty: 35.55 MW.

Item	Shell side	Tube side
Fluid name	Reaction feed	Reaction effluent
Flow rate (kg/h)	184,037	309,726
Allowable pressure drop (kPa)	7	3.5
Temperature (°C)	$230.8 \rightarrow 535.8$	$580.1 \rightarrow 404.2$
Gas phase density (kg/m^3)	$1.73 \rightarrow 0.882$	$0.144 \rightarrow 0.163$
Gas phase viscosity (mN s/m^2)	$0.013 \rightarrow 0.02$	$0.026 \rightarrow 0.021$
Gas phase thermal conductivity (W/(m K))	$0.0292 \rightarrow 0.063$	$0.08 \rightarrow 0.0592$
Gas phase specific heat (kJ/(kg K))	$1.96 \rightarrow 2.58$	$2.46 \rightarrow 2.25$

Table 5.23 Comparison of feed/effluent heat exchanger options in high-temperature section.

Item	Original option	Enhanced option
Equipment size (shell diameter × straight tube length) (mm)	$5000 \times 10,100$	$4000 \times 10,000$
Tube size (outer diameter × wall thickness) (mm)	50.8×2.11	32×2
Pressure drop (kPa)	9.5 (total)	10.5 (total)
Overall heat transfer coefficient (W/(m^2 K))	61.3	81.5
Heat transfer area (m^2)	6685	4871
Total weight of equipment (t)	220	$\sim$180

As can be seen from Table 5.23, the overall heat transfer coefficient of the rod baffle heat exchanger is increased by 33%, the heat transfer area is reduced by 27%, the weight of equipment is reduced by 18%, the equipment cost and site floor area are saved, and the heat transfer enhancement effect is obvious. The feed/effluent heat exchanger in high-temperature section is generally assembled into large equipment in series with a high-pressure steam generator and a feed vaporizer. The equipment operates under high temperature and high pressure, and the shell side flow temperature difference is more than 100°C, and the maximum pressure difference is more than 4.3 MPa; thus there is a big stress problem. Startup, shutdown, and unexpected operation fluctuations may cause damage to the equipment. Heat transfer enhancement can significantly reduce the size of equipment, which helps to reduce the difficulty of strength design.

The reboilers of polyethylbenzene tower bottom and low–pressure circulation tower bottom have the characteristics of high medium viscosity and easy scaling. The use of falling film evaporation structure can effectively enhance heat transfer and prolong the maintenance cycle.

5.5.4.2 Heat transfer enhancement technology for the reboiler in the refining tower of styrene extraction process

The vertical thermosiphon reboiler of styrene refining tower is easy to block and the operation cycle is short, which are the main problems faced by the styrene extraction process in the early stage of operation. The process conditions of the reboiler in the refining tower of a styrene extraction unit are shown in Table 5.24.

Table 5.24 Process conditions for reboiler in refining tower of styrene extraction unit—Duty: 1.091 MW.

Item	Shell side	Tube side
Fluid name	0.15 MPa (gauge pressure) steam	Styrene
Flow rate (kg/h)	1800	37,732
Allowable pressure drop (kPa)	3	Thermosiphon
Temperature (°C)	127.4	$85 \rightarrow 87.58$
Gas phase density (kg/m^3)	–	$\rightarrow 0.461$
Liquid phase density (kg/m^3)	–	$874 \rightarrow 876.9$
Gas phase viscosity (mN s/m^2)	–	$\rightarrow 0.008$
Liquid phase viscosity (mN s/m^2)	–	2.42
Gas phase thermal conductivity (W/(m K))	–	$\rightarrow 0.014$
Liquid phase thermal conductivity (W/(m K))	–	0.131
Gas phase specific heat (kJ/(kg K))	–	$\rightarrow 1.422$
Liquid phase specific heat (kJ/(kg K))	–	$1.89 \rightarrow 1.91$

Because styrene is prone to self-polymerization at high temperature, the refining tower is usually operated under vacuum. When vertical thermosiphon reboiler is used, the hydrostatic column in the heat exchange tube has an obvious influence on the temperature of the bubble point, and the length of sensible heat section can account for more than 30% of that of the heat exchange tube. The sensible heat section has not only a large thermal resistance but also a high metal wall temperature, and higher metal wall temperature will further aggravate the styrene self-polymerization and eventually lead to a rapid decline in the heat transfer effect of the reboiler. In extreme cases, the operating cycle of the reboiler is less than 24 h. The addition of polymer inhibitor can prolong the operation cycle, but it is easy to lead to an increase in product chromaticity and a decline in product quality.

Falling film evaporation is a process of evaporation of the liquid film on the inner surface of the heat exchange tube. The liquid film is thin, the thermal resistance is small, and the heat transfer coefficient is large. The falling film evaporation process eliminates the influence of hydrostatic pressure on the bubble point temperature, effectively utilizes the heat transfer temperature difference, and reduces the metal wall temperature. The falling film flow eliminates the liquid holdup in the heat exchange tube and shortens the residence time of fluid in the high temperature zone, which is beneficial to slowing down the styrene self-polymerization. Actual operation shows that the area of heat exchanger is reduced, the operation cycle is prolonged, and the product quality is improved by using falling–film evaporator. The comparison of vertical thermosiphon and falling film evaporation options in the reboiler is shown in Table 5.25.

Table 5.25 Comparison of reboiler options for refining tower in styrene extraction unit—Duty: 1.091 MW.

Item	Vertical thermosiphon option	Falling film evaporation option
Equipment size (shell diameter × straight tube length) (mm)	1000×2000	800×3500
Number of equipment (piece)	2 (1 standby)	1
Pressure drop (kPa)	Thermosiphon	2.5
Overall heat transfer coefficient (W/(m^2 K))	260	335
Heat transfer area (m^2)	118	89
Total weight of equipment (t)	9.8	3.5

As can be seen from Table 5.25, the falling-film evaporation option has obvious advantages, for it not only reduces the equipment weight by 64% but also reduces the spare equipment by prolonging the operation cycle, saving the investment, land occupation, and operation cost.

In order to improve the problem of styrene self-polymerization scaling, in addition to the above falling film evaporation options, Liu et al.[31] proposed to use internal siphon as a heating element in a vertical thermosiphon reboiler to enhance heat transfer and slow down the fouling growth in the tube. Wu et al.[32] proposed to introduce solid particles into the tube side of vertical thermosiphon reboiler to form a gas-liquid-solid multiphase flow system to enhance heat transfer. In addition, a low-pressure steam is introduced into the inlet of the reboiler to form a gas-liquid two-phase flow in advance in the heat exchange tube, which is helpful to reduce the length of the supercooled section and improve the overall heat transfer coefficient. The practice of introducing low-pressure steam has been applied in styrene extraction unit with a good effect on heat transfer enhancement, but it has a high requirement on the design of steam inlet distributor.

5.5.5 Summary

Styrene is a heat-sensitive material prone to scaling. Selecting falling film evaporator for reboiler can enhance heat transfer, reduce material scaling, and effectively extend the operation cycle. The reaction heat and condensation heat of overhead gas are recovered, and different grades of steam are generated. According to the heat transfer characteristics, appropriate heat transfer elements are selected to enhance heat transfer.

5.6 C$_5$ separation unit

5.6.1 Brief description of process

Cracked C$_5$ is a by-product of ethylene plant, and its main components are isoprene (IP), cyclopentadiene (CPD), and piperylene. These diolefins can be synthesized into many important high–value–added products and are valuable resources for the chemical industry [33,34]. The separation process of cracked C$_5$ fractions mainly includes the complete separation process for separating IP and the simple separation process for separating CPD [35–44]. This section focuses on the typical DMF extractive distillation process and heat transfer enhancement technology.

DMF method was first successfully developed by Zeon Corporation (Japan) [45,46] in 1971. Sinopec's DMF separation technology for cracked C$_5$ was jointly developed by Sinopec Engineering Incorporation (hereinafter referred to as SEI), Sinopec Beijing Research Institute of Chemical Industry, and Sinopec Shanghai Petrochemical Co., LTD., and has formed a proven packaged technology. Sinopec's DMF separation technology for cracked C$_5$ has been developed and improved continuously, such as adding front-end acetylene removal unit and liquid feed extractive distillation process [47,48]. By these, the treating capacity is greatly improved, the operation cycle is further prolonged, the unit consumption of solvent is continuously reduced, and the product quality is steadily improved. The process flow of typical DMF C$_5$ separation process is shown in Fig. 5.22.

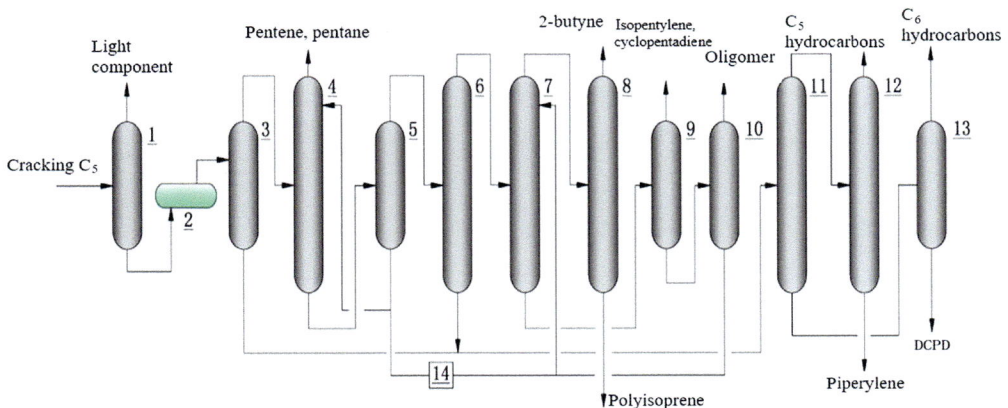

Fig. 5.22 Process flow diagram of typical DMF C$_5$ separation. (1) Predelight fractionator; (2) dimerization reactor; (3) predeheavy fractionator; (4) 1st extractive distillation tower; (5) 1st stripper; (6) deheavy fractionator; (7) 2nd extractive distillation tower; (8) delight fractionator; (9) 2nd stripper; (10) 3rd stripper; (11) De-C$_5$ tower; (12) piperylene tower; (13) dicyclopentadiene (DCPD) tower; (14) solvent recovery.

5.6.2 Analysis of characteristics of thermal energy

The operating temperature and pressure of the whole C_5 separation unit are relatively mild. The main feature of the unit is the easy polymerization of diolefin, which affects the long-term operation of the unit. The factors affecting diolefin polymerization mainly include temperature, oxygen content, and impurity content. High system temperature can greatly improve the polymerization rate, resulting in the formation of rubber-like polymer in the extraction system, which will clog the equipment and affect the separation efficiency of trays and the heat transfer effect of heat exchangers, thus affecting the long-term operation of the unit. The highest temperature in the DMF C_5 separation unit is the bottom temperature of the stripper. In order to ensure the complete removal of all hydrocarbons in the solvent, the operating temperature must be the boiling point of the solvent under the pressure, and medium-pressure steam is used as the heat source. The second highest operating temperature in the unit is the bottom temperature of the 1st and 2nd extractive distillation towers and the 3rd stripper. The operating temperature is about 105–115°C, and 0.4–0.6 MPa (gauge pressure) low-pressure steam is used as the heat source. Diolefin polymerization reaction easily occurs in the bottom of the 1st and 2nd extractive distillation towers. The deheavy fractionator and delight fractionator have the lowest bottom temperature, and low-pressure steam or the self-produced condensate of the unit is used as the heat source. In order to reduce material polymerization as far as possible, according to the operating temperature of the materials, the heating steam of the reboiler in the unit is divided into two grades, i.e., medium pressure and low pressure.

The energy consumption of DMF C_5 separation unit is mainly composed of steam, cooling water, and electricity. Table 5.26 shows the energy consumption of typical DMF C_5 separation unit.

Table 5.26 Energy consumption of typical DMF C_5 separation unit.

Item	Cooling water	Electricity	1.0 MPa (gauge pressure) steam	0.4 MPa (gauge pressure) steam	Total energy consumption
Energy consumption (kgoe/t)	16	20.8	37.6	130	204.4
Proportion (%)	7.8	10.2	18.4	63.6	100

It can be seen from Table 5.26 that in order to reduce the energy consumption of the unit, the consumption of steam, especially low-pressure steam, shall be reduced. Firstly, the process operating conditions shall be optimized. The reflux ratio shall be reduced to reduce the heat load of the reboiler of tower bottom and to reduce the steam consumption. Secondly, the optimization of energy shall be performed. The reuse of high-grade heat source in the unit and the reasonable distribution of heat sources can also reduce steam consumption, thus achieving the purpose of energy saving and consumption reduction.

The thermal energy reuse of C_5 separation includes the heat utilization of circulating solvent and the heat utilization of condensate. The difference between the beginning and ending temperatures of circulating solvent in DMF C_5 separation unit is large, and the heat can be used step by step. The heat utilization of condensate is to make full use of the by-product condensate from the unit as the heat source of the reboiler of distillation tower.

5.6.3 Process energy optimization and heat transfer network optimization

1. Heat transfer enhancement in thermal dimerization reactor

SEI optimized the structure of reactors and other key equipment, including the thermal dimerization reactor, by using software such as CFD, and applied it to the C_5 separation unit of Sinopec [49]. One of the existing thermal dimerization reactors is a horizontal tank with a partition plate. The material is easy to take shortcuts and the mixing during material flow causes the unsatisfactory polymerization effect of some CPD. The other is a hollow reactor, and the problem is that the reaction flow pattern cannot be well controlled, easily leading to complete mixing flow. In order to maintain a high CPD conversion rate, the existing reactor has a large volume and a large floor area and produces a large amount of impurities, which directly affects the economic benefits of the unit.

In view of the shortcomings of the traditional type of reactor, the reactor with improved structure is equipped with a multilattice distributor, and the distributor is designed with one layer or multiple layers. After adding the distributor, the material can flow uniformly and smoothly, and back mixing can be reduced, thus improving the thermal dimerization effect, increasing the conversion rate, and reducing the content of CPD at the reactor outlet. The comparative results before and after the application of the improved structure are shown in Table 5.27.

Table 5.27 Comparison between improved thermal dimerization reactor and conventional reactor.

Item		Unit I	Unit II
C_5 raw material	CPD content, % (mass fraction)	19.8	20.0
	DCPD content, % (mass fraction)	2.0	1.7
	IP content, % (mass fraction)	19.5	20.0
	PD content, % (mass fraction)	16.5	15.9
Reactor outlet	CPD content, % (mass fraction)	4.0	2.8
	X_3 content, % (mass fraction)[a]	1.2	0.6

Continued

Table 5.27   Comparison between improved thermal dimerization reactor and conventional reactor.

Item		Unit I	Unit II
Conversion rate	CPD, % (mass fraction)	82.1	88.2
	IP, % (mass fraction)	2.8	1.5
DCPD yield, % (mass fraction)		79.8	86.8
CPD selectivity, % (mass fraction)		97.3	98.7
Total yield of unit	DCPD, % (mass fraction)	89.0	> 93.0
	IP, % (mass fraction)	90.0	92.0
CPD content in IP (mass fraction)		$\leqslant 3 \times 10^{-6}$	$\leqslant 1 \times 10^{-6}$

[a]X_3 is the copolymer of CPD and IP.

2. Reactive distillation heat transfer enhancement

Neither the traditional structure nor the improved structure fully utilizes the reaction heat released by thermal dimerization. If this reaction heat can be fully utilized, the energy consumption of the unit can be reduced, and reactive distillation can achieve this purpose. In the 1990s, Hu et al.[50] from Sinopec Beijing Research Institute of Chemical Industry proposed the application of reactive distillation technology in the separation of C_5 fractions. Cheng et al.[51] from Sinopec Beijing Research Institute of Chemical Industry simulated the reactive distillation process by using simulation software and proposed process optimization conditions. Ma et al.[52] from Sinopec Engineering Incorporation (SEI) proposed a reactive distillation column for C_5 separation, as shown in Fig. 5.23.

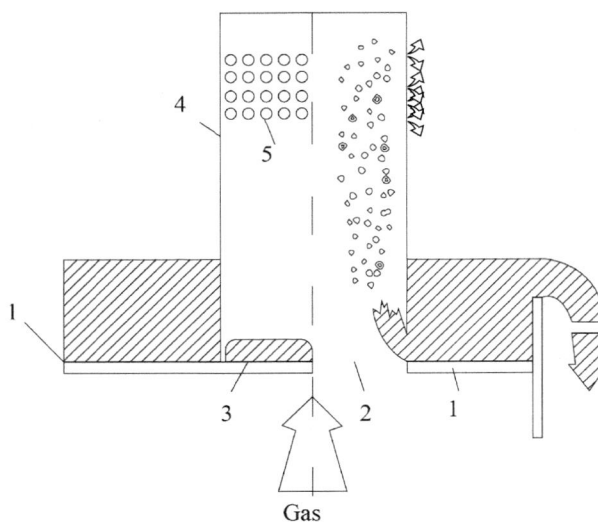

Fig. 5.23 Structure diagram of special tray developed by SEI. (1) Tower plate; (2) plate hole; (3) cover body bottom gap; (4) hood; (5) cover holes.

After using reactive distillation, the reactive distillation process has simple process flow and few equipment, while the process with thermal dimerization has more equipment and the heat released by thermal dimerization is cooled by cooling water, resulting in exergy waste. The reactive distillation technology jointly developed by SEI et al. eliminates the traditional thermal dimerization reactor and makes full use of the reaction heat to enhance heat transfer and improve the conversion rate of the main reaction, which has been proved feasible in practice [53,54]. The IP loss rate of the preseparation unit is reduced from 2%–3% to about 1%. The recovery rate of IP and CPD in the whole unit is increased by more than 2% compared with the traditional technology, and the energy consumption is saved by 5%.

3. Circulating solvent heat transfer enhancement

The bottom temperature of the stripper in C_5 separation unit is 160–165°C, but the temperature of solvents entering the extractive distillation tower is required to be 50–55°C. There is a large temperature difference between them. If this temperature difference is not fully utilized, not only will a large amount of heat be lost, but also a large amount of water will be consumed, which is extremely uneconomical. There are many ways of recycling solvent heat, but most of them have problems such as cross–pinch heat transfer and unreasonable energy utilization. Combined with many years of technological development and engineering experience, SEI has optimized and improved it, and the following advantages have been improved: ① Energy cascade utilization of thermal solvents is achieved, saving steam consumption by 5%–10%. ② Due to the energy supplement of the intermediate reboiler, the gas-liquid phase load of the tower, especially the lower section, is reduced, the gas phase load is reduced by 10%–30%, and the liquid phase load is reduced by 1%–10%. For the tower dominated by gas phase load, the tower diameter is reduced, and the investment is saved [55].

4. Steam condensate system heat transfer enhancement

As shown in Fig. 5.24, the stripper and solvent refining tower in the unit have a high bottom temperature (160–165°C) and uses medium-pressure steam as heat

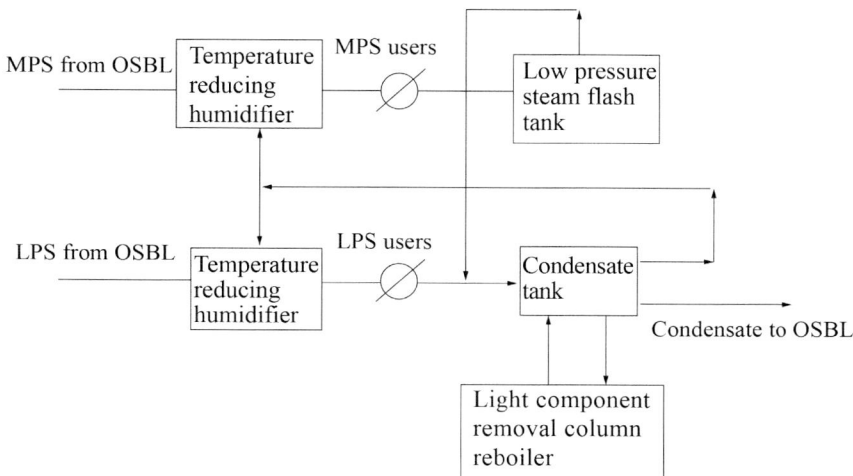

Fig. 5.24 Steam balance diagram of C_5 separation unit.

source, while other steam users use low-pressure steam. Through the self-balance of condensate in the unit, the full utilization of heat is achieved. Taking a 150 kt/a C_5 separation unit as an example, using condensate as heat source in the delight fractionator can save 4 t/h of low-pressure steam.

5. Solvent regeneration system heat transfer enhancement

The solvent of C_5 separation unit by DMF method is recycled. As time goes by, dimers, tar, and chemical polymerization inhibitors are easily enriched in the circulating solvent, so it is necessary to continuously extract small streams of solvent to send to the regeneration system for impurity removal and concentration. The main problem of the traditional solvent recovery process is that the heavy gum and black residue in the circulating solvent accumulate continuously during evaporation in solvent regenerator. When DMF is difficult to steam out, the regenerator is switched. As the gum in the regenerator is thick and flows slowly, it stays on the outer wall of the heat exchange tube, making the heat transfer effect worse. At the same time, the tar cannot be discharged completely during slag removal. The solids formed after cooling adhere to the heat exchange tube of the regenerator and cannot be completely flushed clean even by high-pressure water gun, and special manual cleaning is required so that the foul tar odor is discharged into the atmosphere. On the one hand, frequent cleaning increases the labor intensity of workers and affects the service life of regenerator, thus affecting the economic benefits of the unit. On the other hand, foul-smelling polymer solvent gum pollutes the environment, fails to meet relevant environmental laws and regulations, and affects the health of workers.

The improved solvent recovery process [56,57] achieved the following features:

① The heater is separated from the regenerator and solvent circulating pump is added to enhance the heat transfer effect, reduce the heat transfer area, and save investment.

② The whole process can be closed and continuous, avoiding the exposure of the polymer to air and reducing the emission of pollutants.

③ The indirect stirring of the circulating pump enhances the flow of materials in the flash tank, achieves a better separation of solvent from tar, reduces the possibility of tar accumulation, improves the economic benefits of the unit, and reduces the operation intensity.

The improved process is suitable not only for C_5 separation unit by DMF but also for extraction and separation unit using DMF as solvent.

5.6.4 Heat transfer enhancement technology using elements

Isoprene, piperylene, and cyclopentadiene have high heat sensitivity and the tendency of self-polymerization or copolymerization. This not only reduces the product yield but also increases the heat and mass transfer resistance due to the accumulation of polymers in the system, leading to a decrease in unit capacity or even unit shutdown. Therefore, the heat transfer enhancement of C_5 extraction separation unit must adapt to the easy scaling characteristics of materials and meet the requirements of long-term operation of the unit at full load.

5.6.4.1 Heat transfer enhancement method of reboiler

Taking a 150 kt/a C_5 extraction unit as an example, the process conditions of the reboiler in predeheavy fractionator are shown in Table 5.28. Conventional vertical thermosiphon reboiler is traditionally used. Due to the influence of static pressure head, the inlet of heat exchange tube of the vertical thermosiphon reboiler under vacuum operation usually has a long "supercooled" section, with high metal wall temperature, low liquid flow rate, and long residence time, which aggravate the polymerization of heat-sensitive substances in the material and worsen the heat transfer process. Because the medium is easy to scale and the operation cycle is short, it is generally designed as one in operation and one standby. After improvement, vertical falling film reboiler is selected, which can not only use lower-grade steam as heat source but also achieve the purposes of reducing metal wall temperature, slowing downscaling trend, improving product quality, and prolonging operation cycle, and no spare equipment is needed. The comparison option is shown in Table 5.29.

Table 5.28 Process conditions of reboiler in predeheavy fractionator—Duty: 8.33 MW.

Item	Shell side	Tube side
Fluid name	0.4 MPa steam	C_5
Flow rate (kg/h)	14,030	636,770
Allowable pressure drop (kPa)	3	Thermosiphon
Temperature (°C)	143.6	$101.4 \rightarrow 104.6$
Gas phase density (kg/m³)	–	8.34
Liquid phase density (kg/m³)	–	$683.4 \rightarrow 693.7$
Gas phase viscosity (mN s/m²)	–	0.0091
Liquid phase viscosity (mN s/m²)	–	0.178
Gas phase thermal conductivity (W/(m K))	–	0.0207
Liquid phase thermal conductivity (W/(m K))	–	0.101
Gas phase specific heat (kJ/(kg K))	–	1.767
Liquid phase specific heat (kJ/(kg K))	–	2.17

Table 5.29 Comparison of predeheavy fractionator bottom reboiler option.

Item	Original option	Enhanced option
Equipment size (shell diameter × straight tube length) (mm)	1300 × 4500	1300 × 4500
Specifications of heat exchange tube (outer diameter × thickness) (mm)	25 × 2.5	38 × 2.5
Number of equipment (piece)	2 (one in operation and one standby)	1
Pressure drop (kPa)	Thermosiphon	3.0
Overall heat transfer coefficient (W/(m^2 K))	540	766
Heat transfer area (m^2)	466	263
Total weight of equipment (t)	25.5	8.9

It can be seen that the falling film evaporation option can increase the heat transfer coefficient of the reboiler by 40%, reduce one piece of equipment, and reduce the total weight by 65%. The heat transfer enhancement effect is obvious, and the easy scaling problem of reboiler and the decomposition problem of dicyclopentadiene are solved.

The reboiler of the dicyclopentadiene tower is operated under vacuum, and vertical thermosiphon reboiler is generally selected. Vacuum reboiler is extremely sensitive to installation height. Selecting vertical falling film evaporator is helpful to solve these problems.

5.6.4.2 Heat transfer equipment related to lean solvent heat transfer

According to the temperature level, the hot solvent from the bottom of stripper is successively used as the heating source of the reboiler of the first extraction tower and the second extraction tower, the intermediate reboiler of the first extraction tower and the second extraction tower, and the feed preheater of the first extraction tower and the second extraction tower. As the heat source of reboiler, the heat transfer temperature difference of the subcooled section of cold flow is increased and the length of the subcooled section of the flow is reduced when the solvent and material flow in parallel.

In the stripping process, a large amount of polymer is dissolved in the solvent. When the solvent temperature is low, the polymer may be precipitated from the solvent, reducing the heat transfer effect of the heat exchanger and shortening the operation cycle of the unit. The solvent of the feed preheater of the first extraction tower and second extraction tower goes to the shell side, and the polymerization products are more likely to accumulate in the shell side due to the more dead zones on the shell side. Vertical cut in baffle helps to reduce the accumulation of polymerization products in the heat exchanger.

Heavy components in solvent should be removed before solvent refining and enter first into the solvent refining kettle. Vacuum evaporation is adopted and the heat exchanger adopts kettle evaporator. There is a large gap between the tube bundle and the shell of the solvent refining kettle, and the solvent is not easy to be removed at the gap. The oval tube bundle [58] using Sinopec patented technology is helpful to reduce heat transfer dead zone and timely discharge the bottom-heavy components, to achieve continuous operation of solvent refining kettle and prolong the operation cycle of heat exchanger.

5.6.5 Summary

The IP loss rate of the preseparation unit in a 150 kt/a C_5 separation unit was reduced from 2%–3% to about 1% by using reactive distillation technology. The recovery rate of IP and CPD in the whole unit was increased by more than 2% compared with the traditional technology, and the energy consumption was reduced by 5%. Through the optimization and improvement of solvent heat transfer process, the energy cascade utilization of thermal solvent was achieved, and the steam consumption was saved by 5%–10%. At the same time, due to the energy supplement of the intermediate reboiler, the vapor-liquid phase load of the tower, especially in the lower section, was reduced. The gas phase load was reduced by 10%–30%, and the liquid phase load was reduced by 1%–10%. For the gas-phase load-dominated tower, the tower diameter was reduced accordingly, and the investment was saved. Through the self-balance of the condensate in unit, the heat was fully utilized. Using condensate as heat source in the delight fractionator saved 4 t/h of low-pressure steam. Through the optimization and improvement of the solvent regeneration process, the heat transfer effect was enhanced, the operation cycle of the unit was prolonged, the labor intensity of the workers was reduced, and the economic benefit of the unit was improved.

The materials of C_5 separation unit are easy to polymerize and foul. These factors should be fully considered in the design of heat exchanger. Both theory and practice show that falling film evaporator is suitable for the occasion with evaporation of heat-sensitive medium.

5.7 Butyl rubber unit

5.7.1 Brief description of process

With excellent air tightness, butyl rubber is mainly used to make the inner tube of various tires and the inner liner of tubeless tires. It is an important material for civil and military use. It is the copolymerization product of isobutene and a small amount of isoprene at a low temperature of $-100°C$, and its typical process is the slurry process with aluminum chloride and water as the coinitiators and methyl chloride as the solvent.

Slurry process mainly includes six units, i.e., refrigeration unit, raw material refining unit, raw material and initiator preparation unit, reaction and stripping unit, methyl chloride recovery unit, and rubber particle storage and posttreatment unit. Its flow diagram is shown in Fig. 5.25.

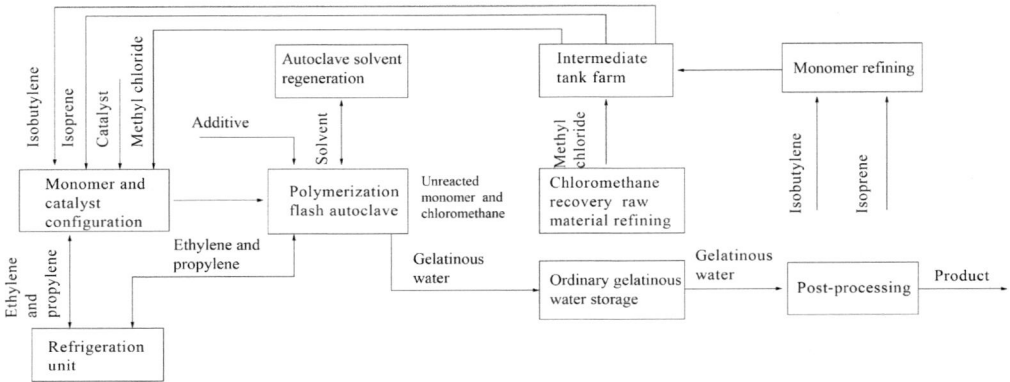

Fig. 5.25 Block diagram of butyl rubber unit.

5.7.2 Analysis of characteristics of thermal energy

5.7.2.1 Analysis of characteristics of cold energy utilization

Butyl rubber polymerization is carried out at a low temperature of $-90°C$. The refrigeration energy consumption accounts for more than 40% of the total energy consumption. The rational utilization of ethylene and propylene cold energy in the unit has become the focus of attention. Typical butyl rubber units use propylene and ethylene cascade refrigeration, and the high-grade cold energy released by LNG gasification can be used for the cooling of butyl rubber unit [59].

1. Characteristics of propylene cold energy utilization

   Propylene compression is divided into two stages. The $-45°C$ atmospheric pressure propylene gas from the secondary user is compressed in the first stage and then mixed with the propylene gas from the secondary inlet tank and compressed to 2.0 MPaA in the second stage. High-pressure propylene is cooled by cooling water, condensed to obtain liquid propylene, and collected in the liquid propylene collection tank. Liquid propylene provides cooling capacity through depressurized evaporation, and its main evaporation temperature is divided into two levels: $-8°C$ and $-45°C$. $-8°C$ propylene is mainly used to produce low-temperature salt water of $0-8°C$, and $-45°C$ propylene is mainly used to cool the reaction materials and initiators to about $-40°C$, which are used to condense high-pressure ethylene refrigerant.

In order to reduce the load of the first stage of propylene compressor, the high-pressure liquid propylene is first flashed to $-8°C$ in the economizer, and the flash vapor is sent to the second stage of the compressor. Then the $-8°C$ liquid propylene is flashed to produce $-45°C$ propylene. The use of economizer reduces the load of the low-pressure section of propylene compressor and saves the energy consumption for propylene compression.

2. Characteristics of ethylene cold energy utilization

Ethylene compression is divided into two stages. The negative pressure ethylene gas is compressed in the first stage and then mixed with ethylene gas from the second stage inlet tank and compressed to 1.9 MPaA in the second stage. The high-pressure hot ethylene gas is cooled by cooling water, cooled by $-8°C$ propylene, condensed by $-45°C$ propylene, and then collected in the liquid ethylene collection tank. Liquid ethylene provides cooling capacity through depressurized evaporation, and its main evaporation temperature is divided into three levels: $-75°C$, $-101°C$, and $-115°C$. $-75°C$ ethylene is generated by economizer, and then $-101°C$ ethylene and $-115°C$ ethylene are generated from $-75°C$ ethylene. $-101°C$ ethylene is mainly used to cool the reaction materials and initiators to about $-98°C$. $-115°C$ ethylene is mainly used for polymerization heat removal. Through the cascade utilization of ethylene cold energy, reasonable allocation of the load of each section of ethylene compressor can be achieved, so as to reduce energy consumption.

In order to reduce the load of the first stage of ethylene compressor, the high-pressure liquid ethylene is first flashed to $-75°C$ in economizer, and the flash vapor is sent to the second stage of the compressor. Then the $-75°C$ liquid ethylene is flashed to generate $-101°C$ ethylene and $-115°C$ ethylene, respectively. The use of economizer reduces the load of the low-pressure section of ethylene compressor and saves the energy consumption for ethylene compression.

Primary flashing of liquid ethylene is carried out to provide cooling capacity for primary users, and the operating temperature is about $-70°C$. The flashed ethylene gas enters the second stage of ethylene compressor. Secondary flashing of the liquid ethylene at the bottom of primary flash tank is carried out to provide cooling capacity for secondary users, and the operating temperature is about $-101°C$ to $-120°C$. The flashed ethylene gas enters the first stage of ethylene compressor.

Butyl rubber unit is usually designed with four kinds of cooling media, namely cooling water, low-temperature salt water, propylene refrigeration, and ethylene refrigeration. The graded utilization of refrigerant is conducive to energy saving and consumption reduction of the unit.

5.7.2.2 Analysis of characteristics of heat energy utilization

In the case that large compressors are driven by electricity, the heat energy of the unit is mainly used in stripping, methyl chloride recovery and raw material refining, and post-treatment. The characteristics of each section are as follows:

The stripping section is a major heat energy user, which directly injects water vapor to disperse the $-100°C$ low-temperature slurry into two parts: the rubber particle water and the mixture of methyl chloride and unreacted monomer. Because the slurry composition is relatively stable during production, the heat energy consumption of this section is basically linear with the yield.

Methyl chloride recovery and raw material refining section uses high-pressure steam, medium-pressure steam, low-pressure steam, and circulating hot water as the heat source according to the operating temperature of materials. High-pressure steam (3.5 MPa, gauge pressure) is used to heat the material with an operating temperature above $250°C$. The steam condensate is first used to flash medium-pressure steam (1.0 MPa, gauge pressure) and then to produce circulating hot water. Medium-pressure steam is mainly used for tower reboiler heating. The steam condensate is first used to flash low-pressure steam (0.35 MPa, gauge pressure) and then to produce circulating hot water. Low-pressure steam is used to heat the tower reboiler with a low operating temperature, and the steam condensate is used to produce circulating hot water. The circulating hot water is mainly used for the heating and tracing of heat-sensitive material reboiler. Through cascade utilization of heat energy, the steam condensate of the unit can be cooled to below $90°C$ and sent out of the unit, effectively utilizing heat energy and reducing steam consumption.

In the posttreatment section, medium-pressure steam is mainly used to heat the air, and the steam condensate is first used to flash low-pressure steam and then to produce circulating hot water.

5.7.3 Process energy optimization and heat transfer network optimization

5.7.3.1 Cold energy system integration

Liquid ethylene provides cooling capacity through depressurized evaporation, and its main evaporation temperature is divided into three levels: $-75°C$, $-101°C$, and $-115°C$. The $-75°C$ ethylene enters the second stage of ethylene compressor, and the depressurized $-101°C$ ethylene enters the first stage of ethylene compressor together with $-115°C$ ethylene. In the traditional process, there is an unreasonable process of first depressurizing the $-101°C$ ethylene to $-115°C$ ethylene and then compressing, which leads to high energy consumption of the ethylene compressor. The new ethylene refrigeration process divides the ethylene compressor into three sections, and its process flow is shown in Fig. 5.26.

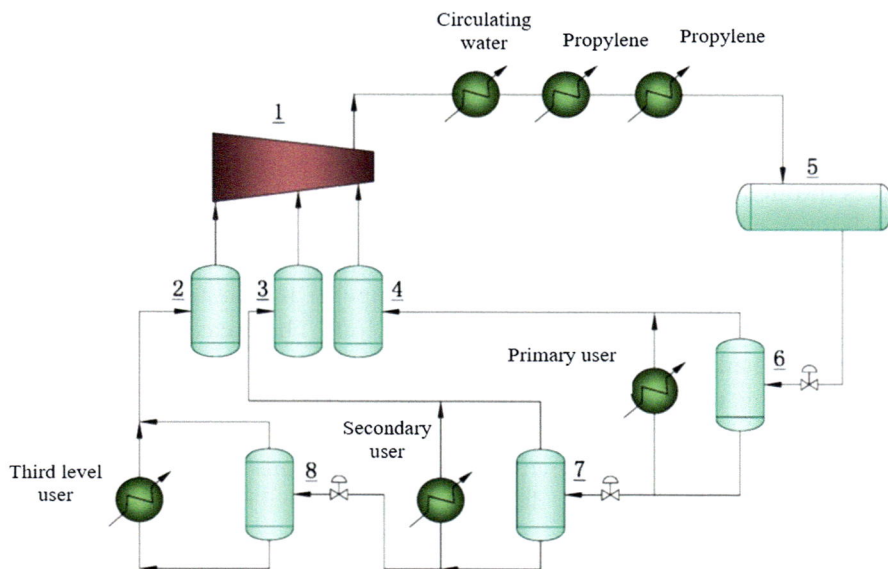

Fig. 5.26 Process flow diagram of new ethylene refrigeration. (1) Ethylene compressor; (2) first-stage inlet tank; (3) second-stage inlet tank; (4) third-stage inlet tank; (5) liquid ethylene collection tank; (6) primary flash tank; (7) secondary flash tank; (8) third-stage flash tank.

Through the above heat transfer process optimization, the unreasonable depressurization process is avoided, and the energy consumption of the compressor is reduced by about 5%–8%. By dividing the ethylene compressor into three sections, the use of negative pressure pipeline is reduced and the investment is saved.

5.7.3.2 Polymerization heat removal process enhancement

Butyl rubber cation polymerization reaction is characterized by rapid initiation, rapid growth, and easy transfer. It is difficult to terminate. Since the chain transfer activation energy is higher than the chain growth activation energy, reducing the reaction temperature is helpful to reduce the chain transfer reaction speed and to obtain polymers with higher molecular weight. The key to ensuring a good polymerization reaction is efficient mass and heat transfer, removal of reaction heat from the reactor, and maintenance of uniform reaction temperature.

There are mainly two types of industrial butyl rubber polymerizer, i.e., axial flow draft tube type and multilayer stirring type [60].

Axial flow agitator has the function of pumping so that the reactant material and slurry circulate in the kettle along the route of "draft tube → kettle bottom → heat exchange tube → kettle top → draft tube." Through the large circulation, the material can fully transfer heat through the tubes, and the gap between tubes can provide cold energy by evaporation of ethylene under negative pressure. The key to ensuring the effect of axial flow draft tube polymerizer is as follows: the conveying capacity of axial flow pump is sufficient to ensure the number of material circulation, the draft tube is set reasonably to avoid backmixing at the top and bottom of the polymerizer and avoid the formation of local dead zone, and the distribution of heat exchange tube is reasonable so that the material passes through the tubes evenly.

The built-in up and down through-shaft multiple impeller of multilayer stirring polymerizer can provide strong stirring for materials. Six groups of inner cool tube bundles are set evenly along the radial direction between the impeller and the tank wall, and ethylene evaporation on the tube side provides cold energy [60]. The mixed feed enters from the lower part of the tank body, and the initiator enters from the side of the tank body. Under high-speed stirring, the material is strongly turbulent. The mass transfer and heat transfer are sufficient, and the polymerization is stable. The material in the polymerizer has high turbulence degree, sufficient heat transfer, multiple heat transfer tube bundles, large heat transfer area, and small temperature difference of the material in the kettle, which is conducive to stable production.

No matter which type of polymerizer is used, the heat is transferred through tubes, so enhancing the uniformity and turbulence of the material flowing through the tubes is the key to enhancing heat transfer. In recent years, the optimization has been mainly focused on baffle and impeller.

5.7.4 Heat transfer enhancement technology using elements
5.7.4.1 Heat transfer enhancement of polymerization reactor
Butyl rubber polymerization conditions are extremely harsh. If the operation temperature is too high or the local temperature is not uniform, rubber particles will soon stick to the heat transfer surface, leading to a decrease in heat removal capacity and the deterioration of product quality. Therefore, butyl rubber polymerization reactor should have the characteristics of rapid heat transfer and mixing [61] in order to maintain a uniform reaction temperature and achieve the purposes of prolonging the operation cycle of reactor, increasing the molecular weight of polymers and reducing the distribution width of products.

As the core equipment of butyl rubber plant, polymerization reactor is the largest user of the highest-grade refrigerant. Enhancing heat transfer and prolonging the operation cycle of polymerizer not only reduce the material consumption and energy consumption of the unit but also have important significance for improving the production capacity.
1. Heat transfer enhancement of axial flow draft tube polymerizer

The length-to-diameter ratio of the reactor is generally between 3 and 5. Increasing the flow rate is helpful to improve the heat transfer coefficient in the tube, improve the shear force between the fluid and the tube wall, reduce the tendency of rubber sticking on the wall of heat exchange tube, and help to prolong the operation time of the polymerizer. It is an effective means to enhance heat transfer. Taking the axial flow draft tube polymerizer of a 10 kt/a butyl rubber plant as an example, the evaporation of $-115°C$ ethylene at 6638 kg/h outside the tube is the control item of heat transfer. Inside the tube, there is about 900,000 kg/h of $-95°C$ slurry. The effective means of heat transfer enhancement are low-finned tubes and external porous high flux tubes. The comparison results are shown in Table 5.30.

Table 5.30 Scheme comparison of axial flow draft tube polymerizer.

Item	Plain tube option	Low-finned tube option	High flux option
Equipment size (shell diameter × straight tube length) (mm)	1300×6000	1100×6000	1100×4500
Film heat transfer coefficient (W/(m² K))	606	1305	2350
Overall heat transfer coefficient (W/(m² K))	305	466	605

It can be seen from Table 5.30 that the overall heat transfer coefficient can be increased by 98% when high flux tube is used instead of plain tube. Using low-finned tube can increase the overall heat transfer coefficient by 50%.

2. Heat transfer enhancement of multilayer stirring polymerizer

Multilayer stirring polymerizer can provide more sufficient agitation to the flow, promote the turbulence degree of the flow, and enhance the heat transfer outside the tube. The heat transfer control item is outside the tube of the multilayer stirring polymerization kettle, and there is a large flow and heat transfer dead zone between the tube bundles, through which it is easy to lead to colloidal particle attachment. Spiral baffle is helpful to reduce the flow dead zone and enhance the heat transfer outside the tube bundle.

5.7.4.2 Heat transfer enhancement of heat exchangers

1. Heat transfer enhancement of heat exchangers in refrigeration unit

In the use of cooling water and chilled water for ethylene gas cooling, the heat transfer coefficient of the water side can reach more than 10 times that of the ethylene gas side. Enhancing heat transfer on the ethylene gas side is helpful to increase the overall heat transfer coefficient. Low-finned tube and high flux condenser tube have the advantages of enhancing the heat transfer coefficient of gas outside the tube and expanding the secondary heat transfer area. The heat transfer characteristics of cooling water/propylene cooler and cooling water/propylene condenser are the same as those of ethylene gas cooler. Low-finned tube or high flux condenser tube can be selected to enhance heat transfer.

In ethylene condensers, the heat transfer coefficient of ethylene condensation is close to that of propylene evaporation. If the equipment is installed vertically, high flux tube with inner sintering and outer longitudinal groove is more suitable for this scenario.

2. Heat transfer enhancement of heat transfer equipment in catalyst preparation and polymerization raw material precooling unit

Polymerization raw materials are cooled to $-98°C$ by cooling water, low-temperature water, $-43°C$ propylene, and $-101°C$ ethylene successively before entering the polymerizer. When propylene and ethylene evaporation is used for refrigeration outside the tube, the use of low-finned tubes or high flux tubes with outer sintering can enhance the boiling heat transfer outside the tube.

3. Heat transfer enhancement of heat transfer equipment in methyl chloride recovery unit

The mixed flow removed from the top of degassing kettle contains about 15% (mass fraction) water vapor, which is condensed by cooling water. The film heat transfer coefficient of shell side of conventional shell and tube heat exchanger is only one-tenth of the heat transfer coefficient of cooling water in the tube. The low-finned tube can meet the requirements of pressure drop and heat transfer enhancement at the same time. The heat load of the condenser in a degassing kettle is 20 MW, and the cooling water cools the medium from 54°C to 43°C. Low-finned tube and plain tube options are compared in Table 5.31.

Table 5.31 Comparison of degassing kettle condenser options.

Item	Plain tube option	Low-finned tube option
Equipment size (shell diameter × straight tube length) (mm)	1900 × 9000	1700 × 7500
Number of equipment (piece)	1	1
Overall heat transfer coefficient (W/(m² K))	530	776
Heat transfer area (m²)	2545	1680
Total weight of equipment (t)	49.9	34.5

The process media of circulating methyl chloride condenser and refined methyl chloride condenser contain noncondensing gas, and the heat transfer characteristics and heat transfer enhancement way of the equipment are the same as that of degassing kettle condenser.

4. Heat transfer enhancement of heat transfer equipment in nitrogen drying unit

The heat transfer in nitrogen drying unit is mainly gas–gas heat transfer or gas–water heat transfer. There are problems such as large volume flow rate of gas, low film heat transfer coefficient of gas side, small allowable pressure drop, and bundle vibration, and the gas–water heat transfer has the problem of large difference in volume flow rate on both sides. The use of low-finned tube is helpful to enhance the heat transfer of gas outside the tube and expand the secondary heat transfer area. The use of fins inside the tube is helpful to enhance the heat transfer of gas inside the tube, but the pressure drop increases significantly.

5.7.5 Summary

Butyl rubber polymerization is carried out at a low temperature of $-90°C$. The energy consumption of refrigeration accounts for more than 40% of the total energy consumption. Through the integration of cold energy, the energy consumption of the unit is effectively reduced. By setting four types of refrigerants for cooling including cooling water, chilled brine, propylene refrigeration, and ethylene refrigeration, the purposes of graded utilization of refrigerants, energy saving, and consumption reduction are achieved. In the propylene refrigeration and ethylene refrigeration systems, the economizer is fully used and the number of compressor stages is reasonably set to reduce the compression energy consumption. Through strong stirring and excellent heat exchange tube design, the heat transfer of the reactor is enhanced, and the reaction is ensured to be carried out at low temperature, so as to stabilize the production process and extend the operation cycle of the reactor. The rapid development of LNG in China has made it possible to use LNG vaporization cold energy to supply cooling to butyl rubber unit.

① The polymerization of butyl rubber is a cationic polymerization reaction at low temperature, which requires strict temperature control. For the polymerizer, it is necessary to enhance the heat transfer on both sides of the ethylene side and the material side and reduce the dead zone on the polymerization side.

② Low-finned tube or high flux condenser tube is suitable for enhancing condensing heat transfer and suitable for heat transfer of refrigerant gas condenser.

③ The low-finned tube is helpful to enhance the heat transfer of gas outside the tube, and the fins inside the tube are helpful to enhance the heat transfer of gas inside the tube. It can be used for the heat exchanger of chlorine and methane condensation and nitrogen drying unit of the plant.

References

[1] Wang SH. Ethylene plant technology and operation. Beijing: China Petrochemical Press; 2009. p. 62–88.
[2] Xiao XJ, He XO. Energy saving technology of cracker in ethylene plant. Petrochem Ind 2013;32 (3):254–7.
[3] He XO. Principle and industrial practice of hydrocarbon steam cracking (I). Ethylene Ind 2008;20 (3):49–55.
[4] Li CHL. Preheating and heat recovery in convection section (I). Ethylene Ind 2009;21(3):58–61.
[5] Li CHL. Preheating and heat recovery in convection section (II). Ethylene Ind 2009;21(4):57–9.

[6] Wu DF, He XO, Sun LL, et al. Numerical simulation of three-dimensional flow field and combustion in radiation section of cracking furnace. Petrochem Ind 2005;34(8):749–53.

[7] Zhang LJ. Improvement of heat transfer enhancement technology of twisted-strips tube. Petrochem Equip Technol 2017;38(4):21–4.

[8] Chen B. Ethylene engineering. Beijing: China Petrochemical Press; 1997. p. 193–210.

[9] Zhao BR, Wang ZW, Wang ZY, et al. Reverse coagulation of hydrogen in supercooling with multiple coolers. Chem Eng 2008;36(7):16–9.

[10] Zhao BR, Li GH. Analysis and countermeasures of light coolant flow channel blockage in ternary refrigeration system. Petrochem Ind 2014;43(8):948–53.

[11] Sheng JY. Application of binary refrigeration technology in ethylene plant. Chem Ind Prog 2002;21 (9):663–7.

[12] Chen YD, Li GF. Technology progress of replacing petroleum ethylene with coal to synthesize ethylene glycol. J Univ Sci Technol China 2009;39(1):1–10.

[13] Zhang JD, Ye JY, Li JH, et al. Analysis of low-temperature heat recovery and utilization in ethylene oxide/ethylene glycol unit. Energy Conserv Emiss Reduction Pet Petrochem Ind 2014;4(1):6–9.

[14] Tang ZQ, Gu YL, Li JB. Advances in ethylene oxide/ethylene glycol production technology. Guangdong Chem Ind 2013;40(4):73–4.

[15] Zhang XZ, Wang SH, Qi YZ. Engineering of ethylene derivatives. Beijing: Chemical Industry Press; 1995. p. 172.

[16] Chen GG. New development of ethylene oxide/ethylene glycol production technology abroad. Chem Ind Prog 1990;3:26–30.

[17] Zhang XZ, Wang SH, Qi YZ. Engineering of ethylene derivatives. Beijing: Chemical Industry Press; 1995. p. 181–3.

[18] Zhang GH. Energy analysis and optimization of ethylene oxide/ethylene glycol plant. Dalian: Dalian University of Technology; 2013.

[19] Wu LJ. Comparative analysis of high purity ethylene oxide technology. Chem Ind Des 2015;25(1):16–9.

[20] Yan W. Application of high efficiency heat exchanger in ethylene glycol distillation. Manag Technol Small Medium-sized Enterprises (Mid-month) 2015;6:170.

[21] Xu Y. Process route selection for industrial propylene oxide production. Petrochem Des 2016;33(1):7–9.

[22] Tang Z, Feng YY, Wang ZQ. Economic and technical evaluation of propylene oxide production process. Prog Fine Petrochem Ind 2000;1(5):47–52.

[23] Yang SC. Design of falling film evaporator. Petrochem Des 1995;12(1):45–57.

[24] Zhang XY. Study on energy saving measures of styrene plant. Utiliz Rubber Plastic Res 2014;6:31–7.

[25] Han YQ. Energy saving approach of ethylbenzene/styrene separation process. Qilu Petrochem Ind 1993;2:131–4.

[26] Shen J. Study on energy saving technology of styrene production process. Chem World 2010;2:98–101.

[27] Wang MF, Qi H. Energy consumption analysis and energy-saving measures of styrene plant. Qilu Petrochem Ind 2007;35(3):194–7.

[28] Liu WJ. Study on energy saving process of styrene plant. Petrochem Des 2008;25(4):47–9.

[29] Tian LS, Ming Z, Tang WC, et al. Composite solvent and method for separation of styrene from hydrocarbon mixtures by extractive distillation. 2009. CN101468938. 2009-7-1.

[30] Zhang H, Ma TZ, Su HC. Analysis of measures affecting the long cycle operation of styrene extraction plant. Petrochem Technol 2017;6:21–2.

[31] Liu GW, Li ZT, Huang HD. Heat transfer enhancement of vertical thermosiphon reboiler. J Chem Eng Chin Univ 1989;3(4):45–52.

[32] Wu JT, Wang SG, Yao PJ. Simulation of energy saving vertical thermosiphon reboiler. J Chem Eng Chin Univ 2002;16(3):252–6.

[33] Zhang XZ. Carbon tetra-carbon pentaolefin engineering. Beijing: Chemical Industry Press; 1998. p. 585–612.

[34] Wu HJ, Guo SZ. Development trend of comprehensive utilization of pyrolysis carbon. Contemp Pet Petrochem Ind 2004;12(6):25–8.

[35] R JY, Michaelson R.C. Isoprene process. 1993. US 5177290. 1993-01-05.

[36] Cheung TtP, Johnson MM. Separation of cyclopentadiene from dicyclopentadiene. 1997. US 5659107. 1997-08-19.

[37] Kulprathipanja S, Chang CH. Process for separating isoprene. 1986. US 4570029. 1986-02-11.

[38] D'sidocky RM. Reduction of cyclopentadiene from isoprene streams. 1983. US 4392004. 1983-07-05.

[39] Arakawa M, Yamanouchi H, Okumura T, et al. Process for purification of isoprene. 1979. US 4147848. 1979-04-03.

[40] Liakumovich AG, Pantukh BI, Lesteva TM. Process for purifying isoprene. 1980. US 4232182. 1980-11-04.

[41] Wu YX, Gu XL, Qiu YX, et al. Effects of water content and polymerization temperature on the cationic polymerization of isobutylene initiated by MeOH/BF_3 system. Acta Polym Sin 2002;4: 498–503.

[42] Throckmorton MC. Process for the removal of cyclopentadiene from unsaturated C5-hydrocarbons. 1984. US 4471153. 1984-09-11.

[43] Ninagawa Y, Yamada O, Renge T, et al. Process for producing isoprene. 1986. US 4593145. 1986-06-03.

[44] Wideman LG. Reduction of cyclopentadiene from isoprene streams. 1983. US 4390742. 1983-06-28.

[45] Takao S, Koide T, Nishitai I. Isoprene purification process. 1970. US 3510405. 1970-05-05.

[46] Yoshiaki W. Utilization of C5 stream and its prospects. Chem Econ Eng Rev 1974;6(8):36–41.

[47] Yu HH, Du CP, Gao JY, et al. A method for separation of five fractions of cleaved carbon by predealkyne process. 2000. CN 1056823C. 2000-09-27.

[48] Yu HH, Lin MX, Du CP, et al. Extraction distillation with liquid feed for separation of carbon pentates from petroleum pyrolysis. 2000. CN 1055281C. 2000-08-09.

[49] Gao S, Ma LG. Design and optimization of C_5 separation dimer reactor. Petrochem Equip Technol 2013;34(4):47–50.

[50] Hu JM, Xu HF, Li X, et al. Reactive distillation technology in cracked C_5 fractions. Petrochem Des 1999;16(2):9–11.

[51] Cheng JM, Li XF, Du CP, et al. Application of reaction distillation in separation of cleaved C5. Chem Ind Ind Prog 2009;28(7):1278–81.

[52] Ma LG, Li M, Sun XJ, et al. A reactive distillation column for C_5 separation. 2013. CN 202740805. 2013-02-20.

[53] Guo SP. Application case analysis of reactive distillation in carbon pentane separation. Shandong Chem Ind 2014;43(7):123–5.

[54] Zhou ZF. Reactive distillation technology in carbon pentane separation process. Ethylene Ind 2014;26 (1):24–7.

[55] Ma LG, Li M, Hou XH, et al. A heat transfer system and method for extractive distillation of carbon penta. 2013. CN 103183578A. 2013-07-03.

[56] Ma LG, Gao D, Li M, et al. A circulating solvent regeneration system and method for a split C_5 separation unit. 2014. CN 103570576A. 2014-02-12.

[57] Ma LG. Process improvement and optimization of solvent recovery system of C_5 separation unit in DMF process. Ethylene Ind 2014;26(2):43–6.

[58] Ruan XQ, Yang L J, Nie YQ, et al. A solvent regeneration kettle. 2013. CN 203264322U. 2013-11-06.

[59] Chen MC, Ding WY. Utilization of cold energy of liquefied natural gas in butyl rubber plant. Petro-chem Ind 2015;44(1):95–102.

[60] Wang B, Zhang PF. Polymerization technology of butyl rubber. Petrochem Technol 2007;14(2):64–8.

[61] Huang Q, Sheng Y, Deng K, et al. The effect of polymerization conditions on crystalline of polybutene-L catalyzed by metallocene catalyst. Chin Chem Lett 2007;18(2):217–20.

CHAPTER 6

Heat transfer enhancement and overall plant energy saving

Contents

In recent years, energy saving and consumption reduction in refining and chemical enterprises have faced a bottleneck constraint in their development. On the one hand, traditional energy-saving technologies are gradually maturing and have been widely implemented. Especially in this century, several large-scale refining and chemical enterprises, with annual capacities exceeding 10 million tons, have adopted energy-saving technologies, progressively reducing the process energy consumption. On the other hand, the continuous upgrading of oil quality requires deeper refining processes, which contribute to increased energy consumption.

In the face of the increasingly severe situation regarding energy saving and emission reduction as well as the new energy consumption challenges of refining and petrochemical enterprises, promoting the systematic design and optimization for energy saving and consumption in these enterprises is an essential approach to overcome the bottleneck of energy saving. This chapter focuses on top-level planning and design-oriented innovative systematic energy-saving design and energy consumption optimization methods for refining enterprises [1]. The content includes the energy-saving design and energy use optimization of units, steam power systems, hydrogen systems, storage and transportation systems, low-temperature heat networks, circulating water systems, etc. Simultaneously, energy-saving technologies and approaches are summarized to provide methodology and technical support for further reasonably reducing the energy consumption level of refining and chemical enterprises, improving energy efficiency, and implementing energy-saving design and energy consumption optimization in refining and chemical engineering enterprises.

6.1 Innovative methodology for systematic energy saving

To implement energy-saving designs and optimize energy consumption in refining and chemical enterprises, it is necessary to understand the factors that are related to energy consumption in these industries.

It is generally believed that the factors affecting the energy consumption of newly built refining and chemical enterprises mainly include crude oil properties, overall process flow, product scheme, and product quality. For refineries that have already been put into operation, attention must be paid to influencing factors such as unit load rate. Among these factors, the overall process flow not only affects the product structure and quality but also significantly impacts the energy consumption of refining and chemical enterprises. The entire processing process of refining and chemical enterprises is determined through multi-objective collaborative optimization, taking into consideration factors such as resources, energy, environment, and profits. Therefore, optimizing the overall process flow to reduce the energy consumption of refining enterprises is a crucial aspect of the implementation of energy-saving design and energy consumption optimization. This optimization encompasses both the processing capability and structure of the plant.

Once the crude oil properties and the overall process flow of refining enterprises are determined, the product scheme, product quality, plant load rate, and other parameters can be essentially determined. This also allows for the determination of energy consumption structure and level. Corresponding to the determined overall process flow, there exists an engineering limit value for energy consumption, primarily constrained by the economic and technological conditions at that time. Based on the specific energy consumption structure of a particular refining and chemical enterprise, the energy allocation system of enterprise can be reasonably planned. With a clear engineering limit of energy consumption, energy-saving goals can be reasonably established, and then various energy-saving technologies can be utilized to improve the energy efficiency of the enterprise, further reducing the energy consumption level and bringing the energy consumption closer to the engineering limit. The engineering limit value assumes that all the currently feasible energy-saving measures have been applied, technically and economically, and that the energy consumption and overall consumption of the enterprise have reached the minimum and optimal value that can be achieved through engineering.

The application of energy-saving technologies includes two parts: energy integration and optimization (energy integration) and process intensification. Therefore, the innovative and systematic energy-saving methodology for refining and chemical enterprises works at three levels: energy planning, energy integration, and process intensification; measures at each level are also optimized. Fig. 6.1 shows the schematic diagram of the innovative and systematic energy-saving method in refining and chemical enterprises.

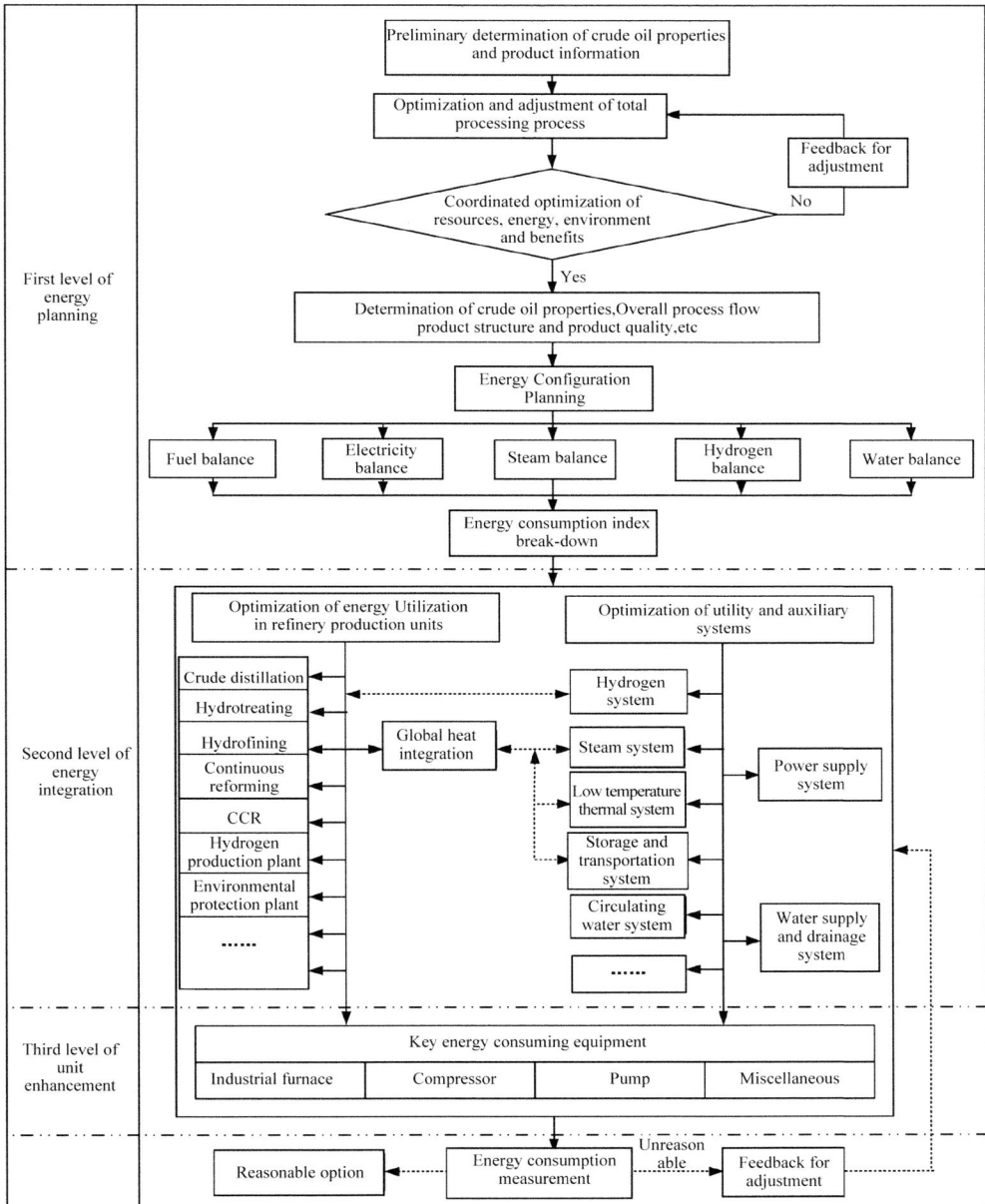

Fig. 6.1 Flow diagram of innovative and systematic energy-saving methods in refining and chemical enterprises.

6.1.1 Energy planning

The first step in implementing energy-saving measures is to optimize and adjust the overall process flow of refining and chemical enterprises, whether they are newly built, renovated, expanded, or in operation.

The energy planning of refining and chemical enterprises based on the overall process flow should be carried out step by step.

Firstly, based on the preliminary information of the enterprise's crude oil properties and products, the overall process flow should be determined through comprehensive trade-off and multiobjective optimization, aiming at the collaborative optimization of resources, energy, environment, and profits.

Secondly, based on the overall process flow, the energy consumption structure of the refining and chemical enterprise is analyzed, and the engineering limit of energy consumption and the demand for fuel, electricity, steam, water, and hydrogen are calculated. The engineering energy consumption limit and the fuel, electricity, steam, water, and hydrogen demand can be calculated and obtained through the refining and chemical energy consumption modeling system.

Next, the energy consumption value of the actual operation (or planning and design) is compared with the energy consumption engineering limit value, and the energy saving target is formulated, along with providing the energy allocation planning suggestions. The energy planning suggestions mainly involve fuel outsourcing, thermal boiler setting, electricity outsourcing, steam balance, circulating water setting, hydrogen production setting, hydrogen pipe network setting, and heat integration.

Finally, combined with the energy-saving goals and energy planning suggestions, the energy consumption indicators are decomposed for each refining and chemical processing unit and utilities and auxiliary system, as a guide and constraint for energy integration and process intensification.

6.1.2 Energy integration

The primary purpose of energy integration is to optimize the energy usage of refining and chemical units, utilities, and auxiliary systems, making it a key component of energy planning for refining and chemical enterprises. The global energy optimization strategy for refining enterprises establishes a logical framework for implementing global energy optimization. Firstly, the energy consumption optimization of the unit is carried out, involving the process improvement and parameter optimization of the core processes such as reaction and separation. On this basis, the optimization of the hydrogen system is conducted on one hand, and on the other hand, the establishment of heat integration between units and the optimization of the heat integration network take place. Subsequently, the energy use optimization of tank farm and auxiliary systems is executed. Following the principle of temperature level match and cascade utilization, the identification and utilization of high-quality heat traps reduce unnecessary steam or other high energy consumption. Then the low-temperature heat of the entire plant is comprehensively recovered and utilized. According to the principle of "long-term, stable and nearby" utilization, combined with the layout of the whole plant, a rational low-temperature thermal system is designed, and the energy optimization of the circulating water system is implemented. Finally, the steam power system is optimized, encompassing the steam pipe network and condensate water system, etc. In conjunction with the steam demand of the entire plant and the gas balance of the refinery, the optimal revamp and operation strategy of the steam power system is proposed.

6.1.3 Unit intensification

Process intensification is a crucial and effective measure to achieve global energy optimization in refining and chemical enterprises, which mainly involves the intensification and optimized utilization of key energy-consuming equipment. In general, key energy-consuming equipment of refining and chemical enterprises includes reactors, waste heat and residual pressure recovery systems or equipment, heating furnaces with power not less than 10 MW, positive displacement compressors with shaft power not less than 1000 kW, turbine compressor with shaft power not less than 2000 kW, and pump with shaft power not less than 200 kW. The major way to realize process intensification is to carry out a special technical revamp or energy-saving application for a single process, unit, or key energy-consuming equipment.

6.1.4 Application case

The crude oil processing capacity of a refinery is 8 Mt/a. The main designed products include liquefied petroleum gas, jet coal, gasoline, diesel, sulfur, fuel oil, benzene, polypropylene, etc. The product quality complies with National V standard. The overall process configuration is as shown in Fig. 6.2, and the design energy consumption data is as shown in Table 6.1.

It can be seen from Table 6.1 that the design value of energy consumption of the refinery is 69.9 kgoe/(t crude oil), the energy consumption factor of the refinery is 8.5, and the design unit factor energy consumption of the refinery is 8.2 kgoe/(t Eff).

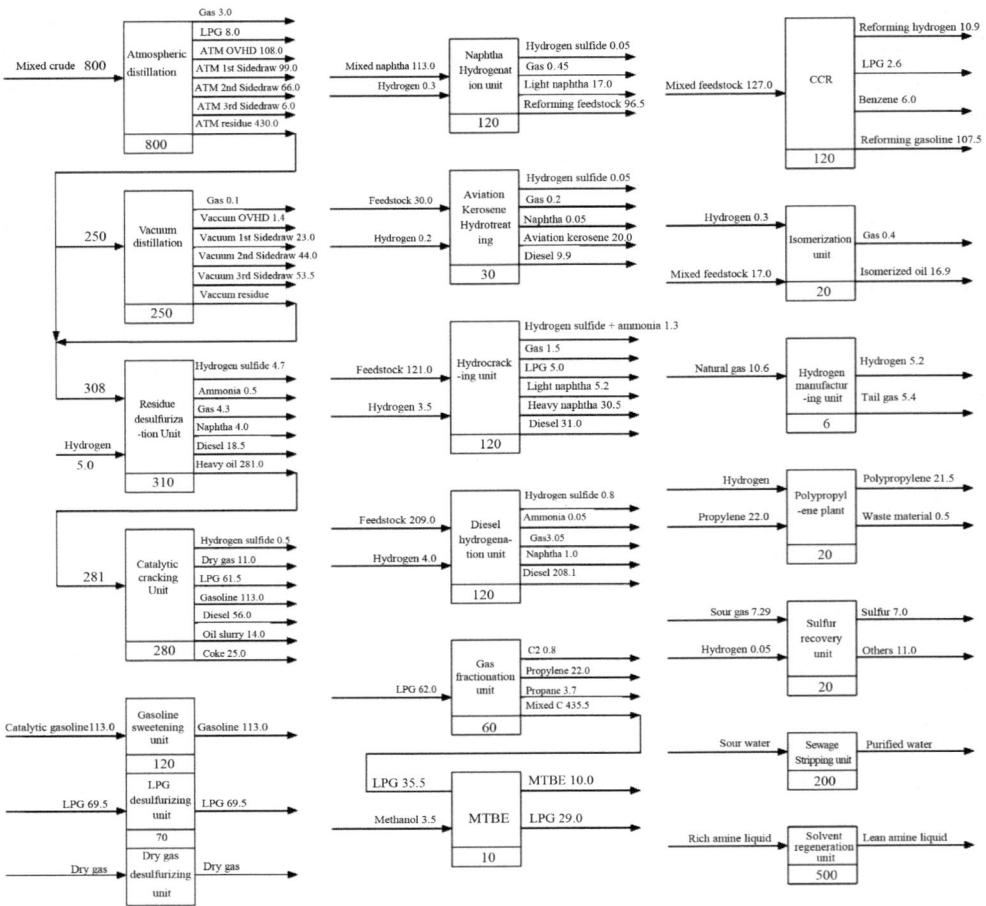

Fig. 6.2 The overall process configuration of an oil refinery (unit: 10 kt/a).

Table 6.1 Design energy consumption of the example refinery.

Item	Index
Consumption of fuel gas (t/h)	27.0
Consumption of fresh water (t/h)	410.0
Consumption of electricity (MW/h)	48.0
Coke burning (t/h)	30.0
Energy consumption of refinery (kgoe/t)	69.9
Energy consumption factor of refinery	8.5
Unit factor energy consumption of refinery (kgoe/(t Eff))	8.2

6.1.4.1 Energy planning

Given that the example enterprise has completed the optimization of the overall process flow in the design process, the energy planning and optimization are based on the established overall process flow. Based on the basic data above, the refinery energy consumption modeling system developed by Sinopec Engineering Incorporation (hereinafter referred to as SEI) is used for energy consumption calculation and the calculation results are summarized in Table 6.2.

Table 6.2 Calculation results of engineering limit energy consumption of the example refining enterprise.

Item	Index	Item	Index
Engineering limit of energy consumption (kgoe/t)	61.0	Electricity demand (MW/h)	40.0
Energy consumption factor of refinery	8.5	Fresh water demand (t/h)	400.0
Engineering limit of unit factor energy consumption of refinery (kgoe/(t Eff))	7.2	3.5 MPa steam demand (t/h)	−50
Pure hydrogen demand (t/h)	6.2	1.0 MPa steam demand (t/h)	−140
Coke burning (t/h)	30.0	0.4 MPa steam demand (t/h)	70
Fuel gas demand (t/h)	20.5	Self-produced fuel gas (t/h)	23.5

It can be seen from Table 6.2 that the engineering limit of the comprehensive energy consumption of the refinery is 61.0 kgoe/t crude oil, the energy consumption factor of the refinery is 8.5, and the engineering limit of the unit faction energy consumption of the refinery is 7.2 kgoe/(t Eff).

Engineering limits lay the foundation for establishing energy-saving goals and clarify the direction of optimization. By comparing the data in Tables 6.1 and 6.2, the difference between the design consumption and the engineering limit of fuel gas and electricity consumption is the main reason for the difference between design energy consumption and engineering limit. Combined with data analysis, planning suggestions for energy and energy transfer media are provided.

1. Optimization of the integration of fuel gas and heat

    Appropriately enhance heat integration, reasonably increase the final temperature of the flow entering the furnace and fractionating column, and appropriately reduce the consumption of fuel gas and heat source steam of fractionating column.

2. Steam system optimization

    The refinery has a surplus of 3.5 MPa, 1.0 MPa, and a shortage of 0.4 MPa steam. According to the type of steam system, it is suggested that after 3.5, 1.0, and 0.4 MPa steam meet the process requirements, the heat of 3.5, 1.0, and 0.4 MPa steam in the process can be properly adjusted to direct heat integration.

3. Power saving measures

Reasonably use the residual pressure of the process to generate electricity, and appropriately use steam as the power source to drive the dynamic equipment to reduce the electricity consumption.

6.1.4.2 Energy integration

According to the energy consumption characteristics of the example refinery, the energy integration suggestions are given in the following aspects.

① Optimize the energy-saving process and catalyst, reduce the reaction severity reasonably, and reduce the total process energy at the source.

② Given the layout of the plant, the heat integration between the crude oil distillation, residual oil hydrogenation, and catalytic cracking unit is mainly considered. The heat integration between hydrofining units is considered.

③ According to the principle of "temperature matching and cascade utilization," the steam power system is optimized to avoid using degraded steam.

④ According to the layout of the plant, a local and whole-plant low-temperature heat recovery network is set up to recover and utilize low-temperature heat resources globally.

⑤ Optimize the circulating water system and properly consider the reuse of circulating water.

⑥ Clarify the pressure level of process flows between plants, avoid repeated rise and fall of pressure between plants, and reasonably save power consumption.

In the energy integration hierarchy, the heat integration between crude oil distillation, residual oil hydrogenation, and catalytic cracking units and the utilization of low-temperature waste heat resources in the whole plant are two key energy-saving measures. Among them, crude oil distillation, residue hydrogenation, and catalytic cracking are the major energy-consuming units. The heat integration of these units mainly includes heat integration within the unit, heat supply/discharge and temperature optimization, and direct heat transfer between material flows in different units to improve the overall energy utilization efficiency. The major heat integration optimization ideas are shown in Fig. 6.3.

Through heat integration, the final temperature of the predistillation bottom oil can reach 290°C after the optimization and revamp of the heat exchange network of the crude oil distillation unit. Then the oil at 290°C and the slurry at 335°C can be heat-exchanged to about 310°C and returned to the atmospheric furnace. It is estimated that 1.5 t/h fuel gas can be saved and the production of 3.5 MPa steam can be reduced by about 24 t/h.

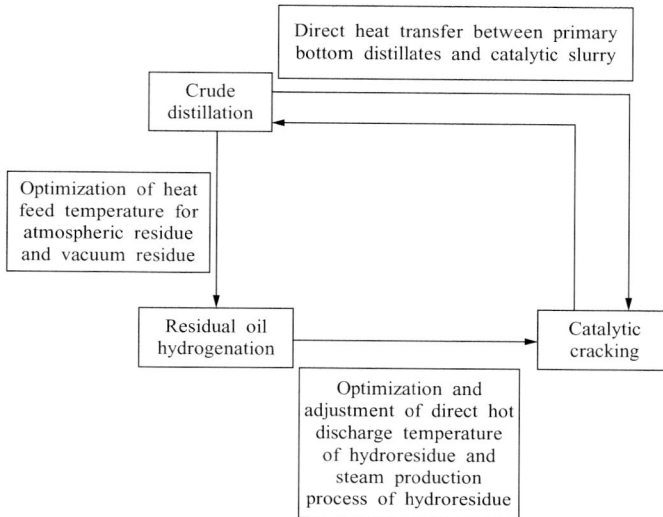

Fig. 6.3 Heat integration renovation ideas of major energy consuming units.

To make rational use of low-temperature heat resources, it is suggested to set up a low-temperature waste heat recovery system for the whole plant based on the design conditions. Hot water at 120°C is produced for low-temperature heat recovery and utilization. Two systems are set up as follows:

① The low-temperature heat utilization system of tank farm temperature maintenance and pipeline heat tracing uses the recovered low-temperature waste heat for tank farm temperature maintenance and pipeline heat tracing, which is expected to recover 1.2 MW of low-temperature heat per hour.

② Low-temperature thermal power generation system uses about 1000 t/h hot water produced by the low-temperature waste heat recovery system for centralized power generation. If the returned hot water temperature is considered 70°C and the thermoelectric conversion efficiency is calculated as 8.0%, the power generation is expected to be 3.5 MW/h.

6.1.4.3 Process intensification

The following suggestions are put forward for process intensification of the example refinery.

① Adopt enhanced combustion and flue gas waste heat recovery technology to reasonably improve the thermal efficiency of the furnace.

② Reasonably improve the efficiency of the compressor included in the key energy-consuming equipment.

③ Reasonably improve the pump efficiency of the key energy-consuming equipment.

The design thermal efficiency of the large furnace of the refinery is 91.0%, and the thermal efficiency of the four-in-one reforming furnace is 92.0%. Combined with the application of combustion enhancement technology and flue gas waste heat recovery technology, the thermal efficiency of the large furnace is optimized to 93.0%, the thermal efficiency of the four-in-one reforming furnace is improved to 94.0%, and the fuel gas is expected to be saved by about 1.0 t/h. The major heating furnaces involved are as shown in Table 6.3.

Table 6.3 Large-scale furnaces in example refinery.

No.	Name	Duty (MW)
1	Atmospheric heating furnace	80.0
2	Vacuum heating furnace	20.0
3	Residue hydrogenation reactor feed heating furnace	25.0
4	Residue hydrogenation fractionation column feed heating furnace	25.0
5	Four-in-one reforming furnace	35.0

6.1.4.4 Energy-saving effect

According to the suggestions of energy planning, energy integration, and process intensification, and combined with the implementation of key energy-saving measures, the energy consumption data of the optimized design is shown in Table 6.4. After optimization, the design energy consumption of the example refinery is 66.0 kgoe/(t Eff), which is about 5.6% lower than the design data. The unit factor energy consumption of the refinery is 7.8 kgoe/(t Eff), which is about 5.0% lower than the design data. It should be noted that there is a certain gap between the optimal value of energy consumption and unit factor energy consumption of the refinery and the engineering limit. This is because some technically feasible energy-saving measures have not been reflected in the actual engineering design because they have not passed the technoeconomic evaluation.

Table 6.4 Energy consumption data of optimized design of the example refinery.

Item	Index
Consumption of fuel gas (t/h)	24.0
Consumption of fresh water (t/h)	410.0
Consumption of electricity (MW/h)	45.0
Coke burning (t/h)	30.0
Design energy consumption of refinery (kgoe/t)	66.0
Design unit factor energy consumption of refinery (kgoe/(t Eff))	7.8

6.2 Global heat integration optimization design

6.2.1 Basis for heat integration of refinery unit

Typical composite curve is shown in Fig. 6.4.

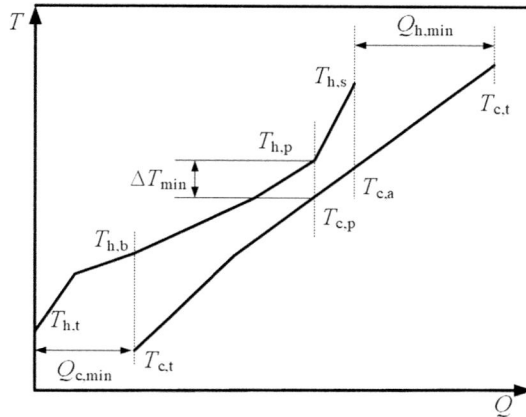

Fig. 6.4 Typical composite curve.

6.2.1.1 Principle of heat output

Basic principles of heat output of unit [2]:

① Above the pinch point, if the temperature is between $T_{h,s}$ and $T_{h,p}$, the flow heat is not exported.

② Under the pinch point, if fluid temperature is between $T_{h,b}$ and $T_{h,p}$, further analysis is needed for whether the flow heat is exported or not.

③ Under the pinch point, if temperature is between $T_{h,t}$ and $T_{h,b}$, the flow heat can be exported.

6.2.1.2 Principle of heat input

Diagnostic principles for rationality of flow heat input in refinery unit [2]:

① Above the pinch point, if temperature is greater than $T_{c,a} + \Delta T_{min}$, further analysis is needed for whether the flow heat is imported or not.

② Above the pinch point, if temperature is between $T_{h,p}$ and $T_{c,a} + \Delta T_{min}$, flow heat is not imported.

③ Under the pinch point, if temperature is less than $T_{h,p}$, flow heat is not imported.

6.2.1.3 Flow heat output and input temperature and heat determination method

For flow heat exportation, when the temperature is under the pinch point and between $T_{h,b}$ and $T_{h,p}$, further analysis is needed to determine whether the flow heat is exported. For flow heat import, when the temperature is above the pinch point and greater than $T_{c,a} + \Delta T_{min}$, further analysis is needed to determine whether the flow heat is imported. Literature [3,4] provides a more intuitive method for determining flow heat output and input temperature.

6.2.2 Heat integration strategy for multiple units in refinery

Wei and Sun suggested the block diagram of the heat integration strategy for multiple units in the refinery [2] as shown in Fig. 6.5. Firstly, the background refinery is selected, and the units with heat integration opportunities are screened and identified. Based on the heat integration analysis strategy, the heat output and heat input flows demand information of each unit is analyzed, and the heat integration flows information database of multiple units is summarized. Based on engineering knowledge, the possible cross–unit heat integration scheme is given considering the constraints of general layout, etc. The economic feasibility analysis of each cross–unit heat integration matching combination is carried out. If the economic efficiency is not reasonable, the heat integration scheme is not supported; if the economic efficiency is reasonable, the heat integration scheme is output. Finally, all the heat integration schemes are summarized to optimize the heat integration network within the unit and between units. The right box in Fig. 6.5 shows the basic analysis strategy of heat integration of units.

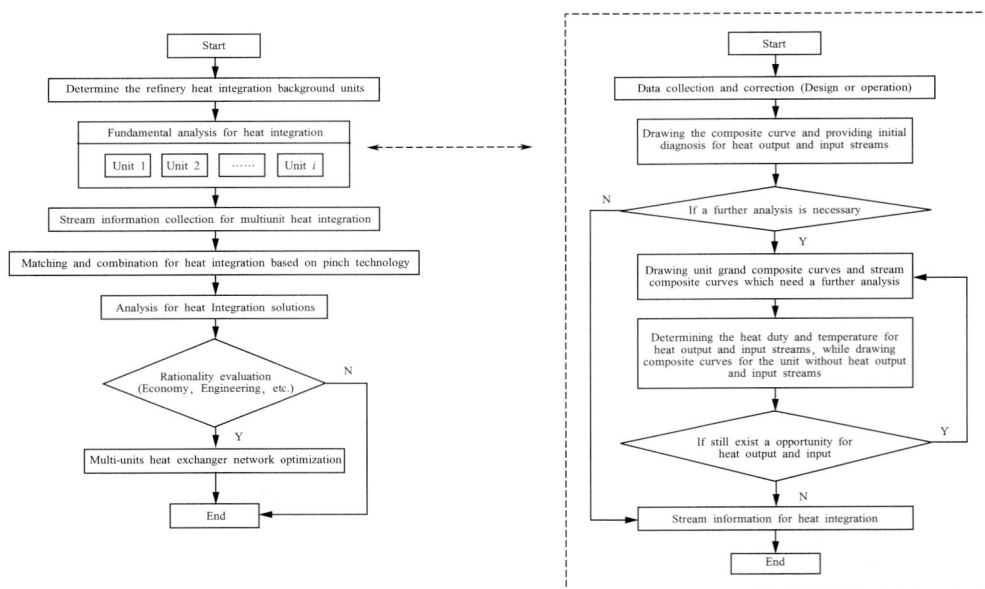

Fig. 6.5 Schematic diagram of heat integration strategy for multiple refinery units [2].

6.2.3 Case study

6.2.3.1 Basic data acquisition and correction

The basic design data of crude oil distillation, catalytic cracking, and light hydrocarbon recovery units were collected for process simulation by Aspen Plus and PRO/II. The design conditions and the cold and hot material flow data are regenerated as shown in Tables 6.5–6.7.

Table 6.5 Cold and hot flow data of crude oil distillation unit.

No.	Flows name	Initial temperature (°C)	Final temperature (°C)	Heat (kW)
C–c1	Crude oil before desalting	40	135	68,380
C–c2	Desalted crude	130	240	93,640
C–c3	Flashed crude	220	370	138,800
C–c4	Atmospheric residue	365	390	20,070
C–h1	Atmospheric top	115	90	16,100
C–h2	Kerosene	170	45	10,650
C–h3	Light diesel	265	60	20,770
C–h4	Heavy diesel	335	60	17,220
C–h5	LVGO	135	60	2480
C–h6	MVGO	245	90	16,440
C–h7	HVGO	300	115	17,870
C–h8	Vacuum residuum	370	150	53,850
C–h9	Atmospheric top pump around	160	100	17,120
C–h10	Atmospheric 1st pump around	215	155	19,060
C–h11	Atmospheric 2nd pump around	310	235	18,280
C–h12	Vacuum top pump around	135	50	11,370
C–h13	Vacuum 1st pump around	245	165	12,030
C–h14	Vacuum 2nd pump around	300	210	28,190

Table 6.6 Cold and hot flow data of light hydrocarbon recovery unit.

No.	Flows name	Initial temperature (°C)	Final temperature (°C)	Heat (kW)
L–c1	Absorber bottom	46	90	1970
L–c2	Rich absorbent	55	130	2460
L–c3	Stripper bottom circulation	105	125	4790
L–c4	Stripper intermediate reboiling circulation	80	100	4580
L–c5	Stripper bottom	125	165	6660
L–c6	Stabilizer bottom circulation	180	190	10,110
L–c7	Naphtha feed	85	120	2110
L–c8	Naphtha column bottom circulation	140	145	7700
L–h1	Compressor exit hydrocarbons	120	40	1500
L–h2	Absorber 1st pump around	45	40	440
L–h3	Absorber 2nd pump around	45	40	510
L–h4	Lean oil	155	40	3440
L–h5	Stabilizer overhead vapor	65	40	9530
L–h6	Stabilizer bottom	140	40	4200
L–h7	Naphtha column overhead vapor	85	40	8810
L–h8	Naphtha column bottom	145	40	10,180

Table 6.7 Cold and hot flow data for catalytic cracking units.

No.	Flows name	Initial temperature (°C)	Final temperature (°C)	Heat (kW)
F–c1	Feed to FCC	170	200	11,950
F–c2	Deethanized gasoline	140	145	3850
F–c3	Sponge absorber rich oil	50	120	5735
F–c4	Debutanizer bottom circulation	180	190	27,150
F–c5	Stripper bottom circulation	140	150	31,050
F–c6	Stripper intermediate reboiling circulation	90	100	8140
F–c7	Condensate oil	40	50	3870
F–h1	Top cycle oil	145	80	29,090
F–h2	Main column LCO pump around	250	185	26,860
F–h3	Main column HCO pump around	320	255	27,150
F–h4	Main column bottom circulation	330	275	64,430
F–h5	Main column bottom product	275	120	2500
F–h6	Light cycle oil	195	90	8030
F–h7	Sponge absorber lean oil	200	40	13,110
F–h8	Main column overhead vapor	125	40	148,610
F–h9	Debutanizer bottoms	180	80	31,220
F–h10	Fresh absorbent	80	40	4480
F–h11	Liquefied petroleum gas	60	40	25,110

6.2.3.2 Heat integration basic analysis

The heat output and heat input supply and demand information of the crude oil distillation unit, light hydrocarbon recovery unit, and catalytic cracking unit are shown in Tables 6.8–6.10, respectively.

Table 6.8 Heat output and heat input of crude oil distillation unit.

Item	Temperature (°C)	Flows name	Note
Heat output	45–91	C–h1, C–h2, C–h3, C–h4, C–h5, C–h6, C–h12	Can export heat
	95–231	C–h1, C–h2, C–h3, C–h4, C–h5, C–h6, C–h7, C–h8, C–h9, C–h10, C–h12, C–h13, C–h14	Need analysis for heat output
	231–369	C–h3, C–h4, C–h6, C–h7, C–h8, C–h11, C–h13, C–h14	Should not export heat
Heat input	40–231	/	Should not import heat
	231–318	/	Should not import heat
	>318	/	Need analysis for heat input

Table 6.9 Heat output and heat input of light hydrocarbon recovery unit.

Item	Temperature (°C)	Flows name	Note
Heat output	38–82	L-h1, L-h2, L-h3, L-h4, L-h5, L-h6, L-h7, L-h8	Can export heat
	82–89	L-h1, L-h4, L-h6, L-h7, L-h8	Need analysis for heat output
	89–165	L-h1, L-h4, L-h6, L-h8	Should not export heat
Heat input	46–89	/	Should not import heat
	89–152	/	Should not import heat
	>152	/	Need analysis for heat input

Table 6.10 Heat output and heat input of catalytic cracking unit.

Item	Temperature (°C)	Flows name	Note
Heat output	40–130	F-h1, F-h5, F-h6, F-h7, F-h8, F-h9, F-h10, F-h11	Can export heat
	130–150	F-h1, F-h5, F-h6, F-h7, F-h8, F-h9, F-h10, F-h11	Need analysis for heat output
	150–330	F-h2, F-h3, F-h4, F-h5, F-h6, F-h7, F-h9	Need analysis for heat output
Heat input	No heat input for hot end threshold problem heat exchange network		

Through the analysis, the integrated flows information of oil distillation, light hydrocarbon recovery, and catalytic cracking heat is summarized.

Table 6.11 Summary of integrated hot flows.

Unit name	Heat output flows information	Heat input flows information
Crude distillation	C-h10 can output heat	Under basic condition, it can receive 318°C + heat input
	In principle, C-h11 no heat output	When C-h11 outputs heat, it can receive 298°C above heat input
Light hydrocarbon recovery	Unsuitable for heat output	Can receive 152°C above heat input
Catalytic cracking	F-h4 can output heat	Unsuitable for heat input

6.2.3.3 Heat integration of crude distillation, light hydrocarbon recovery, and catalytic cracking heat

Based on Table 6.11 and engineering knowledge, three heat integration options are proposed, as shown in Table 6.12. Among them, option 0 is the heat integration scheme adopted in the basic design. The crude oil distillation unit C–h10 and C–h11 both export heat, the light hydrocarbon recovery unit does not consume 3.5 MPa steam, and the catalytic cracking unit F–h4 generates 3.5 MPa steam.

Table 6.12 Summary of heat integration options.

Option	Crude distillation	Light hydrocarbon recovery	Catalytic cracking
0	Both C–h10 and C–h11 export heat	Does not consume 3.5 MPa steam	F–h4 generates 3.5 MPa steam
1	C–h10 exports heat, while C–h11 does not	Consume 3.5 MPa steam	F–h4 generates 3.5 MPa steam
2	C–h10 exports heat, C–h11 does not, and F–h4 exports heat	Consume 3.5 MPa steam	F–h4 exports heat to the crude distillation unit
3	Both C–h10 and C–h11 export heat, while F–h4 imports heat	Does not consume 3.5 MPa steam	F–h4 exports heat to the crude distillation unit

Given the cold and hot utility of the plant, a comparison between the schemes is analyzed. The comparison results are shown in Table 6.13. The cold utilities of each unit are assumed to be circulating water, and the price is 0.03 yuan/kW. The heat utility of the crude oil distillation unit is fuel gas, and the price is 0.43 yuan/kW. The heat utility of light hydrocarbon recovery unit is 3.5 MPa steam, and the price is 0.4 yuan/kW.

As can be seen from Table 6.13, for the three units of crude oil distillation, light hydrocarbon recovery, and catalytic cracking, option 3 has the lowest energy consumption cost and is the optimal heat integration option.

Table 6.13 Comparison of heat integration options.

Option	Crude oil distillation unit		Light hydrocarbon recovery unit		Catalytic cracking unit		Total energy consumption
	$Q_{c,min}$ (kW)	$Q_{h,min}$ (kW)	$Q_{c,min}$ (kW)	$Q_{h,min}$ (kW)	$Q_{c,min}$ (kW)	$Q_{h,min}$ (kW)	C_{op} (yuan/h)
0	10,030	81,060	24,150	0	238,010	0	47,898
1	10,030	68,960	24,150	12,190	238,010	0	42,695
2	10,030	93,270	24,150	12,190	238,010	0	48,272
3	10,030	69,080	24,150	0	238,010	0	37,870

In option 3, the crude oil distillation unit flows C–h10 and C–h11 export heat to the light hydrocarbon recovery unit, and the catalytic cracking unit flow F–h4 imports heat to the crude oil distillation unit. C–h11 is the flow with the temperature above the pinch point of the crude oil distillation unit. According to the principle of "above the pinch point, the temperature is between $T_{h,s}$ and $T_{h,p}$: the flow heat should not be exported," no heat should be exported. Actually, if only C–h11 exports heat, it will inevitably lead to the increase of the minimum hot utility duty of the plant, as can be seen in option 0. Analysis shows that above the pinch point, the temperature is between 230°C and 300°C, and the slope of the total combined curve is negative, indicating that there is a large amount of hot flow, resulting in a relatively small heat transfer driving force in the heat exchange network in this temperature range. If there is heat input above this temperature range, the driving force of heat transfer in the heat exchange network can be increased through partial heat output of the flows in this temperature range. It can be seen that the heat output of C–h11 in option 3 is reasonable. Compared with the basic design heat integration option (option 0), the energy consumption cost can be reduced by about 21.0%.

6.3 Typical cases of heat transfer enhancement across the plant

6.3.1 Direct hot feed of refinery based on "hot flow map"

The heat integration of refinery units includes direct heat integration and indirect heat integration. Direct heat integration includes direct hot feed between units and heat exchange between process flows, etc. Direct hot feed between units avoids the repeated cooling, heating, and pressure changing in the processes of discharge cooling from the upstream unit, intermediate storage tank, and feed heating in the downstream unit, which has been widely implemented in new refineries.

In recent years, due to the upgrading of oil quality and the adjustment of refining product scheme, the composition of refinery units and the direction of material processing has changed, and the traditional direct hot feed mode has also changed accordingly. Therefore, a "hot flow diagram based direct hot feed technology for refinery" is proposed. Based on the total process flow, the "hot flow map" draws information on key units, intermediate storage tanks, key process material and their flow, temperature, etc., which are used to characterize the global direct heat integration of the refinery. Through the "hot flow map," the direct hot feed relationship of the key process flows in the refinery can be easily identified. The "hot flow map" of a typical refinery is shown in Fig. 6.6. From Fig. 6.6, it can be seen that direct hot feed has been put into practice in the refinery, mainly including direct hot feed from crude distillation to VGO hydrogenation, from crude distillation to delayed coking, from crude distillation to kerosene hydrogenation, from crude distillation to hydrocracking, from crude distillation to diesel hydrogenation, from VGO hydrogenation to catalytic cracking, from delayed coking to catalytic cracking, etc. In general, hot feed has been realized in most of the key units. However, there are still some problems, mainly include:

① 20%–40% of the feedstock of VGO hydrogenation, delayed coking, catalytic cracking, and kerosene hydrotreating units passes through the raw material tank, which indicates noncomplete direct hot feed, and there is room for optimization.

② The temperature drop of the direct supply pipeline from crude distillation to VGO hydrogenation and from crude distillation to diesel hydrogenation is large, and the heat preservation and temperature maintenance need to be noted.

③ Process flows such as VGO hydrogenation tail oil, catalytic cracking diesel, and delayed coking wax have realized direct hot feed with regard to corresponding units, but there remains room for feed temperature optimization.

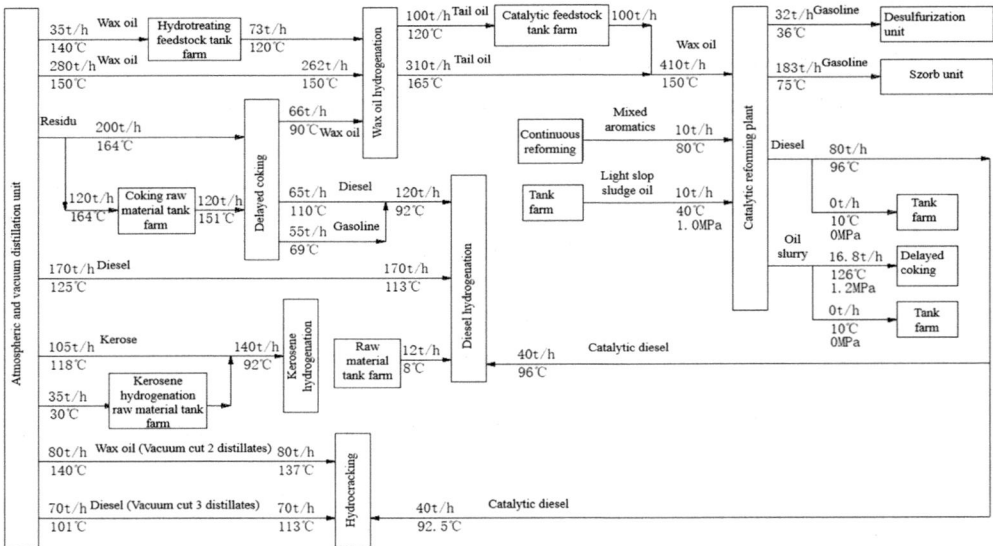

Fig. 6.6 Schematic diagram of hot feed and discharge of refinery units.

6.3.1.1 Direct hot feed from crude distillation/delayed coking unit to VGO hydrogenation unit

The schematic diagram of direct hot feed from crude distillation unit and delayed coking unit to VGO hydrotreating unit is shown in Fig. 6.7.

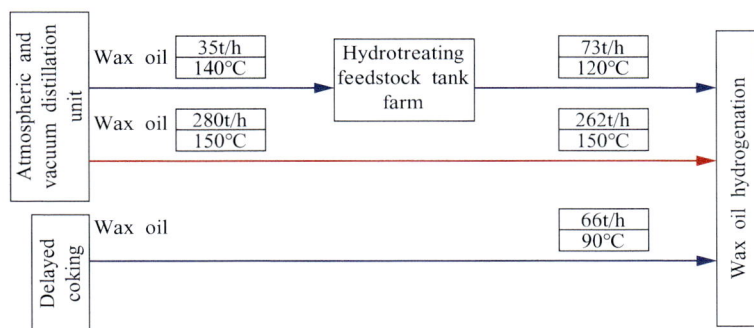

Fig. 6.7 Schematic diagram of direct hot feed from crude distillation unit and delayed coking unit to VGO hydrotreating unit.

1. Existing issues
   ① The direct hot feed has been put into practice between the crude distillation and the VGO hydrotreating unit, but about 20% of the raw material of the VGO hydrotreating unit passes through the tank farm, and the temperature drops from 140°C to 120°C, with a 20°C temperature drop indicating large heat loss.
   ② As for the direct hot feed line from crude distillation to VGO hydrogenation, the temperature drops by about 10°C and there is a certain heat loss.
   ③ The direct hot feed has been put into practice between the delayed coking and the VGO hydrotreating unit, but the temperature of the coking wax is only 90°C, which is lower than that of the crude distillation unit.

2. Revamping measures
   ① All the VGO from the crude distillation enters the VGO hydrotreating unit without passing through the tank farm. If the temperature maintenance equipment in the tank farm cannot meet the temperature requirements of normal conditions, improve the temperature maintenance equipment in the tank farm. For example, if the temperature drop of the pipeline from crude distillation to tank farm and from tank farm to VGO hydrotreating unit is large, repair or replace the insulation material and increase the temperature of heat tracing steam.
   ② Check the thermal insulation of the direct hot feed pipeline from crude distillation to VGO hydrogenation, and repair or replace the thermal insulation material for the pipeline with poor thermal insulation effect.
   ③ Optimize the heat exchange process of coking gas oil, partially canceling the exchangers between VGO and demineralized water and adjusting the output temperature of the VGO to 120–130°C.

3. Energy saving effect
   The feed temperature of the current VGO hydrotreating unit is about 125–130°C, and the temperature is 135–140°C after optimization.

6.3.1.2 Direct hot feed from crude distillation unit to delayed coking unit

The schematic diagram of the direct hot feed from the crude distillation unit to delayed coking unit is shown in Fig. 6.8.

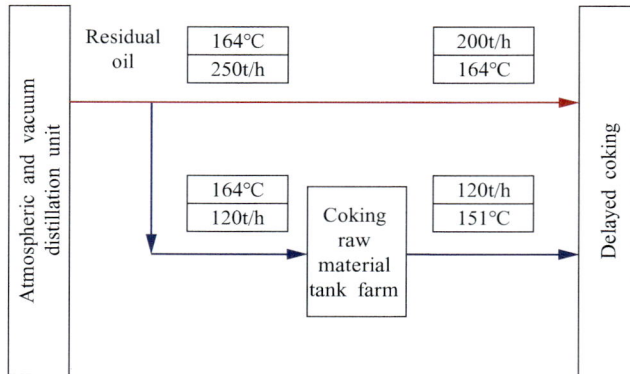

Fig. 6.8 Schematic diagram of direct hot feed from the crude distillation unit to delayed coking unit.

1. Existing issues

    The direct hot feed has been practiced from the crude distillation unit to the delayed coking unit. However, about 40% of the raw material in the delayed coking unit passes through the tank farm, and the temperature drops from 164°C to 151°C, which is about 13°C lower, resulting in a large loss of heat.
2. Revamping measures

    All the residual oil from crude distillation enters the delayed coking unit without passing through the tank farm. If the temperature maintenance equipment in the tank farm cannot meet temperature requirements of normal conditions, the temperature maintenance equipment in the tank farm should be modified. For example, if the temperature drop of the pipeline from the crude distillation to the tank farm and from the tank farm to the VGO hydrotreating unit is large, repair or replace the insulation material and increase the temperature of the heat tracing steam.
3. Energy saving effect

    The current feed temperature of delayed coking unit is about 158–159°C, and the temperature after optimization is 163–164°C.

6.3.1.3 Direct hot feed from crude distillation/delayed coking/catalytic cracking unit to diesel hydrotreating unit

The schematic diagram of direct hot feed from crude distillation/delayed coking/catalytic cracking unit to diesel hydrotreating unit is shown in Fig. 6.9.

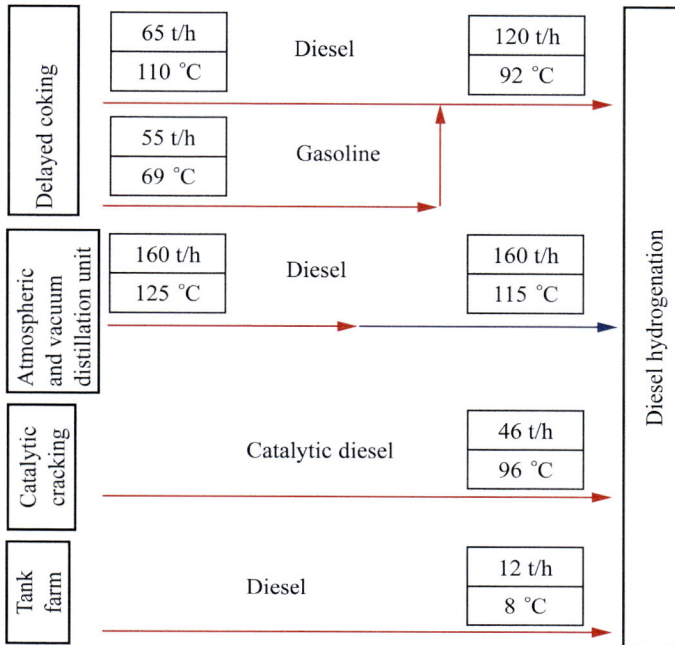

Fig. 6.9 Schematic diagram of direct hot feed.

1. Existing issues

   Direct hot feed has been put into practice between the crude distillation and diesel hydrotreating unit, but the temperature drops from 125°C to 115°C, which is about 10°C lower, with large heat loss.

2. Revamping measures

   Check the thermal insulation of the direct hot feed pipeline from crude distillation to diesel hydrogenation, and repair or replace the insulation material of the pipeline with poor thermal insulation effect.

3. Energy saving effect

   Optimize and increase the feed temperature of diesel hydrotreating unit.

6.3.1.4 Direct hot feed from crude distillation unit to kerosene hydrogenation unit

Refer to Fig. 6.10 for the schematic diagram of direct hot feed from crude distillation to kerosene hydrotreating unit.

Atmolspheric and vacuum distillation unit

| 105t/h | |
| 118°C | Kerosene |

| 35t/h | Kerosene |
| 30°C | |

Kerosene hydrogenation raw material tank farm

| 140t/h | |
| 92°C | |

Kerosene hydrogenation

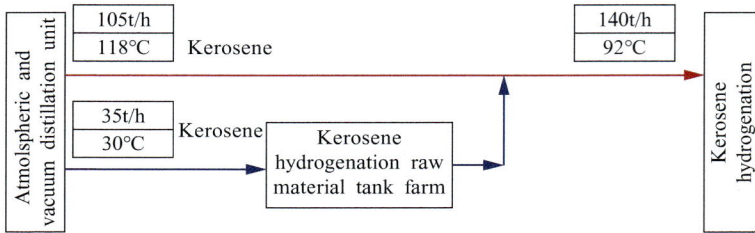

Fig. 6.10 Schematic diagram of direct hot feed from crude distillation unit to kerosene hydrotreating unit.

1. Existing issues

    The direct hot feed from crude distillation to the kerosene hydrotreating unit has been partially realized. About 75% of the raw material of the kerosene hydrotreating unit is fed from the crude distillation unit, and about 25% of the raw material is cooled to 30°C within the crude distillation unit and sent to the tank farm and then to the unit. The cold feed is not reasonable.

2. Revamping measures

    All the kerosene from the crude distillation is directly fed to the kerosene hydrotreating unit without passing through the tank farm. Check the thermal insulation of the feed pipeline from the crude distillation to the kerosene hydrotreating unit, and repair or replace the thermal insulation material of the pipeline with poor thermal insulation effect.

3. Energy saving effect

    The current feed temperature of kerosene hydrotreating unit is about 90–92°C, and the temperature after optimization is 118°C.

6.3.1.5 Direct hot feed from crude distillation/catalytic cracking unit to hydrocracking unit

See Fig. 6.11 for a schematic diagram of direct hot feed from crude distillation/catalytic cracking unit to hydrocracking unit.

1. Existing issues

    The direct hot feed has been put into practice from the FCC to hydrocracking unit, but the temperature of the cracking diesel to diesel hydrocracking unit is about 92.5°C, which is lower than the temperature of other hot feed of the hydrocracking unit.

2. Revamping measures

    Cancel the part of the heat exchanger between cracking diesel and demineralized water, and increase the temperature of cracking diesel to 120°C in the diesel hydrogenation and hydrocracking unit.

3. Energy saving effect

    Optimize and increase feed temperature of hydrocracking unit.

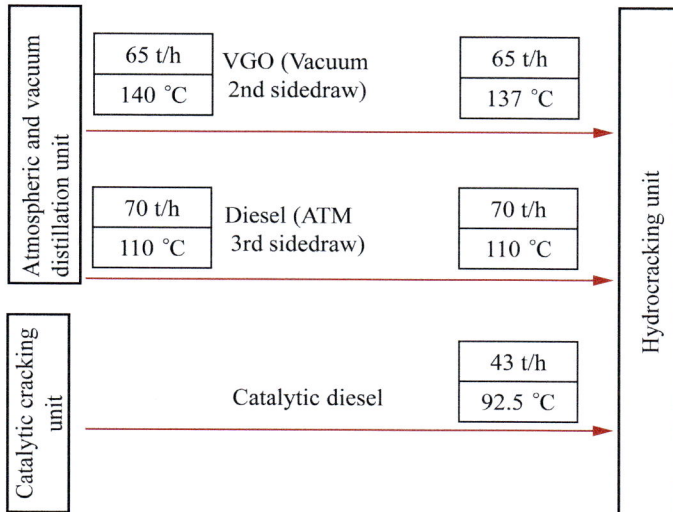

Fig. 6.11 Schematic diagram of hot feed from crude distillation/catalytic cracking unit to hydrocracking unit.

6.3.1.6 Direct hot feed from VGO hydrotreating/continuous reforming unit to catalytic cracking unit

See Fig. 6.12 for the schematic diagram of direct hot feed from VGO hydrotreating/continuous reforming to FCC unit.

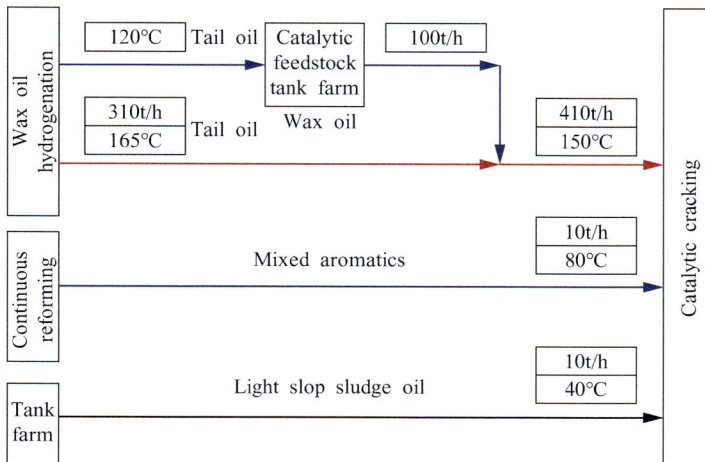

Fig. 6.12 Schematic diagram of direct hot feed from VGO hydrogenation/continuous reforming unit to FCC unit.

1. Existing issues
   ① The direct hot feed is partially realized between the VGO hydrogenation and the FCC unit, but about 25% of the raw material of the FCC unit is mixed with the VGO hydrogenation tail oil after passing through the tank, resulting in the temperature drop of the tail oil from 165°C to 150°C, with large heat loss.
   ② The temperature of heavy aromatics to FCC from continuous reforming is about 80°C, and there is room for optimization and improvement.
   ③ The 1.0 MPa steam generation by hydrogenated gas oil in the VGO hydrotreating unit is suggested to be partially canceled to raise the temperature of the hydrogenated gas oil to that of the FCC unit.

2. Revamping measures
   ① All the hydrogenated gas oil from the VGO hydrotreating unit enters the catalytic cracking unit, without passing through the tank farm. If the temperature maintenance equipment in the tank farm cannot meet the temperature requirements of normal conditions, improve the temperature maintenance equipment in the tank farm. If the temperature drop of the pipeline from the VGO hydrotreating unit to the tank farm and from the tank farm to the catalytic cracking unit is large, repair or replace the insulation material and increase the temperature of the heat tracing steam.
   ② Optimize and increase the temperature of heavy aromatics to 120–140°C in continuous reforming.
   ③ Partially cancel the 1.0 MPa steam generation by hydrogenated gas oil, and raise the temperature of the hydrogenated gas oil to 170–190°C.

3. Energy saving effect
   The feed temperature of the current catalytic cracking unit is about 150°C, and the temperature after optimization is 170–190°C. The catalytic cracking unit can increase 3.5 MPa steam production by 8–10 t/h.

6.3.1.7 Feed for continuous reforming unit

All continuous reforming unit feed comes from the tank farm, and there is no direct feed with related units.

6.3.1.8 Heat integration diagram of the whole plant after improvement

Fig. 6.13 is an optimized "heat flow map" of a refinery. As can be seen from the figure, direct hot feed has been put into practice among the key units of the refinery, which mainly include direct hot feed from crude distillation to VGO hydrogenation, from crude distillation to delayed coking, from crude distillation to kerosene hydrogenation, from crude distillation to hydrocracking, from crude distillation to diesel hydrogenation, from VGO hydrogenation to catalytic cracking, from delayed coking to catalytic cracking, etc. The energy consumption of the refinery can be reduced by about 2.0–4.0 kgoe/ (t crude oil) after the implementation of the across-plant direct hot feed.

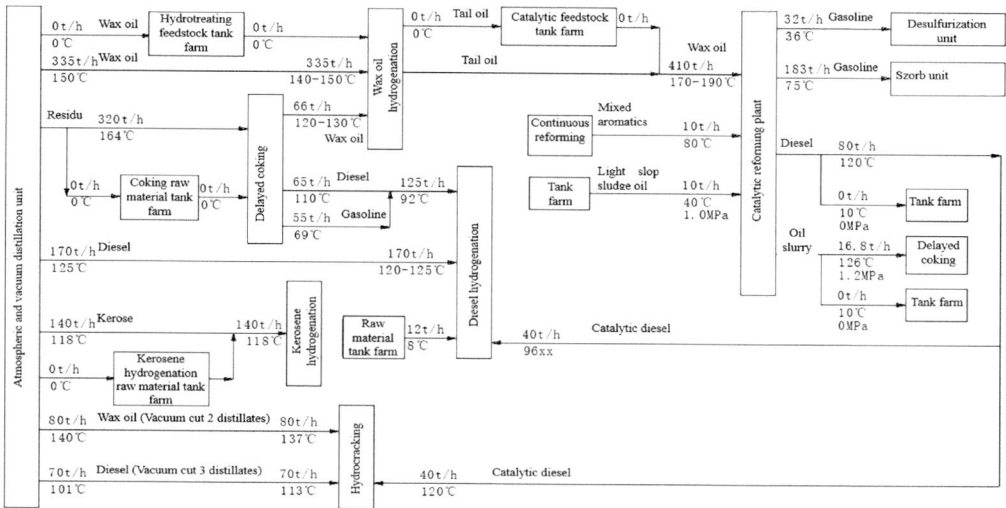

Fig. 6.13 Schematic diagram of hot feed and discharge of units in a refinery.

Atmospheric and vacuum distillation unit

0t/h Wax oil	Hydrotreating feedstock tank farm	0t/h	0°C
0°C		0°C	
335t/h Wax oil		335t/h	140–150°C
150°C			

0t/h Tail oil — Catalytic feedstock tank farm 0t/h — 0°C
Wax oil 410t/h 170–190°C
Tail oil

32t/h Gasoline 36°C — Desulfurization unit
183t/h Gasoline 75°C — Szorb unit

Wax oil hydrogenation

66t/h 120–130°C Wax oil
Continuous reforming — Mixed aromatics 10t/h 80°C
Diesel 80t/h 120°C

Residu 320t/h 164°C

Coking raw material tank farm 0t/h 0°C

Delayed coking

65t/h Diesel 110°C 125t/h 92°C
55t/h Gasoline 69°C

Tank farm — Light slop sludge oil 10t/h 40°C 1.0MPa

0t/h 10°C 0MPa — Tank farm

170t/h Diesel 125°C
170t/h 120–125°C

Diesel hydrogenation

Oil slurry 16.8t/h 126°C 1.2MPa — Delayed coking

0t/h 10°C 0MPa — Tank farm

140t/h Kerose 118°C
140t/h 118°C

Kerosene hydrogenation

Raw material tank farm 12t/h 8°C

40t/h 96xx — Catalytic diesel

0t/h 0°C — Kerosene hydrogenation raw material tank farm

Catalytic reforming plant

80t/h Wax oil (Vacuum cut 2 distillates) 80t/h 137°C
140°C

70t/h Diesel (Vacuum cut 3 distillates) 70t/h 113°C
101°C

Hydrocracking

40t/h 120°C — Catalytic diesel

6.3.2 Centralized layout of units

New refineries in China generally integrate and optimize all units, utilities, and system units as a whole. The main units are generally divided into unit functional areas, such as heavy oil processing, distillate oil processing, gas processing, and environmental protection areas, with relatively centralized layouts in order to facilitate bulk material transportation, reduce pipeline transportation distance, and thus reduce the loss of mechanical energy and heat dissipation.

6.3.2.1 Centralized treatment of refinery gas and expansion of low-temperature heat trap

A lot of gas is produced during refinery processing, which often contains many valuable components, such as light hydrocarbons, hydrogen, ethylene, and ethane. Based on the gas composition, different recovery methods, such as adsorption, absorption, refrigeration, and separation, are taken. It is an important measure and approach to effectively utilize the resources by adopting integrated technology to recover components with low energy consumption. At the same time, the gas recovery process of the refinery is a large system using low-temperature heat to expand a new heat trap. On the one hand, the reboiling temperature of the gas distillation and separation process is generally not high, and hot water with a temperature between 70°C and 90°C is a good heat source for this kind of reboiler. On the other hand, the process of gas refrigeration separation requires a large amount of low-temperature water, and the lithium bromide low-temperature cooling water can fully meet the production needs, which is a good way to utilize the high value-added of hot water in refineries.

1. Centralized light hydrocarbon recovery facilities

    Fuel gas from conventional refineries is processed into high-value–added products. In the past, much of the hydrogen-containing byproduct gas in refineries was burned off as fuel, while large quantities of valuable chemical feedstocks such as naphtha, saturated propane, and butanes were used to produce hydrogen. According to preliminary statistics, currently 90% of raw materials for ethylene cracking are naphtha, kerosene, diesel, and hydrogenated tail oil, and the average yield of ethylene is only about 32%, while the light hydrocarbons and liquefied petroleum gas, such as ethane, propane, and butane, produced by refineries every year, are burned as industrial or civil fuels, resulting in low-value utilization of resources. Therefore, the refinery should set up light hydrocarbon recovery facilities, dry gas recovery facilities, and hydrogen recovery facilities. The liquefied petroleum gas, hydrogen, ethylene, ethane, and other value-added components, which were once burned as fuels, should be recovered. For example, gas fractionation, MTBE, alkylation, and other units can be set up in the processing to separate propane and propylene in liquefied petroleum gas as a chemical feedstock, and the isobutene, isobutene, etc., in liquefied petroleum gas can be used to produce MTBE, alkylation oil, and other gasoline blending components.

2. Management and safe utilization of hydrogen resources

    The consumption of hydrogen in hydrogenation-type refineries is large. Given the different requirements of hydrogen conditions of different units, the use of hydrogen systems in refineries should be reasonably planned, to avoid the energy waste caused by the pressure reduction and boost of hydrogen system, which can improve the utilization of hydrogen and reduce the cost of hydrogen use. The hydrogen source of conventional refineries mainly comes from hydrogen-containing gas from continuous reforming and pure hydrogen from hydrogen production units. The hydrogen pipeline network across the plant is divided into two or more systems according to the hydrogen source and usage. For example, the hydrogen-containing gas from continuous reforming is mainly sent to the hydrotreating units with low hydrogen partial pressure requirement, and the rest enters the hydrogen recovery and purification system of the refinery. The purified hydrogen and the pure hydrogen produced by the hydrogen production unit are supplied to high-pressure hydrotreating units. Connected lines should be set between the pipeline networks to ensure the safe utilization of hydrogen in the hydrotreating units across the plant.

    For a hydrogenation refinery of 10 Mt/a, the hydrogen recovery system can recover about 7000 tons of hydrogen per year. At the same time, the hydrogen-rich PSA tail gas can be used as raw material for hydrogen production units after being boosted. This innovation not only provides a way out for PSA tail gas with low calorific value but also makes reasonable use of resources and saves a lot of raw material for hydrogen production. In addition, it can consume the surplus hot water obtained by the low-temperature heat recovery of the process, reduce the process cooling duty, and save electricity and circulating water.

6.3.2.2 Centralized layout of environmental protection units to improve the energy utilization level of low-pressure steam

Environmental protection units in refineries generally include sour water stripping, solvent regeneration, sulfur recovery, and sewage treatment units. Environmental protection units in traditional refineries are mostly scattered, which is not conducive to energy saving and consumption reduction and also increases potential safety risks. Acidic water stripping and solvent regeneration are also important low-pressure steam consumers in refineries. Taking a 10 Mt/a refinery as an example, 0.4 and 1.0 MPa steam consumed by sour water stripping and solvent regeneration units are about 120–150 t/h, and the steam energy consumption accounts for about 5.0%–10.0% of the energy consumption of the entire refinery. The integrated and refined setting of environmental protection units in refineries is of great significance for optimizing steam power systems and improving the energy utilization level of refineries.

The centralized layout of sour water stripping mainly considers that the sour water produced by hydrogenation and nonhydrotreating units contains different kinds and amounts of ammonia, phenols, cyanide, and other impurities. By treating them separately, on the one hand, it reduces the water pollution by each other, and on the other hand, it reduces the extra energy consumption due to mixed processing, reducing low-pressure steam consumption. The centralized layout of solvent regeneration units also considers the treatment of amine-rich solutions of hydrogenation and nonhydrotreating units, respectively, with regenerated solution sent to different units. To achieve unified management and centralized regeneration, one key is to reduce the consumption of low-pressure steam and reduce the energy consumption of the plant, and the other key is to avoid the long-distance transportation and pressure loss of highly acidic gas in the refinery and reduce the corrosion risk of pipelines.

6.3.3 Steam power system heat transfer enhancement technology

6.3.3.1 Steam power system overview

Steam power system is an important part of petrochemical enterprises. Its task is to convert primary energy into secondary energy, to provide steam, power, electricity, and heat for the production process. The safe and stable operation of steam power systems is the basis for the safe, stable, and long-term operation of petrochemical enterprises. The steam power system of petrochemical enterprises is usually composed of power stations (thermal power stations), steam transportation, distribution and balance facilities, steam and heat users, water deoxygenation and condensate recovery, etc.

1. Steam generation and power generation facilities of power station

    Typical steam generation and power generation facilities in petrochemical power stations (thermal power stations) include the boiler, steam turbine generator set, temperature reducer, deaerator, boiler feed pump, and other auxiliary facilities. In the power station (thermal power station), the boiler utilizes the primary energy (such as coal, fuel oil, natural gas, refinery gas, petroleum coke, etc.) to generate high-pressure or medium-pressure steam. Subsequently, the high-pressure or medium-pressure steam is converted into electric energy through the steam turbine generator, which is transmitted to the whole plant's electricity users. Simultaneously, the power station (thermal power station) will produce steam at various pressure levels for distribution throughout the plant's steam pipe network.

2. Steam transportation, distribution, and balancing facilities

The steam transportation, distribution, and balance facilities of petrochemical enterprises are composed of the steam main network of the whole plant, the branch steam network in each unit, the temperature reduction and decompression facilities as balance and regulation, and the overpressure discharge facilities. Each grade of steam generated in the system is distributed to different users by distribution and balancing facilities such as the main network, branch network, and subcylinder. Temperature reduction and decompression facilities are used as steam duty regulation under startup and various accident conditions, such as the steam supply bypass backup for the extraction steam of the condensate extraction unit or the exhaust steam of the back pressure turbine, which can realize automatic input and rapid start, ensuring that the steam duty requirements of all grades of the plant can be met under startup and various accident conditions.

3. Steam users and steam production facilities in the unit

Steam subsystem in the plant (system unit) of a petrochemical enterprise consists of steam generation equipment such as waste heat boiler and steam generator, steam users such as heat exchangers and reboilers, and process-driven turbines. After the steam conveying subsystem delivers the steam to each unit (unit), the steam is distributed to each steam point within the unit (unit) for process heating or turbine work. In addition, part of the refinery chemical plant can utilize flue gas or the waste heat of process flows to generate steam. The steam generated by the waste heat boiler (or steam generator) in the plant is initially employed within the plant, with any excess entering the steam pipe network of the entire facility.

4. Deoxygenation of water supply and recovery of condensate

The deoxygenation facilities for water supply in petrochemical enterprises include a deaerator, boiler feed pump, and dosing equipment. The demineralized water and qualified condensed water, after treatment, enter the deaerator for deoxygenation, and the deoxygenated water is transported to the steam production equipment, such as the power boiler, waste heat boiler, and steam generator, after being boosted by the boiler feed pump.

The condensate recovery facility will recover all kinds of process condensate and turbine condensate generated by each unit. Subsequently, it will treat them separately based on the quality of the process condensate and turbine condensate. The treated condensates will then be reused as makeup water for the deaerator.

6.3.3.2 Heat transfer enhancement measures for steam power system

In order to provide steam, power, and heat to petrochemical processes, steam power systems need to consume a large amount of primary energy such as coal, natural gas (refinery gas), petroleum coke, and fuel oil. The energy consumption of steam power systems occupies a considerable proportion of petrochemical enterprises. It is key to reduce energy consumption of steam power systems by using heat transfer enhancement technology.

According to the characteristics of steam system energy consumption, the major directions of process intensification and design optimization of steam power system are as follows:

① According to the temperature of the furnace or heat source, the heat transfer enhancement technology is used to improve the steam production parameters.

② Taking advantage of the exergy of flue gas after primary fuel combustion, heat transfer enhancement technology is used to reduce the exhaust temperature of the boiler (waste heat boiler) and improve the effective utilization of fuel.

Process intensification and optimization approaches of the steam generation system are as follows:

1. Improvement of steam production parameters of waste heat boiler in catalytic cracking unit by adopting heat transfer enhancement technology.

   The waste heat boiler, catalyst cooler, and main column bottoms steam generator of the catalytic cracking unit in petrochemical enterprises usually generate 3.82 MPa medium-pressure steam. For large catalytic cracking units with incomplete regeneration technology, the CO in the flue gas needs to be burned in the furnace of the waste heat boiler. In order to improve the reaction rate and control the CO content at the outlet of the boiler to meet the requirements of relevant environmental indicators, the furnace temperature of the waste heat boiler is usually 800–850°C. Exergy loss will be large if the flue gas temperature is used for overheating and generating 3.82 MPa medium-pressure steam. In addition, the temperature of the catalyst in the catalyst cooler is about 690°C, and high-pressure saturated steam can be generated. Therefore, based on the principle that the steam generation system should generate high-grade steam as far as possible corresponding to the heat source temperature level, the steam generation system is optimized by heat transfer enhancement technology. The waste heat boiler of the catalytic cracking unit and the catalyst cooler generates 10.0 MPa high-pressure steam, and the generated high-pressure steam drives 10.0/3.82 MPa backpressure generator set for power generation. It can greatly reduce the energy consumption and operation cost of the unit.

   The steam generation system is optimized by taking a 2.8 Mt/a catalytic cracking unit as an example. The waste heat boiler of the unit is designed with double pressure, and the high-pressure saturated steam generated by the catalyst cooler and the high-pressure saturated steam generated by the boiler is overheated in the high-pressure overheating section of the waste heat boiler, which could generate 192 t/h high-pressure steam with 10.0 MPa and 540°C. The 192 t/h high-pressure steam generated by the waste heat boiler is sent to the 10.0/3.82 MPa backpressure generator set, which can generate 13 MW of power under normal conditions. The slurry temperature in the main column bottom steam generator needs to be cooled from 325°C to 275°C. Given the temperature constraint of the heat source, 71 t/h medium-pressure saturated steam is generated in the main column bottom steam generator. It is overheated to 3.82 MPa, 420°C in the medium-pressure overheating section of the waste heat boiler and sent to the medium-pressure steam pipeline of the entire plant. Through the optimization design of heat transfer enhancement technology, without increased fuel consumption, the unit can generate 13 MW more electricity, and the energy consumption of the unit is reduced by 4.45 kg standard oil/ton raw material. The comparison of 3.82 MPa medium-pressure steam and 10.0 MPa high-pressure steam generated by 2.8 Mt/a catalytic cracking unit is shown in Table 6.14.

Table 6.14 Comparison of steam generation system options in catalytic cracking unit.

Item	Conventional option (medium-pressure steam)	Heat transfer enhancement optimization option (high pressure + medium pressure steam)
Steam generation pressure (MPa)	3.82	10.0/3.82
Steam generation temperature (°C)	420	540/420
Steam flow (t/h)	280	High pressure 192 + medium pressure 71
Backpressure generator set (MW)	0	13
Investment of boiler, generator set, and auxiliary facilities/10,000 yuan	9485	15,102
Demineralized water consumption (t/h)	285	265
Boiler self-consumed power (kW)	1520	2070
Generation unit circulating water consumption (t/h)	0	300
Energy consumption (kgoe/(t raw materials))	44.84	40.39
Payback period for incremental part, year		3.4

2. Use heat transfer enhancement technology to improve steam production parameters of low-pressure steam in the aromatics unit.

The raffinate column and the extract column of the Parex unit in the aromatics complex are normally designed as atmospheric columns. The overhead temperature of the columns ranges from 145°C to 150°C, and the temperature after cooling is between 121°C and 130°C. The overhead temperature of the column varies slightly with pressure. The waste heat from the raffinate column overhead and the waste heat from the extract column overhead of the aromatics complex are two major low-temperature waste heat resources with large heat duty. Taking a 600 kt/a aromatics complex unit as an example, the waste heat duty at these two locations is about 100 MW. Recycling and utilizing this part of low-temperature waste heat appropriately can significantly reduce the energy consumption of the aromatics complex unit. When the raffinate and extract columns are operated at atmospheric pressure, saturated steam of about 0.25 MPa can be generated at the column overhead. After overheating and pipeline transportation, the pressure at the generator is generally below 0.2 MPa. If steam at this grade is used to drive turbines for power generation, the conversion efficiency of steam and electricity is low. In addition, due to the low steam pressure and high heat capacity, the diameter of the steam pipeline is large, the area of the condensing generator set is large, the investment of equipment and pipelines is high, the project investment payback period is long, and the technoeconomic indicators of waste heat utilization are poor.

Through the application of heat transfer enhancement technology, the economy of low-temperature heat recovery at the top of the columns can be improved by appropriately increasing the operating pressure of the two columns. When the pressure of the raffinate column and the extract column is raised to 0.25 MPa and 0.28 MPa, respectively, with the reflux temperature at about 180°C, 180 t/h saturated steam under 0.5 MPa can be generated. After overheating to 180–190°C in the convection section of the reboiler at the bottom of the xylene column, it can be used to generate electricity, and the net power generation can be 18,000 kW. The net thermoelectric conversion efficiency can reach more than 15%. With low-pressure steam power generation technology, the energy consumption of the aromatics unit can be reduced by about 50 kgoe/(t PX).

3. Use of heat transfer enhancement equipment to reduce the exhaust temperature of the boiler

According to statistics, each petrochemical enterprise has a number of power boilers and a greater number of waste heat boilers of various types. The exhaust temperature of these boilers is relatively high. The survey data of exhaust temperature of power boilers and waste heat boilers in petrochemical enterprises are shown in Table 6.15.

Table 6.15 Summary of boiler smoke exhaust temperature.

Type of boiler	Power boiler	FCC unit waste heat boiler	Sulfur recovery unit waste heat boiler
Exhaust temperature range (°C)	120–180	160–220	250–300

It can be observed from Table 6.15 that the exhaust temperature of power boilers and waste heat boilers is high, making the flue gas of these boilers a valuable heat source. Historically, due to the high dew point and strong corrosion of boiler flue gas, there has been a lack of flue gas heat exchange equipment with good corrosion resistance for long-term operation. This has resulted in the inefficient use of flue gas energy. The high exhaust temperature of power boilers and waste heat boilers leads to a waste of valuable waste heat resources and an increase in the consumption of spray cooling water in subsequent desulfurization systems. Whether the boiler exhaust temperature can be further reduced is a key factor for enterprises to save energy, reduce energy consumption, and enhance boiler efficiency.

The use of the innovative nonmetallic flue gas heat extractor against low dew point corrosion can enable the deep utilization of the flue gas waste heat of the boiler so that the exhaust temperature of the boiler can be revolutionarily reduced. By adopting the flue gas deep utilization technology, the expected smoke exhaust temperature of power boilers and waste heat boilers is shown in Table 6.16.

Table 6.16 Expected boiler exhaust temperature after adopting flue gas deep utilization technology.

Type of boiler	Power boiler	FCC unit waste heat boiler	Sulfur recovery unit waste heat boiler
Existing boiler Exhaust temperature range (°C)	120–180	160–220	250–300
Boiler Exhaust temperature range heat transfer enhancement (°C)	80–90[a]	80–90[a]	80–90[a]
Flue gas waste heat duty (MW)	5–50	5–25	2–10

[a]The flue gas waste heat duty of power boilers and waste heat boilers can be accounted for 80°C or more. If there is a heat trap with a lower temperature, such as in-room heating, the exhaust temperature can be further reduced.

Through an in-depth study of boiler flue gas properties, the dew point corrosion problem, which hinders the reduction of exhaust temperature, has been successfully addressed. Pioneering efforts have led to a reduction in the boiler exhaust temperature to less than 100°C. On this basis, a thermal water supply system is set up to send the boiler flue gas waste heat taken out by nonmetallic flue gas heat exchange technology to the waste heat power generator or the waste heat refrigerator for upgraded utilization. The deep utilization technology of boiler flue gas waste heat can be combined with the new technologies of waste heat utilization, such as waste heat power generation and waste heat refrigeration. By lowering the boiler exhaust temperature, the flue gas cooling water consumption of the subsequent desulfurization system can be significantly reduced, and water saving and dust removal can be realized while saving energy.

Taking a set of waste heat boiler of a 2 million tons/year FCC unit as an example, the exhaust temperature of the boiler is 180–200°C, and the flue gas conditions at the outlet of the waste heat boiler are shown in Table 6.17.

Table 6.17 Flue gas parameters of waste heat boiler in FCC unit.

Flue gas composition	O_2	CO_2	N_2	H_2O
Flue gas content (%)	2.25	14.11	73.54	10.1
Flue gas flow (m^3/h)		292,384		
Flue gas temperature (°C)	180–200			

Adopting the deep utilization technology of boiler flue gas waste heat reduces the exhaust temperature of FCC waste heat boiler from 180–200°C to 80–90°C; 12–13 MW of waste heat recovered by the thermal water system is sent to the ORC waste heat generator set. The waste heat generator set can achieve net power generation of 1100–1200 kW, and the energy consumption of the unit is reduced by 1.03–1.12 kgoe/(t raw materials). The total investment, including flue gas heat recovery, ORC unit, and auxiliary facilities, is 20.5 million yuan, and the annual power generation benefit (with the taxed price of electricity being 0.7 yuan/(kW h)) is 6.468–7.056 million yuan. The simple investment payback period of the project is about 3 years.

6.3.4 Heat transfer enhancement in the low-temperature waste heat utilization system

6.3.4.1 Heat transfer enhancement technology in low-temperature waste heat utilization system

The heating and cooling process of material flows in petrochemical production will generate a large amount of waste heat, mainly low-temperature waste heat; the temperature is mostly in the range of 50–200°C.

1. Overview of major waste heat resources in petrochemical enterprises

    The low-temperature waste heat of petrochemical enterprises can usually be subdivided into three temperature ranges: the high-temperature heat source of 150–200°C, the medium-temperature heat source of 80–50°C, and the low-temperature heat source of 50–80°C.

    The waste heat at 150–200°C in petrochemical enterprises has been mostly utilized, and the key issue is the comprehensive and effective utilization of a large amount of waste heat resources at 50–200°C.

Low-temperature waste heat is usually generated in process waste heat, flue gas waste heat, utilities, and auxiliary facilities.

Waste heat resources are mainly generated in separation, transformation, refining, upgrading, polymerization, oxidation, and other processes. Flue gas waste heat is generated in furnaces, power boilers, and waste heat boilers of units. Waste heat in utilities and auxiliary facilities includes waste heat of condensed water, sewage discharge from steam generators, circulating cooling water, steam turbines, deaerators, and waste expansion vessels.

According to incomplete statistics, the amount of waste heat resources at medium and low temperatures, at 80–150°C and with utilization potential, is about 4000–5000 MW in the refining and chemical enterprises within the scope of Sinopec Group. According to the relevant data estimation, in the petrochemical industry of China, the amount of potential waste heat resources of medium and low temperatures at 80–150°C is about 20–30 GW, which has great utilization potential.

The following basic principles should be followed in the cascade utilization of waste heat resources:

① Production units and system units should prioritize low-energy technology, and select high-efficiency equipment and facilities to achieve significant energy savings. Optimize the heat exchange network process within the range of the complex plant or the entire plant, striving to reduce or eliminate the generation of low-temperature waste heat.

② Based on the grade and properties of waste heat resources, system optimization can enable the effective cascade utilization of waste heat and improve the recovery and utilization efficiency of waste heat resources. The high-temperature waste heat is preferably allocated for the high-temperature heat users, while the low-temperature waste heat is prioritized for the low-temperature heat users.

③ Priority should be given to continuous, stable, and same-grade utilization. Based on same-grade utilization, upgraded utilization technologies such as waste heat power generation and waste heat refrigeration can be considered. It is preferred to use low-temperature heat instead of steam and other secondary energy sources consumed by the original production process, such as using waste heat to heat the demineralized water, using hot water as the heat source of the reboiler of the gas separation unit, and using hot water as the heat source of the oil tank heating and temperature maintenance in the storage and transportation system.

④ The utilization of waste heat resources will not affect the normal production of the plant while ensuring the safety and stability of the operation of the plant. The safety production of the plant will not be affected when the heat source, heat trap, and production plan are changed.

⑤ For waste heat utilization, consider heat combination first before recovery, consider industrial use before living use, and consider the use within the plant before use outside the plant (park).

2. Low-temperature waste heat utilization technology approach

From the perspective of process thermodynamics, low-temperature heat utilization technology approach has two categories: one is the same-grade utilization, and the other is upgraded utilization.

For the low-temperature waste heat at 50–200°C in refining and chemical enterprises, the optimized heat transfer network is first used to directly utilize the heat at the same grade. Secondly, it can also be used for space heating and heat tracing. In the northern region, because of the need for space heating and heat tracing in winter, the use of low-temperature waste heat in winter is more adequate, and there will be a surplus in summer. However, in the south, there is a surplus of low-temperature heat in both winter and summer.

On the basis of the same-grade utilization of heat exchange, space heating, and heating tracing, if there is a large amount of excess low-temperature waste heat, low-temperature thermal power generation technology or low-temperature thermal refrigeration technology can be considered for upgraded utilization. There are three main ways to upgrade and utilize low-temperature waste heat: heat pump, refrigeration, and power generation.

① Heat pump is a technology utilizing waste heat for heating.

② Waste heat refrigeration technology is usually used in crude distillation, catalytic cracking, and coking units. Through waste heat refrigeration technology, cold water instead of circulating water cooling is used to absorb and stabilize the system, improving the product yield while saving energy.

③ Waste heat power generation. Low-temperature waste heat power generation is an important approach to energy recovery, which can directly convert low-temperature thermal energy into high-grade electric energy. When it is difficult to find a suitable way to recover a large amount of excess low-temperature heat, low-temperature thermal power generation is an effective method. Low-grade thermal energy can be converted into high-grade energy—electricity through low-temperature thermal power generation units. The generator set can be flexibly configured in accordance with the heat source situation. Compared with the low-temperature thermal refrigeration unit, it is not limited by the cold water cooling transmission distance, cold source matching, and other factors and has greater promotion potential.

3. Low-temperature waste heat power generation technology

Low-temperature waste heat power generation technology is a typical technology path of heat transfer enhancement technology in the utilization of low-temperature waste heat in petrochemical enterprises. The low-temperature thermal power generation technologies used in petrochemical enterprises mainly include flash steam power generation, low-pressure saturated steam power generation, Kalina cycle power generation, and organic Rankine cycle power generation.

① Flash steam power generation technology has been tried and practiced in some projects of petrochemical enterprises such as Changling, Luolian, Jinxi, and Jinzhou since the 1980s. Most of the flash steam power generation units have been dismantled, and only one unit is still in operation. The basic situation of the flash steam turbine generator set of a petrochemical enterprise is as follows.

Heat source: FCC unit hot water; hot water temperature: 120°C/75°C; hot water flowrate: −750 t/h

Unit turbine inlet steam parameters:

$$\text{Primary steam intake}: P = 0.14\,\text{MPa}, T = 114°\text{C}$$
$$\text{Secondary steam intake}: P = 0.049\,\text{MPa}, T = 80.8°\text{C}$$
$$\text{Exhaust steam parameters of turbine}: P = 0.0088\,\text{MPa}, T = 42.5°\text{C}$$
$$\text{Unit design rated power}: 3000\,\text{kW}$$
$$\text{Unit self} - \text{use power rate}: 30\% - 35\% \text{ of the installed scale of the unit}$$

The thermal efficiency and economic benefit of the flash steam turbine generator set are greatly affected by the temperature of the heat source and circulating water temperature. The overall thermal efficiency of the unit is low, the project economic benefit is not high, and the investment payback period is long. Since the 1990s, the flash steam power generation technology has not been used to recover low-temperature heat in new projects of petrochemical enterprises.

② Low-pressure saturated steam power generation technology uses the low-temperature waste heat of units to generate low-pressure saturated steam and uses the generated low-pressure saturated steam as a working medium to drive a steam turbine for power generation or drive compressor. In a 0.6 Mt/a PX unit of an enterprise, the low-temperature waste heat of the process is used to produce 160–170 t/h of 0.4 MPa saturated steam for power generation of 17–18 MW. The plant has changed from being a power consumer of 15–16 MW to a power supplier of about 2–3 MW. Compared with the advanced equipment at home and abroad, the energy consumption of the plant is reduced, which is at the leading level in the world. The cost per ton of PX is lower than that of similar plants, resulting in considerable economic benefits.

③ Kalina cycle power generation technology uses 80%–90% ammonia water as the circulating working medium. The hot water heated by the waste heat of the processing unit enters the evaporator of the Kalina generator set and heats up the circulating working medium. The water in the circulating working medium that leaves the evaporator returns to the unit. The ammonia steam enters the steam turbine to expand and drive the generator to generate electricity. The steam from the steam turbine enters the heat exchanger and condenser to condense into ammonia. Then the ammonia water is pumped to the heat exchanger and evaporator to form a Kalina cycle (Fig. 6.14).

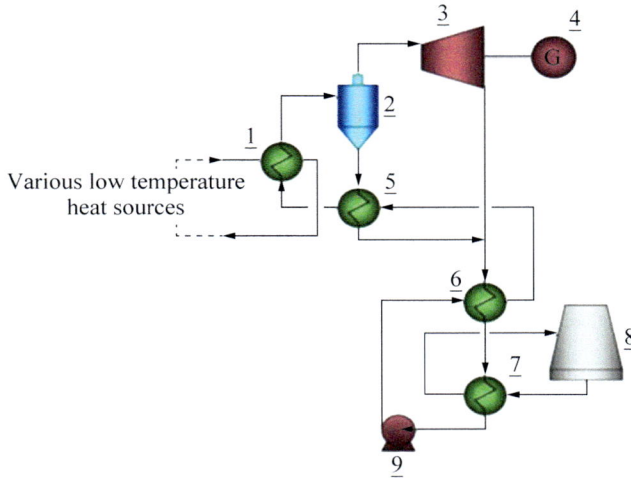

Fig. 6.14 Flow diagram of Kalina cycle power generation process. (1) Evaporator; (2) separator; (3) steam turbine; (4) generator; (5) heat exchanger 3; (6) heat exchanger 2; (7) condenser; (8) cooling tower; (9) ammonia pumps.

This technology is adopted in the HUSAVIK power plant in Iceland and Unterhaching geothermal power station in Munich, Germany, where the heat source is geothermal water and the temperature is 121°C (Table 6.18). HUSAVIK power plant in Iceland has an installed capacity of 2MW, and Unterhaching geothermal power station has an installed capacity of 4MW. In a domestic PX project, China's first Kalina cycle power generator using circulating hot water uses 780 t/h of 70/120°C hot water to generate power. The installed capacity is 4MW, and the net power generation under the design working condition is 3151 kW.

Table 6.18 Performance test results of HUSAVIK power plant in Iceland using Kalina cycle.

Item	Data
Geothermal hot water flow (t/h)	324
Geothermal hot water temperature (°C)	121
Cooling water temperature (°C)	5
Generating power (gross) (kW)	1836
Plant power (kW)	127
Net power generated (kW)	1709
Power generation per ton of hot water (kW)	5.3

Organic Rankine cycle (ORC) power generation technology is a Rankine cycle with a low–boiling–point organic working medium. As shown in Fig. 6.15, the organic working medium absorbs heat from the residual hot flow in the heat exchanger, generating steam with a certain pressure and temperature. The steam enters the turbine to drive the generator or other power machinery. The steam discharged from the turbine condenses into a liquid state in the condenser and then returns to the heat exchanger the pump is boosted, so as to form an ORC cycle.

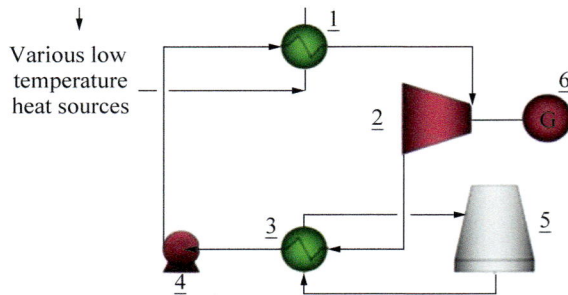

Fig. 6.15 Flow diagram of organic Rankine cycle (ORC) power generation process. (1) Evaporator; (2) expander; (3) condenser; (4) working fluid pump; (5) cooling tower; (6) generator.

ORC power generation technology applications began in the 1950s. It is a widely used industrial technology in Europe, the United States, Japan, and Israel and is mainly used in industrial waste heat recovery and geothermal power generation projects.

ORMAT (Israel), UTC/Pratt & Whitney (United States), OPCON AB (Sweden), Turboden (Italy), and Exergy (Italy) are the main suppliers of ORC power generation technology and equipment in the world. As of 2012, ORC power generation technology has been applied in 2000 low-temperature waste heat and geothermal power generation projects in dozens of countries. In recent years, some leading enterprises have made breakthroughs in the research and development, design, and manufacturing of domestic ORC expanders and have successively put into operation industrial test units.

6.3.4.2 Practical application cases of heat transfer enhancement technology in low-temperature waste heat utilization

1. Energy cascade utilization technology of aromatics unit

The process of the aromatics complex unit is rather long, with many circulating materials, a complicated separation process, and multiple distillation columns. Thus a large amount of low-temperature heat needs to be condensed at the top of the columns. The waste heat has a low temperature, mainly at 90–160°C. Because of its low temperature grade, it is difficult to be utilized within the unit, so air coolers and water coolers are usually used for cooling. Due to the low temperature of the heat source, there exist problems of difficult recovery and low economic benefits. Currently, the conventional method is to cooperate with the overall planning of low-temperature heat utilization of the plant and to use demineralized water or deoxygenated water as the circulating heat medium to recover the low-temperature heat of the unit, which is used for hot water heat tracing, tank farm temperature maintenance, space heating, etc. These methods can only recover part of the low-temperature heat, and there are seasonal variations. Aromatics complex plant is an intense energy consuming plant, and one of the reasons is that the low-temperature heat cannot be effectively used.

Through heat transfer enhancement technology, a large amount of waste heat is recovered and utilized by the deep energy integration of the aromatics unit and the low-temperature waste heat power generation technology, to realize the energy cascade utilization of the aromatics unit. Low-temperature waste heat power generation is an important form of energy recovery, which can directly convert low-temperature thermal energy into high-grade electric energy. When there is a large amount of low-temperature heat surplus that is hard to recover, the use of low-temperature thermal power generation is an effective way.

An aromatics complex with a capacity of 0.6 Mt PX per year recovers more than 100 MW of low-temperature heat using energy cascade utilization technology. The energy cascade utilization technology of the aromatics unit includes two parts: low-pressure steam power generation and hot water power generation:

① Low-pressure steam power generation. It is the first technology in the world to adopt pressurized operation of the raffinate column and the extract column so that the temperature at the top of the columns rises to 195–200°C for 0.5 MPa steam generation.

Raffinate column: with recoverable heat 78.252 MW, it can be used to generate 0.45 MPa (gauge pressure) steam at 121 t/h. Extract column: with recoverable heat 26.066 MW, it can be used to generate 0.45 MPa (gauge pressure) steam at 41 t/h.

The 162 t/h of 0.5 MPa steam generated from the waste heat at the top of the raffinate column and the extract column is overheated by the convection section of the reboiler of the xylene column, and a small amount of the steam provides power for the circulating hydrogen compressor at the toluene disproportionation while most of the steam is used for power generation. The installed capacity of the low-pressure steam generator set is 20 MW.

② Hot water power generation. There are some low-temperature heat sources lower than 130°C in the aromatics unit, which are difficult to use to generate steam. A hot water system is used for series/parallel heat exchange with each process material, and the recovered low-temperature heat is centralized to the hot water generator set for power generation. The hot water power generation can use Kalina cycle power generation technology or organic Rankine cycle power generation technology given the heat source temperature and heat source/trap ratio. The installed capacity of the hot water generator set is 4 MW.

A total of 24 MW waste heat power generation units is set by the energy cascade utilization technology of the aromatics unit, which has achieved a historic breakthrough from electricity consumption to external power supply. The project saves 51,429 tce/a and reduces CO_2 emission by 135,773 t/a. The economic benefits of low-temperature waste heat power generation are remarkable, and the annual cost saving is over 100 million yuan.

2. Application of comprehensive utilization technology of low-temperature waste heat

Based on heat exchange network optimization, the across-plant low-temperature waste heat system is established by fully considering the specific conditions such as low-temperature heat sources and traps, temperature grade, and plant layout of the petrochemical enterprise. According to the principle of "temperature matching, cascade utilization," the heat sources scattered in each unit are concentrated and supplied to heat traps scattered in different places at different temperature grades, so as to enable globally optimized utilization of low-temperature heat. The organic Rankine cycle power generation technology can recover concentrated waste heat at relatively high temperatures. The waste heat refrigeration technology can utilize low-temperature waste heat to drive refrigerators to produce cold water which can be used as a cooling medium for column overhead at units such as FCC, coking, and gas separation. The hot water system can recover excessive low-temperature waste heat for space heating, which saves the cost of heating by steam with high temperature grade. Through the comprehensive utilization of waste heat power generation, refrigeration, heating, and so on, the waste heat of the whole plant can achieve "high energy for high-grade use, low energy for low-grade use" so that the energy utilization can be optimized.

For example, a petrochemical enterprise has many low-temperature waste heat resources in FCC, coking, hydrocracking, diesel hydrogenation, jet hydrogenation, and desulfurization units. The waste heat is not fully utilized, and a large amount of additional electricity and circulating water need to be consumed for air and water cooling. Meanwhile, 1.0 MPa steam is used for low-temperature heating at the gas separation unit as well as for space heating. In addition, electric/ammonia refrigeration units with high energy consumption are still used to provide cold water for the reforming and gas recovery units. The whole plant has high energy consumption with "high-grade energy for lower-grade use." To establish a comprehensive system of low-temperature waste heat cascade utilization across the plant, the waste heat resources of the entire plant are assigned to several regions which help enable the approaches of both same-grade utilization and upgraded utilization. The regions should fully take account of factors such as waste heat sources, heat traps, temperature grades, plant layout, and processes of different units. Besides direct heat exchange between units, the low-temperature waste heat recovery system of the recognized regions should be carefully designed with appropriate hot water flow rate and temperature. By the principle of "temperature matching and cascade utilization," the waste heat resources scattered in various units can be concentrated through the series/parallel heat exchange with hot water and then supplied to heat traps at different places and different temperature levels, so as to realize optimized matching of cold and hot flows and maximized heat recovery and utilization. The schematic flow diagram of the regional hot water system is shown in Fig. 6.16.

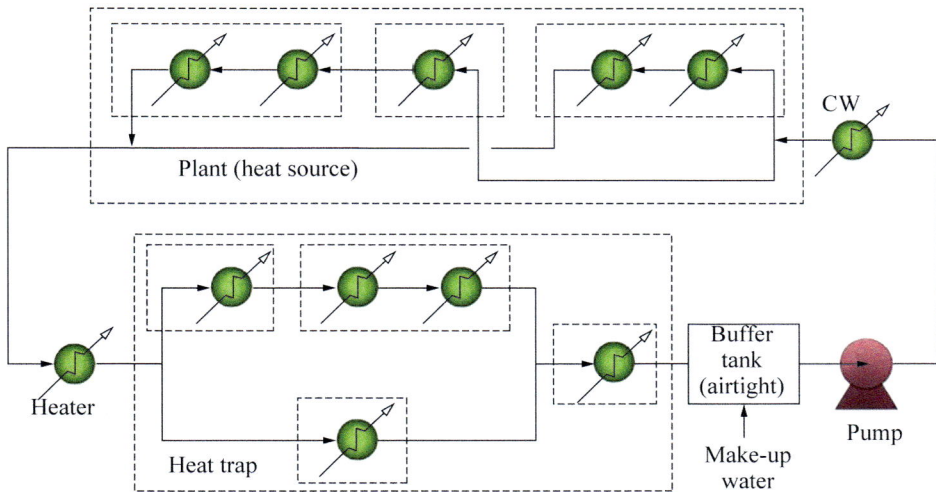

Fig. 6.16 Schematic diagram of regional hot medium water system of a petrochemical enterprise.

Besides passive utilization of waste heat, the waste heat recovery and utilization approaches and requirements of each unit are proposed in accordance with different waste heat utilization technologies, such as waste heat power generation, waste heat refrigeration, heat pump, and waste heat space heating. Through scheme optimization, the comprehensive waste heat cascade utilization project of the enterprise sets up three regional subsystems.

① First comprehensive utilization area of waste heat (including coking, high-pressure hydrocracking and diesel hydrotreating units):

Recover the low-temperature heat of coking, high-pressure hydrocracking, and diesel hydrotreating units. Through comprehensive waste heat utilization technologies such as demineralized water preheating, waste heat power generation, waste heat refrigeration, and heat pump, the deoxygenated steam of the thermoelectric boiler demineralized water preheating is saved, the electric refrigeration unit of the saturated gas recovery unit is replaced, and the circulating water of the coking unit is replaced by cooling water. The operating conditions are optimized and the LPG yield increases.

② Second comprehensive utilization area of waste heat (including 3# FCC, gas separation, and reforming unit):

The waste heat from the main column top pump around, column overheads and light diesel of the FCC unit is used to generate 70–90°C hot medium water, which is supplied to the reboiler of the fractionator of the gas separation unit to replace the 1.0 MPa steam to avoid the waste of high-grade energy. Taking into account the cooling duty and demand of the newly built reforming unit, as well as the cooling demand of various optimization schemes of the absorption and stability section, a waste heat utilization method is proposed for the reforming unit that the waste heat is first used for refrigeration and then for power generation, realizing the combination of the waste heat power generation and waste heat refrigeration of the reforming unit.

③ Third comprehensive utilization area of waste heat (including 2# FCC, gas separation, FCC gasoline desulfurization, and jet coal hydrotreating units):

Considering the heat trap duty of gas separation units and the demand for winter space heating, as well as the technical characteristics of waste heat generation, refrigeration, heat pump, heating, and other low-temperature waste heat utilization, the optimization requirements of waste heat recovery and utilization of eachunit are proposed, so as to realize the integrated and optimized utilization of waste heat resources of each unit in the region. According to the principle of "high energy for high-grade use, low energy for low-grade use" and the principle of "power generation based on heat resource" for the configuration of the waste heat generator, the waste heat resources with relatively high temperature in the FCC gasoline desulfurization and jet coal hydrotreating units in the region are used for waste heat power generation, while the waste heat resources with lower temperature in other units are preferentially used for winter space heating and waste heat refrigeration in nonheating seasons. The comprehensive and effective utilization of waste heat resources in the entire plant is realized through the energy-saving measures of integrated low-temperature heat recovery, integrated heat utilization, and heat cascade utilization.

6.3.5 Energy saving of furnace

6.3.5.1 Enhanced technology to recover low-temperature flue gas waste heat

The furnace is the main heat supplier of the refinery units, and its fuel consumption occupies a considerable proportion of the energy consumption of the unit: at least 20%–30% and up to 80%–90% [5]. Taking the crude distillation unit and catalytic reforming unit as examples, the fuel consumption of the furnace accounts for 85% and 80% of the energy consumption of the entire crude distillation unit and catalytic reforming unit, respectively. Therefore, it is of great significance to improve the fuel thermal efficiency of the heating furnace to reduce the energy consumption of the unit, reduce the greenhouse gas emissions, and improve the economic benefits.

1. The main direction of energy saving of the furnace

    The simplified thermal efficiency formula of the furnace is as follows:

$$\eta = (1 - q_1 - q_2 - q_3) \times 100\% \qquad (6.1)$$

    where,

    η = furnace fuel thermal efficiency;

    q_1 = the ratio of exhaust gas heat loss to total heat input, as a function of exhaust gas temperature and air-gas ratio;

    q_2 = the ratio of incomplete combustion loss to total heat input;

    q_3 = the ratio of radiation loss to total heat input.

    The combustion technology of burners for furnaces is continuously improving. Incomplete combustion basically only occurs occasionally under bad operation, so its heat loss can be neglected. The wall losses of newly built and well-maintained furnaces are not large, generally only accounting for about 1.5%–3% of the total heat input of the furnace. The use of new refractory insulation material and a stacked lining structure of various materials can also reduce the wall loss of the furnace, but the effect is not obvious. The exhaust gas heat loss accounts for a significant proportion of the total heat loss. When the furnace's thermal efficiency is high, the exhaust gas heat loss accounts for 70%–80% of the total loss. When the furnace's thermal efficiency is low, the exhaust gas heat loss can account for up to more than 90%. Therefore, reducing the exhaust gas temperature is the most direct and effective measure to improve the heat efficiency of furnace fuel.

2. Measures to reduce the exhaust gas heat loss

    The following methods are usually adopted to reduce the exhaust gas temperature of the furnaces in refineries to reduce the exhaust gas heat loss and to improve the thermal efficiency of furnaces.

    ① Introduce low–temperature load into the end of the convection chamber for heating. By the comprehensive heat integration across the entire plant, a low-temperature load is introduced into the end of the convection chamber of the furnace for heating, so as to reduce the exhaust gas temperature. This measure is theoretically feasible, but in practice, it is necessary to fully consider whether the flow rate, inlet temperature, and required outlet temperature of the low-temperature load match the duty of the furnace. In addition, the pipeline pressure drop, pipeline insulation, and locations of the introduced load should be considered. Economic analysis should be carried out to determine whether it is reasonable and economical to introduce the low-temperature load into the end of the convection chamber of the furnace. It is also necessary to consider the fuel cost saved and the payback of the one-time construction investment.

② Use a waste heat boiler to generate steam. For furnaces of the catalytic reforming unit and the hydrocarbon–steam hydrogen production unit, the heating of the main process materials and the hydrocarbon–steam conversion only occur in the radiation chamber. The temperature of the flue gas entering the convection chamber is very high, usually around 700–1000°C. There is no suitable low-temperature load to introduce, and the unit consumes a certain amount of steam. Therefore, a flue gas waste heat boiler is installed in the convection chamber to generate medium- and high-pressure steam to recover heat and reduce the flue gas temperature.

In addition to the flue gas waste heat boiler, the convection chamber of the hydrocarbon–steam reformer is also provided with a feed preheating section, a high-temperature air preheater (under some working conditions), and a low-temperature air preheater to further recover flue gas waste heat.

The feed heating furnace of the catalytic reforming unit is limited by the temperature of boiler water supply (usually about 105°C), and currently, the exhaust gas temperature is usually about 145°C.

Due to the increasingly strict environmental regulations, the fuel of the furnaces in the refinery must be pretreated, such as desulfurized, and the exhaust gas temperature of furnaces can be further reduced. However, adopting a waste heat boiler will limit the reduction of exhaust gas temperature. To achieve the goal of further energy saving, an air preheater where the flue gas and air exchange heat is added after the waste heat boiler, and the exhaust gas temperature can be reduced to 90–120°C.

③ Adopt an exhaust gas/air preheater. The temperature of exhaust gas in the convection chamber of the furnace is usually 400–180°C. It is a common and effective way to recover exhaust gas waste heat with an air preheater system. The most important equipment in the recovery system is the exhaust gas/air preheater, where the exhaust gas from the convection chamber exchanges heat with the ambient air. The heated air enters the furnace burner for combustion. The cooled flue gas is discharged into the atmosphere through the stack.

Commonly used air preheater includes tubular air preheater, gravity heat pipe air preheater, plate air preheater, bidirectional finned cast iron plate air preheater, heat medium water air preheater, etc., or a combination of the above.

The tubular air preheater is an air preheater with heat transfer through walls. In order to enhance the heat transfer within the tube, a spoiler is set inside the tube. The role of the spoiler is to make the gas in the tube out of the stable flow state, so as to reduce the boundary layer and corresponding thermal resistance, which can improve the heat transfer. However, the pressure drop will be increased. In order to enhance the heat transfer outside the tube, a finned tube or nail-headed tube can be adapted accordingly. The added surface area can enhance the heat transfer performance.

Gravity heat pipe air preheater takes advantage of the liquid–vapor phase transition of the heat transfer medium. A single heat pipe is a tube closed at both ends, filled with a working medium (usually water in refineries) with a certain vacuum degree. When the hot end is heated, the working medium evaporates and flows to the cold end; the heat is transferred to the cold medium outside the tube while the working medium condenses and flows back to the hot end and thus completes the heat transfer cycle. Due to the large latent heat of vaporization of the working medium, a large amount of heat can be transferred from one end of the tube to the other by a very small temperature difference, which shows a superior heat transfer performance.

In order to intensify the heat transfer performance of the heat pipe, usually the cold and hot ends of the heat pipe adopt a fin or nail head, which can be adjusted accordingly. Furthermore, increasing the vacuum degree in the tube is helpful for the vapor–liquid transition, which enables effective heat transfer. The pipe distance between rows and the number of pipes in each row can be adjusted with correspondence to the flue gas temperature so that the flue gas flow rate outside the tube can be maintained in a high range, which is beneficial for heat transfer.

The heat medium water air preheater is mainly composed of the exhaust gas/water heat exchanger, air/water heat exchanger, a hot water circulating pump, and the corresponding pipes. It is an air-preheating system with softened water circulating in a closed pipe system as the heat transfer medium. In order to enhance heat transfer, fins are adopted for the tubes of the flue gas side and the air side of the heat exchange while the water in the tube always remains liquid with a relatively high flow rate to maintain an efficient heat transfer. On the flue gas side, the high-temperature flue gas transfers heat to the softened water in the tube system, and the hot water transfers heat to the combustion air. The hot water circulation pump maintains the softened water flow for continuous heat transfer.

The original intention of the combined preheater application is to effectively utilize the preheater adapted to a specific flue gas temperature range while taking into account the economy. The part with higher temperature of the flue gas adopts the preheater with excellent heat transfer performance or high temperature resistance, such as tube bundle type and plate type. The heat pipe preheater is used in the middle-temperature section. The low-temperature section adopts the form of a cast iron plate preheater, glass tube or glass plate, graphite bundle type, silicon carbide tube type, or engineering plastic tube type. On the one hand, the heat transfer is strengthened. On the other hand, the corrosion resistance is improved, and the online rate of the equipment is improved, so as to effectively guarantee the thermal efficiency of the heating furnace and achieve the purpose of energy saving and emission reduction.

④ Application of new materials. In order to improve the heat transfer efficiency and reduce dew point corrosion, materials with better corrosion resistance can be used in the heat transfer part of the air preheater, such as ND steel pipe, enamel steel pipe, borosilicate glass pipe and glass plate, graphite pipe, silicon carbide pipe, and engineering plastic pipe.

The tubular air preheater and heat pipe air preheater adopt ND steel pipes and enamel steel pipes in the low-temperature section, which achieves certain antic-orrosion. However, due to differences in the expansion coefficient between enamel and metal pipes, the preheater is prone to cracking, falling, and loss of dew point corrosion resistance. In addition, spots on the pipe, especially the fin, may lack enamel coating, resulting in electrochemical corrosion.

The preheater of borosilicate glass tube and glass plate are new in China, mainly used below 200°C. It shows a good performance of low-temperature dew point corrosion resistance, especially glass plate preheater, which shares the characteristics of good plate heat transfer. However, the glass plate needs to improve the stability of thermal shock resistance.

Graphite tubes and silicon carbide tubes have good corrosion resistance, and their thermal conductivity is better than steel, but currently, they can only be processed into tubes without fins. Due to the high-temperature sintering required for tube processing, the prices are high.

Engineering plastic pipes have good corrosion resistance and poor thermal conductivity compared with steel. It is mainly used below 200°C. At present, there are few industrial applications, and the actual performance needs to be tested.

⑤ Reduce excess air by careful operation. In order to ensure complete combustion of the fuel in the furnace, a certain amount of excess air must always be supplied to the furnace. The ratio of the actual amount of air to the theoretical amount of air required for stoichiometric combustion is called the excess air coefficient. The calculation for excess air coefficient is specified in *Fired heaters for general refinery service* [6], which is detailed in Table 6.19.

Table 6.19 Design excess air coefficient.

Type of burner	Design excess air coefficient	
	Fuel gas	Fuel oil
Natural ventilation	1.20	1.25
Forced ventilation	1.15	1.20

The heating furnace is operated under a minor vacuum. In order to observe the combustion of the burner inside the furnace, the heating of the furnace tubes, the supports and hangers of the furnace tubes and the lining, there are observation holes and inspection doors in different locations of the furnace. In practice, more or less, some air will enter the furnace interior from the furnace body. If the amount of excess air increases, a large amount of air will take away the heat with the exhaust gas. An increase in exhaust gas loss will result in a reduction in thermal efficiency and an increase in fuel consumption. Fig. 6.17 [7] shows the effect of each 0.1 incremental in excess air coefficient α on the decreased value of thermal efficiency at different exhaust temperatures.

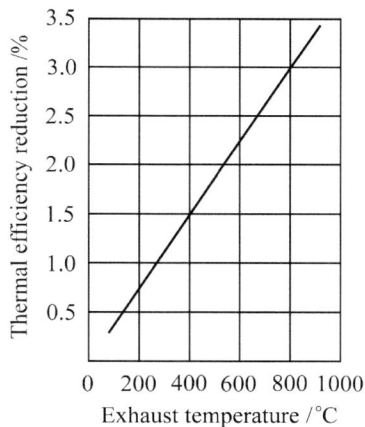

Fig. 6.17 Influence of excess air coefficient on thermal efficiency.

In the operation of the furnace, the burner with good technical performance should be selected first, which can help achieve combustion in the case of low excess air. At the same time, there is a large load regulation ratio and a low NO_X emission level. In addition, the spots that may leak should be reasonably sealed, and the opening of the damper of the burner should be finely adjusted to ensure that the excess air supply is reduced as far as complete combustion in the chamber. Choose the stack baffle with small leakage and which is easy to adjust, and adjust the opening of the stack baffle to ensure that the negative pressure of the furnace top is about -40 to $-10\,Pa$, to avoid the large negative pressure of the furnace chamber and too much ambient air inhaled from the leakage point of the furnace body. Appropriately controlling the amount of air entering the furnace can effectively reduce the exhaust gas loss, improve the thermal efficiency of the furnace, and save fuels to achieve energy saving.

6.3.5.2 Reduce radiation loss

Reduction of the radiation loss of furnace walls, ducts, and preheaters is achieved mainly by reasonable and efficient lining structure.

1. Commonly used lining materials
   ① Ceramic fiber material has the characteristics of low volume density, low heat capacity, low thermal conductivity, ease of construction, excellent thermal and mechanical vibration resistance, no drying requirement, and stable chemical properties. It is one of the mainstream materials used in furnaces in recent years.

      The diameter of the ceramic fiber is 2–$3\,\mu m$. Due to its characteristics of many voids, elasticity, and very low thermal conductivity, it can be prefabricated into various shapes of blocks, felting, blankets, etc., which can effectively shorten the construction period. Hence it is applied to almost all parts of the furnace.
   ② Refractory insulation castable is a dry mixture composed of lightweight aggregate and binder. After mixing with water or other liquid, it is constructed on walls by means of ramming and daubing. The biggest advantage of castables is strong plasticity: it is suitable for parts with a variety of complex shapes and is easy to construct. Compared with ceramic fiber material, it has higher erosion resistance and compactness and is mainly used in the parts with high flue gas flow rate (such as stack, etc.) or the parts prone to dew point corrosion.
   ③ Refractory brick. In the design of the furnace, the furnace bottom, the load-bearing wall, or the hot-face layer, which are easy to be licked by the flame, usually adopts a brick structure. According to the material composition, the commonly used refractory bricks in furnaces can be divided into clay refractory bricks and high aluminum refractory bricks.

2. Technical development of lining structure

   With the widespread use of new technologies and materials in the lining design of furnaces, the theoretical design heat loss can be reduced to 1.5%–3%. After follow-up, it is found that there is still a gap between the actual and theoretical value in practical application. This is closely related to factors such as the performance of lining materials and construction quality. Taking the actual measurement at a plant as an example, the overall thermal insulation effect of the furnace body is good and meets the design requirements. The ambient temperature of the tested furnace is 25°C, basically without wind. The temperature of the furnace wall is in the range of 63.3–79.1°C, with an average of 68°C, except around the doors, holes, and sleeves of the furnace. The hot spot temperature at the fire door is 159.4°C. The local hot spot of the pressure relief door is 162.4°C. Near the bottom of the furnace burner, the hot spot temperature is 274.2°C.

   There are also localized points or regions of above-average temperature in the plant's furnace cluster. The more obvious parts are distributed around the pressure relief doors and the bottom burners. The highest temperature reached over 100°C. In terms of temperature values, there is room for engineering improvement.

3. Adoption of new heat insulation technology

   SEI carried out infrared thermal imaging scanning of key parts of the furnace in a plant. The scanning results show that the temperature of most of the outer wall of the furnace is low, at an average of 68°C, and the heat dissipation is small. It is proved that the design of SEI in recent years has been successful. SEI has developed fire-watching doors and pressure relief doors with patented technologies and improved the design of the furnace lining near the door holes to reduce heat loss. The new heat insulation technology is applied to the furnace of the coking unit in Jinan Refinery, and the temperature is lower than 50°C.

4. Application of new structure and new material on the lining of furnaces

   In order to reduce the wall loss of furnaces, the furnace lining design widely adopted by SEI at present is: the radiation section of the furnace adopts ceramic fiber + castable composite furnace lining structure. The castable acts as a dense layer to prevent flue gas leakage to the furnace wall plate and reduces acidic dew point corrosion. The ceramic fiber on the fire surface is an insulating layer. This structure can effectively reduce the temperature of the furnace wall and reduce the heat dissipation of the furnace body.

   SEI has developed castable with high strength and low thermal conductivity. Under the same insulation layer thickness, the outer wall temperature of the castable is 15–20°C lower than that of ordinary castable.

References

[1] Sun LL, Jiang RX, Wei ZHQ. Research on innovative systematic energy saving method and application scheme. Energy Conserv Emiss Reduct 2015;5(4):1–5.

[2] Wei ZHQ, Sun LL. Research and application of heat integration strategy for multiple units in refining process based on pinch point technology. Acta Pet Sin (Pet Process) 2016;32(2):221–9.

[3] Zhang BJ, Luo XL, Chen QL, et al. Heat integration by multiple hot discharges/feeds between plants. Ind Eng Chem Res 2011;50(18):10744–54.

[4] Zhang BJ, Luo XL, Chen QL. Hot discharges/feeds between plants to combine utility streams for heat integration. Ind Eng Chem Res 2012;51(44):14461–72.

[5] Qian JL. Tube heating furnace. 2nd ed. Beijing: China Petrochemical Press; 2003. p. 520–1.

[6] Fired Heaters for General Refinery Service. SH/T3036-2012. Beijing: China Petrochemical Press; 2012. p. 18.

[7] Qian JL. Tube heating furnace. 2nd ed. Beijing: China Petrochemical Press; 2003. p. 200.

Index

Note: Page numbers followed by *f* indicate figures and *t* indicate tables.